AF389217

Risk Assessment of Nanomaterials Toxicity

Risk Assessment of Nanomaterials Toxicity

Editors

Andrea Hartwig
Christoph van Thriel

MDPI • Basel • Beijing • Wuhan • Barcelona • Belgrade • Manchester • Tokyo • Cluj • Tianjin

Editors
Andrea Hartwig
Karlsruhe Institute of
Technology (KIT)
Germany

Christoph van Thriel
TU Dortmund
Germany

Editorial Office
MDPI
St. Alban-Anlage 66
4052 Basel, Switzerland

This is a reprint of articles from the Special Issue published online in the open access journal *Nanomaterials* (ISSN 2079-4991) (available at: https://www.mdpi.com/journal/nanomaterials/special_issues/Risk_Assessment_Nano).

For citation purposes, cite each article independently as indicated on the article page online and as indicated below:

LastName, A.A.; LastName, B.B.; LastName, C.C. Article Title. *Journal Name* **Year**, *Volume Number*, Page Range.

ISBN 978-3-0365-7812-5 (Hbk)
ISBN 978-3-0365-7813-2 (PDF)

Contents

 *nanomaterials*

Editorial

Risk Assessment of Nanomaterials Toxicity

Andrea Hartwig [1] and Christoph van Thriel [2,*]

[1] Department of Food Chemistry and Toxicology, Institute of Applied Biosciences (IAB), Karlsruhe Institute of Technology (KIT), Gebäude 50.41 (AVG), Adenauerring 20a, 76131 Karlsruhe, Germany; andrea.hartwig@kit.edu

[2] Leibniz Research Centre for Working Environment and Human Factors, Research Group Neurotoxicology and Chemosensation, Department of Toxicology, TU Dortmund, Ardey Str. 67, 44139 Dortmund, Germany

* Correspondence: thriel@ifado.de

Citation: Hartwig, A.; van Thriel, C. Risk Assessment of Nanomaterials Toxicity. *Nanomaterials* **2023**, *13*, 1512. https://doi.org/10.3390/nano13091512

Received: 12 April 2023
Accepted: 24 April 2023
Published: 28 April 2023

The increasing use of nanomaterials in almost every area of our daily life renders toxicological risk assessment a major requirement for their safe handling. Thus, risk assessment strategies ensuring the health of individuals exposed to these types of materials must be adopted and continuously reviewed. Major challenges include the enormous amount of engineered nanomaterials (ENMs) used in workplaces [1], the limited capacity for testing ENMs in long-term animal inhalation studies [2], and the political and societal efforts to reduce animal experiments according to the 3R principles [3]. Against this background, much attention has been paid to grouping of nanomaterials, mainly based on their physicochemical properties and their toxicity in various in vitro models. These new approach methodologies (NAMs) include a detailed characterization of the respective materials in physiologically relevant media, but also more realistic exposure systems, such as co-cultures, at the air–liquid interface, combined with comprehensive cellular investigations providing quite detailed toxicological profiles. These NAM-based approaches have been recently reviewed by the U.S. Federal Agencies and the authors concluded that " . . . two key issues in the usage of NAMs, namely dosimetry and interference/bias controls, . . . " are crucial aspects in ongoing validation processes [4]. In workplaces where inhalation is the major route of exposure, potential toxicity affecting the lungs needs to be considered. Here, advanced in vitro models have documented their predictive capacity for adverse outcomes such as lung fibrosis [5]. Neurotoxicity associated with exposure to nanomaterials is another growing field of scientific investigation [6] and, here, the use of nanocarriers for drug delivery provides a special "route of exposure" [7]. We initiated this Special Issue to further promote scientific progress in the area of nanosafety and are glad to share 13 papers on various topics with the readership of *Nanomaterials*. This Special Issue highlights recent advances in the mechanisms of nanomaterial toxicity as well as approaches for risk assessment, linking nanoparticle characteristics and in vitro toxicity to in vivo observations for advanced risk assessment. Here, the availability of data and the development of databases are important.

With three original articles by Murugadoss, Mülhopt et al., Elje et al., and Meindl et al., addressing various aspects, the respiratory tract toxicities of titanium dioxide, carbon nanotubes, and nanosilver have been described and some assays can be further validated. A link between in vitro screening and results from in vivo testing for lung effects is provided by Creutzenberg et al., describing results from the PLATOX project. Focusing on the aspect of data availability and reproducibility, Krug describes the development of the CoCoN-Database, while Elberskirch et al. describe the results of a round-robin test that includes data science tools to increase comparability among different labs. Another relevant and important aspect is addressed by de Souza Castro et al., comparing 2D and 3D cell culture models of bone mineralization. Here, the 3D model showed improved induction of bone osteointegration by nanoparticles. Mechanisms related to the possible genotoxicity of ENM are described in the papers by Schumacher et al., May et al., and Murugadoss, with

Godderis et al. also addressing the crucial aspect of realistic exposure scenarios in vitro. These papers are also relevant to the key issue of dosimetry, as described by Petersen et al. [4]. The paper by Wall et al. provides new insight into the physico-chemical properties of particulate and fibrous nanomaterials that can modulate their toxicity. Finally, the review by Ruijter et al. highlights various aspects of how in vitro methods can be incorporated into the *Safe-by-Design* concept that is expected to foster the development of safe ENMs before they enter the market.

Author Contributions: Writing—original draft preparation, C.v.T.; writing—review and editing, A.H. and C.v.T. All authors have read and agreed to the published version of the manuscript.

Conflicts of Interest: The authors declare no conflict of interest.

References

1. Schulte, P.A.; Kuempel, E.D.; Drew, N.M. Characterizing Risk Assessments for the Development of Occupational Exposure Limits for Engineered Nanomaterials. *Regul. Toxicol. Pharmacol.* **2018**, *95*, 207–219. [CrossRef] [PubMed]
2. Landsiedel, R.; Ma-Hock, L.; Hofmann, T.; Wiemann, M.; Strauss, V.; Treumann, S.; Wohlleben, W.; Groters, S.; Wiench, K.; van Ravenzwaay, B.; et al. Application of Short-Term Inhalation Studies to Assess the Inhalation Toxicity of Nanomaterials. *Part. Fibre Toxicol.* **2014**, *11*, 16. [CrossRef] [PubMed]
3. Caloni, F.; De Angelis, I.; Hartung, T. Replacement of Animal Testing by Integrated Approaches to Testing and Assessment (IATA): A Call for in Vivitrosi. *Arch. Toxicol.* **2022**, *96*, 1935–1950. [CrossRef] [PubMed]
4. Petersen, E.J.; Ceger, P.; Allen, D.G.; Coyle, J.; Derk, R.; Garcia-Reyero, N.; Gordon, J.; Kleinstreuer, N.C.; Matheson, J.; McShan, D.; et al. U.S. Federal Agency Interests and Key Considerations for New Approach Methodologies for Nanomaterials. *ALTEX* **2022**, *39*, 183–206. [CrossRef] [PubMed]
5. Barosova, H.; Maione, A.G.; Septiadi, D.; Sharma, M.; Haeni, L.; Balog, S.; O'Connell, O.; Jackson, G.R.; Brown, D.; Clippinger, A.J.; et al. Use of EpiAlveolar Lung Model to Predict Fibrotic Potential of Multiwalled Carbon Nanotubes. *ACS Nano.* **2020**, *14*, 3941–3956. [CrossRef] [PubMed]
6. Boyes, W.K.; van Thriel, C. Neurotoxicology of Nanomaterials. *Chem. Res. Toxicol.* **2020**, *33*, 1121–1144. [CrossRef] [PubMed]
7. Mulvihill, J.J.; Cunnane, E.M.; Ross, A.M.; Duskey, J.T.; Tosi, G.; Grabrucker, A.M. Drug Delivery across the Blood-Brain Barrier: Recent Advances in the Use of Nanocarriers. *Nanomedicine* **2020**, *15*, 205–214. [CrossRef] [PubMed]

 *nanomaterials*

Article

Assessment of Carbon Nanotubes on Barrier Function, Ciliary Beating Frequency and Cytokine Release in In Vitro Models of the Respiratory Tract

Claudia Meindl [1], Markus Absenger-Novak [1], Ramona Jeitler [2], Eva Roblegg [2] and Eleonore Fröhlich [1,*]

1   Center for Medical Research, Medical University of Graz, Stiftingtalstr. 24, 8010 Graz, Austria
2   Department of Pharmaceutical Technology and Biopharmacy, Institute of Pharmaceutical Sciences, University of Graz, Universitaetsplatz 1, 8010 Graz, Austria
*   Correspondence: eleonore.froehlich@medunigraz.at; Tel.: +43-31638573011

Abstract: The exposure to inhaled carbon nanotubes (CNT) may have adverse effects on workers upon chronic exposure. In order to assess the toxicity of inhaled nanoparticles in a physiologically relevant manner, an air–liquid interface culture of mono and cocultures of respiratory cells and assessment in reconstructed bronchial and alveolar tissues was used. The effect of CNT4003 reference particles applied in simulated lung fluid was studied in bronchial (Calu-3 cells, EpiAirway™ and MucilAir™ tissues) and alveolar (A549 +/−THP-1 and EpiAlveolar™ +/−THP-1) models. Cytotoxicity, transepithelial electrical resistance, interleukin 6 and 8 secretion, mucociliary clearance and ciliary beating frequency were used as readout parameters. With the exception of increased secretion of interleukin 6 in the EpiAlveolar™ tissues, no adverse effects of CNT4003 particles, applied at doses corresponding to the maximum estimated lifetime exposure of workers, in the bronchial and alveolar models were noted, suggesting no marked differences between the models. Since the doses for whole-life exposure were applied over a shorter time, it is not clear if the interleukin 6 increase in the EpiAlveolar™ tissues has physiological relevance.

Keywords: carbon nanotubes; toxicity; in vitro models; respiratory tract; bronchial epithelium; alveolar epithelium; ciliary beating frequency

Citation: Meindl, C.; Absenger-Novak, M.; Jeitler, R.; Roblegg, E.; Fröhlich, E. Assessment of Carbon Nanotubes on Barrier Function, Ciliary Beating Frequency and Cytokine Release in In Vitro Models of the Respiratory Tract. *Nanomaterials* **2023**, *13*, 682. https://doi.org/10.3390/nano13040682

Academic Editors: Andrea Hartwig and Christoph Van Thriel

Received: 21 December 2022
Revised: 31 January 2023
Accepted: 6 February 2023
Published: 9 February 2023

1. Introduction

Nanoparticles are used in industry, consumer products, health care and medicine, but the safety of some particles is still a matter of discussion [1]. One particle type with unclear safety is carbon nanotubes (CNTs). CNTs have been evaluated for cytotoxicity, genotoxicity, inflammatory potential and carcinogenicity with variable results [2]. The absorption of CNTs through epithelial barriers was found to be low. Among all barriers, the respiratory barrier is the less protected and most permeable. The lung consists of the conducting and the respiratory airways, which have different protection mechanisms and require different models for toxicity testing in vitro [3].

Reported cytotoxicity of multi-walled CNTs (MWCNTs) varied from 5 ng/mL to 10 mg/mL and based on a systematic review, A549 and HUVEC were the most appropriate cells for comparison among research groups [4]. The differences in cytotoxicity by a factor of 10^6 are thought to be due not only to differences in cells, assays, exposure conditions and sample preparation but also to different production of the MWCNTs [4]. To enable better comparison of the results, the use of standardized nanoparticles is suggested. Sponsored by several European Union Joint programs, European Commission's Directorate General Joint Research Centre (JRC) has established the JRC Nanomaterials Repository of industrially manufactured nanomaterials with the aim of providing the scientific and regulatory communities with nanomaterials for testing [5]. Standard Operation Procedures for the dispersion of titanium dioxide, silicon dioxide and MWCNTs in 0.05% bovine serum albumin (BSA) are available due to the NANOGENOTOX European joint action to exclude differences in the preparation of particle suspensions [6]. The CNT4003 particles, formerly termed NM-403, belong to the thin and short MCNTs with reported primary sizes of 12 nm diameter and 443 nm length. They were chosen for this study because measurements of CNT manufacturing workplaces report sizes of highest exposure between 10 and 100 nm [7–9], whereas the thicker and longer CNTs (e.g., 80 nm × 3.7 μm) form particles with aerodynamic size of 260–381 nm [10]. No decrease in viability was observed at 45.7 μg/cm^2 NM-403 for A549 alveolar epithelial cells in conventional testing [11].

Air–liquid interface culture (ALI), where cells grow on transwell membranes and are exposed to medium with nutrients only from the basal side and the apical parts face the air, is the physiologically relevant culture for respiratory cells. Using this technique, Calu-3 cells, established models for the bronchial epithelium, produce mucus and A549 as models for the alveolar epithelium surfactant [12,13]. Calu-3 cells also form epithelial barriers of sufficient tightness identified by transepithelial electrical resistance (TEER) values of >300 Ω × cm^2. The limitation of Calu-3 cells is that effects of toxicants on ciliary function, the important clearance mechanism of the upper airways, cannot be assessed because the cells form only immature and not motile cilia [14]. Reconstructed tissues prepared from human bronchial epithelial cells, EpiAirway™ from MatTek Cooperation and MucilAir™ from Epithelix Sárl, on the other hand, possess cilia and, therefore, allow the assessment of this relevant protection system of the lung. The main limitation of A549 cells, which are routinely used in respiratory toxicity testing, is the fact that despite expression of tight junction proteins, they do not form a functional epithelial barrier, which corresponds to the generally low TEER values in the range of 20–60 Ω × cm^2 [15,16]. EpiAlveolar™ from MatTek is composed of alveolar epithelial cells, fibroblasts and endothelial cells and enables assessment of alveolar barrier function. EpiAlveolar™ enriched with THP-1 macrophages are also available for more realistic testing of the alveolar barrier [17].

To mimic inhalation of MWCNTs, some researchers have used the VITROCELL Cloud system [17–19]. According to the producer, 200 μL of nebulization volume are recommended for the 12-well exposure chambers (https://www.vitrocell.com/inhalation-toxicology/exposure-systems/vitrocell-cloud-system/vitrocell-cloud-alpha-6 accessed on 2 May 2021), indicating that each well of a 12-well plate will receive ~17 μL of saline together with the particles. This solution differs in volume and composition from the lung lining fluid in the lung. The surface of the lung is covered by a dipalmitoyl-glycero-3-phosphocholine (DPPC)-rich lung lining fluid of 70–300 nm thickness [20]. The presence of DPPC is important because nanoparticles in contact with biological fluids will absorb biomolecules, proteins and lipids onto their surface [21]. This so called "corona" will determine the cellular response of the host cells. It has been reported in animal studies that coating with DPPC markedly increased uptake and lung retention of nanoparticles compared to coating with BSA [22,23]. To take the effect of the DPPC binding into account, CNT4003 particles were suspended in simulated lung fluid (SLF), a phosphate buffer containing 0.02% DPPC.

Sauer et al. compared the cytotoxicity of respiratory toxicants in A549 and 3T3 monolayers, MucilAir™ and EpiAirway™ tissues [24], but there is no systematic comparison between models based on cell lines and commercially available reconstructed tissues of bronchial and respiratory part. This comparison is important because the use of the more complex reconstructed tissues is associated with higher costs. The aim of this study was to identify adverse effects of standardized CNTs in physiologically relevant systems and to reveal the role of the testing system. To this end, effects of CNT4003 particles were compared in models of different complexity (monolayers vs. multilayers) and origin (primary vs. cell lines). The focus is not on cytotoxicity but on functional aspects because physiologically relevant exposure doses to CNTs are too low to act acutely toxic. For the bronchial part of the airways Calu-3 cells, EpiAirway™ and MucilAir™ were used. Effects of CNT4003 on the alveolar part of the airways were studied using A549 cells and EpiAlveolar™ tissues with and without THP-1 macrophages.

2. Materials and Methods

2.1. Preparation of Simulated Lung Fluid (SLF)

1,2-Dipalmitoyl-sn-glycero-3-phosphocholine (DPPC, Avanti Polar Lipids, Birmingham, AL, USA) is dissolved in ethanol and added dropwise to prewarmed (37 °C) phosphate-buffered saline with Ca^{2+} and Mg^{2+} (PBS, ThermoFisher Scientific, Vienna, Austria) to reach a concentration of 0.02% (w/v) DPPC.

2.2. Preparation of Particle Suspensions

CNT4003 was obtained from the Joint Research Center (JRC, Ispra, Italy). According to the list of representative nanomaterials status June 2016 of the JRC, CNT4003 are MWCNTs with 189 m^2/g surface area with average length of 443 nm and average diameter of 12 nm [25]. Suspensions were prepared according to the protocol established in the NANOGENOTOX project [26] in 0.05% (w/v) BSA and in the same way in SLF to find out if the same protocol works also for SLF. Briefly, 15.36 mg of CNT4003 powder was prewetted with 30 μL absolute ethanol. Subsequently, 0.97 mL of SLF or of 0.05% BSA in distilled water was added while slowly rotating the glass vessel. To prepare the stock solution, additional 5 mL of SLF or of 0.05% (w/v) BSA was added. Sonication time and amplitude can be optimized, e.g., 16 min at 10% amplitude or 12 min at 20% amplitude can be used [6]. Suspensions were sonicated on ice with a Branson SFX250 Digital Sonifier (Branson Ultrasonic Cooperation, Danbury, USA) equipped with a microtip at 250 W, 20 kHz and 30% amplitude with 10 s impulses (i.e., 10 s pulse on, 1 s pulse off) for 14 min. Pilot experiments showed that the use of a higher amplitude in combination with a shorter sonication time led to a more effective deagglomeration. Suspensions were prepared directly prior to the administration because it was reported that NM-403 showed high polydispersivity and were shown to be stable for 1–2 h.

2.3. Determination of Size and Zeta Potential

Measurements were performed at 25 ± 0.5 °C with a particle concentration of 25.6 μg/mL in triplicates. Hydrodynamic size and polydispersity index (PDI) were measured via photon correlation spectroscopy (PCS) using a Zetasizer Nano ZS (Malvern Instruments, Malvern, United Kingdom) equipped with a 532 nm laser. The scattering angle chosen was 173°, and the refractive indices for the CNTs and the dispersion medium were 2.50 and 1.330, respectively. The zeta potential was measured by electrophoretic light scattering (scattering angle of 173°; Zetasizer Nano ZS, Malvern Instruments) considering the same optical properties of the CNTs and the dispersion media. The calculation was based on the electrophoretic mobility using Henry's equation. The zeta potential was -9.9 ± 0.7 mV for CNT 4003 in BSA and 1.4 ± 0 mV for CNT 4003 in SLF. The hydrodynamic size of particles suspended in 0.05% BSA determined by PCS was 2327 nm with a PDI of 1.0, indicating that PCS was not suitable for accurately determining particle size. In addition, SLF contained DPPS, which formed micelles of 58 ± 7 nm and interfered with the size measurements. Therefore, the sizes of CNT 4003 were additionally measured by laser diffraction (LD, Mastersizer 2000, Malvern Instruments, United Kingdom) by adding the CNT suspension to 18–20 mL SLF or 0.05% BSA until a laser obscuration of 4–6% was achieved. Volume based values considering d (0.1), d(0.5) and d (0.9) were determined in triplicate at 25 °C and CNTs showed d(0.1) 974 ± 32, d(0.5) 2254 ± 44 and d(0.9) 5399 ± 71 nm in BSA and d(0.1) 1290 ± 8, d(0.5) 3829 ± 38 and d(0.9) 6990 ± 43 nm in SLF. These data indicate that the protocol using SLF leads to a particle distribution similar to that observed with BSA.

2.4. Cellular Models of Bronchial and Alveolar Epithelium

Calu-3 cells were obtained from LGC Standards GmbH (Wesel, Germany) and cultured in minimum essential medium (MEM), 2 mM L-glutamine, 1% penicillin/streptomycin, 10% fetal bovine serum (FBS) and 1 mM sodium pyruvate. A549 cells obtained from Deutsche Sammlung für Mikroorganismen und Zellkulturen GmbH (Braunschweig, Germany) were cultured in Dulbecco's modified Eagle's medium (DMEM), 2 mM L-glutamine,

1% penicillin/streptomycin (P/S) and 10% FBS. THP-1 was purchased from Cell Line Services (Eppelheim, Germany) and cultured in RPMI, 2 mM L-glutamine, 1% penicillin/streptomycin and 10% FBS. Cells were passaged at regular intervals.

For the exposures with the CNTs, 0.5×10^6 Calu-3 cells were seeded in 500 µL MEM, 2% L-glutamine and 1% PS + 10% FBS on 12-well polyethylene terephthalate transwell inserts (pore size 0.4 µm, Greiner Bio-one, Kremsmünster, Austria) with 1500 µL of the same medium in the basolateral compartment. Medium in the apical compartment was removed after 24 h, and the medium amount in the basolateral compartment was reduced to 500 µL, which was changed every 2 or 3 days. Cells were used for the experiments when they had reached a transepithelial electrical resistance (TEER) value of >300 $\Omega \times cm^2$.

A549 was seeded at a density of 0.8×10^5 in DMEM + 10% FBS on 12-well polyethylene terephthalate transwell inserts (pore size 0.4 µm, Greiner Bio-one) with 1500 µL of the same medium in the basolateral compartment. Medium in the apical compartment was removed after 24 h as reported previously [27]. According to these experiments, the ratio of 9 alveolar cells to 1 alveolar macrophage observed in vivo [28] is obtained by seeding THP-1 macrophages to the cultured A549 cells in a ratio 1:1. A549 cells were cultured for 9 days in ALI prior to the addition of the THP-1 cells and the average number of A549 cells was determined in duplicates. For the coculture, RPMI + 10% FBS was added in the basal compartment and the medium changed every other day.

Differentiation to THP-1 Macrophages for Coculture with A549 Cells or EpiAlveolar™ Tissues

For coculture with A549 cells, 1.0×10^6 THP-1 cells/mL were seeded in RPMI 1640 containing 10% FBS, 1% L-glutamine and 1% P/S in flasks. Differentiation to macrophages was induced by addition of 10 nM phorbol 12-myristate 13-acetate (PMA, Sigma-Aldrich, Vienna, Austria) to the media for 48 h. The stimulation medium was changed to medium without PMA for another 24 h before use in the cocultures. Cells were harvested by treatment with trypsin/EDTA (0.05%). For addition to EpiAlveolar™ tissues (MatTek Corporation, Ashland, OR, USA), THP-1 cells were stimulated with 8 nM PMA for 72 h, harvested by treatment with trypsin/EDTA and added immediately to the tissues.

2.5. Reconstructed Tissues

EpiAirway™ 3D human respiratory epithelial tissues (AIR-100 PC6.5/PE6.5) were obtained from MatTek Corporation. The tissues were cultivated at the air–liquid interface and medium (MatTek Corporation) changed every other day according to the instructions of the producer. EpiAlveolar™ 3D tissues with and without THP-1 macrophages (MatTek Co-operation) were maintained at the air–liquid interface with medium (MatTek Corporation) changes every other day. Since EpiAlveolar™ 3D tissues with THP-1 macrophages did not tolerate the long delivery, they were added to the EpiAlveolar™ 3D tissues in the laboratory in Graz following the protocol of the company in which 25,000 THP-1 macrophages were added per insert. The cells were seeded in 75 µL media into the apical compartment during feeding and after 24 h the culture was switched back to air-liquid interface condition.

MucilAir™ tissues were purchased from Epithelix Sárl (Geneva, Switzerland) and maintained in MucilAir™ culture medium (Epithelix Sárl) in 24-well format Transwell® cell culture inserts (Sigma-Aldrich) in a humidified incubator (37 °C; 5% v/v CO$_2$). The culture medium was changed every 2–3 days.

2.6. Exposure to Carbon Nanotubes

CNT4003 in different doses were suspended in SLF. A total of 10 µL of the suspension were used for the 0.336 cm^2 tissues and 30 µL for the models with 1.12 cm^2. They were applied on the surface of the cell monolayers or reconstructed tissues. The entire observation time was 14 d. After initial addition of the CNT dose for seven days, cell and tissue surfaces were rinsed with PBS, Transepithelial Electrical Resistance (TEER) measured and new particles or SLF added for another seven days.

2.7. Measurement of Transepithelial Electrical Resistance (TEER) Values

TEER values were determined with an EVOM STX-2-electrode (World Precision Instruments, Berlin, Germany). A total of 0.4 mL MEM was added to the apical and 1.2 mL MEM to the basolateral compartment for TEER measurements for EpiAirway™ and 0.5 mL in the apical and 1.5 mL in the basolateral compartment for EpiAlveolar™ tissues. TEER values were calculated as follows: TEER ($\Omega \times cm^2$) = (Sample–blank resistance, given in Ω) xmembrane area, given in cm^2. Blank resistance is defined as the resistance of the membrane without cells. Membrane area is indicated for EpiAirway™ as 0.336 cm^2 and for EpiAlveolar™ as 1.12 cm^2.

2.8. Determination of Viability (MTS Assay)

Viability was assessed at the end of the exposure time (14 d). CellTiter 96® AQueous Non-Radioactive Cell Proliferation Assay (Promega, Mannheim, Germany) was used according to the manufacturer's instructions. To avoid interference with CNT4003 absorption, inserts were first rinsed with cell culture medium. Subsequently, 100 µL of the combined MTS/PMS solution + 500 µL medium was added in each insert. Plates were incubated for up to 2 h at 37 °C in the cell incubator. The supernatant was transferred into a new plate to remove remaining CNT4003 particles and absorbance was read at 490 nm on a plate reader (SPECTRA MAX plus 384, ServoLab, Kumberg, Austria).

2.9. Interleukin Measurements

Medium from the basolateral compartments of all cultures was collected, and the release of IL-6 and IL-8 was determined using the human IL-6 ELISA set (BD OptEIA™, BD Biosciences, Heidelberg, Germany) and the human IL-8 ELISA set (BD OptEIA™) according to the protocol given by the producers. To induce IL-6 and IL-8 secretion, 24 h prior to the harvesting, 20 µL containing 1, 5 and 10 µg lipopolysaccharide from Escherichia coli 055:B5 (LPS, Sigma-Aldrich) as proinflammatory stimulus was added. The concentration of LPS, which caused the maximum release of the cytokines by A549 cells was already determined in a previous study [27]. Absorbance was read at 450 nm (with correction wavelength of 570 nm) on a SPECTROstar (ServoLab) photometer.

2.10. Mucociliary Clearance

Prior to the measurements, tissues were rinsed in PBS and 10 µL of PBS containing 4×10^5 yellow-green, fluorescent carboxyl polystyrene particles (1.0 µm, Thermo Fisher Scientific, Waltham, MA, USA) was added and migration of the beads monitored for 10–30 s using Ti2-E/confocal system (Nikon CEE GmbH, Vienna, Austria) equipped with heated stage (37 °C) and incubation chamber at 10× magnification. Object tracking was performed by NIS Element software (Nikon CEE GmbH).

2.11. Ciliary Beating Frequency

Prior to the measurements, tissues were rinsed in PBS. Tissues were cut out from the insert and stripes of epithelium were placed on a glass slide and put into the incubation chamber (37 °C) on a fully motorized Ti2-E/confocal system (Nikon CEE GmbH) mounted on an antivibration table. Peripheral and central parts of the stripes were imaged. In pilot experiments, propranolol hydrochloride (100 µM, Sigma-Aldrich) was included as inhibitor and 5 µM forskolin (Sigma-Aldrich) + 100 µM 3-Isobutyl-1-methylxanthin (IBMX, Sigma) as stimulator of CBF. Background noise was determined by measurement of samples fixed with 4% paraformaldehyde. The background was negligible at 0.15–0.75 Hz. Further, the reaction to 1, 5 and 10 µg/insert LPS was recorded. Specimens were examined using 20× magnification. Beating ciliated edges were recorded using a digital high-speed video camera (Andor Zyla VSC-08691, Oxford Instruments, Wiesbaden, Germany) at a rate of 100 frames per second with a frame size of 512 × 512 pixel. Five regions of interest (ROI) were analysed per sample. CBF was calculated by NIS Element software (Nikon CEE

GmbH) and indicated as beats/second (Hz). In the later experiments, only propranolol was used because no prominent increase in CBF was seen.

2.12. Histology

MatTek tissues were fixed in 4% paraformaldehyde and embedded in paraffin using Tissue-Tek®VIP™ 5 (SanovaPharma GesmbH, Gallspach, Austria). Radial sections of 2–5 μm were cut at a rotary microtome, stained with hemalaun and viewed with an Olympus BX51 microscope. Additional sections were cut for immunocytochemical staining.

2.12.1. Immunocytochemical Detection of α-Tubulin

Sections were incubated with DakoCytomation Target Retrieval Solution pH 6.0 (Agilent, Vienna Austria) in decloaking chamber (Biocare Medical, Pacheco, CA, USA) for 10 min at 110 °C and 20 s at 85 °C. Sections remained in the solution for cool down for 30 min. Three rinses with PBS were performed prior to the incubations with blocking serum (1% goat serum in PBS for 30 min at RT) and between each staining step. Antibodies were diluted in antibody diluent (DAKOCytomation, Hamburg, Germany). Incubation with anti-α-tubulin antibody (mouse, 1:500, Thermo Fisher Scientific) or control mouse IgG (negative control, Linaris, Mannheim, Germany) for 1 h at RT was followed by incubation with anti-mouse Alexa 488 antibody (goat, 1:400, Thermo Fisher Scientific) in antibody diluent (DAKOCytomation) for 30 min at RT and counterstain with Hoechst 33342 (1 μg/mL, Thermo Fisher Scientific) for 15 min at RT. After three final washes in PBS, sections were mounted with fluorescence mounting media.

2.12.2. Immunocytochemical Detection of Mucin

Antigen retrieval was performed with DakoCytomation Target Retrieval Solution pH 9.0 (DAKO) in decloaking chamber (Biocare Medical) for 10 min at 110 °C and 20 s at 85 °C. Sections remained in the solution for cool down for 30 min. Three rinses with PBS were performed, cells permeabilized with PBS plus 0.2% Triton X-100 for 2 h at RT. Unspecific antibody binding was blocked with 10% normal goat serum for 30 min at RT. Incubation with anti-mucin 5AC (mouse, 1:200, Abcam, Cambridge, United Kingdom) antibody was performed for 1 h at 37 °C, incubation with anti-mouse Alexa 488 antibody (goat, 1:400, Thermo Fisher Scientific) for 1 h at 37 °C and with Hoechst 33342 (1 μg/mL, Thermo Fisher Scientific) in PBS for 15 min at RT.

2.12.3. Immunocytochemical Detection of Macrophages

After deparaffinization, cells were permeabilized by incubation with 0.1% Triton X-100 in PBS for 60 min at RT and incubation with APC-CY7 anti-CD45 (mouse, 1:500, BD Biosciences, Vienna, Austria) for 1 h at RT and counterstain of nuclei with Hoechst 33342 (1 μg/mL, Thermo Fisher Scientific) for 30 min at RT.

2.12.4. Staining of Whole Mounts with Phalloidin

EpiAlveolar™ tissues were fixed for 15 min with 4% paraformaldehyde. After three washes with PBS, the tissues were permeabilized with 0.1%Triton X100 in PBS, washed three times in PBS again and subsequently incubated with Alexa Fluor™ 488 Phalloidin (Thermo Fisher Scientific, 1:100 in antibody diluent) and Hoechst 33342 (1 μg/mL, Thermo Fisher Scientific) for 30 min at RT.

Images were taken at a Ti2-E/confocal system (Nikon CEE GmbH) with ex/em of 395 nm/414–450 nm for Hoechst 33342 and 470 nm/500–530 nm for α-tubulin, mucin 5AC and phalloidin, and 640 nm/660–850 nm for CD45.

2.13. Statistics

Experiments were performed in triplicates and repeated at least two times. Data from all experiments were analyzed with one-way analysis of variance (ANOVA) followed by

Tukey's HSD post hoc test for multiple comparisons (SPSS 28 software). Results with *p*-values < 0.05 were considered to be statistically significant.

3. Results

In all evaluations, CNT4003 was applied in SLF and the following parameters and assays used as readout: MTS assay for cytotoxicity, TEER measurements for epithelial barrier tightness, interleukin 6 and 8 levels for proinflammatory effects, movement of polystyrene marker particles for mucociliary clearance and high-speed video microscopy for ciliary beating frequency. IL-6 is a marker for acute and chronic inflammation and IL-8 a marker for acute inflammation and a chemoattractant for neutrophils [18]. Doses are indicated as µg/insert and the distribution of CNT4003 on the cell-grown inserts documented. After 14 d of exposure, several CNT4003 agglomerates were visible in the central region of inserts containing mucus-producing cells and rare agglomerates in the inserts containing EpiAlveolar™ tissues (Figure S1, Supplementary Material). This can be explained by the fact that mucus can trap the agglomerates better than surfactant.

3.1. Particle Effect in Models of the Bronchial Part of the Respiratory Tract

Calu-3 cells, EpiAirway™ and MucilAir™ tissues were used as models for the bronchial part and A549 and EpiAlveolar™ tissues with and without macrophages as models for the alveolar part. The models were characterized regarding the stability and ability to react to LPS as an inflammatory stimulus. Evaluation of CNT4003 effects over 14 d corresponded to the recommendation of Behrsing et al. for analysis in EpiAirway™s and MucilAir™ tissues [29].

3.1.1. Calu-3 Cells

No cytotoxicity of CNT4003 was seen up to 50 µg with a viability of 86 ± 15% of untreated controls (Figure S2, Supplementary Material).

As seen in the immunocytochemical staining, Calu-3 monolayers do not form cilia but contain mucus-producing cells (Figure 1). Alpha tubulin is a microtubule marker, and the tubulin-dynein system is a central part of flagellar and ciliary movement [30]. To a lesser extent, tubulin is also expressed in the cytoplasm for structural support, a pathway for transport and force generation in cell division [31].

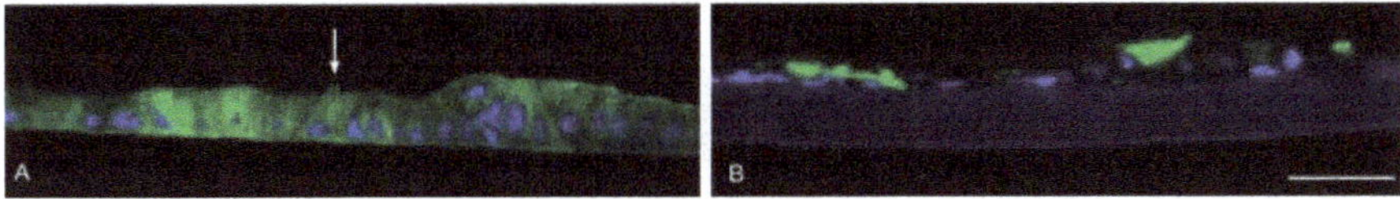

Figure 1. Immunochemical detection of tubulin (**A**) and identification of mucus-producing cells by anti-mucin 5AC staining (**B**). Tubulin is contained in high amounts in cilia and in lower amounts in the cytoskeleton. In the absence of cilia, the staining of the cytoskeleton and of immature cilia (arrow) is visible. Scale bar: 50 µm.

Stability of the Calu-3 model has been shown previously [32]. In this study, TEER values decreased significantly from 1 d to 14 d of culture in all cultures with no difference between untreated and CNT4003-treated cultures (Figure 2). The decrease upon LPS treatment at 7 d was significant compared to 1 d and also significantly more pronounced than that of the control.

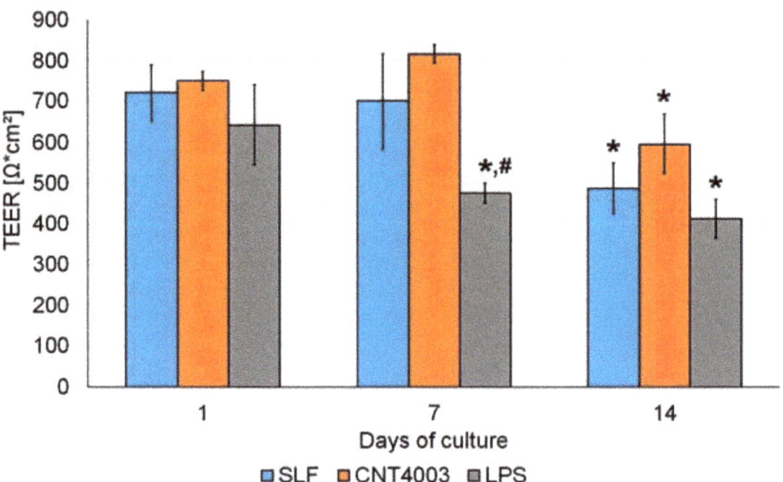

Figure 2. Time-dependent changes of transepithelial electrical resistance (TEER) in Calu-3 cells exposed to simulated lung fluid (SLF) alone, LPS or SLF containing CNT4003. Significant changes compared to 1 d are indicated by an asterisk and significant differences between treated and control cells by a hash.

Stimulation with 10 µg LPS increased IL-6 and IL-8 secretions at all time points significantly (Table 1). Interleukin-6 levels of cultures treated with 25 µg CNT4003 after one and seven days were significantly lower than the levels of the controls (Table 1). Interleukin 8 levels were significantly increased after one day and decreased after seven days. No changes were seen after 14 days.

Table 1. Time-dependent changes (%) of cytokine secretion by Calu-3 upon stimulation with 10 µg LPS or 25 µg CNT4003. Secretion of unstimulated tissues is set as 100%. Significant differences ($p < 0.05$) between treated and untreated samples are indicated by an asterisk.

Cytokine		Day 1	Day 7	Day 14
Interleukin 6	LPS	478 ± 71 *	703 ± 9 *	722 ± 97 *
Interleukin 8	LPS	1504 ± 69 *	367 ± 88 *	266 ± 32 *
Interleukin 6	CNT4003	55 ± 38 *	25 ± 38 *	116 ± 36
Interleukin 8	CNT4003	118 ± 6*	32 ± 4 *	109 ± 2

3.1.2. EpiAirway™ Tissues

EpiAirway™ cultures are composed of columnar epithelium cells arranged as three to five rows of nuclei (Figure 3A). Cilia at the apical surface are seen at low density. The composition of the epithelium of bronchial epithelial cells and goblet cells is identified by anti-α-tubulin staining (Figure 3B) and anti-mucin 5AC staining (Figure 3C), respectively.

TEER values of EpiAirway™ tissues over 28 d were lower than in the other models of the bronchial tract but remained constant for 3 weeks with $259 \pm 124 \; \Omega \times cm^2$ at 1 d and $244 \pm 69 \; \Omega \times cm^2$ at 21 d. Only after 28 d, a significant decline to $110 \pm 67 \; \Omega \times cm^2$ was seen.

CNT4003 particles did not affect TEER values over the entire observation time of 14 d, whereas the initially higher TEER values of the cultures exposed to LPS were significantly decreased compared to 1 d at 7 d and 14 d (Figure 4).

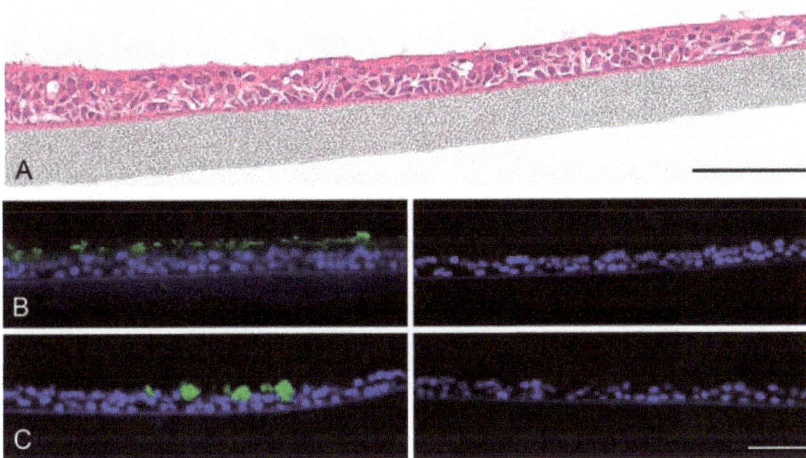

Figure 3. Morphology of EpiAirway™ tissues according to hemalaun staining (**A**) and cellular composition according to immunochistochemical staining. Scale bar: 100 µm. Anti-α-tubulin (**B**) and anti-mucin 5AC (**C**) staining (green) for visualization of bronchial epithelial and goblet cells. Controls with mouse IgG instead of primary antibody are shown on the right side. Nuclei are counterstained with Hoechst 33342 (blue). Scale bar 50 µm.

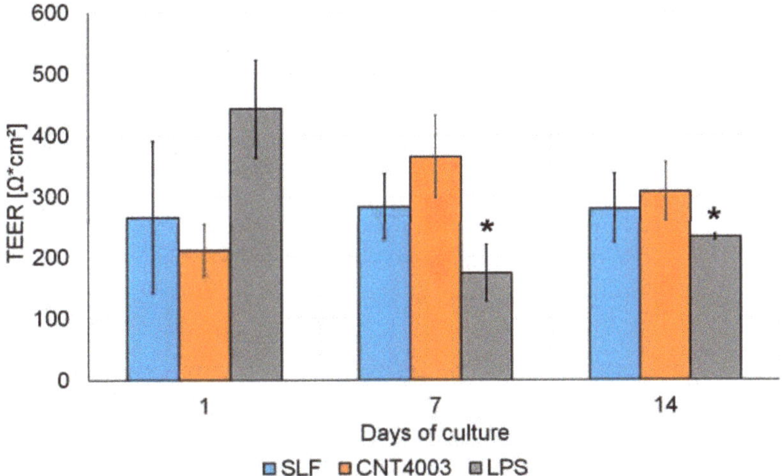

Figure 4. Time-dependent changes of transepithelial electrical resistance (TEER) in EpiAirway™ tissues exposed to simulated lung fluid (SLF) alone or SLF containing CNT4003. Significant decreases compared to 1d are indicated by an asterisk.

EpiAirway™ tissues reacted to stimulation with all concentrations of LPS with significant increases of IL-6 and IL-8 secretions compared to unstimulated tissues at all time points with the exception of IL-6 at day 7 (Table 2).

For assessment of the clearance function of the bronchial epithelium, mucociliary clearance and CBF were evaluated. No transport of the polystyrene indicator beads was detected in the EpiAirway™ tissues, presumably due to the low density of the cilia.

Table 2. Time-dependent changes (%) of cytokine secretion by EpiAirway™ tissues upon stimulation to 1, 5 and 10 µg lipopolysaccharide (LPS). Secretion of unstimulated tissues is set as 100%. Significant differences ($p < 0.05$) between treated and untreated samples are indicated by an asterisk.

Cytokine	LPS	Day 1	Day 7	Day 14
	1 µg	228 ± 83 *	89 ± 7	325 ± 18 *
Interleukin 6	5 µg	368 ± 165 *	163 ± 7 *	211 ± 18 *
	10 µg	847 ± 327 *	240 ± 12 *	383 ± 24 *
	1 µg	133 ± 10 *	255 ± 46 *	291 ± 37 *
Interleukin 8	5 µg	298 ± 18 *	262 ± 22 *	329 ± 26 *
	10 µg	509 ± 12 *	595 ± 7 *	428 ± 4 *

Prior to the assessment of CNT4003, the reaction to LPS was tested. It was found that exposure to LPS (1, 5, 10 µg) increased CBF at all concentrations significantly from 9.9 ± 1.7 Hz (untreated controls) to 15.3–17.8 Hz at 7 d and from 12.7 ± 1.4 Hz (untreated controls) to 15.1–16.2 Hz at 14 d. Basal CBF was significantly lower in this batch of tissues than the frequencies measured in the batch where CNT4003 were tested and no LPS was included.

CBF of untreated tissues was highest at 1 d and significantly lower at all time points (Table 3). The positive control propranolol decreased CBF significantly at all time points. Significant decreases of CBF upon exposure to CNT4003 compared to the respective medium control were not noted. There was a trend of higher CBF in the peripheral regions (15.8 ± 4.2 Hz) than in the central regions of the insert (14.7 ± 2.8 Hz).

Table 3. Ciliary beating frequency (Hz) in EpiAirway™ tissues after exposure to negative control, positive control (propranolol) or CNT4003. Significant differences ($p < 0.05$) between treated and untreated samples are indicated by an asterisk. Significant changes over time of the untreated tissues are marked by a hash.

Treatment	Day 1	Day 7	Day 14
Medium	34 ± 6	27 ± 2 [#]	23 ± 8 [#]
Propranolol	4 ± 0 *	1 ± 0 *	0 ± 0 *
CNT4003	23 ± 10	22 ± 4	16 ± 2

3.1.3. MucilAir™

MucilAir™ tissues are composed of two to three rows of cells. Cilia are already visible in the hemalaun staining (Figure 5A). They form much longer structures at the apical surface (Figure 5B) than in the EpiAirway™ tissue (Figure 3B). Further, rare mucin-producing cells can be seen in the MucilAir™ tissues (Figure 5C).

TEER values of the untreated MucilAir™ cultures significantly decreased over 21 d for cultures from 563 ± 66 Ω × cm^2 at 1 d to 282 ± 24 Ω × cm^2 at 21 d. Therefore, tissues were studied for only 14 d of exposure with LPS and CNT4003. While TEER values of the LPS-treated tissues were significantly decreased compared to 1 d, CNT4003 particles did not affect TEER values over the entire observation time of 14 d (Figure 6).

The reactivity of the tissues to an inflammatory stimulus was tested by incubation with 1, 5 and 10 µg LPS at 1 d, 7 d and 14 d and quantification of IL-6 and IL-8 was performed. IL-6 levels in LPS-stimulated cultures at all time points and all LPS concentrations were significantly lower than that of the untreated controls (Table 4), which is consistent with other studies reporting low IL-6 levels in LPS-stimulated MucilAir™ tissues. In contrast to the action of IL-6, a robust increase in IL-8 secretion was seen after stimulation with LPS at all concentrations. Metz et al. suggested that concentrations of 10 µg/mL LPS were too low to induce a significant increase in IL-6 [33]. IL-6 secretions of the CNT4003-stimulated MucilAir™ tissues were significantly lower than the unstimulated tissues at all time points

(Table 4). IL-8 secretions were also significantly lower than that of the unstimulated controls, except at 1 d.

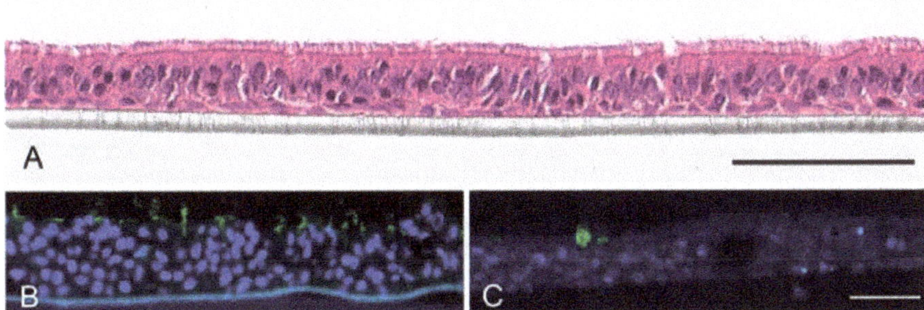

Figure 5. Morphology of MucilAir™ tissues according to hemalaun staining (**A**, scale bar: 100 μm) and cellular composition of the tissues according to immunocytochemistry (**B,C**). Anti-α-tubulin staining (green, (**B**)) visualized cilia of bronchial epithelial cells and anti-mucin 5AC staining (green, (**C**)) visualized goblet cells. Nuclei are counterstained with Hoechst 33342 (blue). Scale bar 50 μm.

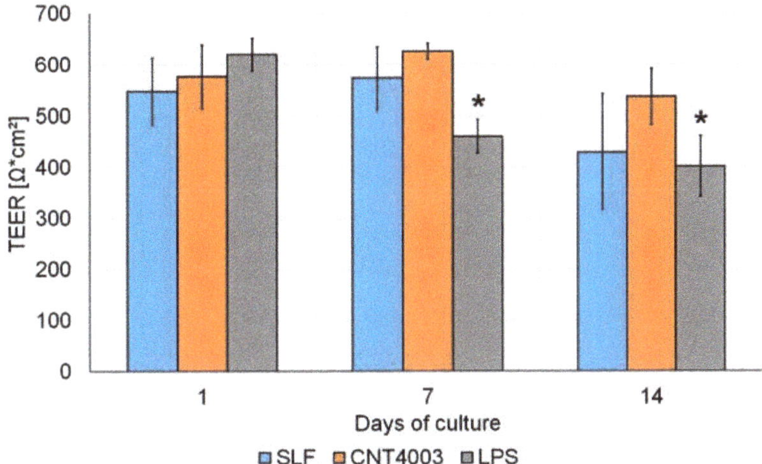

Figure 6. Time-dependent changes in TEER values of MucilAir™ tissues exposed to simulated lung fluid (SLF) alone or with SLF containing CNT4003. Significant decreases compared to 1 d are indicated by an asterisk.

MucilAir™ tissues were able to transport the polystyrene marker beads. However, since there was no coordinated beating like in the in vivo situation, there was no linear transport of the beads. If transport was seen, the beads moved in circles (Video S1, Supplementary Material). Quantification of the effect of the CNT4003 particles was not possible because the diameter and number of the vortexes showed prominent variations between the tissues. The diameter had an influence on the velocity because transport at the periphery of the vortex was faster than at the center.

CBF of untreated tissues was significantly increased at 14 d compared to 1 d (Table 5). Propranolol treatment decreased CBF significantly at all time points. Additionally, the treatment with 10 μg LPS decreased CBF significantly at 14 d. There was no difference between central (19.5 ± 7.1 Hz) and peripheral regions of the insert (20.6 ± 6.0 Hz). CNT4003 exposure did not changes CBF to significant degree.

Table 4. Time-dependent changes (%) of cytokine secretion by MucilAir™ tissues upon stimulation with 1, 5 and 10 µg lipopolysaccharide (LPS) or CNT4003. Secretion of unstimulated tissues is set as 100%. Significant differences ($p < 0.05$) between treated and untreated samples are indicated by an asterisk. Abbreviation: n.a., not available.

Cytokine	Stimulus	Day 1	Day 7	Day 14
Interleukin 6	LPS_1 µg	69 ± 2 *	n.a. [a]	44 ± 3 *
	LPS_5 µg	77 ± 17 *	n.a. [a]	74 ± 15 *
	LPS_10 µg	52 ± 3 *	n.a. [a]	83 ± 5 *
Interleukin 8	LPS_1 µg	164 ± 2 *	214 ± 172 *	250 ± 142 *
	LPS_5 µg	143 ± 2 *	470 ± 131 *	477 ± 37 *
	LPS_10 µg	277 ± 161 *	443 ± 132 *	659 ± 26 *
Interleukin 6	CNT4003	43 ± 42 *	60 ± 26 *	78 ± 26 *
Interleukin 8	CNT4003	54 ± 12 *	108 ± 13	80 ± 17

[a] values deleted due to assay interference.

Table 5. Ciliary beating frequency (Hz) in MucilAir™ tissues after exposure to control, lipopolysaccharide (LPS) or 25 µg CNT4003. Significant ($p < 0.05$) differences in the medium-treated group at different time points are indicated by an asterisk and differences of the treatment to the medium control with a hash.

Treatment	Day 1	Day 7	Day 14
Medium	15 ± 5	21 ± 6	28 ± 3 [#]
Propranolol	5 ± 1 *	5 ± 1 *	8 ± 4 *
CNT4003	12 ± 2	18 ± 8	23 ± 7
10 µg LPS	14 ± 3	19 ± 6	19 ± 4 *

3.2. Assessment of Effects in Models for the Alveolar Part of the Respiratory Tract

3.2.1. A549 Monocultures and Coculture with THP-1 Macrophages

Exposure to 12.5 and 25 µg CNT4003 had no negative effect on the viability of A549 cells in monoculture and coculture with THP-1, but exposure to 50 µg CNT decreased viability significantly in both types of cultures (Figure S3, Supplementary Material).

A549 cells in mono and coculture with THP-1 macrophages secrete low basal levels of IL-6 (A549 monoculture; 0–1.4 pg/mL and 0.04–0.1 pg/mL in coculture; ref. [27]) and were below the limit of detection in this study. They react, however, at lower concentrations than the bronchial models to LPS and, upon stimulation with LPS, IL-6 levels between 140 and 279 pg/mL in A549 monocultures and between 2 and 3 ng/mL in cocultures with THP-1 were measured. IL-8 levels were 21–27 ng/mL and 53–84 ng/mL, respectively (Table 6). This indicates that the cocultures produced approximately 10 times higher amounts of IL-6 and 4 times higher levels of IL-8 upon LPS stimulation than the A549 monocultures. In contrast to LPS, exposure to 25 µg CNT4003 did not stimulate the secretion of IL-6 and IL-8, which remained below the detection limit.

Table 6. Time-dependent changes in interleukin 6 secretion (pg/mL) and interleukin 8 (ng/mL) secretion by A549 mono and A549/THP-1 coculture upon stimulation with 100 ng/mL lipopolysaccharide (LPS). Since cytokine levels of untreated cultures were zero or below zero, normalization to % of control could not be made.

Cytokine	Day 1	Day 7	Day 14
Interleukin 6_A549	140 ± 33	279 ± 91	182 ± 48
Interleukin 8_A549	21 ± 7	27 ± 9	20 ± 2
Interleukin 6_A549/THP-1	2025 ± 152	2137 ± 587	3195 ± 93
Interleukin 8_A549/THP-1	81 ± 11	53 ± 29	84 ± 20

3.2.2. EpiAlveolar™ Tissues with and without THP-1 Macrophages

EpiAlveolar™ tissues consist of alveolar epithelial cells and fibroblasts at the apical site and endothelial cells at the basal site of the membrane [17]. In the hemalaun-stained sections, three to four rows of cells can be seen and tissues with THP-1 (Figure 7B) appear thicker than those without THP-1 (Figure 7A). In tissues with THP-1 cells, the macrophages were identified by CD45-immunoreactivity (Figure 7C).

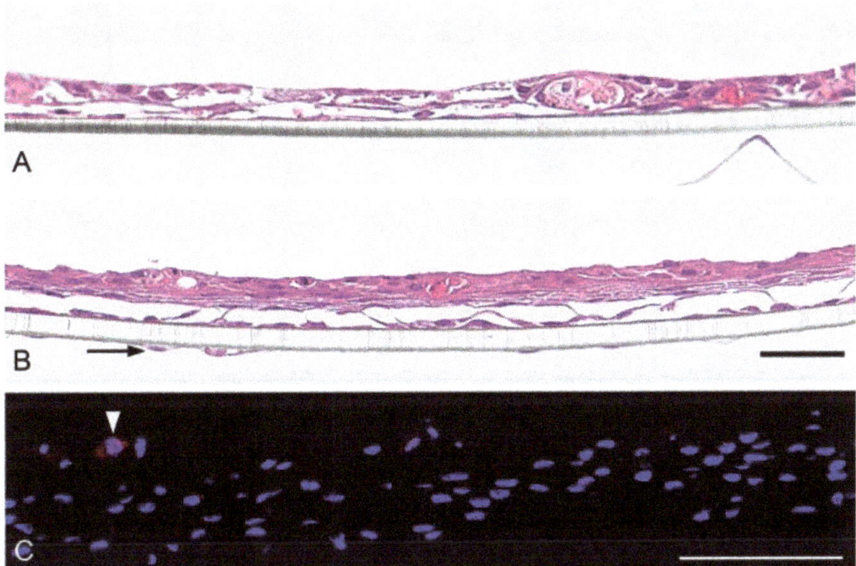

Figure 7. Radial section of EpiAlveolar™ tissues without THP-1 (**A**) and with THP-1 (**B**) macrophages. Endothelial cells at the basal side of the transwell membrane are visible in B (arrow). (**C**) CD45-immunoreactive THP-1 macrophages are rarely seen (arrowhead). Scale bar 50 µm.

Exposure to 25 µg CNT4003 particles did not decrease viability in EpiAlveolar™ tissues with and without THP-1 cells (Figure S4, Supplementary Material).

EpiAlveolar™ tissues formed tight epithelial barriers with TEER values > 850 $\Omega \times cm^2$ (Figure 8). TEER values of LPS-exposed tissues reached 1260 $\Omega \times cm^2$ in the tissues with THP-1 and 1099 $\Omega \times cm^2$ in the tissues without THP-1. There were no differences between tissues with and without THP-1 macrophages and CNT4003 treatment did not influence barrier tightness. At 14 d, prominent variations between tissue replicates (328–974 $\Omega \times cm^2$ in tissues with THP-1 and 322–1701 $\Omega \times cm^2$ in tissues without THP-1) were noted. TEER values may be a good parameter to screen the quality of the tissues because after one delayed delivery, TEER values of the tissues were 97 $\pm$ 137 $\Omega \times cm^2$.

EpiAlveolar™ tissues reacted to stimulation with 1–10 µg LPS with variable and moderate increases in IL-6 and IL-8 secretion. Significant increases in cytokine secretion were mainly seen after stimulation with 5 and 10 µg LPS (Supplementary Material Table S1). In general, more stimulations with significant increases in IL-6 and IL-8 were seen in tissues without THP-1. Treatment with CNT4003 induced a significant increase in IL-6 secretion in EpiAlveolar™ tissues with and without THP-1 macrophages (Table 7). The response in the tissues without THP-1 macrophages was significantly higher than in the tissues with THP-1 cells.

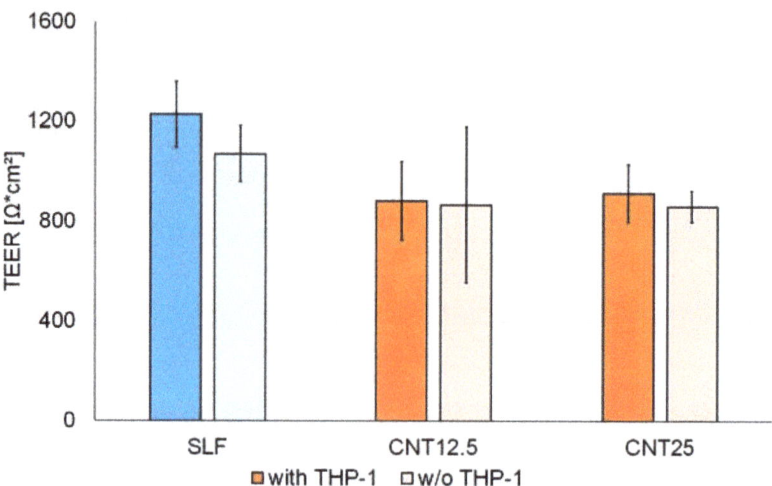

Figure 8. TEER values of EpiAlveolar™ tissues with and without THP-1 treated with control (SLF) or 12.5 and 25 µg CNT4003 particles at 7 d.

Table 7. Time-dependent changes (in %) of cytokine secretion by EpiAlveolar™ tissues without THP-1 (Epi) and with THP-1 (Epi/THP-1) upon stimulation with lipopolysaccharide (LPS) and CNT4003. Secretion of unstimulated tissues is set as 100%. Significant increases ($p < 0.05$) between treated and untreated samples are indicated by an asterisk and significant decreases by a paragraph sign.

Cytokine	LPS_1 µg	LPS_5 µg	LPS_10 µg	CNT_12.5 µg	CNT_25 µg
Interleukin 6_Epi	21 ± 1 §	112 ± 2	174 ± 7 *	750 ± 244 *	618 ± 72 *
Interleukin 8_Epi	102 ± 12	327 ± 36 *	305 ± 41 *	82 ± 42	93 ± 21
Interleukin 6_ Epi/THP-1	65 ± 30	104 ± 30	214 ± 48 *	288 ± 78 *	390 ± 139 *
Interleukin 8_ Epi/THP-1	100 ± 4	107 ± 81	883 ± 29 *	174 ± 53	102 ± 40

4. Discussion

The tightness of the epithelial barrier and low basal levels of proinflammatory cytokines, are crucial properties of the healthy lung, and the used models should be able to assess these parameters. In this comprehensive study, CNTs were evaluated for potential adverse effects on the bronchial and alveolar part of the respiratory system. Further, not only acute cytotoxicity but also epithelial barrier properties, release of cytokines and CBF were used as readout parameters.

4.1. Bronchial Models

The suitability of Calu-3 cells as a model for the respiratory barrier and identification of inflammatory effects has already been reported. In this study, TEER values were 210–365 $\Omega \times cm^2$, which is within the range (~100 $\Omega \times cm^2$, ~250 $\Omega \times cm^2$ and 390 $\Omega \times cm^2$) reported in the literature [34–36]. Variations in TEER values of EpiAirway™ tissues can show prominent interdonor variations from ~300 $\Omega \times cm^2$ to >900 $\Omega \times cm^2$ [34] and values obtained in this study were at the lower end of this range. MucilAir™ tissues in this study showed initial TEER values of 548 ± 65 $\Omega \times cm^2$. Similar to the EpiAirway™ tissues, there were prominent variations depending on the batch ranging from 346 ± 14 to 638 ± 81 $\Omega \times cm^2$ [37]. The difference between different batches of EpiAirway™ tissues suggests differences between donors or prolonged delivery conditions.

Similar to the values in this study, IL-6 and IL-8 values in EpiAirway™ tissues were reported as 100–350 pg/mL and 14,500 ± 3000 pg/mL, respectively [8], and 91 ± 40 pg/mL IL-6 and 5300 ± 220 pg/mL IL-8 [36]. Interbatch levels of IL-6 and IL-8 secretion in MucilAir™ can vary by a factor of two to four (IL-6: 58 ± 26–205 ± 130 pg/mL and IL-8: 4.14 ± 0.61–8.19 ± 3.29 ng/mL) [37]. MucilAir™ tissues in this study secreted up to 358 pg/mL IL-6, while IL-8 were around 100 times higher. The much lower IL-6 than IL-8 secretion by MucilAir™ tissues has also been observed in other studies (100 pg/mL vs. 10,000 pg/mL; [38]). Therefore, most researchers used only IL-8 with basal secretions of ~20 ng/mL, 1.8–5.2 ng/mL and 5–30 ng/mL as indicator for inflammation [39–41].

Normal CBF in human bronchi was reported as 12–15 Hz [42], but great variation from 4 to 19 Hz were reported for human conducting airways [43]. For bronchial epithelial cells of EpiAirway™, a basal CBF of ~18 Hz (17.62–18.02 Hz) has been reported [44], which is slightly lower than the frequencies determined in this study. MucilAir™ tissues in this study had higher CBF than EpiAirway™ tissues. The different ratio of cilia-bearing to mucus-producing cells between the tissues may be the reason for this difference. The average CBF of MucilAir™ tissues in the literature are indicated as 15.8 ± 0.3 Hz and 11.7 ± 1.2 Hz [43,45], and Beyeler et al. observed higher CBF in the peripheral than in the central region [18]. This trend was also to some extent seen in the EpiAirway™ tissues but not in the MucilAir™ tissues of this study. It indicates that it might be good to use similar regions of the insert for the analysis and indicate that in the protocol, e.g., Kim et al. measured 1–2 mm away from the center of the insert [46]. While propranolol reduced CBF, no marked increase in CBF upon administration of forskolin to the tissues was seen. The lack of a strong increase of CBF was also observed in mouse and rat lung slices [47,48]. The authors hypothesized that cells were preactivated by the mechanical manipulation and beat at their maximum frequency. A decrease of CBF has been observed as a reaction to bacteria and diesel exhaust particles and can be interpreted as a toxic response [42]. In this study, 10 µg LPS decreased CBF in the MucilAir™ tissues but increased it in EpiAirway™ tissues. The different reaction may be due to the different basal CBF, which were ~10 Hz in EpiAirway™ and 28 Hz in MucilAir™ tissues. While the reconstructed tissues presented the advantage that detection of CBF was possible, artificial stimulation of CBF by manual manipulation of the insert, differences due to interdonor differences or damage by long shipping times are disadvantages compared to in-house models based on cell lines.

4.2. Alveolar Models

The suitability of A549 cells in mono and in coculture with THP-1 macrophages for the assessment of acute and prolonged effects of particles was shown previously [27]. Assessment of the effect on TEER values is, however, only possible in the reconstructed tissues. Similar to EpiAirway™ and MucilAir™ tissues, marked differences in TEER values were reported for EpiAlveolar™ tissues. We obtained TEER values of 1229 ± 51 $\Omega \times cm^2$ for the EpiAlveolar™ tissues with THP-1 and 1071 ± 111 $\Omega \times cm^2$ for EpiAlveolar™ tissues without THP-1 at 7 d. These data are within the range of the reported values and indicate intact barrier properties. In addition to differences in equipment and operations for the measurements, the shipment of the tissues may impair their viabilities resulting in decreased TEER values compared to those at the producer. TEER values at 7 d and 14 d in the laboratory of the producer were ~1400–1000 $\Omega \times cm^2$ and in another laboratory ~400–600 $\Omega \times cm^2$ [17].

IL-6 levels released by EpiAlveolar™ tissues were reported to vary between 500 pg/mL and 3000 pg/mL in the tissues with THP-1 and between 1500 and 2500 pg/mL in the tissues without THP-1 macrophages [17]. Levels of secreted IL-8 measured over 21 days have been reported as ~5000–15,000 pg/mL in the tissues with THP-1 macrophages and as ~20,000 pg/mL in the tissues without THP-1 macrophages. In this study, IL-6 and IL-8 secretions were in the same order of magnitude as data published by Barosova et al. (e.g., 214 pg/mL–1166 pg/mL in the tissues with THP-1 and 164–1944 pg/mL in those without THP-1 for IL-6 and 5600 pg/mL–26,083 pg/mL in the tissues with THP-1 and from

7476–30,524 pg/mL in EpiAlveolar™ tissues without THP-1 for IL-8). Similar to that study, we did not detect a major influence of THP-1 macrophages on the basal secretion of the interleukins. Upon LPS stimulation, increases were in a similar range for EpiAlveolar™ tissues with and without THP-1 for IL-6 and IL-8 secretions. Coculture with THP-1, by contrast, increased cytokine release markedly upon stimulation with LPS. Increased release of a various proinflammatory cytokines upon LPS stimulation by A549/THP-1 cocultures compared to A549 monocultures has been reported and was reported to be caused by activation of NF-κB [49,50]. The lack of a pronounced response in the EpiAlveolar™ tissues may be due to anti-inflammatory effects by the other cells present in the tissue construct.

4.3. Effects of CNT4003

All models for the bronchial tract were stable over 14 d of evaluation and reacted to the proinflammatory stimulus LPS with significant cytokine increases. They were, therefore, regarded as suitable to detect potential toxic effects of CNT4003. The administered dose was 2×25 µg CNT4003, which is in the range of the lifetime lung exposure of workers to CNTs (12.4–46.5 µg/cm^2 lung surface [51]). Exposure to CNT4003 particles caused no adverse effects on viability, epithelial barrier integrity in bronchial models and inflammation. The significant decrease of IL-6 levels in Calu-3 and MucilAir™ tissues might be explained by the anti-inflammatory action of DPPC bound to the CNTs [52]. Lack of effects on cyto-toxicity, oxidative stress generation and inflammation was also reported after cumulative (5×10 µg/cm^2) MWCNT exposure of human lung explants [19]. They observed, however, a small increase in CBF from ~8 to ~10 Hz, which was not seen in this study. One explana-tion may be that the high basal CBF made a further increase impossible. The model, on the other hand, reacted to LPS exposure with a decrease in CBF, indicating that it is capable of identifying adverse effects. We observed in this study that the distribution of the CNT4003 on the insert was not homogenous (Figure S1, Supplementary Material), and one group reported differences in CBF between central and peripheral areas of the insert [18]. Also due to differences in the deposition of the CNTs (Figure S1, Supplementary material) and the fact that not the entire tissue can be evaluated, small local effects may be overlooked.

The two models for the alveolar part of the respiratory tract indicated adverse effects only in the EpiAlveolar™ tissues with and without THP-1 macrophages by increased IL-6 secretion upon exposure to 12.5 and 25 µg CNT4003 with no dose dependency. The greater cytokine release induced by CNT4003 in the EpiAlveolar™ tissues than in the A549 model is in contrast to the more pronounced response of the A549 model to LPS. Greater toxicity of SDS in A549 than in MucilAir™ tissues has been reported and can be explained by the lack of a mucus layer [37]. Protection of the epithelial cell by greater surface coating may also explain the greater sensitivity of the EpiAlveolar™ tissues to CNT4003 particles. It was reported that the surface tension, as an indication for surfactant production, of EpiAlveolar™ tissues was markedly lower than that of A549 cells [17]. Surfactant is produced by alveolar epithelial type II cells, which represent only a small fraction of the EpiAlveolar™ tissues, whereas the A549 cells resemble these cells and are capable to produce surfactant [13].

5. Conclusions

Cytotoxicity, TEER values, interleukin secretion and CBF show no indication of adverse effects of CNT4003 in the bronchial part of the respiratory tract. The increased IL-6 secretion of the EpiAlveolar™ tissues may indicate adverse effects of CNT4003 on the alveolar part of the lung. Since the doses for whole-life exposure of workers were applied over a shorter time, the physiological relevance is not clear. Although the reconstructed tissues are produced in a standardized way, considerable variations in basal parameters between batches were seen. In addition to the interdonor differences, the duration of the delivery from the producer to the laboratory of the user has an effect on the viability and reaction of the tissues.

Supplementary Materials: The following supporting information can be downloaded at: https://www.mdpi.com/article/10.3390/nano13040682/s1. Figures S1–S4: Distribution of CNT4003 on cell-grown inserts; cytotoxicity of CNT4003 in Calu-3 cells; cytotoxicity of CNT4003 in A549 cells; cytotoxicity of CNT4003 in EpiAlveolar™ tissues; Table S1: LPS-induced cytokine secretion by EpiAlveolar™ tissues; Video S1: Mucociliary clearance in MucilAir™ tissues.

Author Contributions: Conceptualization, E.F.; methodology and data curation, C.M., R.J., M.A.-N.; writing—original draft preparation, C.M. and E.F.; writing—review and editing, C.M., E.R. and E.F.; visualization, M.A.-N. All authors have read and agreed to the published version of the manuscript.

Funding: PETA Science Consortium International e.V. supported this research with an Award from MatTek Life Sciences for 3D human tissue models to assess respiratory toxicity to E.F.

Data Availability Statement: Datasets analyzed or generated during the study are available upon request from the authors.

Conflicts of Interest: The authors declare no conflict of interest.

References

1. Najahi-Missaoui, W.; Arnold, R.D.; Cummings, B.S. Safe Nanoparticles: Are We There Yet? *Int. J. Mol. Sci.* **2020**, *22*, 385. [CrossRef] [PubMed]
2. Dusinská, M.; Collins, A.; Kazimírová, A.; Barancoková, M.; Harrington, V.; Volkovová, K.; Staruchová, M.; Horská, A.; Wsólová, L.; Kocan, A.; et al. Genotoxic effects of asbestos in humans. *Mutat. Res.* **2004**, *553*, 91–102. [CrossRef] [PubMed]
3. Fröhlich, E.; Mercuri, A.; Wu, S.; Salar-Behzadi, S. Measurements of Deposition, Lung Surface Area and Lung Fluid for Simulation of Inhaled Compounds. *Front. Pharmacol.* **2016**, *7*, 181. [CrossRef] [PubMed]
4. Chetyrkina, M.R.; Fedorov, F.S.; Nasibulin, A.G. In vitro toxicity of carbon nanotubes: A systematic review. *RSC Adv.* **2022**, *12*, 16235–16256. [CrossRef]
5. Cotogno, G.; Totaro, S.; Rasmussen, K.; Pianella, F.; Roncaglia, M.; Olsson, H.; Riego Sintes, J.; Crutzen, H. *The JRC Nanomaterials Repository Safe Handling of Nanomaterials in the Sub-Sampling Facility*; JRC Technical Reports; European Union: Ispra, Italy, 2016.
6. Jensen, K.; Kembouche, Y.; Christiansen, E.; Jacobsen, N.; Wallin, H. *The Generic NANOGENOTOX Dispersion Protocol—Standard Operation Procedure (SOP) and Background Documentation*; The National Research Centre for the Working Environment (NRCWE): Copenhagen, Denmark, 2011.
7. Lee, J.H.; Ahn, K.H.; Kim, S.M.; Kim, E.; Lee, G.H.; Han, J.H.; Yu, I.J. Three-Day Continuous Exposure Monitoring of CNT Manufacturing Workplaces. *Biomed Res. Int.* **2015**, *2015*, 237140. [CrossRef]
8. Seagrave, J.; Dunaway, S.; McDonald, J.D.; Mauderly, J.L.; Hayden, P.; Stidley, C. Responses of differentiated primary human lung epithelial cells to exposure to diesel exhaust at an air-liquid interface. *Exp. Lung Res.* **2007**, *33*, 27–51. [CrossRef]
9. Lee, J.H.; Lee, S.B.; Bae, G.N.; Jeon, K.S.; Yoon, J.U.; Ji, J.H.; Sung, J.H.; Lee, B.G.; Lee, J.H.; Yang, J.S.; et al. Exposure assessment of carbon nanotube manufacturing workplaces. *Inhal. Toxicol.* **2010**, *22*, 369–381. [CrossRef]
10. Fujitani, Y.; Furuyama, A.; Hirano, S. Generation of Airborne Multi-Walled Carbon Nanotubes for Inhalation Studies. *Aerosol Sci. Technol.* **2009**, *43*, 881–890. [CrossRef]
11. Di Ianni, E.; Erdem, J.S.; Møller, P.; Sahlgren, N.M.; Poulsen, S.S.; Knudsen, K.B.; Zienolddiny, S.; Saber, A.T.; Wallin, H.; Vogel, U.; et al. In vitro-in vivo correlations of pulmonary inflammogenicity and genotoxicity of MWCNT. *Part. Fibre Toxicol.* **2021**, *18*, 25. [CrossRef]
12. Grainger, C.I.; Greenwell, L.L.; Lockley, D.J.; Martin, G.P.; Forbes, B. Culture of Calu-3 cells at the air interface provides a representative model of the airway epithelial barrier. *Pharm. Res.* **2006**, *23*, 1482–1490. [CrossRef]
13. Öhlinger, K.; Kolesnik, T.; Meindl, C.; Gallé, B.; Absenger-Novak, M.; Kolb-Lenz, D.; Fröhlich, E. Air-liquid interface culture changes surface properties of A549 cells. *Toxicol. In Vitro* **2019**, *60*, 369–382. [CrossRef] [PubMed]
14. Kreft, M.E.; Jerman, U.D.; Lasic, E.; Hevir-Kene, N.; Rizner, T.L.; Peternel, L.; Kristan, K. The characterization of the human cell line Calu-3 under different culture conditions and its use as an optimized in vitro model to investigate bronchial epithelial function. *Eur. J. Pharm. Sci.* **2015**, *69*, 1–9. [CrossRef] [PubMed]
15. George, I.; Vranic, S.; Boland, S.; Courtois, A.; Baeza-Squiban, A. Development of an in vitro model of human bronchial epithelial barrier to study nanoparticle translocation. *Toxicol. In Vitro* **2015**, *29*, 51–58. [CrossRef] [PubMed]
16. Buckley, S.; Kim, K.; Ehrhardt, C. In Vitro Cell Culture Models for Evaluating Controlled Release Pulmonary Drug Delivery. In *Controlled Pulmonary Drug Delivery*; Smyth, H., Hickey, A., Eds.; Springer: New York, NY, USA, 2011; pp. 417–442.
17. Barosova, H.; Maione, A.G.; Septiadi, D.; Sharma, M.; Haeni, L.; Balog, S.; O'Connell, O.; Jackson, G.R.; Brown, D.; Clippinger, A.J.; et al. Use of EpiAlveolar Lung Model to Predict Fibrotic Potential of Multiwalled Carbon Nanotubes. *ACS Nano* **2020**, *14*, 3941–3956. [CrossRef]
18. Beyeler, S.; Chortarea, S.; Rothen-Rutishauser, B.; Petri-Fink, A.; Wick, P.; Tschanz, S.A.; von Garnier, C.; Blank, F. Acute effects of multi-walled carbon nanotubes on primary bronchial epithelial cells from COPD patients. *Nanotoxicology* **2018**, *12*, 699–711. [CrossRef]

19. Chortarea, S.; Barosova, H.; Clift, M.J.D.; Wick, P.; Petri-Fink, A.; Rothen-Rutishauser, B. Human Asthmatic Bronchial Cells Are More Susceptible to Subchronic Repeated Exposures of Aerosolized Carbon Nanotubes At Occupationally Relevant Doses Than Healthy Cells. *ACS Nano* **2017**, *11*, 7615–7625. [CrossRef]
20. Fröhlich, E. Toxicity of orally inhaled drug formulations at the alveolar barrier: Parameters for initial biological screening. *Drug Deliv.* **2017**, *24*, 891–905. [CrossRef]
21. Lima, T.; Bernfur, K.; Vilanova, M.; Cedervall, T. Understanding the Lipid and Protein Corona Formation on Different Sized Polymeric Nanoparticles. *Sci. Rep.* **2020**, *10*, 1129. [CrossRef]
22. Carregal-Romero, S.; Groult, H.; Cañadas, O.; Noelia, A.; Lechuga-Vieco, A.V.; García-Fojeda, B.; Herranz, F.; Pellico, J.; Hidalgo, A.; Casals, C.; et al. Delayed alveolar clearance of nanoparticles through control of coating composition and interaction with lung surfactant protein A. *Biomater. Adv.* **2022**, *134*, 112551. [CrossRef]
23. Kaur, R.; Dennison, S.R.; Burrow, A.J.; Rudramurthy, S.M.; Swami, R.; Gorki, V.; Katare, O.P.; Kaushik, A.; Singh, B.; Singh, K.K. Nebulised surface-active hybrid nanoparticles of voriconazole for pulmonary Aspergillosis demonstrate clathrin-mediated cellular uptake, improved antifungal efficacy and lung retention. *J. Nanobiotechnol.* **2021**, *19*, 19. [CrossRef]
24. Sauer, U.G.; Vogel, S.; Hess, A.; Kolle, S.N.; Ma-Hock, L.; van Ravenzwaay, B.; Landsiedel, R. In vivo-in vitro comparison of acute respiratory tract toxicity using human 3D airway epithelial models and human A549 and murine 3T3 monolayer cell systems. *Toxicol. In Vitro* **2013**, *27*, 174–190. [CrossRef] [PubMed]
25. Rasmussen, K.; Mast, J.; De Temmerman, P.; Verleysen, E.; Waegeneers, N.; Van Steen, F.; Pizzolon, J.; De Temmerman, L.; Van Doren, E.; Jensen, K.; et al. *Multi-Walled Carbon Nanotubes, NM-400, NM-401, NM-402, NM-403: Characterisation and Physico-Chemical Properties*; Publications Office of the European Union: Luxembourg, 2014.
26. Jacobsen, N.; Pojano, G.; Wallin, H.; Jensen, K. *Nanomaterial Dispersion Protocol for Toxicological Studies in ENPRA*; Internal ENPRA Report; National Research Centre for the Working Environment: Copenhagen, Denmark, 2010.
27. Meindl, C.; Öhlinger, K.; Zrim, V.; Steinkogler, T.; Fröhlich, E. Screening for Effects of Inhaled Nanoparticles in Cell Culture Models for Prolonged Exposure. *Nanomaterials* **2021**, *11*, 606. [CrossRef] [PubMed]
28. Jantzen, K.; Roursgaard, M.; Desler, C.; Loft, S.; Rasmussen, L.J.; Møller, P. Oxidative damage to DNA by diesel exhaust particle exposure in co-cultures of human lung epithelial cells and macrophages. *Mutagenesis* **2012**, *27*, 693–701. [CrossRef] [PubMed]
29. Behrsing, H.P.; Wahab, A.; Ukishima, L.; Grodi, C.; Frentzel, S.; Johne, S.; Ishikawa, S.; Ito, S.; Wieczorek, R.; Budde, J.; et al. Ciliary Beat Frequency: Proceedings and Recommendations from a Multi-laboratory Ring Trial Using 3-D Reconstituted Human Airway Epithelium to Model Mucociliary Clearance. *Altern. Lab. Anim.* **2022**, *50*, 293–309. [CrossRef] [PubMed]
30. Mohri, H.; Inaba, K.; Ishijima, S.; Baba, S.A. Tubulin-dynein system in flagellar and ciliary movement. *Proc. Jpn. Acad. Ser. B Phys. Biol. Sci.* **2012**, *88*, 397–415. [CrossRef] [PubMed]
31. Binarová, P.; Tuszynski, J. Tubulin: Structure, Functions and Roles in Disease. *Cells* **2019**, *8*, 1294. [CrossRef] [PubMed]
32. Fröhlich, E.; Meindl, C. In Vitro Assessment of Chronic Nanoparticle Effects on Respiratory Cells. In *Nanomaterials—Toxicity and Risk Assessment*; Soloneski, S., Larramendy, M., Eds.; InTech Open: Rijeka, Croatia, 2015; pp. 69–91.
33. Metz, J.; Knoth, K.; Groß, H.; Lehr, C.M.; Stäbler, C.; Bock, U.; Hittinger, M. Combining MucilAir™ and Vitrocell(®) Powder Chamber for the In Vitro Evaluation of Nasal Ointments in the Context of Aerosolized Pollen. *Pharmaceutics* **2018**, *10*, 56. [CrossRef]
34. Fields, W.; Maione, A.; Keyser, B.; Bombick, B. Characterization and Application of the VITROCELL VC1 Smoke Exposure System and 3D EpiAirway Models for Toxicological and e-Cigarette Evaluations. *Appl. Vitr. Toxicol.* **2017**, *3*, 68–83. [CrossRef]
35. Ren, D.; Nelson, K.L.; Uchakin, P.N.; Smith, A.L.; Gu, X.X.; Daines, D.A. Characterization of extended co-culture of non-typeable Haemophilus influenzae with primary human respiratory tissues. *Exp. Biol. Med.* **2012**, *237*, 540–547. [CrossRef]
36. Zavala, J.; O'Brien, B.; Lichtveld, K.; Sexton, K.G.; Rusyn, I.; Jaspers, I.; Vizuete, W. Assessment of biological responses of EpiAirway 3-D cell constructs versus A549 cells for determining toxicity of ambient air pollution. *Inhal. Toxicol.* **2016**, *28*, 251–259. [CrossRef]
37. Welch, J.; Wallace, J.; Lansley, A.B.; Roper, C. Evaluation of the toxicity of sodium dodecyl sulphate (SDS) in the MucilAir™ human airway model in vitro. *Regul. Toxicol. Pharmacol.* **2021**, *125*, 105022. [CrossRef] [PubMed]
38. Kooter, I.; Ilves, M.; Gröllers-Mulderij, M.; Duistermaat, E.; Tromp, P.C.; Kuper, F.; Kinaret, P.; Savolainen, K.; Greco, D.; Karisola, P.; et al. Molecular Signature of Asthma-Enhanced Sensitivity to CuO Nanoparticle Aerosols from 3D Cell Model. *ACS Nano* **2019**, *13*, 6932–6946. [CrossRef] [PubMed]
39. George, I.; Uboldi, C.; Bernard, E.; Sobrido, M.S.; Dine, S.; Hagège, A.; Vrel, D.; Herlin, N.; Rose, J.; Orsière, T.; et al. Toxicological Assessment of ITER-Like Tungsten Nanoparticles Using an In Vitro 3D Human Airway Epithelium Model. *Nanomaterials* **2019**, *9*, 1374. [CrossRef]
40. Huang, S.; Constant, S.; De Servi, B.; Meloni, M.; Culig, J.; Bertini, M.; Saaid, A. In vitro safety and performance evaluation of a seawater solution enriched with copper, hyaluronic acid, and eucalyptus for nasal lavage. *Med. Devices* **2019**, *12*, 399–410. [CrossRef]
41. Kirkpatrick, D.L.; Millard, J. Evaluation of nafamostat mesylate safety and inhibition of SARS-CoV-2 replication using a 3-dimensional human airway epithelia model. *bioRxiv* **2020**. [CrossRef]
42. Workman, A.D.; Cohen, N.A. The effect of drugs and other compounds on the ciliary beat frequency of human respiratory epithelium. *Am. J. Rhinol. Allergy* **2014**, *28*, 454–464. [CrossRef]

43. Lodes, N.; Seidensticker, K.; Perniss, A.; Nietzer, S.; Oberwinkler, H.; May, T.; Walles, T.; Hebestreit, H.; Hackenberg, S.; Steinke, M. Investigation on Ciliary Functionality of Different Airway Epithelial Cell Lines in Three-Dimensional Cell Culture. *Tissue Eng. Part A* **2020**, *26*, 432–440. [CrossRef]

44. Woodhams, A.; Kidd, D.; Hollings, M. Use of Cilia Beat Frequency in an In Vitro Airway Model (EpiAirway-FT) as a Tool for Assessing Viability. 2018. Available online: https://www.coresta.org/sites/default/files/abstracts/2018_TSRC08_Woodhams.pdf (accessed on 2 November 2020).

45. Phillips, G.; Czekala, L.; Behrsing, H.; Amin, K.; Budde, J.; Stevenson, M.; Wieczorek, R.; Walele, T.; Simms, L. Acute electronic vapour product whole aerosol exposure of 3D human bronchial tissue results in minimal cellular and transcriptomic responses when compared to cigarette smoke. *Toxicol. Res. Appl.* **2021**, *5*, 2397847320988496. [CrossRef]

46. Kim, M.D.; Baumlin, N.; Yoshida, M.; Polineni, D.; Salathe, S.F.; David, J.K.; Peloquin, C.A.; Wanner, A.; Dennis, J.S.; Sailland, J.; et al. Losartan Rescues Inflammation-related Mucociliary Dysfunction in Relevant Models of Cystic Fibrosis. *Am. J. Respir. Crit. Care Med.* **2020**, *201*, 313–324. [CrossRef]

47. Delmotte, P.; Sanderson, M.J. Ciliary beat frequency is maintained at a maximal rate in the small airways of mouse lung slices. *Am. J. Respir. Cell Mol. Biol.* **2006**, *35*, 110–117. [CrossRef]

48. Hayashi, T.; Kawakami, M.; Sasaki, S.; Katsumata, T.; Mori, H.; Yoshida, H.; Nakahari, T. ATP regulation of ciliary beat frequency in rat tracheal and distal airway epithelium. *Exp. Physiol.* **2005**, *90*, 535–544. [CrossRef] [PubMed]

49. Li, J.; Qin, Y.; Chen, Y.; Zhao, P.; Liu, X.; Dong, H.; Zheng, W.; Feng, S.; Mao, X.; Li, C. Mechanisms of the lipopolysaccharide-induced inflammatory response in alveolar epithelial cell/macrophage co-culture. *Exp. Ther. Med.* **2020**, *20*, 76. [CrossRef] [PubMed]

50. Grabowski, N.; Hillaireau, H.; Vergnaud-Gauduchon, J.; Nicolas, V.; Tsapis, N.; Kerdine-Römer, S.; Fattal, E. Surface-Modified Biodegradable Nanoparticles' Impact on Cytotoxicity and Inflammation Response on a Co-Culture of Lung Epithelial Cells and Human-Like Macrophages. *J. Biomed. Nanotechnol.* **2016**, *12*, 135–146. [CrossRef] [PubMed]

51. Gangwal, S.; Brown, J.S.; Wang, A.; Houck, K.A.; Dix, D.J.; Kavlock, R.J.; Hubal, E.A. Informing selection of nanomaterial concentrations for ToxCast in vitro testing based on occupational exposure potential. *Environ. Health Perspect.* **2011**, *119*, 1539–1546. [CrossRef] [PubMed]

52. Tonks, A.; Morris, R.H.; Price, A.J.; Thomas, A.W.; Jones, K.P.; Jackson, S.K. Dipalmitoylphosphatidylcholine modulates inflammatory functions of monocytic cells independently of mitogen activated protein kinases. *Clin. Exp. Immunol.* **2001**, *124*, 86–94. [CrossRef] [PubMed]

 *nanomaterials*

Article

The Effects of Titanium Dioxide Nanoparticles on Osteoblasts Mineralization: A Comparison between 2D and 3D Cell Culture Models

Gabriela de Souza Castro [1,†], Wanderson de Souza [2,†], Thais Suelen Mello Lima [2], Danielle Cabral Bonfim [3], Jacques Werckmann [4], Braulio Soares Archanjo [5], José Mauro Granjeiro [2], Ana Rosa Ribeiro [6] and Sara Gemini-Piperni [1,7,*]

1　Postgraduate Program in Odontology, Unigranrio, Duque de Caxias 25071-202, Brazil
2　Directory of Life Sciences Applied Metrology, National Institute of Metrology Quality and Technology, Rio de Janeiro 25250-020, Brazil
3　LabCeR Group, Federal University of Rio de Janeiro (UFRJ), Rio de Janeiro 21941-901, Brazil
4　Visitant Professor at Brazilian Center for Research in Physics, Rio de Janeiro 22290-180, Brazil
5　Materials Metrology Division, National Institute of Quality and Technology, Rio de Janeiro 25250-020, Brazil
6　NanoSafety Group, International Iberian Nanotechnology Laboratory, 4715-330 Braga, Portugal
7　Laben Group, Federal University of Rio de Janeiro (UFRJ), Rio de Janeiro 21941-901, Brazil
*　Correspondence: sara.gemini@hotmail.com
†　These authors contributed equally to this work.

Abstract: Although several studies assess the biological effects of micro and titanium dioxide nanoparticles (TiO_2 NPs), the literature shows controversial results regarding their effect on bone cell behavior. Studies on the effects of nanoparticles on mammalian cells on two-dimensional (2D) cell cultures display several disadvantages, such as changes in cell morphology, function, and metabolism and fewer cell–cell contacts. This highlights the need to explore the effects of TiO_2 NPs in more complex 3D environments, to better mimic the bone microenvironment. This study aims to compare the differentiation and mineralized matrix production of human osteoblasts SAOS-2 in a monolayer or 3D models after exposure to different concentrations of TiO_2 NPs. Nanoparticles were characterized, and their internalization and effects on the SAOS-2 monolayer and 3D spheroid cells were evaluated with morphological analysis. The mineralization of human osteoblasts upon exposure to TiO_2 NPs was evaluated by alizarin red staining, demonstrating a dose-dependent increase in mineralized matrix in human primary osteoblasts and SAOS-2 both in the monolayer and 3D models. Furthermore, our results reveal that, after high exposure to TiO_2 NPs, the dose-dependent increase in the bone mineralized matrix in the 3D cells model is higher than in the 2D culture, showing a promising model to test the effect on bone osteointegration.

Keywords: TiO_2 NPs; titanium dental implants; 3D spheroids; osteoblasts

Citation: de Souza Castro, G.; de Souza, W.; Lima, T.S.M.; Bonfim, D.C.; Werckmann, J.; Archanjo, B.S.; Granjeiro, J.M.; Ribeiro, A.R.; Gemini-Piperni, S. The Effects of Titanium Dioxide Nanoparticles on Osteoblasts Mineralization: A Comparison between 2D and 3D Cell Culture Models. *Nanomaterials* **2023**, *13*, 425. https://doi.org/10.3390/nano13030425

Academic Editors: Andrea Hartwig and Christoph Van Thriel

Received: 30 December 2022
Revised: 12 January 2023
Accepted: 16 January 2023
Published: 20 January 2023

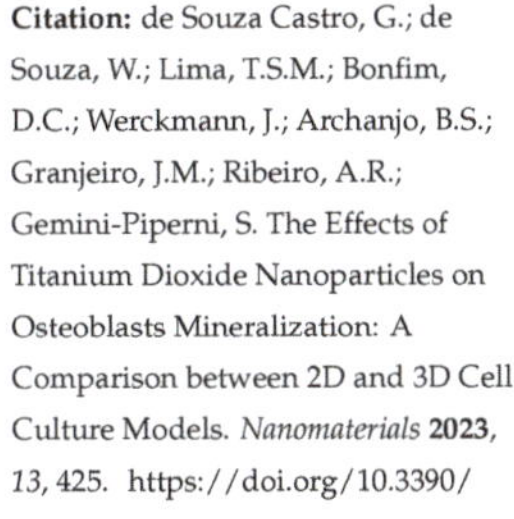

1. Introduction

The development of nanotechnology is increasing exponentially, especially in the production of biomaterials for dental implants [1–3]. This market moved around US$10.87 billion in 2021 and could reach US$12.49 billion in 2022 [1,4]. In this scenario, titanium (Ti) is the metallic material most commonly used for implant applications due to its excellent biocompatibility and osteointegration properties [2,5]; however, the failure of dental implants continues to increase [6]. Although the causes are multifactorial and often related to microbial colonization (biofilms), new questions have been raised about the role of corrosion and/or wear process in the progress of implant failure [7]. Recently, Ti-like particles (mainly titanium dioxide nanoparticles (TiO_2 NPs)) have been found in the peri-implant mucosa and bone cells [6,8,9].

Prothesis degradation processes occur in all material classes, with degradation by-products known to be hazardous. Nonetheless, nanoscale metallic debris released are of tremendous concern since they exhibit enhanced toxicity and dissolving capacities, compared with micron-size polymeric and ceramic debris [7]. Micromovements between the implant abutment and the bone or mucosa unavoidably lead to mechanical wear of the material (formation of nano and microsized debris), which in a corrosive environment result in the release of metallic ions [7,10–12]. Biologically, this event is related to macrophage activation [7,10,11,13] and the onset of pro-inflammatory response, with the release of cytokines such as IL-1β, IL-6, and TNF-α in the peri-implant area, which culminates in reduced osteoblastogenesis, increased osteoclastogenesis, and bone loss in the periprosthetic region [11,13]. Bone loss limits the ability of a prosthesis to withstand physiological loads, giving rise to a need for revision surgery.

Many in vitro and in vivo studies have shown that titanium particles induce pro-inflammatory and toxic effects in the peri-implant environment [7,14–17]. However, some in vitro studies also suggest that TiO_2 NPs can stimulate bone formation [2,18]. Thus, controversy remains about the actual role of TiO_2 NPs on bone cells, likely due to the two-dimensional in vitro models used.

It has been widely reported that 2D osteoblast cell cultures [13–17,19] may lose their original tissue organization and polarity and have limited protein–protein interactions [2,18]. They may exhibit integrins and changes in the cytoskeletal organization that alter their original morphology [2,18,20]. In addition, cells grown in a 2D monolayer often exhibit altered metabolism, phenotype, and gene expression, and the interactions between cells and the extracellular matrix are different from in vivo tissues, which have a three-dimensional architecture [2,20]. In contrast, 3D culture models exhibit greater cell–cell and cell–matrix interactions, which are closer to the in vivo model [21]. Moreover, cellular polarity, which is important for cellular organization and functionality, remains unaltered [2,21,22]. Therefore, surface receptors can bind to extracellular matrix proteins, activate cellular biochemical signals, and influence cell proliferation, differentiation, and mineralization [2,21].

Therefore, in this work, the effects of TiO_2 NPs on the differentiation and production of a mineralized matrix of human osteoblasts cultured in monolayer (2D) and osteoblast-like spheroid culture models (3D) were investigated. Cell viability, morphology, differentiation, mineralization, and nanoparticle internalization were investigated after exposure to NPs. Results demonstrate that TiO_2 NPs lead to a dose-dependent increase in mineralization, although these TiO_2 NPs are clinically known for their possible immune system activation.

2. Materials and Methods

Titanium anatase dispersion: TiO_2 NPs (SIGMA, Kanagawa, Japan) with primary particle size < 25 nm and surface area of 45–55 m^2/g were suspended in ultrapure water (2 mg/mL; pH 4) and dispersed using a direct ultrasound (Q-Sonica) equipped with a 19 mm tip. The sonication was carried out in an ice bath at 32 W of acoustic delivery power for 15 min with 8 s (pulse mode on) and 2 s (pulse mode off), following a protocol previously described by the group [2,5]. After 24 h of stabilization, particle size and particle agglomeration (zeta potential analysis (ζ (mV)) and the polydispersion index (PdI)) were determined by dynamic light scattering (DLS, Zeta-Sizer Nano ZS, Malvern Instruments GmbH, Malvern, UK). DLS measurements were performed at 25 °C using 10 mm polystyrene disposable cuvettes.

To confirm particle size in the medium culture, titanium particles were alternatively suspended in high glucose Dulbecco's Modified Eagle Medium (DMEM, Gibco, Waltham, MA, USA) supplemented with 10% fetal bovine serum (FBS, Gibco) and 1 mg/mL bovine serum albumin (BSA; Sigma-Aldrich, St. Louis, MO, USA) to avoid particle re-agglomeration.

Cell culture: The human osteoblast cell line (SAOS-2) was supplied by the Cell Bank of Rio de Janeiro (BCRJ, Rio de Janeiro, Brazil) packed in frozen ampoules and kept in liquid nitrogen. Cells were thawed and expanded into cell culture flasks (Corning) with DMEM medium supplemented with 10% FBS and 1% penicillin/streptomycin (PS—10,000 units/mL of penicillin and 10,000 µg/mL of streptomycin) (PS, Gibco) in a humidified incubator (5% CO_2, 37 °C). Cell contamination with bacteria, fungi, or mycoplasma was analyzed as previously reported [2,5]. For the 2D model, 10,000 cells/well were seeded in standard flat-bottom 96-well plates for 24 h.

3D culture: For 3D spheroid formation, 96-well U-bottom plates (Corning, Corning, NY, USA) were coated with a thin layer of 1% ultrapure agarose (Sigma-Aldrich), and 10,000 cells were seeded in each well in 200 µL DMEM high glucose medium supplemented with 10% FBS and 1% PS and then incubated for 3 days. Cell growth, shape, and morphology were analyzed on an inverted optical microscope (Nikon Eclipse, Tokyo, Japan), following a protocol previously described [2].

3D cell cytoskeleton: Spheroids were washed with 0.01 M PBS and then fixed with 4% paraformaldehyde (PFA). The cell membrane was then permeabilized by incubation for 2 h with 0.2% BSA + 0.1% Triton X-100 in 0.01 M PBS for 2 h at room temperature (RT). The spheroids were washed three times with a blocking solution containing 50 nM NH_4Cl in 0.01 M PBS. Phalloidin solution (500 ng/mL, stock 1:40, ThermoFisher Lot: F432) diluted in 0.2% BSA + 0.1% Triton X-100 in 0.01 M PBS was then added and incubated for 2 h, and later an additional 30 min with DAPI (1:500) (SIGMA-ALDRICH—Lot: 583-93-7). Cell morphology was visualized using a confocal fluorescence microscope (DMI 6000, Leica, Teaneck, NJ, USA).

NPs exposition: Both 2D and 3D cell cultures were exposed to 0, 5, and 100 µg/mL TiO_2 NPs suspended in incomplete osteogenic medium composed of DMEM supplemented with 10% FBS, 50 µg/mL of ascorbic acid (Sigma), 100 Mm of β-glycerophosphate (Sigma), and antibiotics for 3 and 21 days. Cells without TiO_2 NPs treatment were used as control.

Cytotoxicity assay: After NPs exposition, the cells were washed three times with 0.01 M PBS and then incubated with 0.125% Trypsin (kept in a humidified incubator with 5% CO_2, 37 °C) for 5 min. Trypsin was blocked by adding culture medium with 10% FBS, and 2D adherent cells and 3D spheroids were mechanically dissociated. The cells were centrifuged for 7 min at 500 x g (4 °C), and the pellet was resuspended in 100 µL annexin-binding buffer (Dead Cell Apoptosis Kit for Annexin V; Kit Life and Dead, Life Technologies). The samples were incubated for 15 min (RT) in 3 µL annexin/fluorescein (FITC) solution and 1 µL propidium iodide (according to the manufacturer's instructions). All analyses were performed in a flow cytometer (FACSAria III, BD Biosciences, Franklin Lakes, NJ, USA).

Morphology analysis: Cells were washed with 0.01 M PBS and processed for scanning (SEM) and transmission (TEM) electron microscopy. Briefly, cells were fixed using modified Karnovsky (2% paraformaldehyde, 2.5% glutaraldehyde in 0.1 M sodium cacodylate buffer, pH 7.2) for 2 h at RT and washed with 0.1 M cacodylate buffer. The samples were post-fixed with 1% osmium tetroxide in cacodylate buffer (1:1) for 30 min in the dark, then washed with cacodylate buffer and dehydrated in ethanol (VETEC—1567). Then, for TEM analysis, samples were contrasted in a bloc with 1% of uranyl acetate, dehydrated in acetone, and embedded in Spurr. Ultra-thin sections were also analyzed using EDS in scanning transmission electron microscopy (STEM) mode in a TITAN 80–300 electron microscope (FEI, Netherlands (300 kV).

Alternatively, for SEM analysis, the cells were dried at a critical point (Autosamdri®-815, Series A) and metalized with gold (in a current of 40 mA for 90 sec). The 2D cell samples were analyzed in a scanning electron microscope (JEOL Field Emission Gun-JSM-7401F) with an acceleration voltage of 1 kV. The 3D cells were analyzed under a helium ion beam microscope (HIM) (Carl Zeiss Orion Nanofab—beam current of 0.8 pA, using an electron flood gun to compensate for the positive charge).

Differentiation and analysis of the cell matrix: Cell differentiation was evaluated by alkaline phosphatase histochemistry. The cells were cultured at different times (3, 7, and 21 days). The alkaline phosphatase labeling kit (Sigma-Aldrich Lot: APF-1KT) was used, which is based on the application of 500 µL of diazonium and naphthol salt solution for

30 min in the dark. Afterward, the reaction was stopped with tridistilled water. Positive cells marked in red were photographed under an inverted optical microscope (Nikon Eclipse TS100), using the photo program (Leica Applications Suites—LAS EZ).

To evaluate the production of the mineralized matrix, alizarin red staining was performed after 3, 7, and 21 days of culture. Cells were fixed with 4% PFA and exposed to 1% alizarin red solution (Sigma-Aldrich) at RT for 30 min, then rinsed with ultrapure water. To quantify matrix mineralization, alizarin red-positive nodules were dissolved in a solution of 0.5 N HCl with 5% SDS. The optical density (OD) values of absorbance were quantified spectrophotometrically at a wavelength of 450 nm using a microplate reader (Biotek Synergy 2 multi-mode detection with gen5 software).

Statistical analysis: Data were presented as mean $\pm$ standard deviation (SD). The Gaussian distribution of the samples was tested, and the statistical significance of the data was evaluated using one-way ANOVA or unpaired t-tests. The p values are shown in the figures and statistical significance was considered when $p < 0.05$. Each experiment was performed three times, with triplicates.

3. Results

3.1. Characterization of TiO$_2$ NPs

TiO$_2$ NPs with a primary size of 25 nm were used to mimic the wear particles released by dental implants. The physicochemical characterization of the primary TiO$_2$ NPs was already published [2,5]. TEM micrographs revealed that TiO$_2$ NPs in ultrapure water were agglomerated, requiring the implementation of a dispersion protocol (Figure 1A). Dark-field STEM images show the morphology and agglomeration of TiO$_2$ NPs after dispersion (direct probe sonication), and the STEM/EDS Ti-K map (in blue) confirmed the identity of the TiO$_2$ NPs (Figure 1B). DLS analysis (Figure 1C) showed that the mean diameter (DH (nm)) of the TiO$_2$ NPs was 135 ± 24 nm in water and increased significantly ($p < 0.05$, unpaired t-test) in cell culture medium (156 ± 14 nm), maintaining a polydispersion index (PdI) of less than 0.2. Finally, the zeta potential analysis (ζ (mV)) in water and culture medium showed a significant decrease in the zeta potential value after medium contact, indicating the formation of protein and ionic corona on TiO$_2$ NPs surface ($p < 0.05$, unpaired t-test) (Figure 1C).

3.2. Effect of TiO$_2$ NPs on the Morphology of 2D and 3D Human Osteoblasts

In this study, human osteoblasts (SAOS-2) were cultured as monolayers or spheroids. After 72 h of seeding, cells were exposed for 72 h to 100 µg/mL of TiO$_2$ NPs. Optical microscopy images show the conventional SAOS-2 morphology (Figure 2A, left panel). In monolayers, the cells exhibit an epithelial-like phenotype, which is maintained after exposure to TiO$_2$ NPs. SAOS-2 spheroids have a round shape with a well-organized cytoskeleton (Figure 2B,C), also maintaining their morphology upon titanium exposure (Figure 2C). However, a 29% increase ($p = 0.0151$, unpaired t-test) in diameter and volume was observed after exposition to TiO$_2$ NPs (Figure 2D).

To confirm whether ultrastructural changes occurred after treatment with TiO$_2$ NPs, scanning electron microscopy (SEM) analysis was performed and showed that SAOS-2 in 2D and spheroids (3D) maintained their morphology after 3 days of exposure to TiO$_2$, without changes in their cell–cell contact (Figure 3A,B). Moreover, SEM-EDS analysis confirmed the presence of TiO$_2$ NPs on the surface of both cell models. A detail of interaction of TiO$_2$ NPs with spheroids (Ti-k, marked in blue) can be observed in Figure 3C.

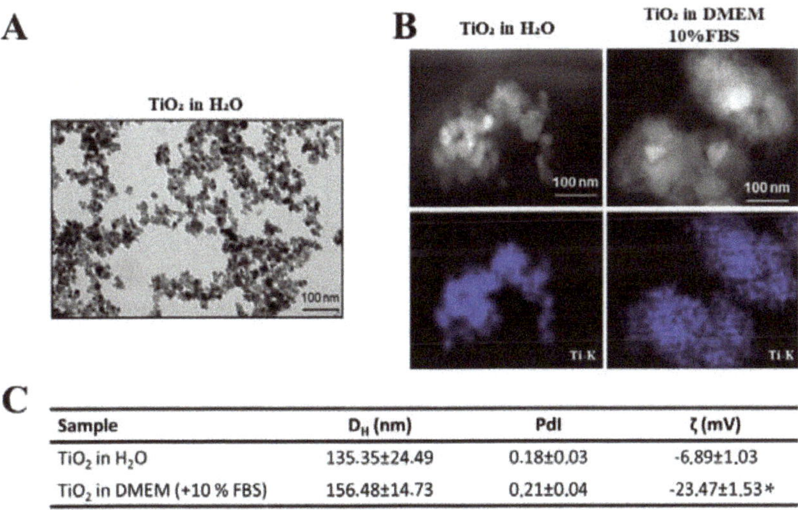

Sample	D_H (nm)	PdI	ζ (mV)
TiO₂ in H₂O	135.35±24.49	0.18±0.03	-6.89±1.03
TiO₂ in DMEM (+10 % FBS)	156.48±14.73	0.21±0.04	-23.47±1.53*

Figure 1. Characterization of TiO₂ NPs: (**A**) Transmission electron micrographs (TEM) of TiO₂ NPs in ultrapure water without dispersion. (**B**) Dark-field STEM micrographs of TiO₂ NPs after dispersion in water and cell culture medium and STEM/EDS Ti-K map confirming titanium presence (in blue). Scale bar: 100 nm. (**C**): Hydrodynamic diameter (D_H (nm)) and polydispersity index (PDI) of TiO₂ NPs after dispersion in ultrapure water and cell culture medium obtained by dynamic light scattering (DLS) and analysis of surface charge by zeta potential of TiO₂ NPs (ζ (mV)) in water and culture medium. The results represent the mean ± standard deviation of three independent experiments performed in triplicate of measurement (* $p < 0.05$ vs. TiO₂ NPs in water).

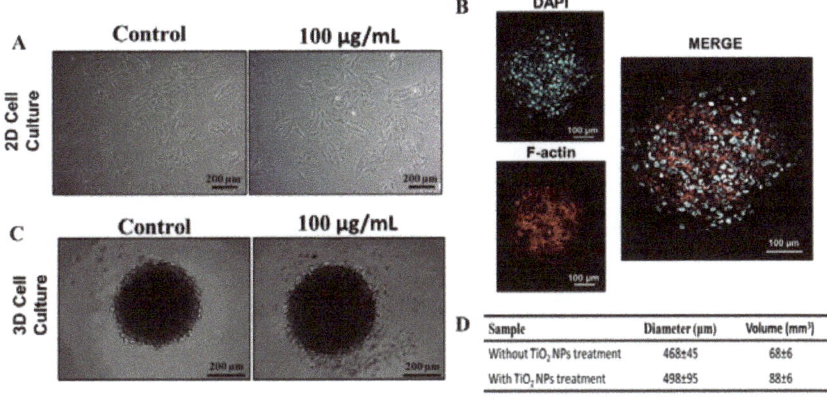

Sample	Diameter (µm)	Volume (mm³)
Without TiO₂ NPs treatment	468±45	68±6
With TiO₂ NPs treatment	498±95	88±6

Figure 2. Treatment of TiO₂ NPs in 2D and 3D cultures of human osteoblasts (SAOS-2): (**A**) Phase contrast micrograph of SAOS-2 in 2D with or without 100 µg/mL of exposition to TiO₂ NPs for 72 h (scale bar: 200 µm) (**B**) Spheroid cytoskeleton (3D), nucleus in blue (stained with DAPI) and the actin filaments in red (stained for F-actin) (scale bar: 100 µm). (**C**) Phase contrast micrograph of SAOS-2 in 3D with or without 100 µg/mL of exposition to TiO₂ NPs for 72 h (scale bar: 200 µm). (**D**) Average diameter and volume of spheroids. The baseline condition was used as a control. The results are representative images of three independent experiments.

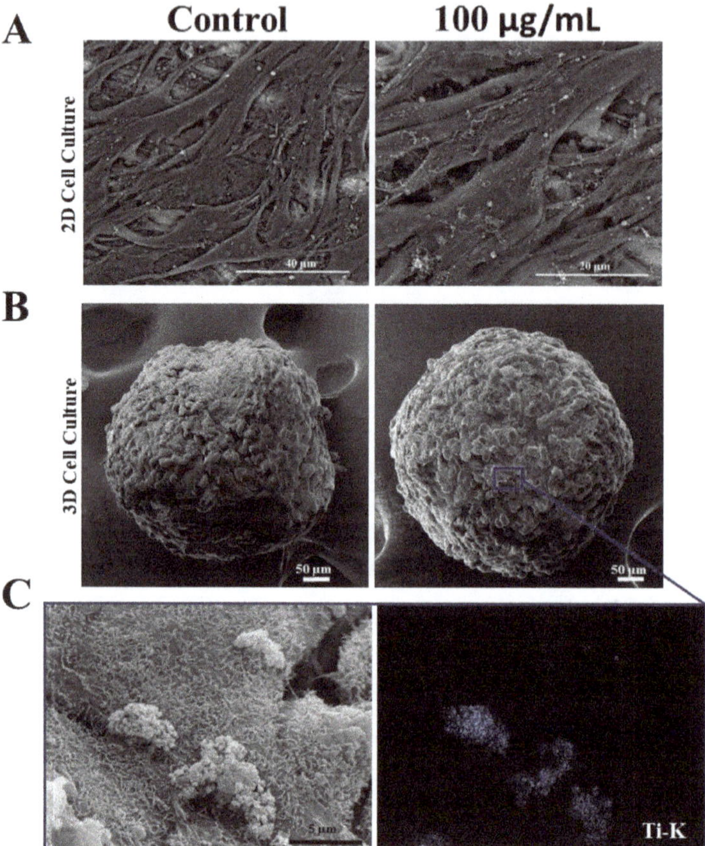

Figure 3. Morphology of human osteoblasts (SAOS-2) cultured in 2D and 3D after exposure to TiO_2 NPs: (**A**) Scanning electron micrographs showing SAOS-2 cultured in 2D and in spheroids (**B**) after treatment with 100 µg/mL of TiO_2 NPs for 3 days. (**C**) STEM/EDS Ti-K map analysis. Ti showed in blue point interacting with cells. Control: cultivation without NPs. The results are representative images of three independent experiments. (scale bar: 40 µm, 20 µm, 50 µm and 5 µm).

3.3. Effect of TiO_2 NPs on Human Osteoblast Viability

Flow cytometry analysis with PI/annexin after 3 and 21 days of culture did not show TiO_2 NPs cytotoxicity, both in the monolayer (Figure 4A) and in the spheroids models (Figure 4B). The levels of apoptosis and necrosis were similar in all conditions evaluated.

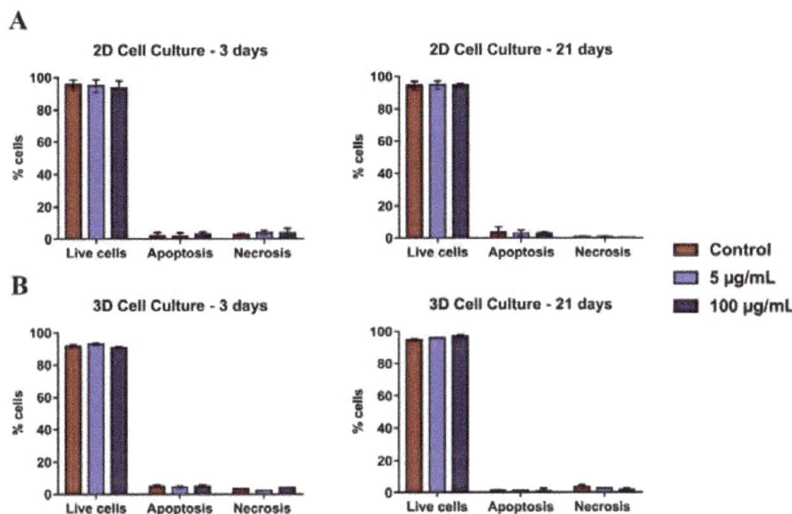

Figure 4. Viability of 2D and 3D human osteoblasts (SAOS-2): SAOS-2 were exposed to TiO$_2$ NPs (100 µg/mL) for 3 and 21 days and the PI/Annexin assay was performed by flow cytometry of cells in (**A**) 2D and (**B**) 3D. The baseline condition was used as a control. The results are the mean ± standard deviation of three independent experiments (no statistical difference).

3.4. Internalization of TiO$_2$ NPs in 2D and 3D Culture of Human Osteoblasts

Transmission electron microscopy (TEM) showed, both in 2D and 3D models, the internalization of TiO$_2$ NPs, that preferentially located in the cell cytoplasm within membrane-like-vesicles or after cell-membrane disruption, possibly in multivesicular bodies (MVBs) or auto-phagolysosomes delimitated by the membrane (Figure 5).

3.5. Differentiation and Mineralization of Human Osteoblasts after Exposure to TiO$_2$ NPs

To understand the influence of TiO$_2$ NPs on the differentiation and mineralization of both cell models (2D and 3D), analyses of alkaline phosphatase (ALP) (differentiation marker) and alizarin red (mineralization marker) were performed. For these analyses, osteoblasts were cultured for up to 14 days, and two exposure concentrations (5 and 100 µg/mL) of TiO$_2$ NPs were used. Previous data in human primary osteoblasts 2D histochemical micrographs showed that the treatment of TiO$_2$ NPs did not enhance the labeling for ALP (marked in red) after 14 days of culture (Figure S1A). However, in the mineralization analysis, there was a dose-dependent increase in alizarin staining after 14 days of treatment with 100 µg/mL (marked in intense red) compared to the control (Figure S1B).

To compare differences in mineralization occurring in 2D and 3D models, we performed alizarin staining after 3, 7, and 14 days after 5 µg/mL or 100 µg/mL TiO$_2$ NPs exposure in both models. Alizarin red results showed a significant dose-dependent increase in mineralization at 14 days compared to the control, both in 2D (Figure 6A) and 3D (Figure 6B). Moreover, when treatment values are normalized by control values, the mineralization increase is higher in the 3D model when compared with the 2D model, suggesting that both models can present different results in the mineralization evaluation (Figure 6C and representative images in Figure 6D).

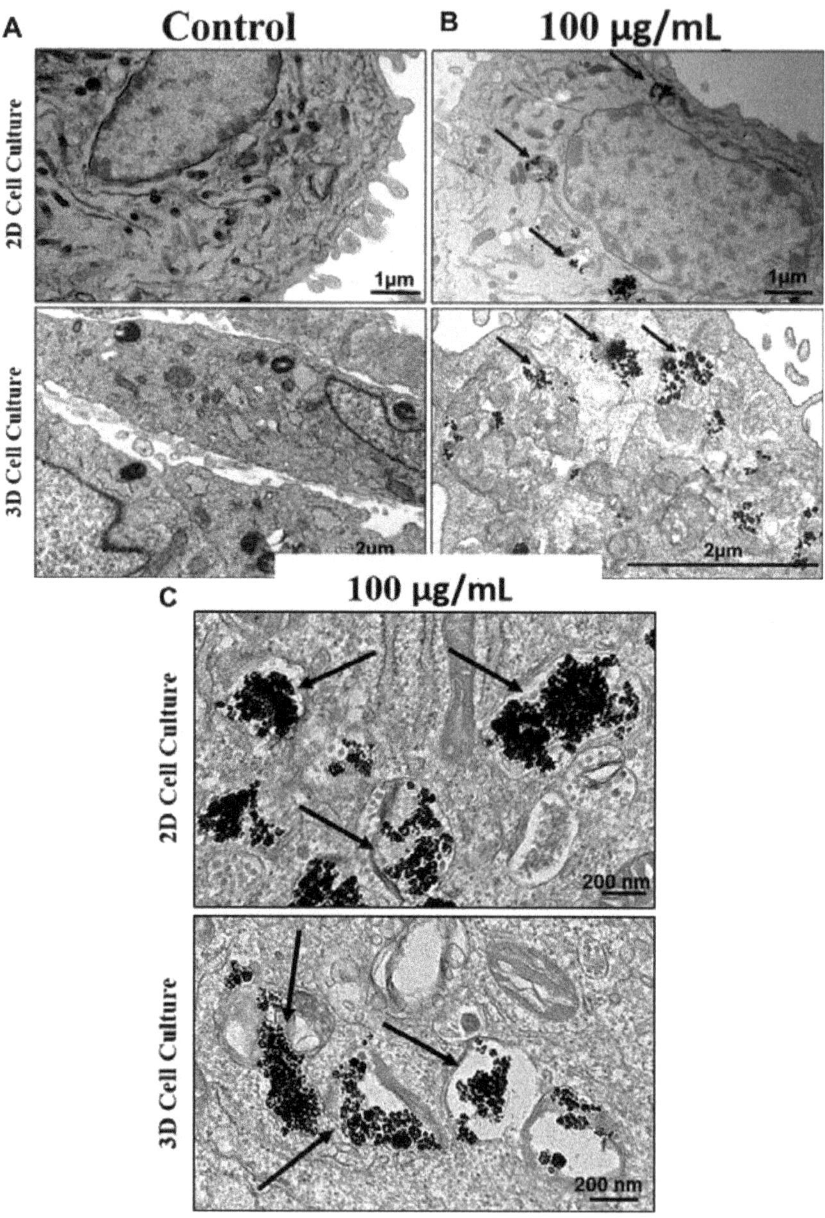

Figure 5. Internalization of TiO$_2$ NPs: Transmission electron microscopy (TEM) micrographs showing TiO$_2$ NPs internalization in human osteoblasts (SAOS-2) 72 h cultured: (**A**) control, cultured without NPs or (**B,C**) treated with 100 µg/mL of TiO$_2$ NPs (black arrow). The results are representative images of three independent experiments. (scale bar: 1 µm, 2 µm, and 200 nm).

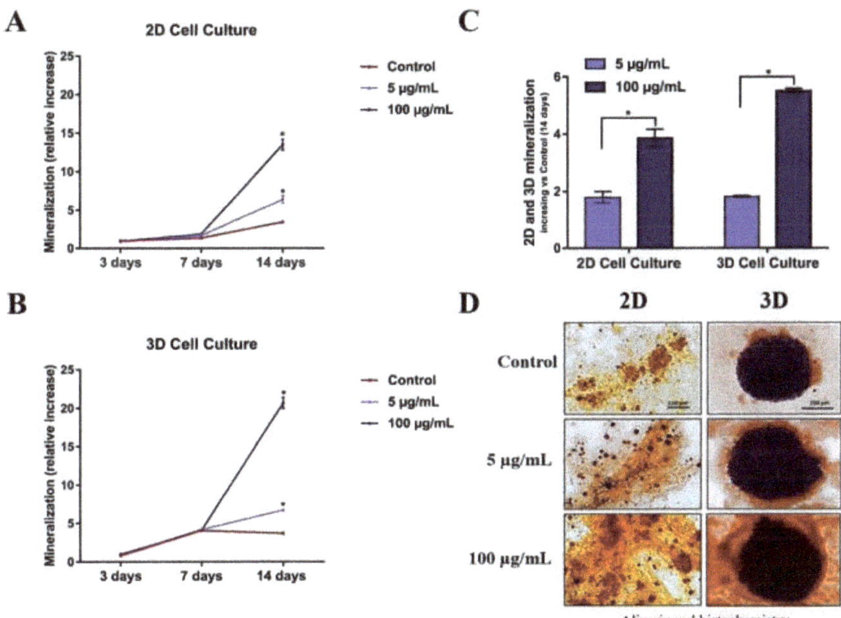

Alizarin red histochemistry

Figure 6. Mineralized matrix produced by human osteoblasts (SAOS-2) in 2D and 3D cells exposed to TiO$_2$ NPs (0, 5 and 100 µg/mL) during 3 and 14 days in osteogenic medium. Alizarin red assay shows a dose-dependent matrix production in (**A**) 2D SAOS-2 model and (**B**) 3D SAOS-2 models. The graphics in (**C**) present a comparison between the inorganic matrix production on both 2D and 3D models at 14 days (final experimental time) normalized *vs* each control to show dose-dependent mineralization fold increase in the different models. The results are average ± standard deviation of three independent experiments * $p < 0.05$; (**D**) representative images obtained through optical microscopy.

4. Discussion

Titanium is the main material employed in the dental implant industry, due to its high mechanical strength, low elastic modulus, corrosion resistance, ductility, and biocompatibility [6,9]. However, tribocorrosion processes at the implant surface lead to accelerated bone loss, compromising osseointegration, and increasing periprosthetic failure [2,5–9,23,24]. The hostile electrolytic environment (oxidation/reduction) together with mechanical action at the interface enables the tribocorrosion phenomena [7,10]. As a consequence, degradation products (released from implants) including metal ions, micrometric, and/or nanometric metallic debris (TiO$_2$ NPs) can be internalized by cells in the bone niche, possibly generating cytotoxic effects [6,9,10]. The adverse effects of TiO$_2$ NPs vary widely in the literature, which raises concern among authorities and physicians due to their high prevalence [5,10,17]. Literature data reveal that inflammatory stimuli associated with cytokine overproduction and increased production of reactive oxygen species are referred to as primary toxic effects that lead to cell death [6,9,13,17].

Some authors explained that this mechanism leads to activation of immunological sentinels and accumulation of antigens such as ions, nanoparticles, microparticles, and bacterial antigens via the functional interface between dental implant and tissue. This leads to immunological cell polarization and follows dental implant loss [6,9].

Most available studies that evaluate osteoblast response to TiO$_2$ NPs use 2D cell culture models, which have shown limitations regarding cell growth and cell–cell and cell–matrix interactions, among others [25–28]. Few studies evaluate the influence of TiO$_2$ NPs on the physiology of bone cells grown in 3D models such as spheroids [2]. Osteoblast spheroids can be considered as a culture model that better mimics living cells in terms of structural and biofunctional properties and provides more reliable results compared to conventional 2D cell cultures (Figure 7) [2,25,29]. Despite this, there are some limitations to spheroid culture, mainly because cellular environments are not similarly exposed to the culture medium. This can lead to the formation of a microenvironment inside the spheroids that can select groups of cells [30,31]. Partial diffusion of nutrients or oxygen can induce necrotic areas in the central area of the spheroids [32]. However, well-characterized multicellular spheroids exhibit different levels of extracellular matrix deposition, growth factor secretion, and gene expression profiles [2]. The viability, morphology, and gene expression of osteoblastic spheroids are contact-dependent, and single or co-culture spheroids have been shown to have an impact on bone cell function [33]. Interestingly, a study reported that primary osteoblasts and pre-osteoblasts MC3T3-E1 can differentiate into osteocytes when grown in 3D cultures [34]. Therefore, 3D culture models can be used to study the pathophysiological reactions of TiO_2 NPs in bone metabolism compared to 2D cultures. A previous study by W. Souza et al., on the cytotoxicity effect of TiO_2 NPs on osteoblast spheroids, revealed that 72 h exposition to TiO_2 NPs can alter the cell cycle, without interfering with osteoblasts' ability to differentiate and mineralize and significantly increase collagen and pro-inflammatory cytokine secretion [2]. In the present study, a longer exposure period (21 days) was assessed to compare 2D with 3D osteoblasts models to better understand their relevance for nanotoxicological studies.

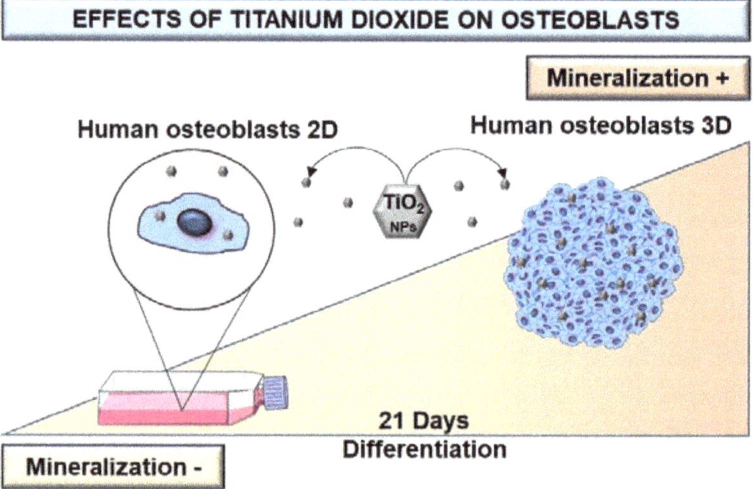

Figure 7. Scheme of differentiation and production of a mineralized matrix of human osteoblasts SAOS-2 cultured in monolayer and 3D spheroid cell models after exposition to TiO_2. After high TiO_2 NPs exposure, the dose-dependent increase of bone-mineralized matrix in the 3D cells model is higher than in monolayer (2D) culture.

TiO_2 NPs are chemically stable, have antibacterial properties, and induce less toxicity than other nanostructures, and, when exposed to the biological environment, blood plasma proteins and ions selectively adsorb on the outer surface of the cell [35]. The complex interface depends on the physical and chemical characteristics of the NPs, as well as the biological characteristics of the environment [36]. In the present study, we observed that TiO_2 NPs had an average size of 150 nm in the culture medium. We can notice an increase in the average size after the addition of the culture medium due to the adsorption of proteins and ions on the TiO_2 NPs surface, which can be correlated with the change in surface charge, identified by zeta potential analysis. Furthermore, in our previous study, we confirmed the adsorption of calcium and phosphate on the surface of TiO_2 NPs, which are important mediators of bone mineralization [5].

To understand the influence of TiO_2 NPs on bone cell mineralization, we used a mature osteoblast line, cultured both in monolayer (2D) and spheroids (3D), the former characterized previously [2]. Spheroids and monolayer cells were treated with 100 μg/mL of NPs for 21 days. In both models (2D and 3D), we observed the internalization of TiO_2 NPs in membrane vesicles (with 72 h). Some studies have shown that NPs can be internalized in a dose-dependent manner, accumulating preferentially in the perinuclear region, and having as their final destination the lysosomes [35,37]. Normally TiO_2 NPs are not observed dispersed in cell cytoplasm [5,35,37]. However, the effect of TiO_2 NPs on cells is directly related to their size distribution, crystal structure, as well as corona formation [35]. Recently, the formation of a bio-camouflage rich in calcium, phosphorus, and hydroxyapatite crystals around TiO_2 NPs was demonstrated, which is known to facilitate the internalization in 2D and 3D osteoblastic models since the detected chemical elements are essential for bone cell metabolism and mineralization [2,5,38].

The present study demonstrated that TiO_2 NPs did not alter the viability of osteoblasts in both cell models (with 21 days). Concomitantly, they did not change the osteoblast morphology or spherical shape of the 3D model upon internalization of the NPs. Interestingly, they were able to stimulate an increase in calcium deposition, which is indicative of the activation of a mineralization process in osteoblastic spheroids. In the present study, results of alkaline phosphatase synthesis and calcium labeling demonstrated that TiO_2 NPs increased osteoblast differentiation that induced greater mineralization in a 3D culture model, suggesting that the 3D architecture possibly increases cell surface interaction with previously reported TiO_2 NPs bio-camouflaged [2]. The mineralization increase in 3D models after exposure to NPs may be related to the greater cell surface capable of contacting NPs when compared to the monolayer (2D), enhancing the stimulatory effects of TiO_2 NPs [39]. This is consistent with previous studies that reveal that 3D osteoblasts models when exposed to TiO_2 NPs, compared to monolayer cells, induce the secretion of vascular endothelial growth factor (VEGF), activating a cascade of events resulting in higher type I collagen production [39–43]. Bone mineralization is the first step for implant osseointegration and begins when collagen I acts as a three-dimensional scaffold for hydroxyapatite deposition [44]. Another study reported greater osteogenic differentiation when using 3D collagen gel culture [34]. Studies also showed that the 2D cell model does not yet seem to be the better model to study interaction with NPs; instead, the spheroids are also promising for application to 3D bioprinting tissue models with biomaterial scaffolds, as an innovative technology to improve bone osteointegration [45].

Unfortunately, there is no consensus in the literature on how to evaluate the biological effect of TiO_2 NPs. Without standardized protocols to assess the biological impacts of NPs, it is necessary to validate safe assessments and mitigate potential health impacts, moving toward the evaluation and development of new cellular study models to better mimic the biological environment [37]. Although osteoblastic spheroids have their advantages compared to monolayers—such as reproducibility, better nutrients, oxygen diffusion gradients, improved cell–cell interactions, matrix deposition, and models with various cell stages (proliferating, quiescent, apoptotic, hypoxic, and necrotic cells) [46,47], 3D spheroid models have not been validated as realistic in vitro models [29,46,48]. One of the main drawbacks of spheroids is that the porosity and mechanical properties is difficult to be studied. Thus, efforts should be made to improve 3D bone cell models to recapitulate the bone microenvironment that is known to be constituted by different cell types and has dynamic and metabolic activity.

TiO_2 NPs released from dental implants are, on the one hand, considered the cause of clinical peri-implant bone loss; on the other hand, they may be able to stimulate the production of a mineralized extracellular matrix in osteoblast spheroids [48]. Another important aspect is that spheroids can respond physiologically better to the stimuli of TiO_2 NPs, which corroborates the development of new studies to create new models applied in clinical studies, to favor the process of bone remodeling and alternative treatment for periodontitis and peri-implantitis. In addition, the spheroids themselves can be applied to high-cell-density tissue models, innovative technology for bone augmentation, and soft tissue replacement procedures [45]. Therefore, the combination of TiO_2 NPs with spheroid cells should be an interesting approach for tissue reconstruction.

Lastly, our results demonstrate that TiO_2 NPs increase calcium deposition in 3D versus 2D cultures. Although this study revealed interesting findings regarding the behavior and role of TiO_2 NPs in generating stimuli for mineralization in 3D models only, it should be noticed that our results are limited to the conditions tested and the experimental setup. Further studies should be encouraged, and further evaluations using the quantification of genes that act on differentiation and mineralization should be performed. However, our results help to better understand the possible impact of 3D culture in dentistry, and also open a discussion about the dual role of TiO_2 NPs, which on one side can activate an inflammatory response that leads to bone resorption. However, on the other hand, it is activating mineralization. Our findings are considered clinically relevant, since, for the first time, we report that at the bone-implant interface, TiO_2 NPs besides the activation of macrophages can also stimulate osteoblasts that play a fundamental role in the mineralization process.

5. Conclusions

In this study, the cells were exposed to TiO_2 NPs at concentrations up to 100 µg/mL in 2D and 3D models for up to 21 days of exposure.

TiO_2 aggregates were dispersed to nanometric size and characterized successfully. Its internalization in both cell models showed no differences in cell morphology or viability and bone mineralization induction in a dose-dependent form in both culture models.

However, the mineralization process was more intense in the 3D spheroid culture compared to the 2D monolayer model.

This brings a new discussion about the possible advantages of TiO_2 NPs on bone mineralization, which may suggest that the action of nanometric particles can contribute to the osseointegration process in titanium dental implants, reducing periprosthetic failures and using 3D cell models as an innovative technology to improve bone osteointegration induced by nanoparticles.

Supplementary Materials: The following supporting information can be downloaded at: https://www.mdpi.com/article/10.3390/nano13030425/s1, Figure S1. Primary osteoblast differentiation and mineralization: alkaline phosphatase (A) and alizarin red (B) staining in primary human osteoblasts exposed during 3, 7, and 14 days after exposure to 5 µg/mL or 100 µg/mL TiO2 NPs. The results are representative images of three independent experiments. (scale bar: 200 µm).

Author Contributions: S.G.-P., A.R.R., W.d.S. and G.d.S.C. designed the study. G.d.S.C., W.d.S., S.G.-P., A.R.R., T.S.M.L. and J.W., performed experimental work and analyses. G.d.S.C., W.d.S., D.C.B., S.G.-P., A.R.R., J.W., J.M.G. and B.S.A. contributed technical support and discussion. G.d.S.C., W.d.S., S.G.-P., A.R.R., T.S.M.L., D.C.B. and J.M.G. wrote/edited the manuscript. All authors have read and agreed to the published version of the manuscript.

Funding: This work was supported by National Council for Scientific and Technological Development (CNPq) with grants 405030/2015-0, 306672/2016-2, and 467513/2014-7. J.M.G. thanks the Cientista do Nosso Estado award from FAPERJ and CNPq/Faperj—National Institute of Science and Technology in Regenerative Medicine (INCT-Regenera, process n. 465656/2014-5). A.R.R. thanks the Sinfonia project H2020 of the European Union (N.857253).

Data Availability Statement: The data that support the findings of this study are available from the corresponding author upon reasonable request.

Acknowledgments: We thank the Physical Research Centre (CBPF), the National Institute of Metrology, Quality and Technology (INMETRO), and the Federal University of Rio de Janeiro (UFRJ), especially Maria Isabel Rossi, for all the support given to the biological experiments. The authors thank Suzana Azevedo dos Anjos for her valuable help in carrying out the sample preparation for SEM, and Mariana Moreira for her help in obtaining micrographs in SEM and TEM. The authors thank the Cell Bank of Rio de Janeiro (BCRJ), especially Radovan Borojevic.

Conflicts of Interest: The authors declare no conflict of interest.

References

1. Awasthi, A.; Awasthi, K.K.; John, P.J. Nanomaterials in Biology. *Environ. Sci. Pollut. Res.* **2021**, *28*, 46334–46335. [CrossRef] [PubMed]
2. Souza, W.; Piperni, S.G.; Laviola, P.; Rossi, A.L.; Rossi, M.I.D.; Archanjo, B.S.; Leite, P.E.; Fernandes, M.H.; Rocha, L.A.; Granjeiro, J.M.; et al. The Two Faces of Titanium Dioxide Nanoparticles Bio-Camouflage in 3D Bone Spheroids. *Sci. Rep.* **2019**, *9*, 9309. [CrossRef] [PubMed]
3. Li, J.; Zhou, P.; Wang, L.; Hou, Y.; Zhang, X.; Zhu, S.; Guan, S. Investigation of Mg–XLi–Zn Alloys for Potential Application of Biodegradable Bone Implant Materials. *J. Mater. Sci. Mater. Med.* **2021**, *32*, 43. [CrossRef]
4. Bayda, S.; Adeel, M.; Tuccinardi, T.; Cordani, M.; Rizzolio, F. The History of Nanoscience and Nanotechnology: From Chemical–Physical Applications to Nanomedicine. *Molecules* **2019**, *25*, 112. [CrossRef] [PubMed]
5. Ribeiro, A.R.; Gemini-Piperni, S.; Travassos, R.; Lemgruber, L.; Silva, C.R.; Rossi, A.L.; Farina, M.; Anselme, K.; Shokuhfar, T.; Shahbazian-Yassar, R.; et al. Trojan-Like Internalization of Anatase Titanium Dioxide Nanoparticles by Human Osteoblast Cells. *Sci. Rep.* **2016**, *6*, 23615. [CrossRef]
6. Delgado-Ruiz, R.; Romanos, G. Potential Causes of Titanium Particle and Ion Release in Implant Dentistry: A Systematic Review. *Int. J. Mol. Sci.* **2018**, *19*, 3585. [CrossRef]
7. Kheder, W.; al Kawas, S.; Khalaf, K.; Samsudin, A.R. Impact of Tribocorrosion and Titanium Particles Release on Dental Implant Complications—A Narrative Review. *Jpn. Dent. Sci. Rev.* **2021**, *57*, 182–189. [CrossRef] [PubMed]
8. Kirmanidou, Y.; Sidira, M.; Drosou, M.-E.; Bennani, V.; Bakopoulou, A.; Tsouknidas, A.; Michailidis, N.; Michalakis, K. New Ti-Alloys and Surface Modifications to Improve the Mechanical Properties and the Biological Response to Orthopedic and Dental Implants: A Review. *Biomed. Res. Int* **2016**, *2016*, 2908570. [CrossRef] [PubMed]
9. Amengual-Peñafiel, L.; Córdova, L.A.; Constanza Jara-Sepúlveda, M.; Brañes-Aroca, M.; Marchesani-Carrasco, F.; Cartes-Velásquez, R. Osteoimmunology Drives Dental Implant Osseointegration: A New Paradigm for Implant Dentistry. *Jpn. Dent. Sci. Rev.* **2021**, *57*, 12–19. [CrossRef] [PubMed]
10. Kim, K.T.; Eo, M.Y.; Nguyen, T.T.H.; Kim, S.M. General Review of Titanium Toxicity. *Int. J. Implant. Dent.* **2019**, *5*, 10. [CrossRef]
11. Romanos, G.; Fischer, G.; Delgado-Ruiz, R. Titanium Wear of Dental Implants from Placement, under Loading and Maintenance Protocols. *Int. J. Mol. Sci.* **2021**, *22*, 1067. [CrossRef] [PubMed]
12. Zhou, Z.; Shi, Q.; Wang, J.; Chen, X.; Hao, Y.; Zhang, Y.; Wang, X. The Unfavorable Role of Titanium Particles Released from Dental Implants. *Nanotheranostics* **2021**, *5*, 321–332. [CrossRef]
13. Christo, S.N.; Diener, K.R.; Bachhuka, A.; Vasilev, K.; Hayball, J.D. Innate Immunity and Biomaterials at the Nexus: Friends or Foes. *Biomed. Res. Int.* **2015**, *2015*, 342304. [CrossRef] [PubMed]
14. Goodman, S.B.; Davidson, J.A.; Song, Y.; Martial, N.; Fornasier, V.L. Histomorphological Reaction of Bone to Different Concentrations of Phagocytosable Particles of High-Density Polyethylene and Ti-6Al-4V Alloy in Vivo. *Biomaterials* **1996**, *17*, 1943–1947. [CrossRef] [PubMed]
15. Bukata, S.V.; Gelinas, J.; Wei, X.; Rosier, R.N.; Puzas, J.E.; Zhang, X.; Schwarz, E.M.; Song, X.R.; Griswold, D.E.; O'Keefe, R.J. PGE2 and IL-6 Production by Fibroblasts in Response to Titanium Wear Debris Particles Is Mediated through a Cox-2 Dependent Pathway. *J. Orthop. Res.* **2004**, *22*, 6–12. [CrossRef] [PubMed]
16. Choi, M.G.; Koh, H.S.; Kluess, D.; O'Connor, D.; Mathur, A.; Truskey, G.A.; Rubin, J.; Zhou, D.X.F.; Sung, K.-L.P. Effects of Titanium Particle Size on Osteoblast Functions in Vitro and in Vivo. *Proc. Natl. Acad. Sci. USA* **2005**, *102*, 4578–4583. [CrossRef]
17. Tuan, R.S.; Lee, F.Y.-I.; Konttinen, Y.T.; Wilkinson, M.J.; Smith, R.L. What Are the Local and Systemic Biologic Reactions and Mediators to Wear Debris, and What Host Factors Determine or Modulate the Biologic Response to Wear Particles? *J. Am. Acad. Orthop. Surg.* **2008**, *16*, S42–S48. [CrossRef]
18. Gutwein, L.G.; Webster, T.J. Increased Viable Osteoblast Density in the Presence of Nanophase Compared to Conventional Alumina and Titania Particles. *Biomaterials* **2004**, *25*, 4175–4183. [CrossRef]
19. Neel, A.; Bozec, L.; Perez, R.A.; Kim, H.-W.; Knowles, J.C. Nanotechnology in Dentistry: Prevention, Diagnosis, and Therapy. *Int. J. Nanomed.* **2015**, *10*, 6371. [CrossRef]
20. Bédard, P.; Gauvin, S.; Ferland, K.; Caneparo, C.; Pellerin, È.; Chabaud, S.; Bolduc, S. Innovative Human Three-Dimensional Tissue-Engineered Models as an Alternative to Animal Testing. *Bioengineering* **2020**, *7*, 115. [CrossRef]

21. Yanagi, T.; Kajiya, H.; Fujisaki, S.; Maeshiba, M.; Yanagi-S, A.; Yamamoto-M, N.; Kakura, K.; Kido, H.; Ohno, J. Three-Dimensional Spheroids of Dedifferentiated Fat Cells Enhance Bone Regeneration. *Regen. Ther.* **2021**, *18*, 472–479. [CrossRef] [PubMed]
22. Gaitán-Salvatella, I.; López-Villegas, E.O.; González-Alva, P.; Susate-Olmos, F.; Álvarez-Pérez, M.A. Case Report: Formation of 3D Osteoblast Spheroid Under Magnetic Levitation for Bone Tissue Engineering. *Front. Mol. Biosci.* **2021**, *8*, 672518. [CrossRef] [PubMed]
23. Schoichet, J.J.; Mourão, C.F.d.A.B.; Fonseca, E.d.M.; Ramirez, C.; Villas-Boas, R.; Prazeres, J.; Quinelato, V.; Aguiar, T.R.; Prado, M.; Cardarelli, A.; et al. Epidermal Growth Factor Is Associated with Loss of Mucosae Sealing and Peri-Implant Mucositis: A Pilot Study. *Healthcare* **2021**, *9*, 1277. [CrossRef] [PubMed]
24. Romanos, G. Current Concepts in the Use of Lasers in Periodontal and Implant Dentistry. *J. Indian Soc. Periodontol.* **2015**, *19*, 490. [CrossRef] [PubMed]
25. Gebhard, C.; Gabriel, C.; Walter, I. Morphological and Immunohistochemical Characterization of Canine Osteosarcoma Spheroid Cell Cultures. *Anat. Histol. Embryol.* **2016**, *45*, 219–230. [CrossRef]
26. Keller, L.; Idoux-Gillet, Y.; Wagner, Q.; Eap, S.; Brasse, D.; Schwinté, P.; Arruebo, M.; Benkirane-Jessel, N. Nanoengineered Implant as a New Platform for Regenerative Nanomedicine Using 3D Well-Organized Human Cell Spheroids. *Int. J. Nanomed.* **2017**, *12*, 447–457. [CrossRef] [PubMed]
27. Maia-Pinto, M.O.C.; Brochado, A.C.B.; Teixeira, B.N.; Sartoretto, S.C.; Uzeda, M.J.; Alves, A.T.N.N.; Alves, G.G.; Calasans-Maia, M.D.; Thiré, R.M.S.M. Biomimetic Mineralization on 3D Printed PLA Scaffolds: On the Response of Human Primary Osteoblasts Spheroids and In Vivo Implantation. *Polymers* **2020**, *13*, 74. [CrossRef]
28. Jensen, C.; Teng, Y. Is It Time to Start Transitioning From 2D to 3D Cell Culture? *Front. Mol. Biosci.* **2020**, *7*, 33. [CrossRef]
29. Duval, K.; Grover, H.; Han, L.-H.; Mou, Y.; Pegoraro, A.F.; Fredberg, J.; Chen, Z. Modeling Physiological Events in 2D vs. 3D Cell Culture. *Physiology* **2017**, *32*, 266–277. [CrossRef]
30. Białkowska, K.; Komorowski, P.; Bryszewska, M.; Miłowska, K. Spheroids as a Type of Three-Dimensional Cell Cultures— Examples of Methods of Preparation and the Most Important Application. *Int. J. Mol. Sci.* **2020**, *21*, 6225. [CrossRef]
31. Holub, A.R.; Huo, A.; Patel, K.; Thakore, V.; Chhibber, P.; Erogbogbo, F. Assessing Advantages and Drawbacks of Rapidly Generated Ultra-Large 3D Breast Cancer Spheroids: Studies with Chemotherapeutics and Nanoparticles. *Int. J. Mol. Sci.* **2020**, *21*, 4413. [CrossRef] [PubMed]
32. Anada, T.; Fukuda, J.; Sai, Y.; Suzuki, O. An Oxygen-Permeable Spheroid Culture System for the Prevention of Central Hypoxia and Necrosis of Spheroids. *Biomaterials* **2012**, *33*, 8430–8441. [CrossRef] [PubMed]
33. Yuste, I.; Luciano, F.C.; González-Burgos, E.; Lalatsa, A.; Serrano, D.R. Mimicking Bone Microenvironment: 2D and 3D in Vitro Models of Human Osteoblasts. *Pharmacol. Res.* **2021**, *169*, 105626. [CrossRef] [PubMed]
34. Sawa, N.; Fujimoto, H.; Sawa, Y.; Yamashita, J. Alternating Differentiation and Dedifferentiation between Mature Osteoblasts and Osteocytes. *Sci. Rep.* **2019**, *9*, 13842. [CrossRef]
35. Buzea, C.; Pacheco, I.I.; Robbie, K. Nanomaterials and Nanoparticles: Sources and Toxicity. *Biointerphases* **2007**, *2*, MR17–MR71. [CrossRef] [PubMed]
36. Corbo, C.; Molinaro, R.; Parodi, A.; Toledano Furman, N.E.; Salvatore, F.; Tasciotti, E. The Impact of Nanoparticle Protein Corona on Cytotoxicity, Immunotoxicity and Target Drug Delivery. *Nanomedicine* **2016**, *11*, 81–100. [CrossRef]
37. Arora, S.; Rajwade, J.M.; Paknikar, K.M. Nanotoxicology and in Vitro Studies: The Need of the Hour. *Toxicol. Appl. Pharmacol.* **2012**, *258*, 151–165. [CrossRef]
38. Ribeiro, A.R.; Mukherjee, A.; Hu, X.; Shafien, S.; Ghodsi, R.; He, K.; Gemini-Piperni, S.; Wang, C.; Klie, R.F.; Shokuhfar, T.; et al. Bio-Camouflage of Anatase Nanoparticles Explored by in Situ High-Resolution Electron Microscopy. *Nanoscale* **2017**, *9*, 10684–10693. [CrossRef]
39. Gurumurthy, B.; Bierdeman, P.C.; Janorkar, A.V. Spheroid Model for Functional Osteogenic Evaluation of Human Adipose Derived Stem Cells. *J. Biomed. Mater. Res. A* **2017**, *105*, 1230–1236. [CrossRef]
40. Yamada, Y.; Okano, T.; Orita, K.; Makino, T.; Shima, F.; Nakamura, H. 3D-Cultured Small Size Adipose-Derived Stem Cell Spheroids Promote Bone Regeneration in the Critical-Sized Bone Defect Rat Model. *Biochem. Biophys. Res. Commun.* **2022**, *603*, 57–62. [CrossRef]
41. Ho, S.S.; Murphy, K.C.; Binder, B.Y.K.; Vissers, C.B.; Leach, J.K. Increased Survival and Function of Mesenchymal Stem Cell Spheroids Entrapped in Instructive Alginate Hydrogels. *Stem Cells Transl Med.* **2016**, *5*, 773–781. [CrossRef] [PubMed]
42. Cuevas-González, M.V.; Suaste-Olmos, F.; Cuevas-González, J.C.; Álvarez-Pérez, M.A. 3D Spheroid Cell Cultures and Their Role in Bone Regeneration: A Systematic Review. *Odovtos Int. J. Dent. Sci.* **2022**, *24*, 44–57. [CrossRef]
43. Lee, H.-J.; Song, Y.-M.; Baek, S.; Park, Y.-H.; Park, J.-B. Vitamin D Enhanced the Osteogenic Differentiation of Cell Spheroids Composed of Bone Marrow Stem Cells. *Medicina* **2021**, *57*, 1271. [CrossRef] [PubMed]
44. Rocha, A.L.; Bighetti-Trevisan, R.L.; Duffles, L.F.; de Arruda, J.A.A.; Taira, T.M.; Assis, B.R.D.; Macari, S.; Diniz, I.M.A.; Beloti, M.M.; Rosa, A.L.; et al. Inhibitory Effects of Dabigatran Etexilate, a Direct Thrombin Inhibitor, on Osteoclasts and Osteoblasts. *Thromb. Res.* **2020**, *186*, 45–53. [CrossRef]
45. Daly, A.C.; Davidson, M.D.; Burdick, J.A. 3D Bioprinting of High Cell-Density Heterogeneous Tissue Models through Spheroid Fusion within Self-Healing Hydrogels. *Nat. Commun.* **2021**, *12*, 753. [CrossRef]
46. Kim, J. bin Three-Dimensional Tissue Culture Models in Cancer Biology. *Semin Cancer Biol.* **2005**, *15*, 365–377. [CrossRef]

47. Chatzinikolaidou, M. Cell Spheroids: The New Frontiers in in Vitro Models for Cancer Drug Validation. *Drug Discov. Today* **2016**, *21*, 1553–1560. [CrossRef]
48. Senna, P.; Antoninha Del Bel Cury, A.; Kates, S.; Meirelles, L. Surface Damage on Dental Implants with Release of Loose Particles after Insertion into Bone. *Clin. Implant. Dent. Relat. Res.* **2015**, *17*, 681–692. [CrossRef]

 *nanomaterials*

Article

Different Sensitivity of Advanced Bronchial and Alveolar Mono- and Coculture Models for Hazard Assessment of Nanomaterials

Elisabeth Elje [1,2,*], Espen Mariussen [1,3], Erin McFadden [1], Maria Dusinska [1] and Elise Rundén-Pran [1,*]

[1] Health Effects Laboratory, Department for Environmental Chemistry, NILU—Norwegian Institute for Air Research, 2007 Kjeller, Norway
[2] Department of Molecular Medicine, Institute of Basic Medical Sciences, University of Oslo, 0372 Oslo, Norway
[3] Department of Air Quality and Noise, Norwegian Institute of Public Health, 0456 Oslo, Norway
* Correspondence: eel@nilu.no (E.E.); erp@nilu.no (E.R.-P.); Tel.: +47-63-89-81-98 (E.E.)

Abstract: For the next-generation risk assessment (NGRA) of chemicals and nanomaterials, new approach methodologies (NAMs) are needed for hazard assessment in compliance with the 3R's to reduce, replace and refine animal experiments. This study aimed to establish and characterize an advanced respiratory model consisting of human epithelial bronchial BEAS-2B cells cultivated at the air–liquid interface (ALI), both as monocultures and in cocultures with human endothelial EA.hy926 cells. The performance of the bronchial models was compared to a commonly used alveolar model consisting of A549 in monoculture and in coculture with EA.hy926 cells. The cells were exposed at the ALI to nanosilver (NM-300K) in the VITROCELL® Cloud. After 24 h, cellular viability (alamarBlue assay), inflammatory response (enzyme-linked immunosorbent assay), DNA damage (enzyme-modified comet assay), and chromosomal damage (cytokinesis-block micronucleus assay) were measured. Cytotoxicity and genotoxicity induced by NM-300K were dependent on both the cell types and model, where BEAS-2B in monocultures had the highest sensitivity in terms of cell viability and DNA strand breaks. This study indicates that the four ALI lung models have different sensitivities to NM-300K exposure and brings important knowledge for the further development of advanced 3D respiratory in vitro models for the most reliable human hazard assessment based on NAMs.

Keywords: NAMs—new approach methodologies; ALI—air–liquid interface; genotoxicity; BEAS-2B; A549; NM-300K; DNA damage; chromosomal damage; cytokines

Citation: Elje, E.; Mariussen, E.; McFadden, E.; Dusinska, M.; Rundén-Pran, E. Different Sensitivity of Advanced Bronchial and Alveolar Mono- and Coculture Models for Hazard Assessment of Nanomaterials. *Nanomaterials* **2023**, *13*, 407. https://doi.org/10.3390/nano13030407

Academic Editors: Saura Sahu and Jean-Marie Nedelec

Received: 25 November 2022
Revised: 3 January 2023
Accepted: 13 January 2023
Published: 19 January 2023

1. Introduction

The production and usage of nanomaterials (NMs) are rising, increasing the risk of human exposure. Inhalation is the most important exposure route for airborne nanomaterials (NMs) and particulate matter (PM) in humans, making the respiratory system a first-target organ [1]. The respiratory tract consists of the tracheobronchial region leading into the alveolar region, where gas exchange with blood occurs across the thin lung–blood barrier (0.4 μm) [1,2]. Besides gas exchange, a main function of the lower respiratory tract is defense against inhaled toxicants [1]. Interaction with and deposition of inhaled NMs are likely to occur in the bronchial and alveolar region. Particle deposition is dependent upon the NMs' physicochemical properties, such as size and solubility [2].

NMs and their dissolved compounds can cause primary effects in the respiratory system, or secondary circulatory effects after crossing the lung–blood barrier and taken up in the blood. A human study has shown the translocation of inhaled gold NM or its dissolved species into the circulatory system and accumulation at sites of vascular disease [3]. Gold was detected in blood and urine up to three months after inhalation exposure to gold NMs [3,4]. Translocation of silver NMs has been seen in in vivo studies in rodents [5].

In order to comply with the 3R′s principle to reduce, refine and replace animal experiments, new advanced in vitro models are developed to better simulate the complexity of human lungs. Reliable in vitro models of the airway system are of critical importance for the risk assessment and governance of NMs and other environmental pollutants [6,7]. Human cells cultured on a microporous membrane at the air–liquid interface (ALI) with cell culture medium only at the basolateral side, represent a highly relevant model for inhalation toxicity studies [8]. Human lung cell lines such as A549 and BEAS-2B are commonly used as model cells in respiratory toxicology. A549 cells are alveolar type-II carcinoma cells, while BEAS-2B cells are immortalized cells from normal human bronchial epithelia. Both A549 and BEAS-2B cells form monolayers when cultivated at the ALI [9,10]. In order to further advance the models, cocultures with other cell types, such as macrophages, dendritic cells, or endothelial cells, can be established. The ALI exposure model aims to better mimic the physiology of the respiratory system and is regarded as a more relevant in vitro model compared to submerged exposure. Aerosolized exposure to the particles on top of the cells introduces less changes in the physicochemical properties of the test substance compared with submerged exposure [8].

Inhalation exposure to NMs, PM or other compounds may lead to adverse human effects. Genotoxicity is a critical endpoint in the hazard assessment of chemicals, including NMs, and should be assessed both at the level of DNA/genes and chromosomes. The comet assay is a widely used assay for determining DNA damage as DNA strand breaks (SBs), and as oxidized or alkylated bases by the inclusion of a repair enzyme such as formamidopyrimidine DNA glycosylase (Fpg) [11]. For the detection of chromosomal damage, the most-used test is the micronucleus assay (OECD test guideline 487), which detects the formation of micronuclei from chromosomes, chromatid fragments or whole chromosomes that lag behind in cell division [12,13]. So far, a very limited number of studies have addressed several genotoxicity endpoints in ALI models. Our approach, combining advanced and more physiologically relevant in vitro respiratory models and exposure systems with genotoxicity testing (by both comet and micronucleus assays), will support the hazard characterization of NMs for risk assessment and safe use.

NMs can induce DNA damage by direct contact with DNA, or indirectly via NM-induced oxidative stress or intermediate molecules and processes in cells (primary genotoxicity). Secondary genotoxicity can be driven by an inflammatory response [14]. The airway epithelium is an integrated part of the inflammatory defense response after inhalation exposure to toxicants. Pro-inflammatory cytokines are considered biomarkers of NM-induced toxicity and can be linked with adverse effects. The pro-inflammatory cytokines IL-6 and IL-8 are among the cytokines predominately secreted by monocytes, and both are coupled to lung injury and considered biomarkers of lung disease [15–17]. IL-8 can also act as a chemokine [17]. The bronchial epithelium serves as a first-line defense system against inhaled pathogens mainly by the release of chemokines, such as IL-8 [18]. The cytokines IL-6 and IL-8 have been shown to be secreted by airway epithelial, including BEAS-2B cells, and endothelial cells and be involved in lung inflammation responses [18–21]. IL-6 has been shown to be released from BEAS-2B cells after exposure to particulate matter below 1 μm in size (PM1), and both IL-6 and IL-8 were induced in BEAS-2B cells after exposure to the PM2.5 fraction [22,23]. Endothelial EA.hy926 cells were shown to release IL-6 and IL-8 after exposure to silica NMs [19].

The A549 cell line has frequently been used in coculture lung models and has been shown to be useful in a range of applications for hazard assessment of NMs [24–35]. The non-cancerous origin of BEAS-2B cells may make the cell line more relevant for use in risk governance of NMs, particularly as a bronchial respiratory model. Coculture models with BEAS-2B in ALI conditions for hazard assessment are, however, much less characterized than those with A549. The main aim of this study was to characterize an advanced respiratory model with BEAS-2B bronchial cells cultivated in ALI models, after exposure to an aerosolized reference silver NM, NM-300K. The cells were cultivated both as monocultures and in cocultures with human endothelial EA.hy926 cells. Cells from ALI cultures were analyzed for cytokine secretion, cytotoxicity, barrier integrity, DNA damage by the comet assay and chromosomal damage by the cytokinesis-block micronucleus assay. Importantly, the responses obtained with the bronchial BEAS-2B model were compared with the A549 alveolar model. The experimental design brilliantly allows for the comprehensive analysis of several endpoints from the same sample, facilitating increased throughput, better comparability, reduced costs, and sustainability by design to support the development of new approach methodologies (NAMs) and next-generation risk assessment (NGRA) of NMs.

2. Materials and Methods

2.1. Experimental Design

An experimental design combining the analysis of several endpoints from the same sample was developed. The same inserts with cells at the ALI were used for the analysis of cytokine secretion in basolateral media (enzyme-linked immunosorbent assay, ELISA), Ag permeation (inductively coupled plasma mass spectrometry, ICP-MS), cell viability (alamar-Blue assay), cell proliferation, DNA damage and oxidized base lesions (enzyme modified version of the comet assay), and chromosomal damage (micronucleus assay) (Figure 1). In parallel, additional experiments on ALI cultures were included to further characterize the models, and experiments with traditional submerged cultures were performed for comparisons. For each exposure condition, 1–2 culture inserts were included from both mono- and cocultures, and at least 3 independent experiments were performed, in order to allow for appropriate biological variation to be included in the results and analysis.

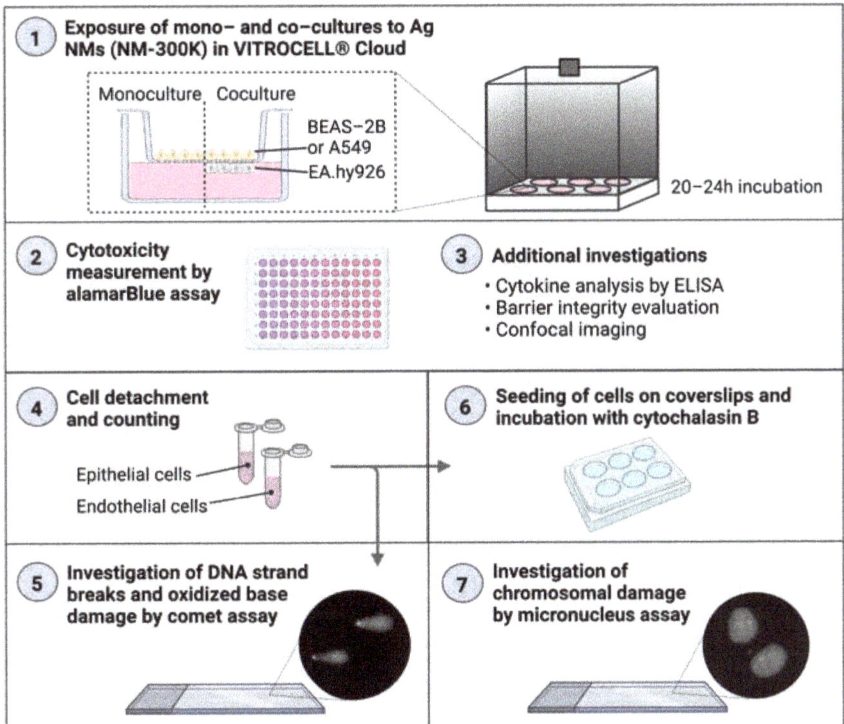

Figure 1. Experimental design for the four different respiratory models exposed at the air–liquid interface (ALI) in the VITROCELL® Cloud system. Created with BioRender.com.

2.2. Nanomaterials

The Ag NM NM-300K is listed on the representative manufactured NMs list of the European Commission Joint Research Centre (JRC, Brussels, Belgium) and was selected for this study based on its toxicity in our previous work [36–39]. NM-300K was provided by the Fraunhofer Institute for Molecular Biology and Applied Ecology (Schmallenberg, Germany). NM-300K is a silver colloidal dispersion with a nominal silver content of 10% *w/w*. The NMs were dispersed in an aqueous solution with stabilizing agents, consisting of 4% *w/w* each of polyoxyethylene glycerol trioleate and polyoxyethylene (20) sorbitan mono-laurate (Tween 20). The pristine diameter of NM-300K is about 15 nm, and the size distribution is narrow, where >99% of particles (by number) have a size below 20 nm. A second peak of smaller NMs of about 5 nm has also been reported. The majority of the NMs have a spherical shape [40].

Dispersed NMs were received in vials of approximately 2.0 g each, sealed under argon. The vials were stored at room temperature (RT) in the dark before use. The dispersion medium, NM-300K DIS, contained the aqueous solution with stabilizing agents at the same concentrations as NM-300K, but without Ag. This was used as a solvent control.

2.3. Nanomaterial Dispersion and Characterization

Stock dispersions of NM-300K were prepared in accordance with the Nanogenotox protocol [41]. The original vial of NM-300K was vortexed (>10 s), before approximately 1 g was added to a scintillation vial (Wheaton Industries, Millville, NJ, USA). To this, water with 0.05% bovine serum albumin (BSA) was added to yield a final nominal concentration of 10 mg/mL Ag-NMs, in order to obtain a high enough concentration of Ag-NMs in the ALI exposure system. The total Ag and dissolved Ag species (<3 kDa fraction) were measured from the same samples in Camassa and Elje et al. (2022), revealing a silver concentration of 7.2 ± 0.9 mg/mL with 3.6 ± 0.1% dissolved silver species [39].

The dispersion was sonicated in an ice bath using a calibrated Q500 sonicator with a 6 mm microtip probe (Qsonica L.L.C, Newtown, CT, USA), with amplitudes of 30–40% for 7–13 min. The energy output of the sample was 1030–1285 J/mL dispersion ($n = 10$), similarly to what is recommended by the Nanogenotox protocol (1176 J/mL [41]). Additional stock dispersions were sonicated using lower energy (95–720 J/mL, $n = 5$), and were included in the study as similar results were seen compared with the other dispersions. The NM stock dispersions were kept on ice for 10 min before use, to let the NMs settle. Before use, the vial was vortexed for approximately 10 s. The dispersion was kept on ice throughout the experiment. The dispersion medium NM-300K DIS (without Ag) was prepared following the same protocol as that for NM-300K.

The NMs were previously tested for endotoxins, with endotoxin contents below the limit of detection [39]. Stock dispersions for use in the submerged exposure experiments were diluted to 2.56 mg/mL in BSA-water before further dilution in culture medium (Section 2.4) in order to ensure consistency with other studies on the same NM.

2.4. Physicochemical Characterization of Nanomaterials in Dispersion

NM-300K was subjected to measurement of hydrodynamic diameter and zeta potential in a Zetasizer Ultra Red (Malvern Panalytical Ltd., Malvern, United Kingdom) immediately after preparation and after 24 h. The hydrodynamic diameter was determined using dynamic light scattering (DLS) by the particles in suspension. The measured particle size is the diameter of a sphere that diffuses at the same speed as the particle being measured, which is determined by measuring the Brownian motion of the particles by DLS and then interpreting the size using the Stokes–Einstein equation.

The NM stock dispersion was vortexed and diluted 1:100 in sterile filtered MilliQ water, and a 1 mL dispersion was transferred to a disposable cuvette (DTS0012) for size analysis. The hydrodynamic diameter was measured by non-invasive back scatter at 174.7° with 3–5 steps. Analysis was performed at 25 °C with 120 s equilibration time, automatic attenuation, and no pause between steps. Data were processed in the ZS Explorer software (version 2.0.0.98, Malvern Panalytical Ltd.), using general purpose model, refractive index 1.59 and absorption 0.01.

Measurement of size distribution of NMs diluted in culture medium was performed directly after preparation and after 24 h incubation at 37 °C, 5% CO_2. First, the stock dispersion was vortexed, and mixed with serum-free LHC-9 medium (article no. 12680013, ThermoFisher Scientific, Waltham, MA, USA) to give the highest tested concentration (141 µg/mL or 100 µg/cm^2 in submerged exposure). Then, the sample was diluted 1:10 in sterile filtered MilliQ water, transferred to a disposable cuvette and measured as described above.

Results were presented as Z-average (Z-ave), which is the intensity-weighted mean hydrodynamic size of the ensemble collection of particles, the polydispersity index (PDI), and hydrodynamic diameter (by intensity) of individual peaks in the size distributions.

For zeta potential analysis, the NM stock dispersion was vortexed and diluted 1:100 in sterile filtered MilliQ water, and 1 mL dispersion was transferred to a pre-wetted disposable folded capillary cell (DTS1070). The zeta potential was measured at 25 °C using mixed-mode measurement phase analysis light scattering (M3-PALS).

2.5. Cell Culture

BEAS-2B cells, an Ad12-SV40 hybrid virus-transformed human bronchial epithelial cell line [42,43], were purchased from ATCC (Manassas, VA, USA) (SV40 immortalized, CRL-9609, LN: 62853911). The cells were cultured in serum-free LHC-9 medium without supplements, and they were maintained in an incubator with humidified atmosphere at 5% CO2 and at 37 °C. The cells were passaged two times a week at 80–85% confluency. To facilitate detachment, the cells were incubated with trypsin-EDTA (0.25%, Sigma-Aldrich, Saint-Louis, MO, USA) with polyvinylpyrrolidone (PVP, 0.5% wt/vol) for 3–5 min. Medium was added, and the suspension was centrifuged to remove the trypsin/PVP before cells were seeded at 1.3×10^4 cells/cm^2 in Corning CellBind® cell culture flasks (Corning, Corning, NY, USA). The cells were used at passages (P) 3–14 (details in Table S1).

The human alveolar type II lung epithelial A549 cells [44] were provided by ATCC, and they were cultured in Dulbecco's Modified Eagle's Medium, DMEM, with low glucose (D6046, Sigma-Aldrich) supplemented with 9% v/v fetal bovine serum, FBS (prod.no. 26140079, ThermoFisher Scientific), and 1% v/v penicillin–streptomycin (100 U/mL pen and 100 µg/mL strep) (catalog no. 15070063, ThermoFisher Scientific). Human endothelial EA.hy926 cells [45] were provided from ATCC and were cultured in DMEM with high glucose (catalog no. 11960, ThermoFisher Scientific), supplemented with 9% v/v FBS, 1% v/v penicillin-streptomycin, sodium pyruvate (1 mM) and L glutamine (4 mM). The cells were maintained in an incubator with a humidified atmosphere at 5% CO_2 and at 37 °C. The cell lines were passaged two or three times a week at 85–90% confluency, using phosphate-buffered saline (PBS, catalog no. 14190094, ThermoFisher Scientific) for washing and semi-dry trypsinization using trypsin-EDTA (0.25%) incubation at 37 °C for 3 min. The cells were seeded at 1.3×10^4 cells/cm^2 in standard cell culture flasks. A549 cells were used at P2–15 and EA.hy926 cells were used at P3–19 (details in Table S1). All cell lines were regularly tested for mycoplasma contamination and found negative.

2.6. Cell Cultures at the Air–Liquid Interface

The seeding of mono- and cocultures were performed in a similar manner as previously described [46,47] with some modifications. All cell types, epithelial A549 (P3–15) and BEAS-2B (P3–14), and endothelial EA.hy926 (P3–16) (details on p numbers in Table S1), were seeded at a density of 1.1×10^5/cm^2. Mono- and cocultures were cultivated on permeable cell culture inserts in 6-well plates with a porous membrane of polyethylene terephthalate (PET) with a 1 µm pore diameter. Two insert types were used, with similar properties and cell attachment results: Millicell (catalog no. MCRP06H48, Sigma-Aldrich) or ThinCert™ (catalog no. 392-0128, Greiner Bio-One, Kremsmünster, Austria). The same insert type was used for all samples within an experiment.

First, the basolateral side of the membrane was pre-wetted (dipped in media), and the insert was placed upside down in the lid of a Falcon 6-well plate for inserts (catalog no. 353502, Corning). Then, 250 µL of EA.hy926 cell suspension was added to the basolateral side to reach a cell density of $1.1 \times 10^5/cm^2$. The lid with inserts was gently tilted to all sides to ensure even distribution of the cell suspension to the whole membrane surface before incubation for 3.5 h at 37 °C, 5% CO_2. After incubation, the plate was turned in a quick movement back to the original position, and 3 mL media (for EA.hy926 cells) was added to the basolateral side. To the apical side, 1 mL of A549 or BEAS-2B cell suspension (in their own media) was added to reach a cell density of $1.1 \times 10^5/cm^2$. Monocultures of BEAS-2B or A549 were prepared in the same way, where the basolateral compartment was filled with 3 mL of media without cells. The medium volumes were optimized in pilot experiments in order to avoid too great a pressure on the cells and the insert. The cultures were incubated for 2–3 days (48–72 h) to let the cells grow to confluency. Two days' incubation was performed only for A549 mono- and cocultures for alamarBlue and comet assay.

Epithelial and endothelial cells were seeded in their respective media. After 2–3 days of incubation of mono- and cocultures, the basolateral medium was replaced by 1.5 mL of fresh media, and the apical media was removed to place the cells in ALI conditions. For BEAS-2B/EA.hy926 a 1:1 mixture of LHC-9 and DMEM high glucose with supplements was used in the basolateral compartment, and for A549/EA.hy926 a 1:4 mixture of DMEM low glucose and DMEM high glucose with supplements was used (Table S2). The mono- and cocultures were incubated for 20–24 h in order to let the cells adapt to ALI conditions before exposure (Section 2.7).

2.7. Exposure of ALI Cultures in the VITROCELL® Cloud System

The VITROCELL® Cloud system (6-well format) (VITROCELL®Systems GMBH, Waldkirch, Germany), was used for the aerosol exposure of mono- and cocultures at ALI conditions to NM-300K and controls. A small volume of NM dispersion or control solution was added to the Aeroneb Pro® vibrating membrane nebulizer, which generates a dense cloud of droplets with a median aerodynamic diameter of 4–6 µm inside an exposure chamber. After some minutes, the humid aerosol will deposit at the bottom of the exposure chamber (area 145 cm^2) with cell inserts [48].

Aerosol exposure was performed by aerosolizing 2×150 µL of sample, followed by 150 µL PBS (details below), to the mono- and cocultures positioned at ALI in the VITROCELL® Cloud system at 37 °C. After 8 min, the aerosol cloud had settled, and the chamber was opened to transfer the cell inserts to 6-well plates (Falcon) with 1.5 mL fresh culture media (for mono- or cocultures). The exposure of ALI cultures was performed in the same sample order for all experiments: PBS (2×150 µL), NM-300K dispersion medium (2×150 µL), NM-300K low concentration (2×150 µL of stock dispersion diluted $10\times$ in PBS or NM-300K dispersion medium), and NM-300K high concentration (2×150 µL of stock dispersion). In order to reduce the amount of NMs left in the nebulizer, all samples were immediately exposed to additional 150 µL PBS. Thus, all samples were exposed to a cloud with a total volume of 450 µL. All solutions and dispersions were vortexed directly before use. The nebulizer was rinsed with PBS between all exposures, and the cloud system and chamber were wiped with a tissue with ethanol.

The relative amount of nebulized solution that is deposited on top of the cells, the deposition efficiency, can be measured by comparing the amount of deposited substance on the insert to the original solution, either by using a fluorescent compound or elemental analysis. As the deposition efficiency can vary between different nebulizers, the same nebulizer was used for all experiments in this study. We previously measured the deposition efficiency of this nebulizer to be 53% [39]. Additionally, the deposition of Ag in NM-300K was measured giving similar results [39]. This information was used to choose the exposure volumes needed for achieving the intended nominal concentrations for cell exposure.

2.8. Positive Control Exposures

Exposure of ALI cultures to positive controls were performed via the basolateral culture media below the inserts in 6-well plates, for 20–24 h. First, stock solutions were prepared and stored for use within all experiments before being diluted in sterile filtered H_2O directly before use and further diluted in culture medium. Chlorpromazine hydrochloride (catalog no. C8138, Sigma-Aldrich) was used as a positive control for cytotoxicity in the alamarBlue assay, with a stock solution at 5 mM in H_2O stored at 4 °C and exposure concentration of 50–100 µM. Mitomycin-C (catalog no. A2190.0002, PanReac AppliChem [VWR/Avantor]) was used as a positive control for micronuclei induction in the micronucleus assay, with stock solution at 0.2 mg/mL in dimethyl sulfoxide (DMSO) stored at −20 °C, and exposure concentration at 0.15 µg/mL, similarly to Reference [49]. In one experiment with A549/EA.hy926 and BEAS-2B/EA.hy926 cocultures, a higher concentration (0.30 µg/mL) was additionally included.

2.9. Characterization of the ALI Cultures by Microscopy Analysis

Daily evaluation of cell density and proliferation was performed with a Leica DM-IL microscope. More detailed characterization was performed by confocal microscopy. For confocal microscopy, cells in separate culture inserts were stained, fixed, and mounted between two glass coverslips, as described in Reference [39]. In brief, the plasma membranes were stained with CellMask Deep Red Plasma Membrane Stain (Invitrogen, 1:750 dilution in serum-free medium, 15 min at 37 °C), and the cells were fixed in formaldehyde (4%, 15 min, RT), before the nuclei were counterstained with DAPI (4′,6-diamidino-2-phenylindole, ProLong Gold Antifade reagent with DAPI, ThermoFischer Scientific). Confocal microscopy was performed using a Zeiss LSM 700 (lasers 405 and 639 nm; objective 40×). Image acquisition and processing were performed with the Zeiss Software ZEN. Z-stack acquisition was performed with 12–52 µm thickness with 7–53 images for each stack.

2.10. Barrier Function of the ALI Cultures by Elemental Analysis and Fluorescence

The barrier function of the BEAS-2B monocultures and BEAS-2B/EA.hy926 cocultures at the ALI, simulating the human lung–blood barrier, was tested by measuring the permeation of Ag or a fluorescent hydrophilic molecule into the basolateral medium. The procedures were performed similarly to as described in Reference [39] for A549 and A549/EA.hy926 cultures.

The permeation of Ag through the barrier was measured by analyzing the total Ag in the basolateral medium of the ALI cultures after 20–24 h exposure, and it was compared with the deposited Ag on empty culture inserts. The basolateral medium was collected in Eppendorf tubes, stored at −20 °C, and used for cytokine analysis by ELISA (Section 2.11) and Ag analysis by inductively coupled plasma mass spectrometry (ICP-MS). Culture medium was thawed on ice, vortexed for 10 s and 250–500 µL was transferred to a Teflon container (18 mL) vial. Sample preparation and ICP-MS analysis was performed as described in our recent study [39]. The ICP-MS results were analyzed to give the total Ag mass per insert, which was then divided by the deposited mass (low: 0.8 µg/cm^2, high: 6.0 µg/cm^2 [39], multiplied by insert area), and multiplied by 100% to give the percentage of Ag permeation through the ALI cultures.

Breakthrough of fluorescein sodium salt was measured on separate cell cultures in order to avoid interference between fluorescein and alamarBlue solution. After exposure to PBS (Section 2.7), 150 µL of fluorescein sodium salt (10 µg/mL in PBS) was nebulized and deposited on top of the cells for 3.5 min. In parallel, fluorescein was deposited on empty inserts in order to estimate the maximum leakage through the insert without cells and on inserts filled with 1 mL PBS in order to measure the maximum deposited fluorescein in the apical side. The leakage samples were transferred to 6-well plates with 1.5 mL medium on the basolateral side, and samples for deposition efficiency were transferred to empty wells. After 22–24 h of incubation, the fluorescence of fluorescein was measured in the basolateral medium or in apical PBS, related to a seven-point fluorescein standard curve (1.6–50 ng/mL) and blank in the respective medium or PBS. Fluorescence was read in triplicate (90 μL/well) in a black 96-well plate on a FLUOstar OPTIMA microplate reader (BMG Labtech, Ortenberg, Germany) with excitation 480 nm and emission 525 nm. Two independent experiments ($n = 2$) were performed, each with 1–2 culture inserts for deposition and breakthrough.

2.11. Cytokine Measurement

The basolateral medium from the exposed mono- and cocultures were transferred to Eppendorf tubes and frozen at $-20\ ^\circ$C. The amounts of cytokines present in the basolateral media was measured by the enzyme-linked immunosorbent assay (ELISA), which is a commonly used colorimetric immunological assay. The target molecule in the sample will bind to a specific antibody immobilized at the bottom of the microplate well. Through the addition of the second antibody, a sandwich complex is formed. A substrate solution binds to this complex and produces a measurable signal, which is directly proportional to the concentration of target present in the original sample.

ELISA was performed using kits for the human cytokines IL-6 (prod.no. 88-7066, Invitrogen) and IL-8 (prod.no. 88-8086, Invitrogen). The manufacturer's recommended procedures were followed. The samples were thawed on ice and vortexed before use and diluted 1–200 times in the assay buffer to fit within the measurement region. A standard curve was included in all plates. Duplicate measurements from each sample were performed. Plate washing was performed on a Hydroflex (TECAN, Grödig, Austria) microplate washer. Absorbance was read at 450 nm on an Infinite 200 Pro M Nano (TECAN) plate reader.

Potential interference between NM-300K and the performance of the ELISA was investigated. NM-300K was prepared as described above and diluted in cell culture media for BEAS-2B cells, A549 cells, or EA.hy926 cells, in order to achieve concentrations of 30, 3 and 0.3 μg/mL. The NMs in media were added to the ELISA plate in duplicates and mixed with reagent buffer or standard (final concentrations 25 pg/mL IL-6 or 31 pg/mL IL-8) provided in the kit. Further steps in the assay were run as described for the other samples.

2.12. Cell Viability Assessed by alamarBlue Assay

Cell viability was determined by the alamarBlue assay, which is based on the metabolic activity of cells and is commonly used for the quantitative analysis of cell viability and proliferation. The active ingredient in alamarBlue reagent (Sigma-Aldrich) is resazurin, which is a blue non-toxic, cell-permeable compound with low fluorescence. In living cells, resazurin is reduced to resorufin which is red and highly fluorescent, and the color change is detected on a plate reader.

AlamarBlue assay was performed 20–24 h after cell exposure. First, the basolateral media of ALI cultures was removed and saved for cytokine analysis. For monocultures, 1 mL alamarBlue reagent 10% v/v in cell culture media was added to the apical compartment, and 1.5 mL alamarBlue-free media was added to the basolateral compartment. For ALI cocultures, coculture media with alamarBlue 10% v/v was used in both compartments with the same volumes as for the monocultures. The plates were incubated for approximately 1 h. The plates were gently swirled to ensure even distribution of the alamarBlue solution, and 40 μL aliquots were transferred in triplicate to black 96-well plates, before fluorescence (excitation 530 nm, emission 590 nm) was measured on a FLUOstar OPTIMA microplate reader. Blank values (alamarBlue medium without cells present) were subtracted from the fluorescence intensity, which was further normalized by the average measurement of negative control (incubator control) set to 100% relative viability. Potential interference of NM-300K with the alamarBlue assay was investigated as described in Supplementary Materials.

2.13. Cell Detachment and Counting

Directly after performing the alamarBlue assay (Section 2.11), both sides of the insert were washed with PBS. Cells were detached by trypsin-EDTA incubation and subsequent mixing/washing of insert. For the apical compartment with BEAS-2B or A549 cells, 300 µL trypsin-EDTA was used (with PVP for BEAS-2B). For the basolateral compartment, PBS was used for monocultures and 1–1.5 mL trypsin-EDTA 0.05% (Sigma-Aldrich) was used for EA.hy926 cells. Inserts were trypsinized for 3–5 min at 37 °C, and the apical suspension was mixed with a pipet to facilitate detachment before medium for each cell type was added (1 mL in apical side, 3 mL in basolateral side). BEAS-2B cells were centrifuged at 200 g for 5 min and resuspended in 1 mL fresh culture medium to remove trypsin-EDTA/PVP which was not neutralized by the serum-free cell culture medium.

The cell suspension was mixed 1:1 with trypan blue (0.4%, Invitrogen) for the staining of cells with compromised cell membrane. The cells were counted in an automated cell counter (Countess® C10227, Invitrogen) in order to determine the total number of live cells and viability (%). The cell density was calculated by dividing the total number of live cells by the membrane insert area. Immediately after counting, the cell suspensions were further diluted to approximately 200,000 live cells/mL and used for genotoxicity studies (Sections 2.14 and 2.15).

2.14. DNA Damage Assessed by the Comet Assay

Cell suspensions from ALI and submerged cultures (Section 2.16) were subjected to DNA damage evaluation by the enzyme-modified version of the comet assay. Briefly, in the comet assay, cells are embedded in gels, lysed, and the remaining nucleoids are subjected to an electrophoretic field. The movement of damaged DNA causes comet formations, wherein the relative amount of DNA in the comet tail is proportional to the number of DNASBs.

Reagents used for the comet assay were provided by Sigma-Aldrich unless otherwise stated. A cell suspension with approximately 10,000 cells in 50 µL was mixed with 200 µL low melting point (LMP) agarose (0.8% in PBS) in a 96-well plate, yielding a final concentration 0.64% LMP agarose. Mini-gels (10 µL) were placed on coded microscopy slides precoated with 0.5% standard melting point agarose, and the slides were submerged in lysis solution (2.5 M NaCl, 0.1 M EDTA, 10 mM Tris, 1% v/v Triton X-100, pH 10, 4 °C) overnight.

For the detection of oxidized bases, the bacterial repair enzyme formamidopyrimidine DNA glycosylase (Fpg, gift from NorGenoTech, Oslo, Norway), which converts oxidized bases to SBs, was used [11]. After lysis, gels for Fpg treatment were washed twice for 8 min in buffer F (40 mM HEPES, 0.1 M KCl, 0.5 mM EDTA, 0.2 mg/mL BSA, pH 8, 4 °C), before being placed in a humid box and covered with Fpg (200 µL/slide) and polyethylene foil. Fpg incubation was performed for 30 min at 37 °C.

All slides were placed in the tank and submerged with electrophoresis solution (0.3 M NaOH, 1 mM EDTA, pH > 13, 4 °C), and DNA was allowed to unwind for 20 min. The electrophoresis was run for 20 min (25 V, 1.25 V/cm, Consort EV202). Gels were neutralized in PBS, washed in ultrapure H_2O, and air-dried overnight. Staining of DNA was performed with SYBR gold (1:2000), and scored in Leica DMI 6000 B (Leica Microsystems), equipped with a SYBR® photographic filter (ThermoFischer Scientific) using the software Comet assay IV 4.3.1 (Perceptive Instruments, Bury St Edmunds, UK). Comets were scored semi-blindly by two operators, where all slides within one experiment were scored by the same operator. Median DNA tail intensity was calculated from 50 comets per gel as a measure of DNA SBs. Medians were averaged from 2–6 gels per sample per n = 3–7 independent experiments.

Hydrogen peroxide (H_2O_2) was used as a positive control for DNA SBs. Cells from negative control inserts were embedded in gels and submerged in 13–100 µM H_2O_2 in PBS for 5 min at 4 °C. The samples were washed twice for 2 min in PBS (4 °C) and then submerged in a separate Coplin jar of lysis solution. The short time between H_2O_2 treatment and lysis limits the process of damage repair. The H_2O_2 exposure experiments with BEAS-2B ALI mono- and cocultures were conducted by placing all cell types (BEAS-2B from monoculture, BEAS-2B from coculture, and EA.hy926 cells from coculture) on the same slide in order to minimize variation. For experiments with submerged cultures (Section 2.16), all cell types and exposure conditions were placed on the same slide. As a negative control for the H_2O_2 exposure experiments, a separate slide with gels was exposed to PBS in parallel with H_2O_2 exposure.

As a positive control for the function of the Fpg enzyme, A549 cells were exposed to a photosensitizer, Ro 19–8022 (Hoffmann La Roche, Switzerland), and irradiated with visible light before embedding in gels, as described in Elje et al. 2019 [50]. The photosensitizer Ro 19-8022 induces with light oxidized purines, mainly 8-oxoG, which is detected by the Fpg [11]. The function of Fpg was controlled on a regular basis and was not included in all experiments. The positive control had an expected effect compared to the historical control data, with a net Fpg (level of SBs + Fpg minus level of SBs) of >20% DNA in tail.

2.15. Chromosomal Damage Assessed by the Cytokinesis-Block Micronucleus Assay

In parallel to the comet assay, cells from ALI cultures were seeded for detection of chromosomal damage by the cytokinesis-block micronucleus assay. The micronucleus assay measures the ability of the test substance to induce structural chromosome damage (clastogenic effect) or numerical chromosome alterations (aneugenic effect). Micronuclei are formed from chromosome or chromatid fragments or from whole chromosomes that lag behind in cell division. The addition of the active polymerization inhibitor cytochalasin B allows for analysis of the micronuclei frequency in cells that have completed one mitosis after treatment with the test substance, as such cells which are binucleated because the cytochalasin B prevents the separation of daughter cells after mitosis [13,51].

After the treatment and detachment of cells from ALI cultures, approximately $0.5–1 \times 10^5$ cells were seeded on flame-sterilized coverslips placed in 6-well plates with 1.5 mL media to a final concentration of 6 µg/mL cytochalasin B (prod.no. C6762, Sigma-Aldrich). The coverslips were incubated with culture media for 1–3 h before adding cells in order to facilitate cell adhesion. Coverslips and plates used for BEAS-2B cells were first coated with collagen IV (prod.no. 804592, Sigma-Aldrich). To each well with a coverslip, 1 mL collagen IV (30 µg/mL) diluted in Hanks' Balanced Salt solution (HBSS) (prod.no. 14175046, ThermoFisher Scientific) was added, before overnight incubation at 4 °C. The coated plates were washed with PBS before medium with cells was added. Cells were incubated at 37 °C and 5% CO_2 for 26–33 h (1–2 cell cycles). The same incubation time was used for all samples within the same experiment.

Cells were washed with PBS and fixed with methanol and acetic acid (3:1) in two steps, first for 15 min at RT, and then for up to 4 days at 4 °C. Coverslips were dried in the fume hood for 3–5 min and mounted on coded standard microscopy slides cleaned with ethanol, with a drop of the mounting medium ProLong Gold Antifade reagent with DAPI (ThermoFischer Scientific).

Cells were imaged in Zeiss Imager-Z2 microscope with a Metafer camera (MetaSystems Hard & Software GmbH, Altlussheim, Germany) and analysis system for micronuclei scoring. Scoring was performed in a semi-automatic manner, using $10\times$ and $40\times$ objectives. More details on the system and settings used for scoring can be seen in Section S1.2 in the Supplementary Materials, including the percentages of analyzed binucleated cells (Figure S1). The selected settings for scoring and analysis did not identify all binucleated cells with micronuclei, giving some false negative and false positive cells. Thus, to avoid this, all identified binucleated cells were manually accepted or rejected, and cells with possible micronuclei were checked with a $40\times$ objective.

2.16. Statistical Analysis

At least three independent experiments were performed for each test method, unless otherwise stated. In each experiment, 1–2 parallel culture inserts were included (Table S4). Results are presented as mean with standard deviation (SD) calculated from the average results from n experiments. Normal distribution of data was assumed. In order to evaluate the statistical significance of the results, one-way ANOVA was performed followed by Dunnett's multiple comparisons test using GraphPad Prism version 9.3.1 for Windows (GraphPad Software, San Diego, CA, USA). Results were compared to the PBS control for

ALI cultures and unexposed control for submerged cells. Fluorescein permeation results were compared to empty inserts with no cells. Ag permeation results were analyzed by one-way ANOVA with post-test Sidak to allow for multiple comparisons between low and high concentrations and between models and empty inserts (16 comparisons in total). The level of significance was set to $p < 0.05$. Calculation of EC_{50} values, the concentration giving 50% response, was performed using non-linear regression analysis with the Hill equation, in GraphPad Prism. Mathematical calculations were performed in Excel (Microsoft 365).

3. Results

3.1. Nanomaterial Dispersion Quality and Physicochemical Characterization

The hydrodynamic diameter of NM-300K in stock dispersion and diluted in LHC-9 culture medium was measured by DLS directly after preparation and after 24 h (Table 1). The NMs in stock dispersion had a hydrodynamic diameter (Z-ave) of 149 nm and were stable in dispersion for 24 h. The size distribution of the NMs had two or three peaks, where most particles were within the peak with mean size 110–265 nm, and with some smaller (5–40 nm) and larger (>3000 nm) particles. The zeta potential was -17.1 ± 2.8 mV, indicating that the NM dispersions were semi-stable.

Table 1. Characterization of NM-300K directly after preparation (T0) and after 24 h (T24h). The stock dispersion (10 mg/mL) was used for aerosolized exposure and was diluted in LHC-9 culture medium to make a medium dispersion (141 µg/mL, 100 µg/cm^2) for submerged exposure. Results are presented as mean $\pm$ standard deviation or interval (lower value–higher value) from n = 3–15 independent experiments. PDI: polydispersity index, h: hours.

Sample	Time	Z-Ave (nm)	PDI (a.u.)	Main Peak (nm)	n
Stock	T0	149.3 ± 42.5	0.331 ± 0.062	110–265	15
dispersion	T24h	129.6 ± 15.4	0.347 ± 0.022	130–147	3
Stock diluted	T0	378.2 ± 109.8	0.393 ± 0.032	148–280	5
in medium	T24h	1838.8 ± 2344.2	0.884 ± 0.739	60–150	5

NM-300K diluted in LHC-9 culture medium had a higher hydrodynamic diameter than the stock dispersion, with a Z-ave of 378 nm and main peak with slightly higher size as for stock dispersion. After 24 h, larger NMs were detected in some of the measurements, giving a Z-ave between 119–5180 nm and a high PDI. Medium without NM-300K had a Z-ave of 15.4 $\pm$ 1.0 nm with PDI 0.349 $\pm$ 0.054 (n = 2).

3.2. Characterization of the Advanced Models

The advanced models of BEAS-2B/EA.hy926 and A549/EA.hy926 cells, cultured at the ALI, were characterized for cell density, viability and barrier integrity. The ALI-cultured apical cells were moist with a shiny surface, though some cultures occasionally showed a drier appearance (Figure S2). BEAS-2B cells were grown in dense structures on the porous membranes, as seen by confocal microscopy (Figure 2A,B). Higher density was observed for BEAS-2B in coculture with EA.hy926 cells compared to monocultures. Some holes in the apical cell layers of both mono- and cocultures were observed, and the fewest holes were seen in A549 cocultures (details not shown). The confocal images of BEAS-2B cells also indicated slightly higher thickness of the apical cell layer in cocultures, with cells growing in multilayers, compared to the BEAS-2B cells in monocultures (Figure 2 and Figure S3). Cell counting after detachment of cells confirmed a higher density of BEAS-2B cells in coculture compared to BEAS-2B in monocultures, indicating a higher proliferation of the cells in this condition (Table 2). The opposite result was seen for A549 cells, where the density was higher in monoculture compared to coculture with EA.hy926 cells. The endothelial cells had a similar density of both types of cocultures (Table 2) and were growing in a confluent monolayer (Figure 2C). The collected cells had high viability, though some cells were lost during the detachment process and during washing before trypsinization, and some cells still remained on the insert.

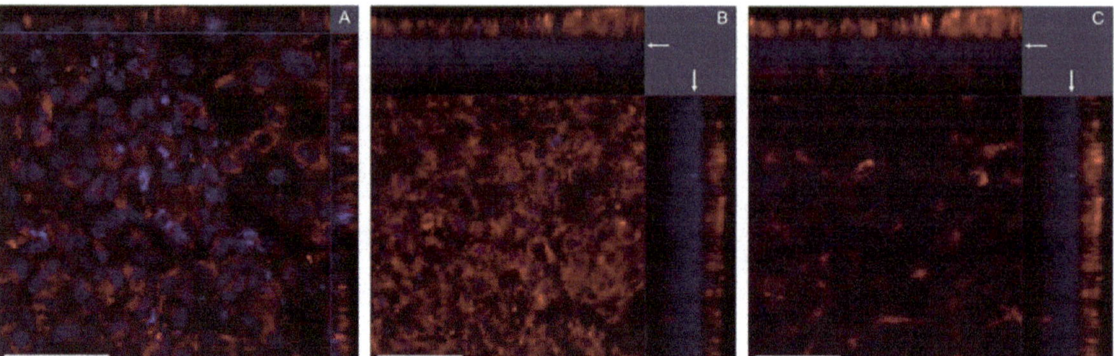

Figure 2. Confocal images of advanced bronchial BEAS-2B models. Z-stack image series (2D x–y view and respective side views) showing the distribution of BEAS-2B and EA.hy926 cells on the opposite sides of a transwell membrane insert (arrow). (**A**) BEAS-2B cells in monocultures (z-stack thickness 12 µm). (**B**) BEAS-2B cells in cocultures (z-stack thickness 52 µm). (**C**) EA.hy 926 cells in cocultures (z-stack thickness 52 µm). Red: cellular membranes stained with Cell Mask red dye; blue: nuclei counterstained with DAPI. Magnification: 40×. Scale bars 50 µm.

Table 2. Number of live cells, cell density and cell viability from ALI cultures at the end of the cultivation period, evaluated by cell counting with trypan blue staining after detaching the cells from the inserts. Results are presented as mean with standard deviation (SD) from 2–3 independent experiments (*n*) and 2–7 replica culture inserts in each experiment. ALI: air–liquid interface.

ALI Model		Cell Line	Live Cells ($\times 10^6$)	Cell Density ($\times 10^5/cm^2$)	Viability (%)	*n*
BEAS-2B	Monoculture	BEAS-2B	2.9 ± 0.3	6.8 ± 0.8	95 ± 2	2
	Coculture	BEAS-2B	4.7 ± 0.6	11.1 ± 1.4	97 ± 1	2
		EA.hy926	0.9 ± 0.3	1.6 ± 0.4	83 ± 15	3
A549	Monoculture	A549	2.4 ± 0.7	5.7 ± 1.5	97 ± 2	2
	Coculture	A549	1.4 ± 0.8	3.3 ± 1.9	94 ± 4	2
		EA.hy926	0.8 ± 0.1	1.8 ± 0.1	93 ± 3	2

The barrier integrity of the advanced models was investigated by measuring the permeation of the water-soluble fluorescein sodium salt and Ag (NMs or dissolved species) from NM-300K after aerosol exposure through the cellular layer by quantification in the basolateral media after 24 h (Table 3). A high permeation of fluorescein, at the same level as empty inserts without cells, was found in BEAS-2B/EA.hy926 cocultures (70%) (Table 3). Strongly reduced fluorescein permeation was seen in BEAS-2B monocultures (20%) and in A549/EA.hy926 cocultures (9%). No difference was seen in the permeation of fluorescein between incubator control and PBS-exposed cultures.

Table 3. Comparison of the barrier integrity of mono- and cocultures of BEAS-2B/EA.hy926 and A549/EA.hy926 cells, by measurement of fluorescein sodium salt and Ag permeation. Results are presented as mean permeation with SD of n = 2–3 independent experiments with single or duplicate inserts. Permeation is presented as the basolateral concentration (μM) and as the basolateral concentration relative to the deposited apical concentration (%). Results for A549/EA.hy926 cultures are based on our previous publication [39]. NC: negative control. PBS: phosphate-buffered saline, NM-300K DIS: dispersant control, NM-300K low: nominal 1 μg/cm^2, NM-300K high: nominal 10 μg/cm^2.

Experiment Type	Treatment	Fluorescein or Ag Permeation				Empty Inserts
		BEAS-2B/EA.hy926 Cultures		A549/EA.hy926 Cultures *		
		Monoculture	Coculture	Monoculture	Coculture	
Fluorescein (% of deposited apical concentration)	NC	21 ± 2% (n = 2) [a]	69 ± 2% (n = 2)	-	-	74 ± 11% (n = 2)
	PBS	22 ± 1% (n = 2) [a]	71 ± 7% (n = 2)	-	9 ± 5% (n = 3) [a]	
Ag (μM and % of deposited apical concentration) [b]	NM-300K DIS	0.017 ± 0.003 μM (n = 2)	0.026 ± 0.013 μM (n = 2)	0.069 ± 0.025 μM (n = 2)	0.030 ± 0.017 μM (n = 2)	-
	NM-300K low	11.9 ± 1.5 μM 57.1% (n = 3) [c,d]	7.7 ± 1.3 μM 37.1% (n = 3) [e]	1.9 ± 0.6 μM 8.1% (n = 2) [f]	3.3 ± 0.6 μM 14.3% (n = 2) [f]	-
	NM-300K high	11.9 ± 0.4 μM 7.7% (n = 3)	8.3 ± 2.2 μM 5.3% (n = 3) [e]	14.5 ± 1.4 μM 8.7% (n = 3)	15.4 ± 1.3 μM 9.2% (n = 2)	11.40 ± 0.03 μM 7% (n = 2)

* Results on A549/EA.hy926 cultures are based on [39]. [a] Statistically significant difference from empty inserts, evaluated by one-way ANOVA with post-test Dunnett ($p < 0.05$). [b] Ag permeation results were analyzed by one-way ANOVA with post-test Sidak ($p < 0.05$). A total of 16 comparisons were made: between low and high concentration for each model, low and low concentration for BEAS-2B and A549 monocultures, high and high concentration for BEAS-2B and A549 monocultures, low and low concentration for BEAS-2B and A549 cocultures, high and high concentration for BEAS-2B monocultures and A549 cocultures, and high concentration for all models and empty inserts. [c] Statistically significant difference from BEAS-2B coculture at the same concentration. [d] Statistically significant difference from A549 monoculture at the same concentration. [e] Statistically significant difference from A549 coculture at the same concentration. [f] Statistically significant difference from the same model at high concentration.

The permeation of Ag was higher in BEAS-2B monocultures compared with cocultures (Table 3). This difference was highest after exposure to the low concentration of NM-300K, giving a permeation of 57% in monocultures and 37% in cocultures. In the basolateral medium of cultures exposed to the high concentration of NM-300K, the Ag concentrations of both mono- and cocultures were similar to the maximum permeation through empty inserts (11 μM) (no statistically significant difference by one-way ANOVA with post-test Sidak, $p > 0.05$). The barrier integrity of A549 cultures differed from the BEAS-2B cultures. A slightly higher Ag permeation was seen after exposure to low concentration of NM-300K in A549 cocultures (14%) compared to monocultures (8%) although the difference was not statistically significant. However, the permeability of both A549 mono- and cocultures was lower than for BEAS-2B mono- and cocultures ($p < 0.05$). After exposure to the high concentration of NM-300K, similar results were seen for both A549 and BEAS-2B models, where the permeability was about the same as for the maximum permeation through empty inserts ($p > 0.05$). A low concentration of Ag was found in the basolateral medium of cultures exposed to NM-300K DIS, and the concentration was similar for all models.

3.3. Toxic Responses after NM-300K Exposure in Advanced Respiratory BEAS-2B or A549 Models

3.3.1. Cytotoxicity

The mono- and cocultures of BEAS-2B/EA.hy926 or A549/EA.hy926 cells were exposed to aerosolized NM-300K and control solutions in the VITROCELL® Cloud. After 20–24 h, cytotoxicity was investigated by the alamarBlue assay following the experimental design in Figure 1. The measured deposited concentrations of NM-300K (low and high), with nominal concentrations 1 and 10 $\mu g/cm^2$, were measured in our recent study to be 0.8 and 6.0 $\mu g/cm^2$, respectively [39].

Cell viability is presented relative to incubator control (NC, set to 100%) and statistically analyzed against the PBS control. A reduction in the relative cell viability was seen after NM-300K exposure at high concentration in BEAS-2B monocultures (57%, Figure 3A), but not in the cocultures compared with the PBS exposure control. The viability of BEAS-2B cells in cocultures was significantly reduced after aerosol exposure to PBS, with a relative viability of 67%, compared with the incubator control. This effect of PBS was not seen in the monocultures (Figure 3B). The viability of cells exposed to NM-300K DIS was similar to that of cells exposed to PBS, in both models. The viability of EA.hy926 cells was not affected by aerosol exposure to NM-300K or PBS. The positive control, 50–100 μM chlorpromazine hydrochloride in basolateral media, strongly reduced the viability in all cultures, as expected.

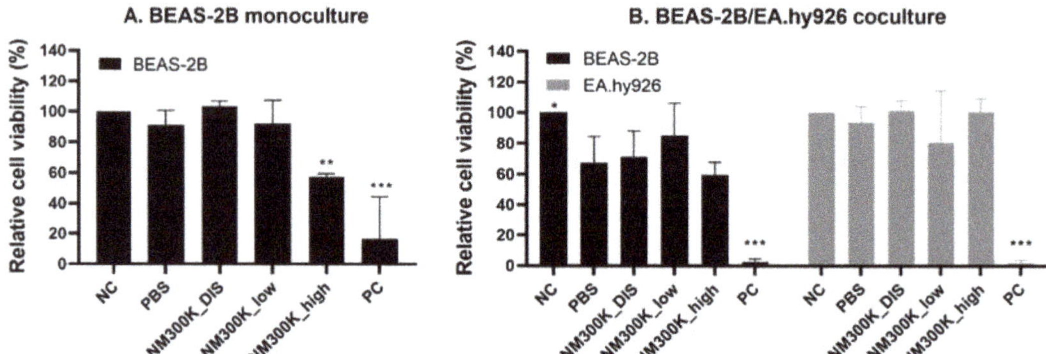

Figure 3. Relative cell viability of BEAS-2B and EA.hy926 cells after exposure to aerosolized NM-300K and control solutions at the air–liquid interface, evaluated by alamarBlue assay. The response of cells in monocultures (**A**) was different compared to cocultures (**B**). Cell viability is presented relative to NC, which is set to 100%. Results are presented as the mean with standard deviation from n= 5–7 (**A**) and n = 4 (**B**) independent experiments (where the results from each experiment are averaged in the case of two replica inserts). Statistically significant different effects on cell viability compared to control inserts with PBS-exposed cells were analyzed by one-way ANOVA followed by Dunnett´s post-hoc test (* $p < 0.05$, ** $p < 0.01$, *** $p < 0.001$). NC: negative control, PBS: phosphate-buffered saline, NM-300K DIS: dispersant control, NM-300K low: nominal 1 $\mu g/cm^2$, NM-300K high: nominal 10 $\mu g/cm^2$, PC: positive control (chlorpromazine hydrochloride 50–100 μM in basolateral medium for 24 h).

Similar results as with the BEAS-2B models were seen with A549 monocultures and A549/EA.hy926 cocultures after NM-300K exposure (Table 4, details in [39]). The viability of EA.hy926 cells was reduced after aerosol exposure when in coculture with A549 but not with BEAS-2B (Table 4).

Table 4. Comparison of the relative cell viability in bronchial BEAS-2B and alveolar A549 advanced models after aerosol exposure to PBS and NM-300K. Results are presented as the mean relative cell viability (compared to NC, set to 100%) with standard deviation from 4–9 independent experiments (*n*), each with 1–2 replica culture inserts. Statistically significant differences compared to control exposed to PBS were analyzed by one-way ANOVA with Dunnett´s multiple comparisons post-test, and they are indicated by * $p < 0.05$. Data for A549 mono- and cocultures are based on our results from [39]. NC: negative control, PBS: phosphate-buffered saline, NM-300K low: nominal 1 $\mu g/cm^2$, NM-300K high: nominal 10 $\mu g/cm^2$.

	Relative Cell Viability (%)					
	BEAS-2B/EA.hy926 Cultures			A549/EA.hy926 Cultures [a]		
	Monoculture	Coculture		Monoculture [a]	Coculture [a]	
Treatment	BEAS-2B	BEAS-2B	EA.hy926	A549 [a]	A549 [a]	EA.hy926 [a]
NC	100 ± 0	100 ± 0 *	100 ± 0	100 ± 0	100 ± 0	100 ± 0
PBS	91 ± 9	67 ± 17	94 ± 10	93 ± 23	62 ± 22	69 ± 19
NM-300K low	92 ± 15	85 ± 21	81 ± 34	72 ± 13	57 ± 20	67 ± 21
NM-300K high	57 ± 2 *	60 ± 8	100 ± 9	59 ± 26	59 ± 29	68 ± 21
n	5–8	4	4	6–9	4–5	4–5

[a] Data based on the results from [39].

For the comparison of the new advanced models with the corresponding traditional cell models, the cytotoxicity of NM-300K was also tested with submerged exposure of monocultured cells by alamarBlue assay. NM-300K was cytotoxic in BEAS-2B cells at concentrations above 10 $\mu g/cm^2$ and at 10 $\mu g/cm^2$ for submerged and ALI exposure, respectively (Figure 3 and Figure S4). BEAS-2B cells were more sensitive to NM-300K exposure compared to A549 and EA.hy926 cells in submerged conditions (Figure S4 and Table S5, [52,53]). No interference with the alamarBlue assay was detected for NM-300K (Figure S5).

3.3.2. Secretion of Pro-Inflammatory Cytokines IL-6 and IL-8

Upon NM exposure, pro-inflammatory cytokines can be secreted by the airway epithelium and endothelium in order activate the immune system. The concentrations of the pro-inflammatory cytokines IL-6 and IL-8 secreted from the ALI cultures into the basolateral medium during exposure were measured by ELISA. The results are presented as absolute concentrations (Figure 4 and Tables S6 and S7) and relative to NC (Figure S6).

An apparent trend towards increased levels of IL-8 was seen after NM-300K exposure in all cell models; however, a statistically significant increase was measured only for BEAS-2B in mono- and coculture for the lowest concentration of NM-300K compared with untreated incubator control. A similar effect was seen on IL-6 levels in BEAS-2B cocultures only. In BEAS-2B mono- and cocultures, the concentrations of both IL-6 and IL-8 were higher for low-concentration NM-300K compared with the high concentration. There was a significant effect of PBS exposure on the levels of IL-6 in monocultures of BEAS-2B compared to NC (Figure 4 and Figure S6).

IL-6 and IL-8 concentrations were found to be higher (3× and 9×, respectively) in cocultures of BEAS-2B/EA.hy926 compared with BEAS-2B monocultures (Figure 4 and Tables S6 and S7). For the monocultures, the increase in IL-6 was about the same for low-concentration NM-300K and NM-300K DIS. However, no increase was detected after exposure to the high concentration of NM-300K, for which the concentrations of NM-300K DIS was matching. For the A549 models, the level of IL-8 was about 4× higher in monocultures than in cocultures. For IL-6 level, there was no difference between mono- and cocultures of A549 cells. No interference between the NM-300K and the assay was found (Figure S7).

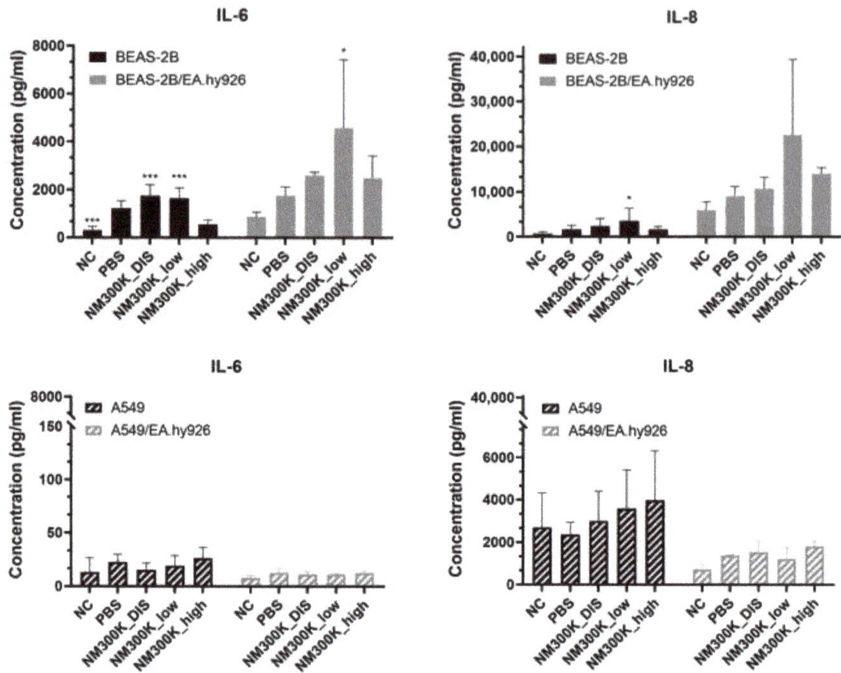

Figure 4. Concentrations of IL-6 and IL-8 in mono- and cocultures of BEAS-2B/EA.hy926 cells (top panel) and A549/EA.hy926 cells (bottom panel) after exposure to aerosolized NM-300K and control solutions at the air–liquid interface, evaluated by ELISA. Results are presented as the mean with standard deviation from single or duplicate inserts from n = 2–6 independent experiments (n = 6 for BEAS-2B monocultures, n = 3 for BEAS-2B/EA.hy926 cocultures and A549 monocultures, and n = 2 for A549/EA.hy926 cocultures). Statistically significant different effects on cytokine concentration compared to the negative control inserts with PBS-exposed cells (PBS) were analyzed by one-way ANOVA followed by Dunnett´s post-hoc test (* $p < 0.05$, *** $p < 0.001$). NC: negative control, PBS: phosphate-buffered saline, NM-300K DIS: dispersant control, NM-300K low: nominal 1 µg/cm^2, NM-300K high: nominal 10 µg/cm^2.

3.3.3. Genotoxicity by DNA and Chromosomal Damage in ALI Cultures

After viability analysis by alamarBlue assay and cytokine secretion analysis by ELISA, cells from the same samples were analyzed for DNA damage by the comet assay and cytokinesis-block micronucleus assay.

Different sensitivities on induction of DNA SBs and oxidized base lesions after exposure to NM-300K were measured by the enzyme-modified comet assay when comparing the different cell types and models. NM-300K exposure at low concentrations induced an increase in DNA SBs and SBs + Fpg in BEAS-2B cells in monoculture, with 26 ± 18% DNA in the tail (Figure 5A), but no effect in cocultures (Figure 5B). No significant effect was measured in BEAS-2B or in EA.hy926 cells (Figure 5) after exposure to high-concentration NM-300K. The levels of DNA SBs were similar in the incubator control (NC) and in samples exposed to PBS. No effect of NM-300K DIS was seen. However, the background of DNA SBs (in NC) was slightly higher in BEAS-2B cells from cocultures (5.9 ± 3.6% DNA in tail, n = 3) compared with monocultures (1.4 ± 1.6% DNA in tail, n = 6). For comparison, submerged NM300-K exposure of BEAS-2B cells did not induce any genotoxicity, as an increase in SBs was detected only at cytotoxic concentrations (from 10 µg/cm^2) (Figure S8).

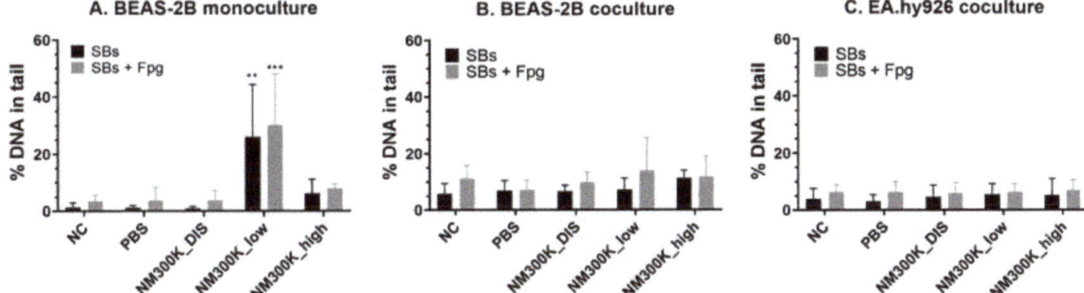

Figure 5. DNA damage by strand breaks (SBs) and oxidized base lesions (Fpg) of BEAS-2B and EA.hy926 cells after exposure to aerosolized NM-300K and control solutions at the air–liquid interface, evaluated by the comet assay. The response of cells in monocultures (**A**) was different compared with cocultures (**B,C**). Results are presented as the mean with standard deviation from single or duplicate inserts from n = 3–6 (**A**), n = 3 (**B**), and n = 4 (**C**) independent experiments. Statistically significant different effects on DNA damage compared to control inserts with PBS-exposed cells (PBS) were analyzed by one-way ANOVA followed by Dunnett´s post-hoc test (** $p < 0.01$, *** $p < 0.001$). NC: negative control, PBS: phosphate-buffered saline, NM-300K DIS: dispersant control, NM-300K low: nominal 1 µg/cm^2, NM-300K high: nominal 10 µg/cm^2.

H_2O_2 exposure induced a concentration-related induction of SBs in BEAS-2B cells from monocultures and EA.hy926 cells from cocultures (Figure 6 and Table 5). BEAS-2B cells in cocultures were found to be more sensitive, with a high level of SBs also at the lowest concentrations of H_2O_2 (80 $\pm$ 11% DNA in tail at 13 µM H_2O_2). Different media compositions were used in the BEAS-2B monocultures (LHC-9) and in the BEAS-2B/EA.hy926 cocultures (DMEM and LHC-9, 1:1). In submerged cells, which showed less sensitivity to H_2O_2 exposure than ALI cultures, the different media compositions did not affect the viability (Figure S9) or H_2O_2 sensitivity (Table 5 and Figure S10). Cells from A549 mono- and cocultures had a high response to 100 µM H_2O_2, as expected, and only one concentration was tested.

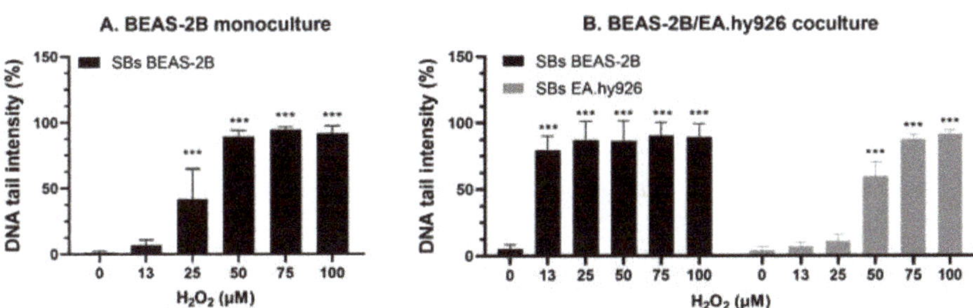

Figure 6. DNA damage by strand breaks (SBs) in BEAS-2B and EA.hy926 cells after exposure to H_2O_2 in gels, evaluated by the comet assay. Cells from unexposed inserts of (**A**) BEAS-2B monocultures and (**B**) BEAS-2B/EA.hy926 cocultures were embedded in gels before H_2O_2 exposure. Results are presented as the mean with standard deviation from single or duplicate inserts from n = 7 (**A**) and n = 3 (**B**) independent experiments. Statistically significant different effects of DNA damage compared to negative control cells without H_2O_2 exposure (0 µM in PBS) were analyzed by one-way ANOVA followed by Dunnett´s post-hoc test (*** $p < 0.001$). NC: negative control; H_2O_2: hydrogen peroxide; PBS: Phosphate-buffered saline.

Table 5. DNA damage response after the hydrogen peroxide (H_2O_2) exposure of BEAS-2B cells from different culturing conditions. Results are presented as the mean with standard deviation from $n = 3$ independent experiments (except ALI monocultures with $n = 6$), each with 1–2 culture inserts or culture wells per treatment. Medium type is presented as the ratio of medium for BEAS-2B cells (LHC-9) and medium for EA.hy926 cells (DMEM). ALI: air–liquid interface, SBs: strand breaks.

| Culture Conditions | | | H_2O_2 Treatment | | | |
| | | | DNA SBs (% DNA in Tail) | | | EC_{50} (μM) |
Culture Type	Cell Line	LHC-9:DMEM	12.5 μM	25 μM	50 μM	
ALI monoculture	BEAS-2B	1:0	7 ± 4	41 ± 23	89 ± 5	28 ± 6
ALI coculture	BEAS-2B	1:1	80 ± 11	87 ± 14	86 ± 15	4 ± 5
	EA.hy926	1:1	7 ± 3	11 ± 5	60 ± 10	45 ± 5
Submerged monoculture	BEAS-2B	1:0	7 ± 3	24 ± 12	84 ± 14	33 ± 6
		1:1	7 ± 3	18 ± 6	77 ± 10	37 ± 6
		0:1	4 ± 3	27 ± 6	85 ± 5	31 ± 1
	EA.hy926	1:0	18 ± 7	20 ± 5	42 ± 17	70 ± 31
		1:1	16 ± 1	19 ± 2	46 ± 19	>100
		0:1	19 ± 3	12 ± 6	34 ± 5	>100

No significant effect on micronuclei induction was found after exposure to PBS, NM-300K DIS, or NM-300K, on any of the cultures, compared to the PBS control (Figure 7). A high level of micronuclei was induced by the positive control (0.15 μg/mL mitomycin-C in basolateral media) in BEAS-2B and EA.hy926 cells from mono- and cocultures (Figure 7A,B), and slightly lower for A549 in mono- and cocultures (Figure 7C,D). The effect of mitomycin-C in A549 monocultures was significantly different from that of the unexposed NC control only ($p = 0.04$), and the increase in micronuclei formation was not statistically significant from the PBS control ($p = 0.08$). The proportion of binucleated cells in the samples for micronuclei investigation was estimated to be 21% for BEAS-2B and 19% for EA.hy926 cells. Corresponding numbers for A549/EA.hy926 cocultures were about 21% for A549 and 8% for EA.hy926 cells (Figure S1).

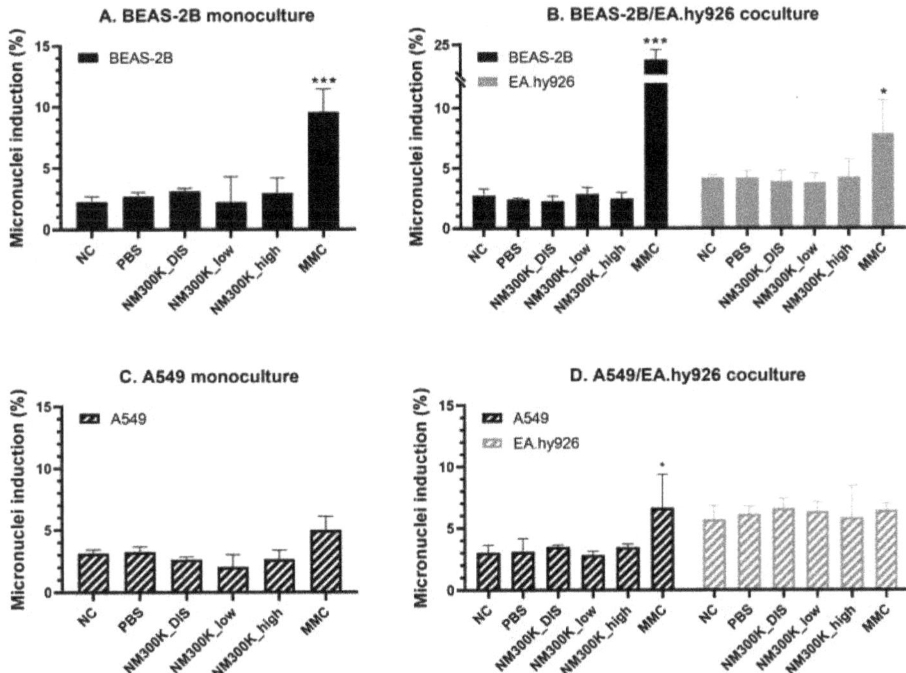

Figure 7. Micronuclei induction in BEAS-2B, EA.hy926 and A549 cells after exposure to aerosolized NM-300K and control solutions at the air–liquid interface, evaluated by the cytokinesis block micronuclei assay. Micronuclei induction was analyzed in cells from BEAS-2B monoculture (**A**), BEAS-2B/EA.hy926 coculture (**B**), A549 monoculture (**C**), and A549/EA.hy926 coculture (**D**). Results are presented as the mean with standard deviation from single or duplicate inserts from n = 3 independent experiments. Statistically significant different effects on micronuclei induction compared to control inserts with PBS-exposed cells (PBS) were analyzed by one-way ANOVA followed by Dunnett´s post-hoc test (* $p < 0.05$, *** $p < 0.001$). NC: negative incubator control, PBS: phosphate-buffered saline, NM-300K DIS: dispersant control, NM-300K low: nominal 1 µg/cm^2, NM-300K high: nominal 10 µg/cm^2, MMC: mitomycin-C at 0.15 µg/mL in basolateral medium for 24 h.

4. Discussion

An important part of NAMs, which are essential for NGRA, is the development and characterization of advanced in vitro models. Advanced respiratory in vitro models are of high importance for the hazard assessment of NMs after inhalation exposure. Cells cultured and exposed at the ALI represent a more physiological scenario than cells in submerged conditions. In order to develop the most realistic NAMs, the characterization, testing and validation of models is needed. Of importance to this is comparison of the effects of reference NMs on different advanced models for the same target, as well as to benchmark the effects of the tested NMs against the effects in traditional 2D in vitro models. This study focused on the characterization and application of the immortalized human bronchial epithelial cell line BEAS-2B cultivated at ALI in monoculture or in cocultures with endothelial EA.hy926 cells for the testing of different toxicity endpoints. It is, to our knowledge, the first study to successfully apply several genotoxicity endpoints, including the micronucleus assay, in advanced BEAS-2B cocultures at the ALI, and to perform a comparison of the effects of a reference NM in mono- and cocultures, and also with the more extensively used human alveolar epithelial cell line A549. Further, effects on the advanced models at the ALI were compared with responses in traditional corresponding submerged cultures in monocultures.

Cell growth, barrier integrity, and confluency differed between mono- and cocultures of BEAS-2B cells. In cocultures with EA.hy926, the BEAS-2B cells had lower confluency compared with monocultures, and they showed a multilayer growth, which was not seen with the A549 cells. Thus, more holes were seen under confocal microscopy in the BEAS-2B epithelial layer, which was also thicker. We found a higher density of BEAS-2B cells in cocultures compared with monocultures, indicating higher cellular growth and stimulated cell proliferation in the cocultures. The opposite was found with A549 cells, where the cells had lower density in cocultures compared to monocultures. The stimulated proliferation of BEAS-2B cells in cocultures with EA.hy926 cells was found not to be due to different media compositions of mono- and cocultures. Rather, the increased proliferation of BEAS-2B cells in coculture might be related to cell signaling from the endothelial cells. The multilayer growth of BEAS-2B cells in coculture indicates that the advanced model facilitates conditions similar to tissue physiology in the lungs, as has been shown also in previous studies [10,54].

In monoculture, BEAS-2B and A549 had similar cell numbers at the end of the cultivation period, despite the longer doubling time (4 h) of BEAS-2B cells compared to A549 [55,56]. The endothelial cells showed a similar density and appearance in both coculture types, with BEAS-2B and A549, respectively.

The barrier function appeared to be less in BEAS-2B cocultures than in monocultures, as indicated by the higher permeability of fluorescein, which was measured in the basolateral medium. This finding is in line with the higher frequency of observed holes in the apical cell layers of cocultures of BEAS-2B. In contrast, Ag permeation was higher in BEAS-2B monocultures than in cocultures. It is known that Ag has a high affinity for sulfur and that it may form toxic complexes with sulfur-containing proteins in the cells [57]. The lower permeation of Ag in the coculture may be explained by the additional barrier constituted by the endothelial cells and the measured higher epithelial cell density in cocultures, and thus increased interaction with or dissolution of the NMs. Ag permeation after exposure to NM-300K at high concentrations was in all culture types similar to the permeation through empty inserts. As total silver was measured, it included both particles and dissolved Ag-ions diffusing from the apical side down the concentration gradient. However, the permeation of Ag was similar after exposure to low and high concentrations, and it was not increased as may be expected by the larger gradient. This low permeability might be due to the agglomeration/aggregation of the nanoparticles at high density, making them less likely to penetrate into the pores of the insert membrane, the pores being blocked, a saturation of the medium, or a combination of these factors.

A549 mono- and cocultures showed less permeation of fluorescein and also of Ag after exposure to low-concentration NM-300K, than BEAS-2B cultures. This is in accordance with previous studies showing that A549 cells form tight junctions [39,46] and thus a stronger barrier, measured as trans-epithelial resistance (TEER), at an earlier stage than BEAS-2B cells [54]. Different bronchial cell lines, including BEAS-2B, cultivated at ALI, were evaluated for barrier functions by He et al., (2021), and only Calu-3 cells were found to sustain a strong TEER for up to 21 days. Also, the immortalized cell line 16HBE formed tight junctions and developed a strong TEER, though the TEER dropped considerably in ALI conditions [10].

In order to compare the responses of the different models to a toxic insult, the cells were exposed to Ag NM-300K, which is a reference NM commonly used in many projects related to the safety of NMs (such as the European Commission FP7 project NanoReg, and H2020 projects NanoReg2, PATROLS and RiskGONE). The relative cell viability of BEAS-2B monocultures at the ALI was reduced by NM-300K exposure at the highest concentration. This is similar to what we observed for A549 monocultures [39]. The viability was reduced at lower concentrations after ALI exposure compared with submerged exposure, which may indicate higher sensitivity of the ALI models. However, direct comparisons between the models are difficult due to large differences in the experimental conditions including aerosol or submerged exposure, cell number, and cell density.

Significant reduction in cell viability was only measured after exposure to high concentrations of NM-300K in monocultures. In cocultures, we observed that PBS aerosol exposure reduced cell viability of the BEAS-2B cells, compared with incubator control, and thus no significant difference in viability could be detected between PBS and NM exposures. The same effect of PBS was also seen previously with A549 cocultures [39]. This indicates that the coculture conditions at the ALI sensitize the apical epithelial cells to the aerosol exposure both of non-toxic and toxic compounds. One could speculate that the lack of direct basolateral contact with the cell culture media in the cocultures could influence this, due to the confluent layer of the endothelial cells. Additionally, the interaction and interplay between the endothelial cells in itself may play a role. Interestingly, the viability of the endothelial cells was not reduced after PBS exposure in coculture with BEAS-2B but with A549 [39], despite the demonstrated stronger barrier of the A549 epithelial layer.

When genotoxicity was tested, the models showed different sensitivities. Genotoxicity measured as DNA SBs was found only in BEAS-2B monocultures at low concentrations of NM-300K. The lack of genotoxic effects at high concentrations could be due to loss of damaged cells during the washing steps, as cytotoxicity was measured at this concentration. No genotoxic effect was seen in BEAS-2B or in the EA.hy926 cells in coculture. The lack of an effect in the BEAS-2B cells in coculture may be due to the higher cell density and thereby relatively lower number of particles per cell. One may also speculate as to whether the co-cultivation with the endothelial cells may increase the emergency preparedness of the BEAS-2B cells towards toxic insults. The increased level of pro-inflammatory cytokines in the cocultures (Figure 4) may support such a theory, and it has been suggested previously that increased expression of cytokines, such as IL-6, can promote DNA repair [58]. In cocultures of A549 and EA.hy926, genotoxicity measured as SBs was only found in the endothelial cells. However, we previously showed that in triculture with addition of differentiated THP-1 cells, no SBs were detected [39]. These results are not false negative, as NMs were internalized in the A549 cells [39]. A genotoxic response can be secondary due to the induction of an inflammatory response. Thus, we measured IL-6 and IL-8 levels in mono- and cocultures. No significant increase in IL-6 or IL-8 levels was detected in the A549 models after NM-300K exposure. In general, there was a trend towards increased levels after ALI exposure. However, the basal level of IL-8 was much higher in A549 cocultures than in monocultures, which was the opposite as seen with BEAS-2B. For BEAS-2B cocultures, the level of both IL-6 and IL-8 was increased after exposure to low concentration of NM-300K, although statistical significance was reached only for IL-6. The level of IL-6 was also increased in BEAS-2B monocultures at low concentrations of NM-300K, but it was about $9\times$ lower than for cocultures. In contrast, significant induction of SBs was measured only in monocultures at low concentrations of NM-300K. As the toxic response was changed when coculturing different cell types, as compared with monocultures, further studies are merited to elucidate the interplay between the cell types and the importance of coculturing multiple cell types for hazard identification.

Genotoxicity was further tested at the chromosomal level by the micronucleus assay. Few studies have applied the micronucleus assay on advanced human lung models at the ALI; at the time of writing, we found only one publication on monocultured A549 [49], one on monocultured BEAS-2B and other bronchial cell types [59], and four studies on nasal epithelial cells from donors [60–63]. We successfully employed this assay to both mono- and cocultures of BEAS-2B cells exposed at the ALI. The effect of the positive control mitomycin-C (0.15 µg/mL in basolateral medium) was found to be more pronounced in the cocultures. The effect of mitomycin-C on A549 mono- and cocultures was lower than on BEAS-2B, and the effect in A549 monocultures was only significantly different from the unexposed control and not from the PBS control. Mitomycin-C also induced micronuclei in the underlying endothelial cells in coculture with the BEAS-2B-cells, but not with A549 cells, and the proportion of binucleated cells was lower when cocultured with A549 cells. The epithelial and endothelial cells were cultured with cytochalasin B for the same duration; however, a longer incubation time was used for BEAS-2B/EA.hy926 cocultures compared to A549/EA.hy926 cocultures due to the differences in cell doubling times. Further investigations are needed in order to determine whether the differences in response are due to lower sensitivity towards mitomycin-C when in coculture with A549 or if experimental optimization is needed to produce a higher proportion of binucleated cells.

No aneugenic or clastogenic effects were detected after NM-300K exposure at the ALI, which is in contrast to previous studies with significant micronuclei induction by NM-300K in submerged BEAS-2B cells [64,65]. ALI exposure was performed when the cells had developed a confluent cell layer, which may have affected the cell cycle and proliferation rate. In our study the cells from ALI cultures were directly seeded with cytochalasin B on coverslips after NM treatment, at a lower density to initiate DNA replication and MN formation, before the cell fixation and analysis of MN in binucleated cells. The discrepancy in the effects between submerged and ALI may, therefore, be due to their different stages in division cycles during NM exposure, which merits further experiments to optimize the method. The agglomeration state of the NMs would also influence their toxicity, and this might differ between submerged and ALI exposure. A previous study on titanium dioxide NMs in BEAS-2B cells showed that changes in the composition of exposure medium affected the induction of MN due to differential states of agglomeration [66].

The experimental design presented in this study, enabling several endpoints to be measured from each insert, allows for increased throughput, reduced costs, time and materials and thus better sustainability, compared to measuring all endpoints in separate inserts. This is an important aspect of the NAMs and crucial for an integrated approach to testing and assessment (IATA) in NGRA. Further, more direct comparisons are enabled, as the different endpoints are measured from the same exposure, thus reducing the variability induced by distinct exposures. This is an essential issue, as variability may be expected to increase with the increasing complexity of the model, in line with variability in human responses between individuals. Our results, which showed the toxicity of PBS at the ALI in cocultures compared with an incubator control not exposed at the ALI, point to the importance of always including an incubator control in ALI experiments.

The higher complexity of the advanced models makes them more laborious and maybe less applicable for screening purposes. However, this is significantly improved by the efficient experimental design we here present, which also contributes to more robust mechanistical data, as several endpoints are measured from the same insert. One argument that has been used for the application of more complex cell models is that they may increase the sensitivity towards toxic insults. An approach to test the sensitivity towards the induction of SBs is to expose cells with H_2O_2 on gels directly before lysis. A striking observation was that the BEAS-2B cells in coculture appeared to be much more sensitive to H_2O_2 exposure. Cells in monoculture at ALI showed slightly higher sensitivity to H_2O_2 exposure compared to cells in submerged conditions. Also, the endothelial cells showed higher sensitivity to H_2O_2 when in coculture with BEAS-2B than as submerged monoculture. The culture media formulation appeared to have no influence on the outcome either in terms of viability or sensitivity towards SBs. The reasons for the higher sensitivity to H_2O_2 in coculture are not known and merits further investigations, but it may be related to the above-mentioned interaction between cells in coculture making the DNA more prone to damage.

The higher sensitivity of the BEAS-2B cells in coculture to NM-300K was observed in cell viability by the alamarBlue assay but not on SBs by the comet assay. NM-300K has a high cytotoxic potential and a quite narrow concentration window between non-cytotoxic and cytotoxic effects. During the technical procedure in the sample preparation for comet assay, the cells go through several washing steps, and there are reasons to believe that the more damaged cells are less attached to the insert and can be lost. In the MN assay, the effect of mitomycin-C was most pronounced in the cells from the coculture, which could be an indication that the BEAS-2B cells in coculture may be a more sensitive model for the hazard assessment of genotoxic compounds, and thus a promising advanced model to be adapted for risk assessment based on in vitro data and an IATA approach. The differences in sensitivity between mono- and cocultures, and between the application of lung epithelial A549 cells or bronchial BEAS-2B cell models, for the various endpoints measured emphasizes the importance of carrying out proper characterization of the emerging advanced models, as well as developing robust SOPs. Further, it points to the importance of developing advanced coculture in vitro models, allowing for intercellular signaling, to better mimic tissue organization and enhance the prediction of human hazard.

5. Conclusions

An important step in finding the best predictive model for human adverse effects is to increase complexity and thereby obtain more tissue- and organ-like structures, use human cells, characterize the models, and compare responses in different models. This work indicates that the bronchial mono- and coculture models of BEAS-2B and EA.hy926 cells have different sensitivities to NM-300K exposure as measured by cytotoxicity and genotoxicity at DNA and chromosomal levels, and they are different from the alveolar models of A549 and EA.hy926 cells. This is important knowledge to provide more robust reproducible and reliable results, for the further development of advanced 3D respiratory in vitro models relevant to inhalation exposure, and to obtain the most reliable hazard identification and prediction of effects on humans based on non-animal studies. This study provides important knowledge for the further development of advanced 3D respiratory in vitro models for the most reliable hazard identification and prediction of the effects from inhalation on human-based NAMs for NGRA.

Supplementary Materials: The following supporting information can be downloaded at: https://www.mdpi.com/article/10.3390/nano13030407/s1, Figure S1: Percentage of binucleated cells in the samples for micronucleus assay; Figure S2: Images of A549/EA.hy926 cocultures at the ALI; Figure S3: Confocal images of advanced bronchial models; Figure S4: Relative cell viability of submerged BEAS-2B cells after exposure to NM-300K and NM-300K DIS; Figure S5: AlamarBlue interference test; Figure S6: Relative concentrations of IL-6 and IL-8 in ALI mono- and cocultures; Figure S7: ELISA interference test of NM-300K in cell culture media; Figure S8: DNA damage by strand breaks and oxidized base lesions in submerged BEAS-2B cells after exposure to NM-300K evaluated by Fpg-modified comet assay; Figure S9: Relative cell viability of submerged BEAS-2B cells after exposure to DMEM and LHC-9 culture media; Figure S10: DNA strand breaks evaluated by the comet assay after exposure to hydrogen peroxide of submerged monocultures of BEAS-2B and EA.hy926 cells with different media compositions. Table S1: Passage numbers; Table S2: Medium types used in mono- and cocultures; Table S3: NM-300K concentrations applied for exposure; Table S4: Number of cell culture inserts per experiment; Table S5: EC50 values for cytotoxic effect of NM-300K on submerged monocultures; Table S6: Concentrations of interleukin 6; Table S7: Concentrations of interleukin 8.

Author Contributions: Conceptualization, E.E., E.M. (Espen Mariussen) and E.R.-P.; formal analysis, E.E. and E.M. (Erin McFadden); investigation, E.E. and E.M. (Erin McFadden); writing—original draft preparation, E.E.; writing—review and editing, E.E., E.M. (Erin McFadden), E.M. (Espen Mariussen), M.D. and E.R.-P.; visualization, E.E.; supervision, E.M. (Espen Mariussen), M.D. and E.R.-P.; project administration, E.M. (Espen Mariussen), M.D. and E.R.-P.; funding acquisition, E.M. (Espen Mariussen), M.D. and E.R.-P. All authors have read and agreed to the published version of the manuscript.

Funding: This research was funded by The Norwegian Research Council, NanoBioReal project: 288768; The Norwegian Research Council Ph.D. project (E.E.): 272412/F40; TEPCAN project funded by the Program "Applied research" under the Norwegian Financial Mechanisms 2014–2021/POLNOR 2019 (EEA and Norway Grants), Thematic areas: welfare, health and care: NCBR Funding No. NOR/POLNOR/TEPCAN/0057/2019-00; European Union's Horizon 2020 research and innovation program RiskGONE: 814425; European Union's Horizon 2020 research and innovation program TWINALT: 952404—H2020-WIDESPREAD-2020-5.

Data Availability Statement: Data are available from the researchers on request.

Acknowledgments: The authors would like to thank Mihaela R. Cimpan and Ivan Rios Mondragon (University of Bergen, Norway) for the loan of LHC-9 medium, Marit Vadset (NILU) for performing ICP-MS analysis, Tatiana Honza (NILU) for assistance with comet scoring, Rudolf Dreschler (Metasystems) and Naouale El Yamani (NILU) for training on/assistance with the use of the Metafer system and expert opinions, Andrew Collins and Sergey Shaposhnikov (NorGenoTec AS, Norway) for providing the Fpg, and Reidun Torp (University of Oslo, Norway) for a critical revision of the manuscript. The graphical abstract was created with BioRender.com.

Conflicts of Interest: The authors declare no conflict of interest.

References

1. Bonner, J.C. Molecular Mechanisms of Respiratory Toxicity. In *Molecular and Biochemical Toxicology*; Smart, R.C., Hodgson, E., Eds.; Wiley: Hoboken, NJ, USA, 2018.

2. Oberdörster, G.; Oberdörster, E.; Oberdörster, J. Nanotoxicology: An Emerging Discipline Evolving from Studies of Ultrafine Particles. *Environ. Health Perspect.* **2005**, *113*, 823–839. [CrossRef] [PubMed]

3. Miller, M.R.; Raftis, J.B.; Langrish, J.P.; McLean, S.G.; Samutrtai, P.; Connell, S.P.; Wilson, S.; Vesey, A.T.; Fokkens, P.H.B.; Boere, A.J.F.; et al. Inhaled Nanoparticles Accumulate at Sites of Vascular Disease. *ACS Nano* **2017**, *11*, 4542–4552. [CrossRef] [PubMed]

4. Raftis, J.B.; Miller, M.R. Nanoparticle Translocation and Multi-Organ Toxicity: A Particularly Small Problem. *Nano Today* **2019**, *26*, 8–12. [CrossRef] [PubMed]

5. Hadrup, N.; Sharma, A.K.; Loeschner, K.; Jacobsen, N.R. Pulmonary Toxicity of Silver Vapours, Nanoparticles and Fine Dusts: A Review. *Regul. Toxicol. Pharmacol.* **2020**, *115*, 104690. [CrossRef]

6. Zavala, J.; Freedman, A.N.; Szilagyi, J.T.; Jaspers, I.; Wambaugh, J.F.; Higuchi, M.; Rager, J.E. New Approach Methods to Evaluate Health Risks of Air Pollutants: Critical Design Considerations for In Vitro Exposure Testing. *Int. J. Environ. Res. Public Health* **2020**, *17*, 2124. [CrossRef]

7. Pfuhler, S.; van Benthem, J.; Curren, R.; Doak, S.H.; Dusinska, M.; Hayashi, M.; Heflich, R.H.; Kidd, D.; Kirkland, D.; Luan, Y.; et al. Use of in Vitro 3D Tissue Models in Genotoxicity Testing: Strategic Fit, Validation Status and Way Forward. Report of the Working Group from the 7th International Workshop on Genotoxicity Testing (IWGT). *Mutat. Res. Genet. Toxicol. Environ. Mutagen.* **2020**, *850–851*, 503135. [CrossRef]

8. Lacroix, G.; Koch, W.; Ritter, D.; Gutleb, A.C.; Larsen, S.T.; Loret, T.; Zanetti, F.; Constant, S.; Chortarea, S.; Rothen-Rutishauser, B.; et al. Air-Liquid Interface in Vitro Models for Respiratory Toxicology Research: Consensus Workshop and Recommendations. *Appl. In Vitro Toxicol.* **2018**, *4*, 91–106. [CrossRef]

9. Heijink, I.H.; Brandenburg, S.M.; Noordhoek, J.A.; Postma, D.S.; Slebos, D.J.; van Oosterhout, A.J.M. Characterisation of Cell Adhesion in Airway Epithelial Cell Types Using Electric Cell-Substrate Impedance Sensing. *Eur. Respir. J.* **2010**, *35*, 894–903. [CrossRef]

10. He, R.W.; Braakhuis, H.M.; Vandebriel, R.J.; Staal, Y.C.M.; Gremmer, E.R.; Fokkens, P.H.B.; Kemp, C.; Vermeulen, J.; Westerink, R.H.S.; Cassee, F.R. Optimization of an Air-Liquid Interface in Vitro Cell Co-Culture Model to Estimate the Hazard of Aerosol Exposures. *J. Aerosol Sci.* **2021**, *153*, 105703. [CrossRef]

11. Collins, A.R. The Use of Bacterial Repair Endonucleases in the Comet Assay. In *Methods in Molecular Biology*; Humana Press Inc.: Totowa, NJ, USA, 2017; Volume 1641, pp. 173–184. [CrossRef]

12. Kohl, Y.; Rundén-Pran, E.; Mariussen, E.; Hesler, M.; el Yamani, N.; Longhin, E.M.; Dusinska, M. Genotoxity of Nanomaterials: Advanced In Vitro Models and High Throughput Methods for Human Hazard Assessment—A Review. *Nanomaterials* **2020**, *10*, 1911. [CrossRef]

13. OECD. Test No. 487: In Vitro Mammalian Cell Micronucleus Test. In *OECD Guidelines for the Testing of Chemicals, Section 4*; OECD Publishing: Paris, France, 2016. [CrossRef]

14. Magdolenova, Z.; Collins, A.; Kumar, A.; Dhawan, A.; Stone, V.; Dusinska, M. Mechanisms of Genotoxicity. A Review of in Vitro and in Vivo Studies with Engineered Nanoparticles. *Nanotoxicology* **2014**, *8*, 233–278. [CrossRef]

15. Butt, Y.; Kurdowska, A.; Allen, T.C. Acute Lung Injury: A Clinical and Molecular Review. *Arch. Pathol. Lab. Med.* **2016**, *140*, 345–350. [CrossRef] [PubMed]

16. Ekstrand-Hammarström, B.; Akfur, C.M.; Andersson, P.O.; Lejon, C.; Österlund, L.; Bucht, A. Human Primary Bronchial Epithelial Cells Respond Differently to Titanium Dioxide Nanoparticles than the Lung Epithelial Cell Lines A549 and BEAS-2B. *Nanotoxicology* **2012**, *6*, 623–634. [CrossRef] [PubMed]

17. Borish, L.C.; Steinke, J.W. 2. Cytokines and Chemokines. *J. Allergy Clin. Immunol.* **2003**, *111*, S460–S475. [CrossRef]

18. Zou, J.; Zhou, L.; Hu, C.; Jing, P.; Guo, X.; Liu, S.; Lei, Y.; Yang, S.; Deng, J.; Zhang, H. IL-8 and IP-10 Expression from Human Bronchial Epithelial Cells BEAS-2B Are Promoted by Streptococcus Pneumoniae Endopeptidase O (PepO). *BMC Microbiol.* **2017**, *17*, 187. [CrossRef]

19. Napierska, D.; Thomassen, L.C.J.; Vanaudenaerde, B.; Luyts, K.; Lison, D.; Martens, J.A.; Nemery, B.; Hoet, P.H.M. Cytokine Production by Co-Cultures Exposed to Monodisperse Amorphous Silica Nanoparticles: The Role of Size and Surface Area. *Toxicol. Lett.* **2012**, *211*, 98–104. [CrossRef] [PubMed]

20. Khair, O.A.; Davies, R.J.; Devalia, J.L. Bacterial-Induced Release of Inflammatory Mediators by Bronchial Epithelial Cells. *Eur. Respir. J.* **1996**, *9*, 1913–1922. [CrossRef]
21. Adler, K.B.; Fischer, B.M.; Wright, D.T.; Cohn, L.A.; Becker, S. Interactions between Respiratory Epithelial Cells and Cytokines: Relationships to Lung Inflammation. *Ann. N. Y. Acad. Sci.* **1994**, *725*, 128–145. [CrossRef]
22. Longhin, E.; Holme, J.A.; Gualtieri, M.; Camatini, M.; Øvrevik, J. Milan Winter Fine Particulate Matter (WPM2.5) Induces IL-6 and IL-8 Synthesis in Human Bronchial BEAS-2B Cells, but Specifically Impairs IL-8 Release. *Toxicol. In Vitro* **2018**, *52*, 365–373. [CrossRef]
23. Tian, G.; Wang, J.; Lu, Z.; Wang, H.; Zhang, W.; Ding, W.; Zhang, F. Indirect Effect of PM 1 on Endothelial Cells via Inducing the Release of Respiratory Inflammatory Cytokines. *Toxicol. In Vitro* **2019**, *57*, 203–210. [CrossRef]
24. Herzog, F.; Loza, K.; Balog, S.; Clift, M.J.D.; Epple, M.; Gehr, P.; Petri-Fink, A.; Rothen-Rutishauser, B. Mimicking Exposures to Acute and Lifetime Concentrations of Inhaled Silver Nanoparticles by Two Different in Vitro Approaches. *Beilstein J. Nanotechnol.* **2014**, *5*, 1357–1370. [CrossRef]
25. Klein, S.G.; Cambier, S.; Hennen, J.; Legay, S.; Serchi, T.; Nelissen, I.; Chary, A.; Moschini, E.; Krein, A.; Blömeke, B.; et al. Endothelial Responses of the Alveolar Barrier in Vitro in a Dose-Controlled Exposure to Diesel Exhaust Particulate Matter. *Part. Fibre Toxicol.* **2017**, *14*, 7. [CrossRef]
26. Di Ianni, E.; Erdem, J.S.; Møller, P.; Sahlgren, N.M.; Poulsen, S.S.; Knudsen, K.B.; Zienolddiny, S.; Saber, A.T.; Wallin, H.; Vogel, U.; et al. In Vitro-in Vivo Correlations of Pulmonary Inflammogenicity and Genotoxicity of MWCNT. *Part. Fibre Toxicol.* **2021**, *18*, 25. [CrossRef]
27. Meindl, C.; Öhlinger, K.; Zrim, V.; Steinkogler, T.; Fröhlich, E. Screening for Effects of Inhaled Nanoparticles in Cell Culture Models for Prolonged Exposure. *Nanomaterials* **2021**, *11*, 606. [CrossRef]
28. Barosova, H.; Karakocak, B.B.; Septiadi, D.; Petri-Fink, A.; Stone, V.; Rothen-Rutishauser, B. An In Vitro Lung System to Assess the Proinflammatory Hazard of Carbon Nanotube Aerosols. *Int. J. Mol. Sci.* **2020**, *21*, 5335. [CrossRef] [PubMed]
29. Cappellini, F.; di Bucchianico, S.; Karri, V.; Latvala, S.; Malmlöf, M.; Kippler, M.; Elihn, K.; Hedberg, J.; Wallinder, I.O.; Gerde, P.; et al. Dry Generation of CeO2 Nanoparticles and Deposition onto a Co-Culture of A549 and THP-1 Cells in Air-Liquid Interface—Dosimetry Considerations and Comparison to Submerged Exposure. *Nanomaterials* **2020**, *10*, 618. [CrossRef] [PubMed]
30. Wang, Y.; Adamcakova-Dodd, A.; Steines, B.R.; Jing, X.; Salem, A.K.; Thorne, P.S. Comparison of in Vitro Toxicity of Aerosolized Engineered Nanomaterials Using Air-Liquid Interface Mono-Culture and Co-Culture Models. *NanoImpact* **2020**, *18*, 100215. [CrossRef] [PubMed]
31. Loret, T.; Peyret, E.; Dubreuil, M.; Aguerre-Chariol, O.; Bressot, C.; le Bihan, O.; Amodeo, T.; Trouiller, B.; Braun, A.; Egles, C.; et al. Air-Liquid Interface Exposure to Aerosols of Poorly Soluble Nanomaterials Induces Different Biological Activation Levels Compared to Exposure to Suspensions. *Part. Fibre Toxicol.* **2016**, *13*, 58. [CrossRef]
32. Herzog, F.; Clift, M.J.D.; Piccapietra, F.; Behra, R.; Schmid, O.; Petri-Fink, A.; Rothen-Rutishauser, B. Exposure of Silver-Nanoparticles and Silver-Ions to Lung Cells in Vitro at the Air-Liquid Interface. *Part. Fibre Toxicol.* **2013**, *10*, 11. [CrossRef]
33. Friesen, A.; Fritsch-Decker, S.; Hufnagel, M.; Mülhopt, S.; Stapf, D.; Weiss, C.; Hartwig, A. Gene Expression Profiling of Mono- and Co-Culture Models of the Respiratory Tract Exposed to Crystalline Quartz under Submerged and Air-Liquid Interface Conditions. *Int. J. Mol. Sci.* **2022**, *23*, 7773. [CrossRef]
34. Kaur, K.; Mohammadpour, R.; Ghandehari, H.; Reilly, C.A.; Paine, R.; Kelly, K.E. Effect of Combustion Particle Morphology on Biological Responses in a Co-Culture of Human Lung and Macrophage Cells. *Atmos. Environ.* **2022**, *284*, 119194. [CrossRef] [PubMed]
35. Friesen, A.; Fritsch-Decker, S.; Hufnagel, M.; Mülhopt, S.; Stapf, D.; Hartwig, A.; Weiss, C. Comparing α-Quartz-Induced Cytotoxicity and Interleukin-8 Release in Pulmonary Mono- and Co-Cultures Exposed under Submerged and Air-Liquid Interface Conditions. *Int. J. Mol. Sci.* **2022**, *23*, 6412. [CrossRef] [PubMed]
36. Gábelová, A.; el Yamani, N.; Alonso, T.I.; Buliaková, B.; Srančíková, A.; Bábelová, A.; Pran, E.R.; Fjellsbø, L.M.; Elje, E.; Yazdani, M.; et al. Fibrous Shape Underlies the Mutagenic and Carcinogenic Potential of Nanosilver While Surface Chemistry Affects the Biosafety of Iron Oxide Nanoparticles. *Mutagenesis* **2017**, *32*, 193–202. [CrossRef] [PubMed]
37. El Yamani, N.; Collins, A.R.; Rundén-Pran, E.; Fjellsbø, L.M.; Shaposhnikov, S.; Zielonddiny, S.; Dusinska, M. In Vitro Genotoxicity Testing of Four Reference Metal Nanomaterials, Titanium Dioxide, Zinc Oxide, Cerium Oxide and Silver: Towards Reliable Hazard Assessment. *Mutagenesis* **2017**, *32*, 117–126. [CrossRef]
38. Elje, E.; Mariussen, E.; Moriones, O.H.; Bastús, N.G.; Puntes, V.; Kohl, Y.; Dusinska, M.; Rundén-Pran, E. Hepato(Geno)Toxicity Assessment of Nanoparticles in a HepG2 Liver Spheroid Model. *Nanomaterials* **2020**, *10*, 545. [CrossRef]
39. Camassa, L.M.A.; Elje, E.; Mariussen, E.; Longhin, E.M.; Dusinska, M.; Zienolddiny-Narui, S.; Rundén-Pran, E. Advanced Respiratory Models for Hazard Assessment of Nanomaterials—Performance of Mono-, Co- and Tricultures. *Nanomaterials* **2022**, *12*, 2609. [CrossRef]
40. Klein, C.L.; Comero, S.; Stahlmecke, B.; Romazanov, J.; Kuhlbusch, T.A.J.; van Doren, E.; De Temmerman, P.-J.; Mast, T.J.; Wick, P.; Krug, H.; et al. *NM-Series of Representative Manufactured Nanomaterials NM-300 Silver Characterisation, Stability, Homogeneity*; Publications Office of the European Union: Luxembourg, 2011. [CrossRef]
41. Jensen, K.A.; Clausen, P.A.; Birkedal, R.; Kembouche, Y.; Christiansen, E.; Jacobsen, N.R.; Levin, M.; Koponen, I.; Wallin, H.; de Temmerman, P.-J.; et al. Towards a Method for Detecting the Potential Genotoxicity of Nanomaterials. Deliverable 3. Final Protocol for Producing Suitable Manufactured Nanomaterial Exposure Media. The Generic NANOGENOTOX Dispersion

Protocol. Standard Operating Procedure (SOP) and Background Documentation. Available online: https://www.anses.fr/en/system/files/nanogenotox_deliverable_5.pdf (accessed on 17 June 2022).

42. Ke, Y.; Reddel, R.R.; Gerwin, B.I.; Miyashita, M.; McMenamin, M.; Lechner, J.F.; Harris, C.C. Human Bronchial Epithelial Cells with Integrated SV40 Virus T Antigen Genes Retain the Ability to Undergo Squamous Differentiation. *Differentiation* **1988**, *38*, 60–66. [CrossRef]

43. Reddel, R.R.; Ke, Y.; Kaighn, M.E.; Malan-Shibley, L.; Lechner, J.F.; Rhim, J.S.; Harris, C.C. Human Bronchial Epithelial Cells Neoplastically Transformed by V-Ki-Ras: Altered Response to Inducers of Terminal Squamous Differentiation. *Oncogene Res.* **1988**, *3*, 401–408.

44. Lieber, M.; Todaro, G.; Smith, B.; Szakal, A.; Nelson-Rees, W. A Continuous Tumor-Cell Line from a Human Lung Carcinoma with Properties of Type II Alveolar Epithelial Cells. *Int. J. Cancer* **1976**, *17*, 62–70. [CrossRef]

45. Suggs, J.E.; Madden, M.C.; Friedman, M.; Edgell, C.J.S. Prostacyclin Expression by a Continuous Human Cell Line Derived from Vascular Endothelium. *Blood* **1986**, *68*, 825–829. [CrossRef]

46. Rothen-Rutishauser, B.M.; Kiama, S.C.; Gehr, P. A Three-Dimensional Cellular Model of the Human Respiratory Tract to Study the Interaction with Particles. *Am. J. Respir. Cell. Mol. Biol.* **2005**, *32*, 281–289. [CrossRef] [PubMed]

47. Klein, S.G.; Serchi, T.; Hoffmann, L.; Blömeke, B.; Gutleb, A.C. An Improved 3D Tetraculture System Mimicking the Cellular Organisation at the Alveolar Barrier to Study the Potential Toxic Effects of Particles on the Lung. *Part. Fibre Toxicol.* **2013**, *10*, 31. [CrossRef] [PubMed]

48. Ding, Y.; Weindl, P.; Lenz, A.G.; Mayer, P.; Krebs, T.; Schmid, O. Quartz Crystal Microbalances (QCM) Are Suitable for Real-Time Dosimetry in Nanotoxicological Studies Using VITROCELL®Cell Exposure Systems. *Part. Fibre Toxicol.* **2020**, *17*, 44. [CrossRef] [PubMed]

49. Binder, S.; Cao, X.; Bauer, S.; Rastak, N.; Kuhn, E.; Dragan, G.C.; Monsé, C.; Ferron, G.; Breuer, D.; Oeder, S.; et al. In Vitro Genotoxicity of Dibutyl Phthalate on A549 Lung Cells at Air–Liquid Interface in Exposure Concentrations Relevant at Workplaces. *Environ. Mol. Mutagen.* **2021**, *62*, 490–501. [CrossRef]

50. Elje, E.; Hesler, M.; Rundén-Pran, E.; Mann, P.; Mariussen, E.; Wagner, S.; Dusinska, M.; Kohl, Y. The Comet Assay Applied to HepG2 Liver Spheroids. *Mutat. Res. Genet. Toxicol. Environ. Mutagen.* **2019**, *845*, 403033. [CrossRef] [PubMed]

51. Dusinska, M.; Mariussen, E.; Rundén-Pran, E.; Hudecova, A.M.; Elje, E.; Kazimirova, A.; el Yamani, N.; Dommershausen, N.; Tharmann, J.; Fieblinger, D.; et al. In Vitro Approaches for Assessing the Genotoxicity of Nanomaterials. In *Methods in Molecular Biology*; Humana Press Inc.: Totowa, NJ, USA, 2019; Volume 1894, pp. 83–122. [CrossRef]

52. El Yamani, N.; Mariussen, E.; Gromelski, M.; Wyrzykowska, E.; Grabarek, D.; Puzyn, T.; Tanasescu, S.; Dusinska, M.; Rundén-Pran, E. Hazard Identification of Nanomaterials: In Silico Unraveling of Descriptors for Cytotoxicity and Genotoxicity. *Nano Today* **2022**, *46*, 101581. [CrossRef]

53. Rundén-Pran, E. Hazard assessment of spherical and rod-shaped silver nanomaterials by an advanced lung model at the air-liquid interface. Manuscript in Preparation. 2023.

54. Schlinkert, P.; Casals, E.; Boyles, M.; Tischler, U.; Hornig, E.; Tran, N.; Zhao, J.; Himly, M.; Riediker, M.; Oostingh, G.J.; et al. The Oxidative Potential of Differently Charged Silver and Gold Nanoparticles on Three Human Lung Epithelial Cell Types. *J. Nanobiotechnol.* **2015**, *13*, 1. [CrossRef]

55. A549—CCL-185 | ATCC. Available online: https://www.atcc.org/products/ccl-185 (accessed on 10 October 2022).

56. Haniu, H.; Saito, N.; Matsuda, Y.; Tsukahara, T.; Maruyama, K.; Usui, Y.; Aoki, K.; Takanashi, S.; Kobayashi, S.; Nomura, H.; et al. Culture Medium Type Affects Endocytosis of Multi-Walled Carbon Nanotubes in BEAS-2B Cells and Subsequent Biological Response. *Toxicol. In Vitro* **2013**, *27*, 1679–1685. [CrossRef]

57. Le Ouay, B.; Stellacci, F. Antibacterial Activity of Silver Nanoparticles: A Surface Science Insight. *Nano Today* **2015**, *10*, 339–354. [CrossRef]

58. Centurione, L.; Aiello, F.B. DNA Repair and Cytokines: TGF-β, IL-6, and Thrombopoietin as Different Biomarkers of Radioresistance. *Front. Oncol.* **2016**, *6*, 175. [CrossRef]

59. Platel, A.; Privat, K.; Talahari, S.; Delobel, A.; Dourdin, G.; Gateau, E.; Simar, S.; Saleh, Y.; Sotty, J.; Antherieu, S.; et al. Study of in Vitro and in Vivo Genotoxic Effects of Air Pollution Fine (PM2.5-0.18) and Quasi-Ultrafine (PM0.18) Particles on Lung Models. *Sci. Total Environ.* **2020**, *711*, 134666. [CrossRef] [PubMed]

60. Koehler, C.; Ginzkey, C.; Friehs, G.; Hackenberg, S.; Froelich, K.; Scherzed, A.; Burghartz, M.; Kessler, M.; Kleinsasser, N. Aspects of Nitrogen Dioxide Toxicity in Environmental Urban Concentrations in Human Nasal Epithelium. *Toxicol. Appl. Pharmacol.* **2010**, *245*, 219–225. [CrossRef]

61. Koehler, C.; Ginzkey, C.; Friehs, G.; Hackenberg, S.; Froelich, K.; Scherzed, A.; Burghartz, M.; Kessler, M.; Kleinsasser, N. Ex Vivo Toxicity of Nitrogen Dioxide in Human Nasal Epithelium at the WHO Defined 1-h Limit Value. *Toxicol. Lett.* **2011**, *207*, 89–95. [CrossRef] [PubMed]

62. Koehler, C.; Thielen, S.; Ginzkey, C.; Hackenberg, S.; Scherzed, A.; Burghartz, M.; Paulus, M.; Hagen, R.; Kleinsasser, N.H. Nitrogen Dioxide Is Genotoxic in Urban Concentrations. *Inhal. Toxicol.* **2013**, *25*, 341–347. [CrossRef] [PubMed]

63. Wiest, F.; Scherzad, A.; Ickrath, P.; Poier, N.; Hackenberg, S.; Kleinsasser, N. Studies on Toxicity and Inflammatory Reactions Induced by E-Cigarettes: In Vitro Exposure of Human Nasal Mucosa Cells to Propylene Glycol at the Air–Liquid Interface. *HNO* **2021**, *69*, 952–960. [CrossRef] [PubMed]

64. Cervena, T.; Rossnerova, A.; Zavodna, T.; Sikorova, J.; Vrbova, K.; Milcova, A.; Topinka, J.; Rossner, P. Testing Strategies of the in Vitro Micronucleus Assay for the Genotoxicity Assessment of Nanomaterials in Beas-2b Cells. *Nanomaterials* **2021**, *11*, 1929. [CrossRef]
65. García-Rodríguez, A.; Kazantseva, L.; Vila, L.; Rubio, L.; Velázquez, A.; Ramírez, M.J.; Marcos, R.; Hernández, A. Micronuclei Detection by Flow Cytometry as a High-Throughput Approach for the Genotoxicity Testing of Nanomaterials. *Nanomaterials* **2019**, *9*, 1677. [CrossRef]
66. Prasad, R.Y.; Wallace, K.; Daniel, K.M.; Tennant, A.H.; Zucker, R.M.; Strickland, J.; Dreher, K.; Kligerman, A.D.; Blackman, C.F.; Demarini, D.M. Effect of Treatment Media on the Agglomeration of Titanium Dioxide Nanoparticles: Impact on Genotoxicity, Cellular Interaction, and Cell Cycle. *ACS Nano* **2013**, *7*, 1929–1942. [CrossRef]

Article

Impact of Nano- and Micro-Sized Chromium(III) Particles on Cytotoxicity and Gene Expression Profiles Related to Genomic Stability in Human Keratinocytes and Alveolar Epithelial Cells

Paul Schumacher [1], Franziska Fischer [1], Joachim Sann [2,3], Dirk Walter [4,5] and Andrea Hartwig [1,*]

[1] Department of Food Chemistry and Toxicology, Institute of Applied Biosciences (IAB), Karlsruhe Institute of Technology (KIT), Adenauerring 20a, 76131 Karlsruhe, Germany; paul.schumacher@kit.edu (P.S.); franziska.fischer@kit.edu (F.F.)

[2] Institute of Physical Chemistry, Justus-Liebig-University Giessen, Heinrich-Buff-Ring 17, 35392 Giessen, Germany; joachim.sann@phys.chemie.uni-giessen.de

[3] Center for Materials Research (LaMa/ZfM), Justus-Liebig-University Giessen, Heinrich-Buff-Ring 16, 35392 Giessen, Germany

[4] Laboratories of Chemistry and Physics, Institute of Occupational and Social Medicine, Justus-Liebig-University Giessen, Aulweg 129, 35392 Giessen, Germany; dirk.walter@arbmed.med.uni-giessen.de

[5] Institute of Inorganic and Analytical Chemistry, Justus-Liebig-University Giessen, Heinrich-Buff-Ring 17, 35392 Giessen, Germany

* Correspondence: andrea.hartwig@kit.edu

Citation: Schumacher, P.; Fischer, F.; Sann, J.; Walter, D.; Hartwig, A. Impact of Nano- and Micro-Sized Chromium(III) Particles on Cytotoxicity and Gene Expression Profiles Related to Genomic Stability in Human Keratinocytes and Alveolar Epithelial Cells. *Nanomaterials* **2022**, *12*, 1294. https://doi.org/10.3390/nano12081294

Academic Editor: Miguel Gama

Received: 2 March 2022
Accepted: 5 April 2022
Published: 11 April 2022

Publisher's Note: MDPI stays neutral with regard to jurisdictional claims in published maps and institutional affiliations.

Abstract: Exposure to Cr(VI) compounds has been consistently associated with genotoxicity and carcinogenicity, whereas Cr(III) is far less toxic, due to its poor cellular uptake. However, contradictory results have been published in relation to particulate Cr_2O_3. The aim of the present study was to investigate whether Cr(III) particles exerted properties comparable to water soluble Cr(III) or to Cr(VI), including two nano-sized and one micro-sized particles. The morphology and size distribution were determined by TEM, while the oxidation state was analyzed by XPS. Chromium release was quantified via AAS, and colorimetrically differentiated between Cr(VI) and Cr(III). Furthermore, the toxicological fingerprints of the Cr_2O_3 particles were established using high-throughput RT-qPCR and then compared to water-soluble Cr(VI) and Cr(III) in A549 and HaCaT cells. Regarding the Cr_2O_3 particles, two out of three exerted only minor or no toxicity, and the gene expression profiles were comparable to Cr(III). However, one particle under investigation released considerable amounts of Cr(VI), and also resembled the toxicity profiles of Cr(VI); this was also evident in the altered gene expression related to DNA damage signaling, oxidative stress response, inflammation, and cell death pathways. Even though the highest toxicity was found in the case of the smallest particle, size did not appear to be the decisive parameter, but rather the purity of the Cr(III) particles with respect to Cr(VI) content.

Keywords: Cr_2O_3 particles; Cr(VI) release; cytotoxicity; gene expression profiles; DNA damage signaling; DNA repair proteins; oxidative stress; cell death pathways

1. Introduction

Chromium (Cr) is a naturally occurring element, with three thermodynamically stable forms, namely, Cr(0), Cr(III), and Cr(VI). From a toxicological perspective, the distinction between hexa- and trivalent chromium is of major importance. Exposure to various Cr(VI) compounds has been consistently associated with elevated incidences of respiratory cancers in humans and experimental animals. In contrast, there is no evidence of a carcinogenic action in case of trivalent chromium compounds [1–3]. This difference is explained by the so-called uptake-reduction model originally described by Wetterhahn [4]. Cr(VI) ions travel easily through the anion channels of the plasma membrane, and are reduced by intracellular electron donors in three one-electron steps via Cr(V) and Cr(IV), to the stable form of Cr(III). Whereas the anionic chromate is unable to react with DNA directly, Cr(III) forms stable binary (Cr(III)-DNA) and ternary (ligand-Cr(III)-DNA) DNA adducts in Cr(VI) treated cells, where the ligand can be ascorbic acid (Asc), glutathione (GSH), cysteine, or histidine [5]. One proposed outcome of processing the respective DNA lesions is the induction of microsatellite and chromosomal instability [6]. Furthermore, reactive oxygen species are generated in the course of the intracellular reduction of Cr(VI) to Cr(III), leading not only to oxidative stress associated with oxidative DNA damage, but also the activation of redox-regulated signal pathways [7,8]. In addition, epigenetic changes, both on the level of DNA methylation as well as post-translational histone modifications, appear to be associated with Cr(VI) induced carcinogenicity (for a recent review see [9]). It is very likely that a combination of all these mechanisms is involved in Cr(VI)-induced carcinogenicity.

In contrast, the absence of toxic effects in Cr(III) complexes results from their poor ability to enter cells, their lack of intracellular accumulation, and their high stability of coordinated multidentate ligands, which prevent binding to cellular macromolecules (for review see [5]). Nevertheless, contradictory results have been published concerning particulate Cr(III) compounds, which may enter the cell via endocytosis, thereby circumventing the cell membrane barrier reported for water-soluble Cr(III) compounds. Horie and coworkers [10], in particular, demonstrated that in human lung carcinoma A549 cells and human keratinocyte HaCaT cells, Cr_2O_3 nanoparticles show severe cytotoxicity, an increase in intracellular reactive oxygen species (ROS) levels, and an activation of antioxidant defense systems and apoptosis; cellular responses were stronger in the Cr_2O_3 nanoparticle-exposed cells when compared to cells exposed to micro-sized Cr_2O_3 particles or $CrCl_3$ [10]. The authors proposed extracellular and/or intracellular Cr(VI) release from nano-sized Cr(III) particles, which needs further clarification, since usually Cr(VI) is reduced in biological media as well as intracellularly to Cr(III). Therefore, the oxidation of Cr(III) particles to Cr(VI) would contradict the current understanding of chromium-induced toxicity, but would be quite important for the toxicological risk assessment of Cr(III) compounds.

Within the present study, we compared two nano-sized and one micro-sized Cr_2O_3 particles with respect to cytotoxicity and cellular effects related to genomic stability, and compared it to both $K_2Cr_2O_7$ and $CrCl_3$. Two different cell lines were used, namely the human keratinocyte cell line HaCaT, and A549 human alveolar lung carcinoma cells. The particles were characterized with respect to size, oxidation state, as well as in relation to the release of Cr(III) and Cr(VI) ions in ultrapure water and artificial lysosomal fluid (ALF). Furthermore, besides cytotoxicity, special attention was given to gene expression profiles related to genomic stability, including the genes coding for proteins involved in metal homeostasis, specific DNA repair factors, DNA damage response, oxidative stress response, cell cycle control, and cell proliferation. To this end, a high-throughput RT-qPCR approach was applied as described previously [11]. We observed cytotoxicity and pronounced gene expression alterations, typical for Cr(VI)-induced cellular damage, in the case of water-soluble Cr(VI) and one nano-sized particle. There were very minor, or no effects, in the case of the other two particles under investigation.

2. Materials and Methods

2.1. Materials

The Cr_2O_3 particles were purchased from Nanostructured & Amorphous (Katy, TX, USA) Lot: 1910-091918 (particle A), and Sigma Aldrich (Steinheim, Germany) Lot: 634239 (particle B). Particle C (Lot: CHC 2018-19) was kindly provided by Lanxess (Cologne, Germany). $CrCl_3$ hexahydrate ($\geq$97%) and $K_2Cr_2O_7$ ($\geq$99.5%) were purchased from Carl Roth (Karlsruhe, Germany).

Dimethyl sulfoxide ($\geq$99.9%) and 1,5-diphenylcarbazide ($\geq$97.0%) were purchased from Sigma Aldrich (Steinheim, Germany). A CellTiter-Glo® Luminescent Cell Viability Assay was purchased from Promega (Madison, WI, USA). All PCR consumables, including PCR tubes, strips, reaction tubes, and tubules, as well as cell culture dishes and flasks, were obtained from Sarstedt (Nuembrecht, Germany). The primer pairs were synthesized by Eurofins Genomics (Ebersberg, Germany) or Fluidigm (San Francisco, CA, USA). The DNA suspension buffer, PCR-certified water, and TE buffer were obtained from Teknova (Hollister, CA, USA). The 2X Assay Loading Reagent and 20X DNA Binding Dye Sample Loading Reagent were purchased from Fluidigm (San Francisco, CA, USA). Bio-Rad (Munich, Germany) provided the 2X SsoFastTM EvaGreen® Supermix with Low ROX and the 2X SYBR Green Supermix. The 2X TaqMan® PreAmp Master Mix was obtained from Applied Biosystems (Darmstadt, Germany) and exonuclease I from New England Biolabs (Frankfurt am Main, Germany).

2.2. Physicochemical Characterization of Cr_2O_3 Particles

2.2.1. TEM

The Cr_2O_3 particles were suspended in either sterile ultrapure water or complete media at different concentrations. After sonification, the particle suspensions were applied on copper grids (Plano, Wetzlar, Germany), and then dried prior to analyses. To characterize primary core size, size distribution, and morphological shape, the particles were examined using transmission electron microscopy (CM 200 FEG/ST, Philips, Eindhoven, The Netherlands). ImageJ 1.52d software (U.S. National Institutes of Health, Bethesda, MD, USA) was used to analyze the diameter of individual, non-overlapping particles, and their size distribution was calculated by counting 300 to 500 particles.

2.2.2. Hydrodynamic Size and Polydispersity Index (PDI)

The hydrodynamic size and PDI were determined for particles A and B (Supplementary Tables S1 and S2). Particle C exerted a very high PDI and sedimented rapidly, which did not allow for respective measurements.

2.2.3. XPS

The XPS measurements were conducted by applying a PHI VersaProbe II system (Physical Electronics PHI/ULVAC-PHI, Chanhassen, MN, USA) equipped with an Al Kα anode (1486.6 eV). For survey spectra, a pass energy of 93.9 eV was used. For all XP detail spectra, the pass energy was set to 23.50 eV. The X-ray power was 100 W and the spot was scanning over an area of 1400 μm × 100 μm. The powder samples were pressed into PTFE cups to achieve a compact and smooth surface. During measurement, a PHI dual beam charge neutralization with low energy Argon ions (~10 eV) and electrons (~2 eV) was used, ensuring a uniform potential without charging effects. Data evaluation was performed using CasaXPS (version 2.3.22, Casa Software Ltd.). All XP spectra in this work were calibrated in relation to the signal of adventitious carbon at 284.8 eV. For the signal fitting, a Shirley background and GL(30) line shapes were used.

2.2.4. Solubility Measurement/Oxidation State

The release of soluble chromium from the Cr_2O_3 particles was determined under neutral pH conditions or under acidic pH conditions, the latter resembling conditions in the lysosomes, as previously described [12]. Briefly, stock solutions of 1 mg/mL Cr_2O_3 particles were prepared by weighing them into 1.5 mL polystyrene reaction tubes, followed by dilution in 50 mL sterile snap-on lid glasses with either sterile ultrapure water or with artificial lysosomal fluid (ALF), pH 4.5 (composed of sodium chloride (3.210 g/L), sodium hydroxide (6.000 g/L), citric acid (20.800 g/L), calcium chloride dihydrate (0.1285 g/L), disodium hydrogen phosphate (0.0710 g/L), sodium sulphate (0.0390 g/L), magnesium chloride (0.0476 g/L), glycine (0.0590 g/L), sodium citrate dihydrate (0.0770 g/L), sodium tartrate dihydrate (0.0900 g/L), sodium lactate (0.0850 g/L), or sodium pyruvate (0.0860 g/L). The tubes were ultrasonicated for 10 min in a water bath. After 0, 24, 48, or 120 h at room temperature, 1 mL solutions were centrifuged at 16,000× g and 4 °C for 1 h. The chromium content was either quantified by the 1,5-diphenylcarbazid (DPC) method or by graphite furnace atom absorption spectrometry (GF-AAS). For the DPC method, 50 μL of reaction mix consisting of 8 μL DPC (1% DPC in acetone), 15 μL sulfuric acid (1 M), 15 μL phosphoric acid (1 M), and 15 μL ultrapure water were added to 200 μL of the sample in a 96-well plate. After shaking for 3 min and incubating for 17 min, the absorption was measured at 540 nm using a multiplate reader TECAN® Infinite M200 Pro (TECAN Group, Maennedorf, Switzerland). Freshly prepared solutions in concentrations of 0, 0.1, 0.2, 0.5, 1.0, and 2.0 mg/L hexavalent chromium ($K_2Cr_2O_7$ in ultrapure water or ALF, respectively) were used for the calibration.

For the GF-AAS measurement (PinAAcle 900 T, Perkin Elmer, Rodgau, Germany) of the total soluble chromium, 1 mL of the supernatant was heated stepwise to 95 °C to dry up. The remnants were further digested with 1:1 HNO_3 (69%)/H_2O_2 (30%) (v/v) by repeated stepwise heating to 95 °C. The residue was then solubilized for measurement in 1 mL HNO_3 (0.2%). The following AAS temperature program was applied: drying at 120 °C for 30 s, and 140 °C for 45 s, 30 s pyrolysis at 1500 °C, atomization at 2300 °C for 5 s, and cleaning for 3 s at 2450 °C.

2.3. Cell Culture Experiments

2.3.1. Cr_2O_3 Particle Suspensions and $CrCl_3$ as Well as $K_2Cr_2O_7$ Incubation Dilutions

The Cr_2O_3 suspensions, as well as soluble Cr(III) and Cr(VI) dilutions, were freshly prepared for each experiment. Particles, received as dry powder, were aliquoted by weighing them into 1.5 mL sterile polystyrene reaction tubes. Watery stock solutions of 1 mg/mL Cr_2O_3 were prepared in an endotoxin-free snap-on lid glass by ultrasonication for 10 min. Dilutions in the range of 2, 10, 20, and 50 μg/mL were prepared by adding aliquots of the stock solution into 15 mL sterile falcon tubes filled with adequate volumes of fresh complete medium. Incubation volumes of 200 μL/cm^2 were chosen to receive particle doses of 0.4, 2.0, 4.0, and 10 μg/cm^2. Stock solutions of water-soluble $CrCl_3$ (200 mM) and $K_2Cr_2O_7$ (20 mM) were prepared and diluted accordingly, skipping the sonication step. For comparison purposes, the particle suspension of 10 μg/mL (=2 μg/cm^2) Cr_2O_3 was considered equimolar to 132 μM Cr(III) or Cr(VI).

2.3.2. Cell Culture and Incubation

The human adenocarcinoma cell line A549 (ATCC CCL-185) was kindly provided by Dr. Roel Schins (Leibniz Research Institute for Environmental Medicine, Düsseldorf, Germany). The A549 cells were cultured as a monolayer in RPMI-1640, supplemented with 10% heat-inactivated fetal bovine serum (FBS, Invitrogen, Darmstadt, Germany), 100 U/mL penicillin, and 100 μg/mL streptomycin (both Sigma Aldrich, Steinheim, Germany) at 37 °C in a humidified atmosphere containing 5% CO_2 (HeraSafe, Thermo Scientific, Langenselbold, Germany). Human keratinocytes HaCaT cells (CLS 300493) were kindly provided by Prof. Dr. Brunhilde Bloemeke (Trier University, Department of Environmental Toxicology, Trier, Germany). The cells were cultured at 37 °C, 5% CO_2 in a humidified atmosphere in Dulbecco's modified Eagle's medium (DMEM, Sigma Aldrich, Steinheim Germany), supplemented with 10% FBS (Invitrogen, Darmstadt, Germany), 100 U/mL penicillin, 100 μg/mL streptomycin (both Sigma Aldrich, Steinheim, Germany), and 2 mM Gluta-MAX™ (Gibco, Carlsbad, CA, USA). Both cell lines were grown up to 80% confluency and routinely split three times per week. Passage numbers from 8 to 25 (HaCaT) and from 14 to 35 (A549) were used for experiments. The measurement of the ATP content was carried out in white-walled optical-bottom 96-well plates (ThermoFisher, Dreieich, Germany). Seeding density was either 1×10^4 cells/well for HaCaT, or 3×10^4 cells/well for A549 cells. For gene expression analyses 0.5×10^6 were seeded in 6 cm cell culture dishes (Sarstedt, Nuembrecht, Germany). After 24 h (A549) or 48 h (HaCaT) the supernatant was removed from the logarithmically growing cells and was replaced by the particle suspension or Cr(III)/(VI) dilution. For consistent particle deposition, the incubation volume for each experiment was set at 0.2 mL per square centimeter growth area.

2.3.3. Cytotoxicity Assay

A Promega CellTiter-Glo® ATP assay was used to analyze cell viability and cell proliferation. Logarithmically growing cells were incubated for 24 h with 0.4, 2.0, 4.0, or 10.0 μg/cm^2 Cr_2O_3 particles; 26.4, 66.0, 132, 264, 660, or 1320 μM (not shown) $CrCl_3$; or 1.32, 2.64, 6.6, 13.2, 26.4, or 66 μM $K_2Cr_2O_7$. The incubation solution was removed after 24 h and the cells were washed twice with PBS. The CellTiter-Glo® Luminescent Cell Viability Assay was carried out according to the manufacturer's protocol. Briefly, complete medium and the equal volume of CellTiter-Glo® reagent were added to the cavities. The 96-well plate was transferred on an orbital shaker for 2 min to induce cell lysis. The plate was incubated for 40 min in the dark at room temperature to stabilize the luminescent signal, which was then recorded with a microplate reader TECAN® Infinite M200 Pro (TECAN Group, Maennedorf, Switzerland). Data were analyzed with Excel 2010 (Microsoft, Redmond, CA, USA). The ATP content as a measure for cell viability was expressed as percentage normalized to non-treated control cells. To test if soluble Cr(III) or Cr(VI) interfered with the ATP assay, the relevant concentrations from above were added to a standard curve of ATP (0.5, 1.0, 2.5, 5.0 μM).

2.3.4. Gene Expression Analyses

Gene expression analyses via high-throughput RT-qPCR using the Fluidigm dynamic array on the BioMark™ System were performed as described by Fischer et al. [11]. Briefly, 0.5×10^6 cells were treated with different concentrations of Cr_2O_3 particles, $CrCl_3$, or $K_2Cr_2O_7$ in complete medium. After 24 h, the cells were washed in PBS, trypsinized, and resuspended in ice-cold PBS containing 10% FBS, and collected by centrifugation. RNA isolation was performed using MN NucleoSpin® RNA Plus KIT (Macherey-Nagel, Dueren, Germany), according to the manufacturer's protocol. Then, 1 μg of RNA was reverse transcribed in duplicates into complementary DNA (cDNA) using qScript™ cDNA Synthesis Kit (Bio-Rad, Munich, Germany). Subsequently, a specific target amplification (STA) and an exonuclease I digestion (EXO) were performed prior to qPCR. Then, 5 μL of the STA and EXO mix containing 1.25 μL cDNA mix from cDNA synthesis, 0.5 μL pooled primer mix (PPM), 2.5 μL 2X TaqMan® PreAmp Master Mix, and 0.75 μL PCR-certified water were applied on the Fluidigm dynamic array. Controls, such as a no-template control (NTC-STA) and a non-reverse transcribed RNA control (NoRT) were considered. All pipetting steps, until reverse transcription, were performed under a sterile RNA hood. Pipetting the cDNA was carried out under a DNA/DNase-free hood, to prevent cross-contamination during the entire qPCR experiment. Any details regarding various temperature profiles for RT, STA, EXO, qPCR, and melting curve analyses can be found in the original publication from Fischer et al. [11]. Preparation and loading of Fluidigm 96.96 Dynamic Array IFC (integrated fluidic circuit) were performed according to the manufacturer's instructions. After priming, the chip was loaded with samples and primer reaction mixes within 1 h to prevent unfavorable evaporation effects and loss of pressure. Samples and primer reaction mixes were loaded into the chip by running the Load Mix (136×) script of the IFC Controller HX (Fluidigm, San Francisco, CA, USA). The chip was transferred into the BioMark™ System (Fluidigm, San Francisco, CA, USA) immediately, and qPCR and subsequent melting curve analyses were performed. Data analysis was executed with the Fluidigm Real-Time PCR Analysis tool and GenEx™ 5 software (MultiD Analyses AB, Gothenburg, Sweden). Transcription levels of five reference genes (*ACTB*, *B2M*, *GAPDH*, *GUSB*, and *HPRT1*) were used for normalization. Alterations in transcript levels of the target genes were displayed as a $\log_2$ fold change compared to a control group by calculating relative quantities corresponding to the $\Delta\Delta$Cq method [13,14].

2.3.5. Statistics

If not stated otherwise, all data are displayed as the mean of three independently performed experiments, each of which was conducted at least in duplicates. For cell viability, differences on the cellular level between the negative control (non-treated cells) and the metal compound treatment were analyzed using one-way ANOVA followed by a Dunnett's T post hoc test.

3. Results

3.1. Particle Characteristics

Three different Cr_2O_3 particles were included, two in the nano-sized range and one in the micro-sized range. Particle A was nano-sized and was obtained from the same supplier as stated in the study of Horie and coworkers [10].

No differences in morphology were detected between the particles. Particle size was determined using transmission electron microscopy (TEM). Representative TEM images of the particles are shown in Figure 1(1). Particles A and B are nano-sized, with average diameters of 40 nm and 80 nm, respectively. Particle C is micro-sized, with an average diameter of 150 nm (Figure 1(2)).

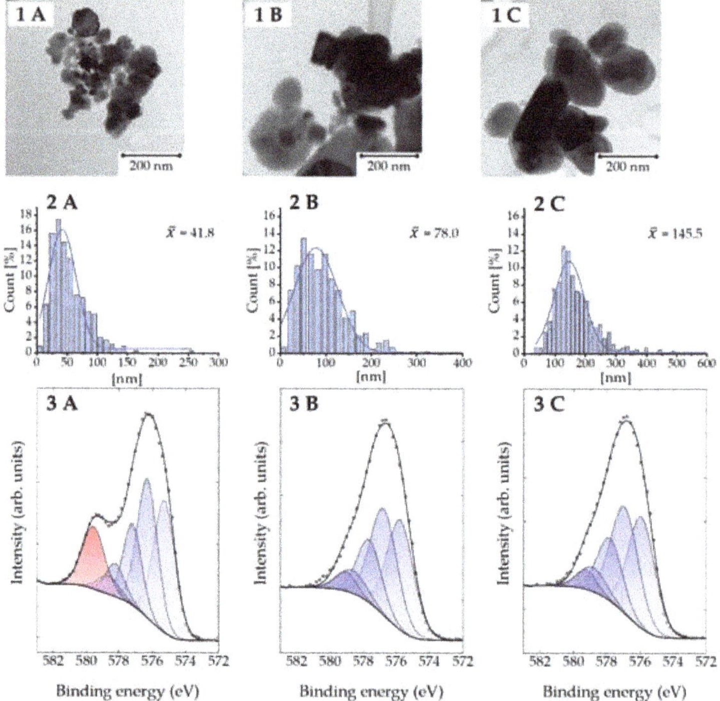

Figure 1. Physicochemical characterization of different Cr_2O_3 particles. Size distribution was measured in ultrapure water at a concentration of 10 µg/cm^2. **1A–1C**: representative transmission electron microscopy (TEM) images of the particles. **2A–2C**: average diameters of particle A (42 nm), B (78 nm) and C (146 nm). **3A–3C**: X-ray photoelectron spectroscopy (XPS) spectra of the particles. The red signal at 579.6 eV only present in Figure 1(**3A**) is characteristic for Cr(VI), while the blue multiplet signal between 575.3 and 578.6 eV is characteristic for Cr(III).

The oxidation state of the Cr_2O_3 particles was measured using X-ray photoelectron spectroscopy (XPS), and detailed spectra of chromium $2p_{3/2}$ are shown for the three different particles (Figure 1(3)). Particles B and C show relatively pure Cr(III) multiplets spectra (shown in blue), as expected for Cr_2O_3 [15]. Compared to values in the literature [15], the peaks are broadened, which is most likely due to the powder nature of the sample. However, in contrast to particles B and C, particle A clearly shows a contribution of Cr(VI) to the spectrum (red signal), accounting for about 15% of the total chromium content.

3.2. Cytotoxicity

In the next step, the cytotoxicity of all three particles was investigated, as well as water-soluble Cr(III) and Cr(VI), by applying the CellTiter-Glo® luminescent cell viability assay (Promega). This assay is based on the quantification of ATP after complete cell lysis, via the reaction of beetle luciferin to oxyluciferin and photons. The luminescence generated correlates with the ATP content, and therefore, the number of viable cells in culture.

The cytotoxicity of $K_2Cr_2O_7$, $CrCl_3$, and the three different Cr_2O_3 particles was measured after 24 h incubation. Results for both cell lines are shown in Figure 2A,B. Treatment with $K_2Cr_2O_7$ exerted pronounced dose-dependent cytotoxicity starting in the low micromolar concentration range; the ATP content was reduced by 11% (A549) or 18% (HaCaT) after treatment with 6.6 µM, and by 42% (A549) or 49% (HaCaT) after treatment with 26.4 µM. Severe cytotoxicity was observed after treatment with Cr(VI) concentrations of 66 µM and above. In case of $CrCl_3$, neither differences in confluency nor a reduction in ATP content were evident at concentrations up to 264 µM. Treatment with 660 µM showed a minor reduction of the ATP content by 20% in HaCaT cells and 10% in A549 cells. Concerning the three different Cr_2O_3 particles, only particle A showed cytotoxic effects at doses above 1 µg/cm². As displayed in Figure 2A, the viability of the A549 cells decreased by 35% after incubation with 4 µg/cm², and by 58% after incubation with 10 µg/cm². ATP depletion was even more pronounced in HaCaT cells (Figure 2B). Here, viability was reduced to 81%, 65% and 21% at 2, 4 and 10 µg/cm² Cr_2O_3, respectively. In contrast, no significant cytotoxicity was observed in either cell line for particles B or C.

A549 cells

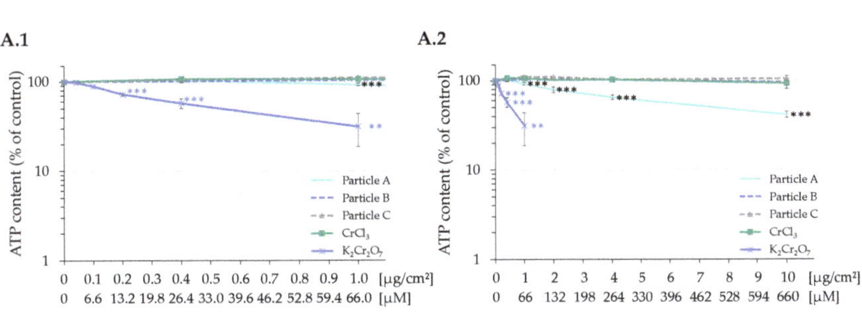

HaCaT cells

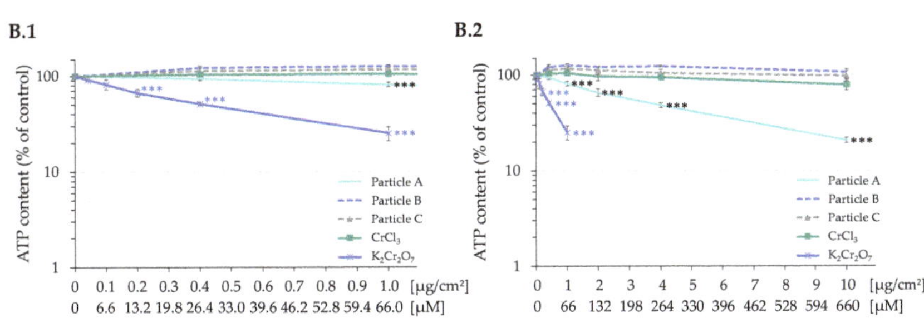

Figure 2. ATP content of A549 (**A**) and HaCaT (**B**) cells after 24 h treatment with Cr(III) oxide particles Cr_2O_3, $CrCl_3$, or $K_2Cr_2O_7$. Treatments below 66 µM (corresponding to 1 µg/cm² chromium) are shown in (**A.1**) and (**B.1**). Treatments covering the entire dose range investigated are depicted in (**A.2**) and (**B.2**). Mean values ± standard deviations derived from three independent experiments are shown. Statistics were performed using ANOVA followed by Dunnett's T post hoc test: ** $p \leq 0.01$, *** $p \leq 0.001$.

3.3. Release of Soluble Chromium from Cr_2O_3 Particles

Since particle A showed considerable cytotoxicity, while neither $CrCl_3$, nor particles B and C were cytotoxic, we investigated whether this may be due to Cr(VI) release as indicated by our XPS studies. As a first step, chromium release from the particles was determined via AAS to quantify the content of all soluble chromium species in supernatants. For this purpose, 1 mg/mL particles were either suspended in ultrapure water (pH 7.0) or artificial lysosomal fluid (ALF) with a pH of 4.5 (both at room temperature), to simulate potential intracellular chromium release after endocytosis within the lysosomes. Supernatants were centrifuged at $16,000\times g$ for 1 h before measurement. To investigate a potential time dependency of chromium release, the measurements were performed immediately after the preparation of the suspension (0 h), after 24 h, 48 h, or 120 h as depicted in Figure 3A. The results obtained by AAS demonstrated an immediate chromium release from particle A of around 4% in ultrapure water, with no considerable increase in time. This fraction was a little less pronounced in ALF, reaching around 3.5%. Chromium release from particle B was significantly lower, resulting in 0.18% in ultrapure water and 0.15% in ALF. Chromium release from particle C was not quantifiable in a reproducible manner.

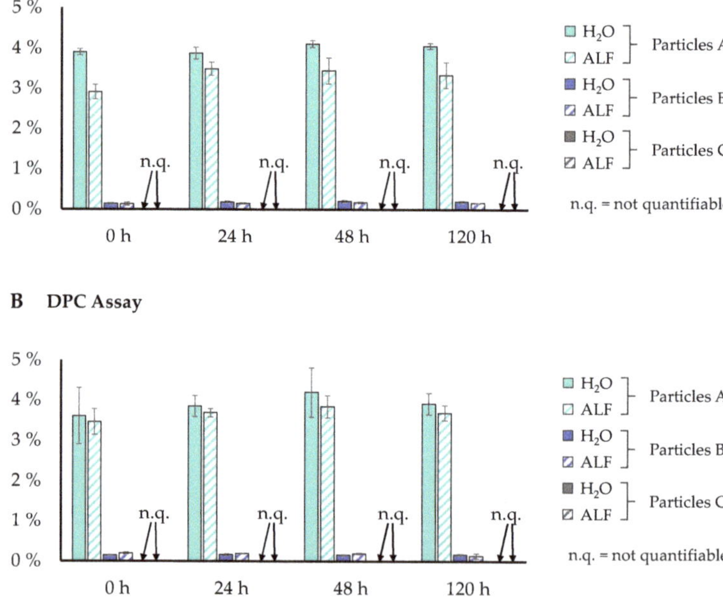

Figure 3. Release of chromium from Cr_2O_3 particles in different media. Cr_2O_3 particles measuring 1.0 mg/mL were either incubated in ultrapure water (pH 7.0) or artificial lysosomal fluid (ALF) (pH 4.5) for 0 h, 24 h, 48 h, or 120 h. The remaining particles were removed from the supernatant by centrifugation as described in Materials and Methods. Total chromium release was quantified by atomic absorption chromatography (AAS) (**A**) or Cr(VI) release by a colorimetric DPC assay (**B**) by applying the chromogenic dye 1,5-diphenylcarbazone. Mean values of 3 independent determinations ± SD are shown.

To differentiate between the release of Cr(VI) and Cr(III), a colorimetric assay was applied. Due to the tetrahedral structure of Cr(VI), it is prone to form complexes with chromogenic dyes, such as 1,5-diphenylcarbazone (DPC), which forms red–violet products, proportional to the amount of Cr(VI) present in the sample. As shown in Figure 3B, a Cr(VI) content of 3.9% for particle A, and 0.16% in case of particle B, was detected in ultrapure water after 24 h. For particle C, the chromium content was below the limit of quantification of 0.038 mg/L. As previously, particle suspensions were measured after 0 h, 24 h, 48 h, and 120 h to examine a time-dependent dissolution of the particles, and a potential release of Cr(VI) in ultrapure water and ALF. As depicted in Figure 3A,B, chromium was detectable immediately after contact with either ultrapure water or ALF, while no significant pH or time dependency was observed.

3.4. Gene Expression Analysis

As described in the introduction, Cr(VI) has been shown consistently to induce DNA damage as well as oxidative stress, while Cr(III) is considered to be largely non-toxic due to a very limited uptake in cells. Nevertheless, whether or not this applies also to Cr(III) oxide particles, which can enter the cells via endocytosis, remains to be elucidated. In the present study, therefore, we aimed to obtain toxicity profiles for all three particles and to compare them with water-soluble Cr(VI) and Cr(III). To this end, we applied gene expression analyses using high-throughput RT-qPCR, established previously in our group using the BioMark HD system [11], and which has been used successfully for metal-based nanomaterials, both after submersed and air-liquid interface (ALI) exposure [16–18]. This method enables the parallel investigation of 96 samples with regard to their impact on 95 genes of interest. Within our system, the genes were selected to yield expression profiles related to genomic stability, and can be grouped into six gene clusters: xenobiotic metabolism, metal homeostasis, (oxidative) stress response and inflammation, DNA damage response and repair, cell cycle regulation, and apoptosis. A complete list of genes and their encoding proteins is provided in Supplementary Table S3. For better visualization, changes in transcription were calculated as fold$_2$ changes of expression levels. A reduction of at least 50% (log$_2$ fold change ≤ -1) or a doubling (log$_2$ fold change ≥ 1) were considered relevant when compared to the respective control [16]. Within these investigations, the impact of K$_2$Cr$_2$O$_7$, CrCl$_3$, and the three Cr$_2$O$_3$ particles on gene expression profiles was examined in both cell lines after 24 h incubation. Based on the cytotoxicity data shown in Figure 2, particle doses of 0.4, 2.0, 4.0, and 10.0 µg/cm^2 were chosen and compared to toxicity profiles at 2.64, 6.6, 26.4, and 66 µM Cr(VI) and 66, 264, and 1320 µM CrCl$_3$. The heatmaps in Figures 4 and 5 summarize expression patterns of those genes for which expression alterations were considered relevant. A complete overview of the results obtained for the entire gene set is provided in Supplementary Figures S1 and S2. For both cell lines, a strong impact of Cr(VI) was observed in the gene clusters of DNA damage response, cell cycle regulation, and apoptosis.

3.4.1. Water Soluble Cr(VI) Treatment

Within the applied dose range of 2.64 to 66 µM, treatment with K$_2$Cr$_2$O$_7$ elicited some distinct changes in gene expression. As a general picture, both cell lines showed similar gene expression profiles, with some differences for individual genes and dose-dependencies (Figures 4 and 5).

In A549 cells, the strongest impact on gene expression was seen after treatment with 26.4 µM Cr(VI). Genes related to DNA damage response and repair were affected the most. Thus, the DNA damage response gene *GADD45A* (growth arrest and DNA damage-inducible gene α) showed a pronounced dose-dependent increase starting at the lowest, non-cytotoxic concentration, and reached induction levels up to a 28-fold expression change. This was also the case for the HaCaT cells, although to a slightly lesser extent. Interestingly, the expression of specific DNA repair proteins was downregulated in both cell lines, particularly in case of *ATR*, *BRCA1*, *BRCA2*, *ERCC4/XPF*, *MLH1*, *MSH2*, *LIG3*, *RAD50*, *RAD51*, and *XPA*, which are involved in all major DNA repair pathways. Once again, effects started at the lowest dose and transcription levels were reduced by up to 80% at cytotoxic concentrations. Even though the gene expression pattern was similar in both cell lines, respective alterations occurred at lower concentrations in the A549 cells.

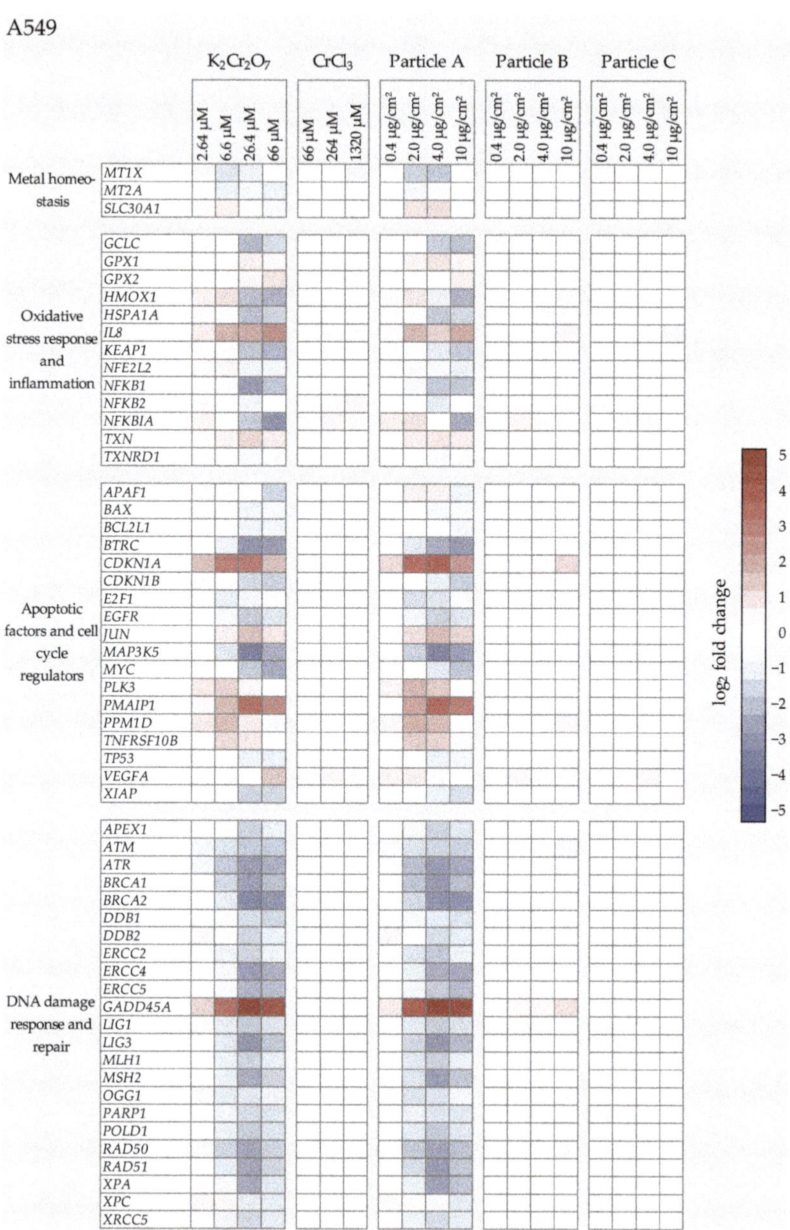

Figure 4. Overview of the impact of $K_2Cr_2O_7$, $CrCl_3$, or three different Cr(III) oxide particles on human lung epithelial cells (A549) using a high-throughput RT-qPCR approach with a custom-designed gene set. The genes under investigation have been clustered into groups associated with metal homeostasis, oxidative stress response, inflammation, apoptosis, and cell cycle regulation as well as DNA damage response and repair. A549 cells were treated with the respective chromium compound for 24 h. Displayed are the $\log_2$ fold changes of relative gene expression as a heatmap. Red colors indicate an enhanced expression, and blue colors indicate a down-regulation. The mean values of at least three independently conducted experiments are shown.

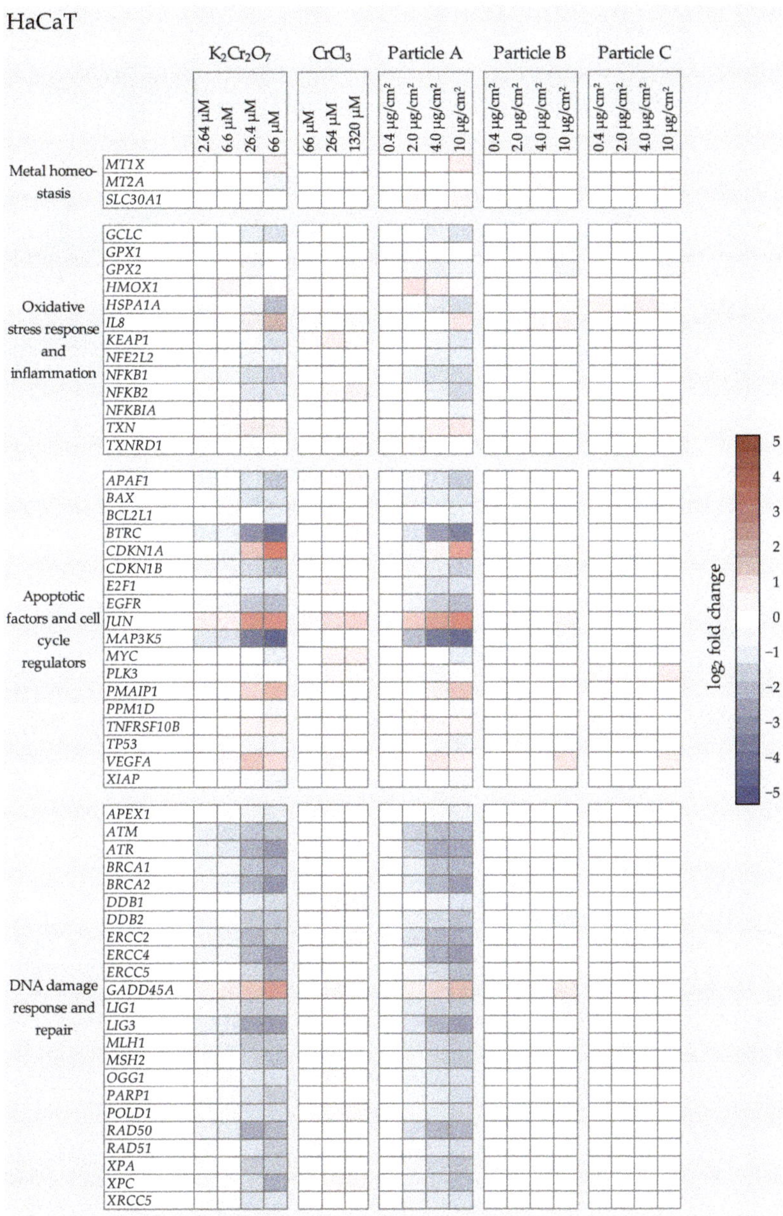

Figure 5. Overview of the impact of $K_2Cr_2O_7$, $CrCl_3$, or three different Cr(III) oxide particles on human keratinocytes (HaCaT) using a high-throughput RT-qPCR approach with a custom-designed gene set. The genes under investigation have been clustered into groups associated with metal homeostasis, oxidative stress response, inflammation, apoptosis, and cell cycle regulation as well as DNA damage response and repair. HaCaT cells were treated with the respective chromium compound for 24 h. Displayed are the $\log_2$ fold changes of relative gene expression as a heatmap. Red colors indicate an enhanced expression, and blue colors indicate a down-regulation. The mean values of three independently conducted experiments are shown.

In the cluster of oxidative stress response and inflammation, in both cell lines a concentration-dependent and relevant induction was only evident for *IL8* (Interleukin 8), which is involved in the inflammatory response. Several other genes such as heme oxygenase 1 (*HMOX1*) were slightly up-regulated at low concentrations, but down-regulated at higher concentrations. Again, effects were more pronounced in the A549 cells when compared to the HaCaT cells. $K_2Cr_2O_7$ further down-regulated the expression of antioxidant responsive genes such as *GCLC* (γ-Glutamyl cysteine synthetase), *NFkB1,* and *NFkB1A* in the A549 cells, as well as *NFkB1* and *NFkB2* in the HaCaT cells.

With regard to cell cycle regulators, one of the most striking effects was the dose-dependent up-regulation of *CDKN1A* in both cell lines, which encodes the protein p21. In addition to its involvement in cell cycle arrest upon DNA damage, it plays an important role in DNA repair, DNA replication, and apoptosis.

Regarding genes related to specific cell death pathways, in A549 and to a lesser extent HaCaT cells, the induction of *PMAIP*, which encodes the proapoptotic BCL-2 protein Noxa, was observed, predominantly at higher concentrations. In contrast, the strongest gene repression was exhibited by the gene *MAP3K5* coding for Ask1 (apoptosis signal-regulating kinase 1); low dose treatment with 6.6 µM decreased the transcription rate of this gene by more than 50% in A549 and HaCaT cells. Similarly, *BTRC* (beta-transducin repeat containing E3 ubiquitin-protein ligase) was strongly repressed in A549 and HaCaT; it also plays a key role in apoptosis. Finally, *JUN* coding for the transcription factor c-Jun was up-regulated in both cell lines; after incubation with 26.4 µM, the transcription increased three-fold (A549) and five-fold (HaCaT).

3.4.2. Water Soluble Cr(III) Treatment

The gene expression patterns of $CrCl_3$ differed strongly from those of $K_2Cr_2O_7$. In general, the applied Cr(III) treatments exhibited no significant modulations of the genes under investigation over the complete concentration range.

3.4.3. Cr(III) Oxide Particle Treatment

In direct comparison with the gene expression results of the soluble Cr(VI) compound, it was noticeable that the gene expression patterns of the $K_2Cr_2O_7$ treatment and those of the Cr(III) oxide particle A treated cells were almost identical, suggesting that the cellular effects were solely due to the release of Cr(VI).

Particle B exhibited almost no changes in gene expression. Only in the A549 cells were *GADD45A* transcript levels very slightly elevated, signaling low levels of DNA damage, as well as *CDKN1A* involved in cell cycle arrest, both apparent at the highest dose level. No relevant effects were observed in the HaCaT cells.

Finally, particle C showed no relevant alterations in gene expression profiles in either cell line within the applied dose range of 0.4 µg/cm^2 to 10 µg/cm^2.

4. Discussion

Based on the publication of Horie et al. [10] we aimed to investigate whether Cr(III) particles—especially those in the nano-sized range—exerted properties of Cr(VI), due to either intracellular Cr(III) release with the subsequent induction of DNA damage, or via the release of Cr(VI) either extracellularly or intracellularly. This question is of utmost importance for toxicological risk assessment, since Cr(VI) is considered to be carcinogenic, and diverse mechanisms, including the induction of DNA damage, have been identified to contribute to carcinogenicity. On the other hand, Cr(III) is considered to be far less toxic and/or genotoxic, but clarification is still needed whether this also applies to nano-sized Cr(III) particles.

In addition to the determination of cytotoxicity, we used a very sensitive high-throughput RT-PCR method [11,16,17] as a central approach, to establish the toxicological fingerprints of water-soluble Cr(VI) and Cr(III), as well as those of three different Cr_2O_3 particles differing in size and manufacturer. While gene expression profiles of Cr(VI) clearly identified the induction of DNA damage, and to some extent also the induction of oxidative stress, water soluble Cr(III) provoked no changes in gene expression profiles up to millimolar concentrations, in agreement with its low toxicity. Regarding the Cr_2O_3 particles under investigation, our results confirm in principle the cellular damage provoked by particle A, which was apparent also in the Horie study [10]. The cause for the observed damage appears to be the Cr(VI) present in the particles, as well as its subsequent release. Only very minor effects were observed in the case of the nano-sized particle B, whereas no corresponding effects were apparent for particle C, which was micro-sized.

Within the present study, $K_2Cr_2O_7$ exerted the highest toxicity, causing severe depletion of ATP content in both cell lines in the low micromolar concentration range. $CrCl_3$, on the other hand, had no significant impact on the metabolic activity and did not decrease the ATP levels in either the A549 or HaCaT cells. A study by Hininger and coworkers revealed similar results for the cytotoxicity of water-soluble Cr(III) and Cr(VI) in HaCaT cells, determined by MTT conversion and LDH release. Even though the toxicity endpoints differ and a direct comparison of the cytotoxic effect requires some adjustment, Cr(VI) was found to be strongly cytotoxic at low micromolar concentrations, whereas Cr(III) chloride exerted toxicity only in the millimolar range [19].

These differences are due to the considerably higher permeability of cell membranes in relation to Cr(VI) when compared to Cr(III). As described extensively in the literature, Cr(VI), due to its tetrahedral configuration, enters the cell via sulfate or phosphate ion channels, whereas the permeability of cell membranes in relation to Cr(III) is much lower [4,20]. The toxicity of Cr(VI) is attributed to its intracellular reduction to Cr(III). In the course of reduction, highly reactive chromium intermediates Cr(V) and Cr(IV) are generated, concurrent with the formation of reactive oxygen species (ROS), that mediate the oxidation of molecular targets, including membrane-associated phospholipids, proteins, and DNA [21,22]. The final intracellular reduction product is Cr(III), which forms stable binary (Cr(III)-DNA) and ternary (ligand-Cr(III)-DNA) adducts, the latter with ascorbic acid (Asc), glutathione (GSH), cysteine, or histidine as ligand, depending on the availability of intracellular reductants [5].

With regard to the three different Cr_2O_3 particles under investigation, only particle A exerted toxicity in the applied dose range, whereas there was no reduction in ATP content detectable for either the nano-sized particle B, nor for the micro-sized particle C. Therefore, even though particle A was the smallest, the size of the particles is not the effective parameter explaining the toxicity. When comparing both cell lines, the HaCaT cells showed a more pronounced reduction in ATP content, and therefore, in cell viability after exposure towards particle A and soluble Cr(VI). This is consistent with results published by Horie and coworkers. In their study, the same particles were investigated, also using HaCaT and A549 cells, even though different toxicological endpoints (MTT conversion and LDH release), higher Cr_2O_3 concentrations, and shorter incubation times were applied [10].

These differences in cytotoxicity of the three particles under investigation were further elucidated. Since neither particle B, which is nano-sized like particle A, nor particle C in the micro-sized range was cytotoxic, the toxicity was not related to the size of the particles. Therefore, differences in particle toxicity were most likely related to the Cr(VI) content. As a first step, the oxidation state of the Cr_2O_3 particles was measured using X-ray photoelectron spectroscopy (XPS.) While particles B and C exerted relatively pure Cr(III) multiplets spectra as expected for Cr_2O_3, particle A clearly revealed a considerable Cr(VI) content of about 15%, based on the total chromium content. This raised the question whether Cr(VI) is released from the particles, and if so, whether or not Cr(VI) is expected to be released extracellularly at a neutral pH, or whether it is more likely to be released intracellularly, for example after endocytic uptake of the particles into lysosomes, under acidic pH conditions. To discriminate between these possibilities, the release of total chromium under both conditions was investigated via AAS, as a first step. In the second step, a colorimetric Cr(VI)-specific assay was applied to distinguish between the two chromium species. The Cr_2O_3 particles were dispersed for up to 120 h in either ultrapure water or in ALF (pH 4.5). After centrifugation at 16,000× *g*, the supernatants were analyzed accordingly. Soluble chromium released from particle C was neither quantifiable by AAS, nor within the Cr(VI) specific DPC assay. Particle B showed a moderate chromium release of 0.18%, which was also detected in its entity by the colorimetric approach (0.19%), and thus, identified as Cr(VI). Particle A, however, showed a pronounced chromium release of around 4%, detected by AAS immediately after dissolution in water, with no time-dependent increase. Furthermore, experiments with the chromogenic dye DPC revealed that the chromium released from particle A was almost exclusively Cr(VI), amounting to 3.6%. Compared to the total chromium content of 15%, as determined by XPS (Figure 1(3)), it is evident that Cr(VI) was not released completely. This fraction was also not increased in artificial lysosomal fluid when compared to water. These data provide evidence that particle A immediately released considerable amounts of Cr(VI) extracellularly under neutral pH conditions.

Our results indicate that the release of chromium strongly depends on the specific particles under investigation. This confirms published results, even though only few studies discriminated between Cr(VI) and Cr(III). Even studies conducted with particles from the same manufacturer have shown large deviations when assessing chromium release, also depending on the respective method of quantification. Horie et al., for example, observed 1.0–1.5% soluble chromium with 0.4% Cr(VI) in supernatants of particle A suspensions after ultrafiltration and centrifugation in complete media containing FBS [10]. Another group using nano-sized Cr_2O_3 from the same manufacturer detected a total chromium release of up to 0.15% with 0.09% Cr(VI) from particles suspended in artificial wastewater (100 mg/L) [23]. Nanoparticles from other suppliers also differed with respect to chromium release. While Kumar et al. [24] could not detect any release of chromium via AAS under their test conditions, Peng and colleagues found 4.1% soluble chromium after 24 h and 5.4% after 48 h in a saline buffered solution [25]. Both studies used Cr_2O_3 NP from different manufacturers, and did not distinguish between the chromium species. In another study conducted by Costa et al. particle suspensions of 1.0 and 10.0 g/L were analyzed, reporting a total chromium release of approximately 0.15%, with a Cr(VI) content of up to 0.08%. [26].

These findings raise the question why in some cases Cr(VI) is released from Cr_2O_3 particles, while in other cases no chromium release is observed at all. This phenomenon could be caused by differences in production processes of the respective Cr_2O_3 particles. They are produced by the reduction of alkali chromates. To obtain pure Cr_2O_3, a complete reduction of chromates, with a subsequent purification process, is mandatory. Production processes with incomplete reductions of alkali chromates, or insufficient purification steps, result in detectable amounts of Cr(VI). This may be more pronounced in the case of nano-sized particles [27]. Therefore, under toxicological considerations, a characterization of the particles including Cr(VI) release prior to use is of utmost importance.

Within this study, to the best of our knowledge, a quantitative high-throughput RT-qPCR method was applied for the first time to elucidate the toxicological impact of the three Cr_2O_3 nano- and micro-sized particles in comparison with water-soluble Cr(VI) and Cr(III). Since Cr(VI) is carcinogenic, via the induction of DNA damage as the primary mode of action, special emphasis was given to genes involved in DNA damage signaling, DNA repair, cell cycle control, and apoptosis. Furthermore, potential effects on the oxidative stress response were considered. In principle, cellular alterations of gene expression patterns may result from Cr(VI) released by the particles, but could also be the consequence of uptake via endocytosis, followed by intracellular chromium release and the formation of DNA adducts typical for Cr(III). The investigations of water-soluble chromium compounds clearly shows significant differences in the effect of Cr(III) and Cr(VI). Even low micromolar concentrations of Cr(VI) resulted in altered gene expression in the clusters of DNA damage response, cell cycle regulation, and cell death pathways, as well as in—even though less pronounced—oxidative stress response. Cr(III), on the other hand, showed no relevant changes in gene expression patterns, even after treatment with millimolar concentrations.

As a predominant mode of action, Cr(VI) induces DNA damage after intracellular reduction. Besides the formation of highly reactive Cr(V) and Cr(IV) intermediates, which may lead to the formation of ROS, stable binary (Cr(III)-DNA) and especially ternary Cr(III)-DNA adducts are generated, involving cellular reducing agents such as ascorbate or glutathione [5,9,28]. One proposed outcome of processing the respective DNA lesions is the induction of microsatellite and chromosomal instability [5]. With regard to the gene expression profiles obtained for Cr(VI) within the present study, the most eminent outcome was the induction of the DNA damage signaling gene *GADD45A* in both cell lines. Furthermore, the cell cycle regulator *CDKN1A* was induced, coding for p21, and stabilizing p53 upon the induction of DNA damage, with a subsequent cell cycle arrest [29]. In contrast, the expression of several specific DNA repair factors was down-regulated. The inhibited transcription was particularly pronounced in case of *ATR, BRCA1, BRCA2, ERCC4/XPF, MLH1, MSH2, LIG3, RAD50, RAD51*, and *XPA*, coding for enzymes and proteins involved in DNA double-strand break (DSB) repair (*ATR, BRCA1*, and *BRCA2, LIG3, RAD50, RAD51*), nucleotide excision repair (*XPA* and *ERCC4/XPF*) or mismatch repair (*MLH1* and *MSH2*).

The down-regulation of DNA repair genes appears to be an important mechanism in order to modulate cellular DNA repair capacity. In this context, epigenetic changes play an important role in gene regulation, and likely contribute to carcinogenicity [30]. In some studies, the interaction of Cr(VI) with proteins related to transcription, or its interference with epigenetic modulators, was also demonstrated and associated with genomic instability [9,30,31]. Thus, many of the not yet fully understood toxic effects of Cr(VI), such as Cr(VI)-induced cellular DNA repair deficiencies or Cr(VI)-induced genomic instability, are currently thought to be due to epigenetic changes [6,32]. Impaired transcription can occur due to different reasons. Thus, promotor regions of genes can accumulate 5-methyl cytosine (5-mC) in guanine- and cytosine-rich regions (CpG islands), which are predominantly found in close distance to the transcription starting site. For instance, in Cr(VI)-exposed lung epithelial cells, a sequence-specific increase of 5-mC led to the down-regulation of DNA repair genes [31,33]. Furthermore, hypermethylation of the *MLH1* gene has been linked to a defective mismatch repair, resulting in microsatellite instability in lung tissue of Cr(VI)-exposed workers [34,35]. Also, recently published transcriptome studies support the hypothesis that Cr(VI) may have a pronounced effect on the transcriptional response via a variety of epigenetic modifications, contributing to carcinogenesis [32]. In addition to altered DNA methylation patterns, there is some evidence that chromates lead to a decreased transcription rate via changes in histone modifications, by interference with enzymes involved in the acetylation and methylation of histone side chains. Thus, in a study conducted by Sun et al. exposure of A549 and BEAS2B cells to Cr(VI) resulted in a decrease in histone modification H3K27Me3, and an increase in modifications H3K4Me3, H3K9Me2, and H3K9Me3 [36]. In particular, methylation at H3K9 or H3K27 is closely associated with transcriptional repression [37].

One other postulated mechanism of Cr(VI) induced cellular toxicity is the generation of ROS. Thus, after the anion carrier-mediated uptake of Cr(VI) [38,39], it is reduced intracellularly to the highly reactive chromium intermediates Cr(V) and Cr(IV), causing oxidative stress [5,9,37]. However, our data do not seem to support the extensive generation of ROS. At low concentrations, only marginal effects were observed in the gene cluster of oxidative stress response. One exception was *IL8*, signaling inflammation, which was induced in both cell lines. At higher concentrations, Cr(VI) ions instead caused a repression of oxidative stress-related genes, such as *NFKB1* and *NFKB2*. These observations appear to contradict results reported by Ye and Shi [40], who described elevated transcription levels of glutathione peroxidase (*GPx*), copper-zinc superoxide dismutase (*SOD*), and metallothionein 2A (*MT2A*), as well as the metal-regulatory transcription factor 1 (*MTF1*) in A549 cells. However, in their study, two hour, high-dose (K$_2$Cr$_2$O$_7$, 300 μM) treatments were applied. With respect to particle A, the data of Horie and colleagues also indicated an increased formation of ROS. However, in this case, ROS were not detected at the transcriptional level, but were measured directly by the colorimetric DCFH assay. By this approach, a clear increase of the intracellular ROS was observed in the A549 and HaCaT cells after treatment with Cr_2O_3 particle A or $K_2Cr_2O_7$. However, more than a 10-fold higher dose of particle A (88 µg/mL), and up to a 20-fold higher concentration in the case of $K_2Cr_2O_7$ (1 mM), as well as far shorter treatment times, were applied [10]. Therefore, we assumed that after 24 h of treatment, the initial chromium intermediates responsible for the generation of ROS may have already been vanished due to their instability and short biological half-life [41,42].

Transcriptional alterations were also observed with respect to apoptosis factors. Pronounced repressions were observed for the genes *MAP3K5* and *BTRC*. In case of *MAP3K5*, a downregulation was associated with an increase in cellular stress levels and dysfunction of apoptotic regulatory pathways. Dysregulation can lead to inhibition of programmed cell death, and thus, a shift toward uncontrolled, necrotic cell death mechanisms, some of which result in severe inflammatory responses [43]. However, toxic concentrations of Cr(VI) also resulted in the induction of pro-apoptotic genes, such as *PMAIP*, *PPM1D*, and *TNFRSF10B* in both cell lines. Therefore, the impact on cell death pathways on the transcriptional level appears to be controversial and needs to be further elucidated.

5. Conclusions

In conclusion, the results of the present study show that gene expression profiles obtained using high-throughput RT-qPCR provide valuable toxicity profiles for different chromium compounds, reflecting the mode of action of chromium in different oxidation states, especially with respect to the higher toxicity and genotoxicity of Cr(VI) vs. Cr(III). Regarding the Cr_2O_3 particles, the cytotoxicity, as well as the gene expression profiles, indicate that the toxicity of the particles resemble either completely (particle 3) or mostly (particle 2) respective profiles obtained for water soluble Cr(III), i.e., showing no or only mild effects, respectively. Therefore, Cr_2O_3 nanoparticles or microparticles as such are neither cyto- nor genotoxic. However, this statement holds only for particles not releasing Cr(VI). If the latter is the case, as shown for particle A and—to a much lesser extent—for particle B, observed effects resemble Cr(VI), both qualitatively and even quantitatively. Since the same solubility was observed in distilled water and in artificial lysosomal fluid, Cr(VI) is expected to be released extracellularly, and to be taken up via anion channels, as is known for water soluble Cr(VI). Our findings also have an important impact on the toxicological risk assessment of Cr(III) oxide particles, either in the nano- or in the micro-sized range. Intracellular conversion to Cr(VI) with subsequent reduction, and the subsequent induction of DNA damage, appears to be absent or negligible; also, the same applies to the intracellular release of Cr(III) and the induction of DNA damage. Therefore, the cellular damage depends not on particle uptake, but solely on whether or not Cr(VI) is released from the particles, which needs to be elucidated on a case-by-case basis for risk assessment.

Supplementary Materials: The following supporting information can be downloaded at: https://www.mdpi.com/article/10.3390/nano12081294/s1, Table S1: Hydrodynamic size of the particles; Table S2: Polydispersity index of the particles; Table S3: Name of coding proteins for the selected genes; Figure S1: complete $\log_2$ fold gene expression data A549 cells; Figure S2: complete $\log_2$ fold gene expression data HaCaT cells.

Author Contributions: Conceptualization, P.S., F.F., D.W. and A.H.; investigation, P.S., F.F. and J.S.; methodology, P.S., F.F., J.S., D.W. and A.H.; validation, P.S., F.F., D.W. and A.H.; formal analysis, P.S., F.F., J.S., D.W. and A.H.; resources, A.H.; data curation, P.S., F.F. and A.H.; writing—original draft preparation, P.S.; writing—review and editing, P.S., F.F., D.W., J.S. and A.H.; visualization, P.S., F.F., J.S., D.W. and A.H.; supervision, A.H.; project administration, A.H. All authors have read and agreed to the published version of the manuscript.

Funding: This research received no external funding.

Nanomaterials **2022**, 12, 1294

Data Availability Statement: The data presented in this study are available on request from the first (P.S.) and corresponding author (A.H.) for researchers of academic institutes who meet the criteria for access to the confidential data.

Acknowledgments: We would like to thank Roel Schins (IUF, Düsseldorf, Germany) and Brunhilde Bloemeke (University Trier, Trier, Germany) for providing A549 cells and HaCaT cells, respectively. Furthermore, we would like to thank Uwe Hempelmann, (Lanxess, Cologne, Germany) for providing the micro-sized Cr_2O_3 particles (Particles 3). Lastly, we would like to thank the Laboratories for Electron Microscopy at KIT, especially Heike Störmer, for taking the TEM images.

Conflicts of Interest: The authors declare no conflict of interest.

References

1. International Agency for Research on Cancer. *Chromium, Nickel and Welding-IARC Monographs on the Evaluation of Carcinogenic Risks to Humans*; World Health Organization: Lyon, France, 1990; Volume 49.
2. Wilbur, S.A.H.; Fay, M.; Yu, D.; Tencza, B.; Ingerman, L.; Klotzbach, J.; James, S. *Toxicological Profile for Chromium*; U.S. Department of Health and Human Services: Atlanta, GA, USA, 2012.
3. International Agency for Research on Cancer. *A Review of Human Carcinogens-IARC Monographs on the Evaluation of Carcinogenic Risks to Humans*; World Health Organization: Lyon, France, 2012; Volume 100 F.
4. Wetterhahn, K.E.; Hamilton, J.W. Molecular basis of hexavalent chromium carcinogenicity: Effect on gene expression. *Sci. Total Environ.* **1989**, *86*, 113–129. [CrossRef]
5. Zhitkovich, A. Chromium in drinking water: Sources, metabolism, and cancer risks. *Chem. Res. Toxicol.* **2011**, *24*, 1617–1629. [CrossRef]
6. Wise, S.S.; Aboueissa, A.E.; Martino, J.; Wise, J.P., Sr. Hexavalent Chromium-Induced Chromosome Instability Drives Permanent and Heritable Numerical and Structural Changes and a DNA Repair-Deficient Phenotype. *Cancer Res.* **2018**, *78*, 4203–4214. [CrossRef] [PubMed]
7. Myers, J.M.; Antholine, W.E.; Myers, C.R. The intracellular redox stress caused by hexavalent chromium is selective for proteins that have key roles in cell survival and thiol redox control. *Toxicology* **2011**, *281*, 37–47. [CrossRef]
8. Hartwig, A. Metal interaction with redox regulation: An integrating concept in metal carcinogenesis? *Free Radic. Biol. Med.* **2013**, *55*, 63–72. [CrossRef]
9. Chen, Q.Y.; Murphy, A.; Sun, H.; Costa, M. Molecular and epigenetic mechanisms of Cr(VI)-induced carcinogenesis. *Toxicol. Appl. Pharmacol.* **2019**, *377*, 114636. [CrossRef] [PubMed]
10. Horie, M.; Nishio, K.; Endoh, S.; Kato, H.; Fujita, K.; Miyauchi, A.; Nakamura, A.; Kinugasa, S.; Yamamoto, K.; Niki, E.; et al. Chromium(III) oxide nanoparticles induced remarkable oxidative stress and apoptosis on culture cells. *Environ. Toxicol.* **2011**, *28*, 61–75. [CrossRef]
11. Fischer, B.M.; Neumann, D.; Piberger, A.L.; Risnes, S.F.; Koberle, B.; Hartwig, A. Use of high-throughput RT-qPCR to assess modulations of gene expression profiles related to genomic stability and interactions by cadmium. *Arch. Toxicol.* **2016**, *90*, 2745–2761. [CrossRef]
12. Semisch, A.; Ohle, J.; Witt, B.; Hartwig, A. Cytotoxicity and genotoxicity of nano—and microparticulate copper oxide: Role of solubility and intracellular bioavailability. *Part. Fibre Toxicol.* **2014**, *11*, 10. [CrossRef]
13. Bustin, S.A. Absolute quantification of mRNA using real-time reverse transcription polymerase chain reaction assays. *J. Mol. Endocrinol.* **2000**, *25*, 169–193. [CrossRef]
14. Pfaffl, M.W.; Hageleit, M. Validities of mRNA quantification using recombinant RNA and recombinant DNA external calibration curves in real-time RT-PCR. *Biotechnol. Lett.* **2001**, *23*, 275–282. [CrossRef]
15. Biesinger, M.C.; Brown, C.; Mycroft, J.R.; Davidson, R.D.; McIntyre, N.S. X-ray photoelectron spectroscopy studies of chromium compounds. *Surf. Interface Anal.* **2004**, *36*, 1150–1163. [CrossRef]
16. Hufnagel, M.; Schoch, S.; Wall, J.; Strauch, B.M.; Hartwig, A. Toxicity and Gene Expression Profiling of Copper- and Titanium-Based Nanoparticles Using Air-Liquid Interface Exposure. *Chem Res. Toxicol.* **2020**, *33*, 1237–1249. [CrossRef] [PubMed]
17. Hufnagel, M.; Neuberger, R.; Wall, J.; Link, M.; Friesen, A.; Hartwig, A. Impact of Differentiated Macrophage-Like Cells on the Transcriptional Toxicity Profile of CuO Nanoparticles in Co-Cultured Lung Epithelial Cells. *Int. J. Mol. Sci.* **2021**, *22*, 5044. [CrossRef] [PubMed]
18. Wall, J.; Seleci, D.A.; Schworm, F.; Neuberger, R.; Link, M.; Hufnagel, M.; Schumacher, P.; Schulz, F.; Heinrich, U.; Wohlleben, W.; et al. Comparison of Metal-Based Nanoparticles and Nanowires: Solubility, Reactivity, Bioavailability and Cellular Toxicity. *Nanomaterials* **2021**, *12*, 147. [CrossRef] [PubMed]
19. Hininger, I.; Benaraba, R.; Osman, M.; Faure, H.; Marie Roussel, A.; Anderson, R.A. Safety of trivalent chromium complexes: No evidence for DNA damage in human HaCaT keratinocytes. *Free Radic. Biol. Med.* **2007**, *42*, 1759–1765. [CrossRef] [PubMed]
20. Barceloux, D.G. Chromium. *J. Toxicol. Clin. Toxicol.* **1999**, *37*, 173–194. [CrossRef]
21. Shi, X.; Dalal, N.S. Chromium (V) and hydroxyl radical formation during the glutathione reductase-catalyzed reduction of chromium (VI). *Biochem. Biophys. Res. Commun.* **1989**, *163*, 627–634. [CrossRef]

22. Stohs, S.J.; Bagchi, D.; Hassoun, E.; Bagchi, M. Oxidative mechanisms in the toxicity of chromium and cadmium ions. *J. Environ. Pathol. Toxicol. Oncol.* **2001**, *20*, 77–88. [CrossRef]

23. Wang, X.; Zheng, Q.; Yuan, Y.; Hai, R.; Zou, D. Bacterial community and molecular ecological network in response to Cr_2O_3 nanoparticles in activated sludge system. *Chemosphere* **2017**, *188*, 10–17. [CrossRef]

24. Kumar, D.; Rajeshwari, A.; Jadon, P.S.; Chaudhuri, G.; Mukherjee, A.; Chandrasekaran, N.; Mukherjee, A. Cytogenetic studies of chromium (III) oxide nanoparticles on Allium cepa root tip cells. *J. Environ. Sci.* **2015**, *38*, 150–157. [CrossRef] [PubMed]

25. Peng, G.; He, Y.; Zhao, M.; Yu, T.; Qin, Y.; Lin, S. Differential effects of metal oxide nanoparticles on zebrafish embryos and developing larvae. *Environ. Sci. Nano* **2018**, *5*, 1200–1207. [CrossRef]

26. Costa, C.H.D.; Perreault, F.; Oukarroum, A.; Melegari, S.P.; Popovic, R.; Matias, W.G. Effect of chromium oxide (III) nanoparticles on the production of reactive oxygen species and photosystem II activity in the green alga Chlamydomonas reinhardtii. *Sci. Total Environ.* **2016**, *565*, 951–960. [CrossRef] [PubMed]

27. Buxbaum, G.; Pfaff, G. *Industrial Inorganic Pigments*, 3rd ed.; Wiley: Weinheim, Germany, 2005. [CrossRef]

28. De Loughery, Z.; Luczak, M.W.; Zhitkovich, A. Monitoring Cr Intermediates and Reactive Oxygen Species with Fluorescent Probes during Chromate Reduction. *Chem. Res. Toxicol.* **2014**, *27*, 843–851. [CrossRef]

29. Abbas, T.; Dutta, A. p21 in cancer: Intricate networks and multiple activities. *Nat. Rev. Cancer* **2009**, *9*, 400–414. [CrossRef]

30. Christmann, M.; Kaina, B. Epigenetic regulation of DNA repair genes and implications for tumor therapy. *Mutat. Res.* **2019**, *780*, 15–28. [CrossRef]

31. Hu, G.; Li, P.; Cui, X.; Li, Y.; Zhang, J.; Zhai, X.; Yu, S.; Tang, S.; Zhao, Z.; Wang, J.; et al. Cr(VI)-induced methylation and down-regulation of DNA repair genes and its association with markers of genetic damage in workers and 16HBE cells. *Environ. Pollut.* **2018**, *238*, 833–843. [CrossRef]

32. Rager, J.E.; Suh, M.; Chappell, G.A.; Thompson, C.M.; Proctor, D.M. Review of transcriptomic responses to hexavalent chromium exposure in lung cells supports a role of epigenetic mediators in carcinogenesis. *Toxicol. Lett.* **2019**, *305*, 40–50. [CrossRef]

33. Wang, Z.; Wu, J.; Humphries, B.; Kondo, K.; Jiang, Y.; Shi, X.; Yang, C. Upregulation of histone-lysine methyltransferases plays a causal role in hexavalent chromium-induced cancer stem cell-like property and cell transformation. *Toxicol. Appl. Pharmacol.* **2018**, *342*, 22–30. [CrossRef]

34. Imai, K.; Yamamoto, H. Carcinogenesis and microsatellite instability: The interrelationship between genetics and epigenetics. *Carcinogenesis* **2008**, *29*, 673–680. [CrossRef]

35. Takahashi, Y.; Kondo, K.; Hirose, T.; Nakagawa, H.; Tsuyuguchi, M.; Hashimoto, M.; Sano, T.; Ochiai, A.; Monden, Y. Microsatellite instability and protein expression of the DNA mismatch repair gene, hMLH1, of lung cancer in chromate-exposed workers. *Mol. Carcinog* **2005**, *42*, 150–158. [CrossRef] [PubMed]

36. Sun, H.; Zhou, X.; Chen, H.; Li, Q.; Costa, M. Modulation of histone methylation and MLH1 gene silencing by hexavalent chromium. *Toxicol. Appl. Pharmacol.* **2009**, *237*, 258–266. [CrossRef] [PubMed]

37. Arita, A.; Costa, M. Epigenetics in metal carcinogenesis: Nickel, arsenic, chromium and cadmium. *Metallomics* **2009**, *1*, 222–228. [CrossRef] [PubMed]

38. Codd, R.; Dillon, C.T.; Levina, A.; Lay, P.A. Studies on the genotoxicity of chromium: From the test tube to the cell. *Coord. Chem. Rev.* **2001**, *216–217*, 537–582. [CrossRef]

39. Valko, M.; Morris, H.; Cronin, M.T.D. Metals, Toxicity and Oxidative Stress. *Curr. Med. Chem.* **2005**, *12*, 1161–1208. [CrossRef]

40. Ye, J.; Shi, X. Gene expression profile in response to chromium-induced cell stress in A549 cells. *Mol. Cell. Biochem.* **2001**, *222*, 189–197. [CrossRef]

41. Suzuki, Y.; Fukuda, K. Reduction of hexavalent chromium by ascorbic acid and glutathione with special reference to the rat lung. *Arch. Toxicol.* **1990**, *64*, 169–176. [CrossRef]

42. Levina, A.; Zhang, L.; Lay, P.A. Structure and reactivity of a chromium(v) glutathione complex. *Inorg. Chem.* **2003**, *42*, 767–784. [CrossRef]

43. Ryuno, H.; Naguro, I.; Kamiyama, M. ASK family and cancer. *Adv. Biol. Regul.* **2017**, *66*, 72–84. [CrossRef]

 *nanomaterials*

Article

PLATOX: Integrated In Vitro/In Vivo Approach for Screening of Adverse Lung Effects of Graphene-Related 2D Nanomaterials

Otto Creutzenberg [1,*], Helena Oliveira [2], Lucian Farcal [3], Dirk Schaudien [1], Ana Mendes [2], Ana Catarina Menezes [2], Tatjana Tischler [1], Sabina Burla [3,4] and Christina Ziemann [1,*]

[1] Fraunhofer Institute for Toxicology and Experimental Medicine ITEM, 30625 Hannover, Germany; dirk.schaudien@item.fraunhofer.de (D.S.); tatjana.gripp@gmx.de (T.T.)

[2] Department of Biology & CESAM, University of Aveiro, 3810-193 Aveiro, Portugal; holiveira@ua.pt (H.O.); anamendes@live.com.pt (A.M.); catarinamenezes@msn.com (A.C.M.)

[3] BIOTOX SRL, 407280 Cluj-Napoca, Romania; lucian.farcal@biotox.ro (L.F.); sabina.burla@list.lu (S.B.)

[4] Department of Environmental Research and Innovation, Luxembourg Institute of Science and Technology, 4422 Belvaux, Luxembourg

[*] Correspondence: otto.creutzenberg@item.fraunhofer.de (O.C.); christina.ziemann@item.fraunhofer.de (C.Z.); Tel.: +49-511-5350-461 (O.C.); +49-511-5350-203 (C.Z.)

Abstract: Graphene-related two-dimensional nanomaterials possess very technically promising characteristics, but gaps exist regarding their potential adverse health effects. Based on their nano-thickness and lateral micron dimensions, nanoplates exhibit particular aerodynamic properties, including respirability. To develop a lung-focused, in vitro/in vivo screening approach for toxicological hazard assessment, various graphene-related nanoplates, i.e., single-layer graphene (SLG), graphene nanoplatelets (GNP), carboxyl graphene, graphene oxide, graphite oxide and Printex 90$^{®}$ (particle reference) were used. Material characterization preceded in vitro (geno)toxicity screening (membrane integrity, metabolic activity, proliferation, DNA damage) with primary rat alveolar macrophages (AM), MRC-5 lung fibroblasts, NR8383 and RAW 264.7 cells. Submerse cell exposure and material-adapted methods indicated material-, cell type-, concentration-, and time-specific effects. SLG and GNP were finally chosen as in vitro biologically active or more inert graphene showed eosinophils in lavage fluid for SLG but not GNP. The subsequent 28-day inhalation study (OECD 412) confirmed a toxic, genotoxic and pro-inflammatory potential for SLG at 3.2 mg/m^3 with an in vivo-ranking of lung toxicity: SLG > GNP > Printex 90$^{®}$. The in vivo ranking finally pointed to AM (lactate dehydrogenase release, DNA damage) as the most predictive in vitro model for the (geno)toxicity screening of graphene nanoplates.

Keywords: graphene; 2D; nanoplates; lung; inhalation; toxicity; genotoxicity; in vitro; inflammation; hazard assessment

Citation: Creutzenberg, O.; Oliveira, H.; Farcal, L.; Schaudien, D.; Mendes, A.; Menezes, A.C.; Tischler, T.; Burla, S.; Ziemann, C. PLATOX: Integrated In Vitro/In Vivo Approach for Screening of Adverse Lung Effects of Graphene-Related 2D Nanomaterials. *Nanomaterials* **2022**, *12*, 1254. https://doi.org/10.3390/nano12081254

Academic Editors: Andrea Hartwig and Christoph Van Thriel

Received: 28 February 2022
Accepted: 5 April 2022
Published: 7 April 2022

Publisher's Note: MDPI stays neutral with regard to jurisdictional claims in published maps and institutional affiliations.

1. Introduction

Graphene-related two-dimensional (2D) nanomaterials (nanoplates; GRNP) belong to the group of graphene-based engineered nanomaterials that are currently subject to an accelerated toxicological characterization. Before its first isolation from graphite by Novoselov and Geim in 2004, graphene was only a scientific model and known as a building block of already well-known carbon-based three-dimensional (3D) nanomaterials, such as graphite, carbon nanotubes (CNT) and fullerenes [1]. Graphene emerged as a 2D allotrope of carbon nanomaterials characterized by a single flat atom layer as a monocrystalline structure. Graphene is believed to be composed of benzene rings stripped of their hydrogen atoms. The carbon atoms are sp2-hybridized and hexagonally arranged, similar to a lattice honeycomb structure, with a C-C bond length of 0.142 nm [2,3]. Graphene belongs, as a pristine, defect-free, single-atom layer carbon plate, to the graphene-based family of nanomaterials (GFN), which also comprises graphene oxide (GO) as the most popular member, reduced graphene oxide (rGO), exfoliated graphene flakes, and graphene nanoplatelets (GNP) as well as the respective chemically functionalized versions [4]. Graphene nanoplates include single-layer (SLG), bilayer, trilayer and few-layer graphene (FLG; 3–10 well-defined stacked graphene layers) as well as graphene nanoplatelets (GNP) with more graphene layers [5,6].

GRNP are unique compared to spherical nanoparticles or one-dimensional (1D) carbon nanomaterials such as CNT or nanorods, owing to their physicochemical (PC) properties. GRNP possess unique features, e.g., high surface area per unit mass, excellent thermal and electric conductivity, and outstanding mechanical properties and strength. Their very promising PC properties have led to manifold applications in electronics, photonics, composite materials, catalysts, energy generation/storage, sensors, computer memory, metrology, and biomedicine [7,8]. However, the unique PC properties of GRNP can also influence interaction with cells and cellular substructures and, thus, their biocompatibility, with the probability to pose human health risks [9].

Given the large-scale use of GRNP in different industries (currently 1 to 10 tons per annum [10]), occupational exposure, but also exposure to consumers and the general population might occur. Appropriate risk assessment is, therefore, of high importance, in order to draw qualified conclusions upon the delicate balance between technology advancements and hazards to human health and the environment. However, according to the notifications provided by companies to ECHA during REACH registrations, no hazards have been classified for graphene [11]. Since large-scale production and commercialization of GRNP are imminent, the study of real hazards is indispensable. Even if the health effects associated with potential adversity of graphene have already been studied at the cellular level (in vitro) and in animal models (in vivo), as shown by the increasing number of research papers published on graphene toxicity and genotoxicity [2,5,12–14], the risk posed to humans remains nearly unexplored.

Despite the popular image of GRNP, resulting from their PC properties and the promising applications in various fields, many forms can be hazardous to human health, as they are processed as dry powders, thus posing a significant exposure risk at workplaces, in particular, through the inhalation route [15]. Humans can be exposed to graphene from various sources, mostly during manufacturing, with the potential risk of inhalation toxicity [16]. Therefore, the measurement of airborne dust levels in research laboratories and pilot and full-scale manufacturing facilities should be performed and integrated with aerodynamic diameters and deposition profiles from published theories on thin-plate aerodynamics [9]. Nanomaterials up to 25 µm in diameter can deposit ahead of the ciliated airways upon inhalation [17]. It is very difficult to figure out appropriate and real-life exposure scenarios leading to significant toxicity, as the PC properties of graphene are novel and have not been explored in depth in inhalation studies. However, a recent publication provides important data on human exposure during GNP production for research and development purposes, including real-time measurements and personal sampling. The publication finally identified the most critical production phases [18].

A review on inhalation testing of graphene for hazard assessment demonstrated that the existing database, especially for the inhalation route, is still too scarce to allow proper risk assessment [19]. Although important toxicological analogies may be available from other carbonaceous nanomaterials such as carbon black (CB) and CNT, direct extrapolation from those material classes to GRNP is not appropriate, as new engineered nanomaterials may pose unexpected risks [20]. In a 4-week inhalation test with rats at aerosol concentrations up to 1.9 mg of GNP per m^3, no signs of inflammation were noticed at any time point [21]. Due to their nano-thickness, GNP and, more generally, GRNP exhibit particular aerodynamic properties, allowing deposition in the lung alveolar region. GRNP are nano-scaled in one dimension, whereas the other two dimensions are often micron-scaled. The specific aerodynamic consequences have been discussed in detail by [9]. Calculation of the deposition of GNP with 0.5, 5 and 25 µm lateral dimensions resulted in unexpectedly high deposition efficiencies of 45–10%. Compared to multiwalled carbon nanotubes (MWCNT, tangled type), GNP showed a lower toxic potential in a 5-day inhalation test, whereas no relevant toxicity was detected for CB and graphite nanoplates. Rigid MWCNT, however, showed a stronger toxic response than GNP [22]. After intravenous injection into mice, no hematotoxicity and no pro-inflammatory potential of highly water-dispersed FLG in lymph node and spleen cells were observed, thus pointing to high biocompatibility [13].

Due to the limited availability of in vivo data for pristine graphene nanoplates in particular and predictive toxicological in vitro screening approaches for GRNP in general, the development of integrated in vitro/in vivo screening approaches and filling of data gaps is inevitable to establish safe exposure levels. In-depth knowledge about graphene-based materials in terms of biocompatibility as well as comprehensive toxicological evaluation and risk analysis is imperative to ensure the safe application of these versatile materials. Risk analysis of graphene should be done by completing the following steps: in vitro and in vivo hazard assessment, risk assessment, risk characterization, risk communication, risk management, and policy relating to risk in the context of risks of concern to individuals, to the public- and private-sector organizations, and to society at a local, regional, national, or global level [23]. The main objective for human health risk assessments is to finally establish acceptable exposure levels for humans to specific GRNP.

In the EU-funded project PLATOX, eight mostly commercially available GRNP species were, therefore, selected, i.e., two SLG, two GNP, carboxyl graphene (CG), graphene oxide (GO) and graphite oxide, supplemented by technical soot (Printex 90®; non-platelet reference), to develop an integrated, lung-focused approach for proper hazard assessment of GRNP and to fill toxicological data gaps. Based on the 3R principles (Replacement, Reduction and Refinement) regarding animal experiments, it is also highly desirable to define in vivo-validated, predictive in vitro screening approaches for GRNP. As such, we screened different lung-relevant cell types, i.e., primary rat alveolar macrophages (AM), human MRC-5 lung fibroblasts, murine RAW 264.7 and rat NR8383 macrophages, using different endpoints/screening methods regarding their ability to predict in vivo lung toxicity of GRNP and, particularly, of pristine graphene nanoplates. In vitro results were finally validated by a 28-day (with 4 weeks of recovery) inhalation study with the pristine graphene nanoplate species that was least biologically active in vitro (GNP) and one of the two that were highly active in vitro (SLG).

When considering the in vivo results (with a clear cytotoxic, genotoxic, and pro-inflammatory potential of SLG, but less activity of GNP after 28 days of inhalation) AM as the primary rat cell type directly derived from the respiratory tract showed the highest reliability in predicting in vivo adverse lung effects of the pristine graphene nanoplates tested in the present study, with lactate dehydrogenase (LDH) release and induction of DNA strand breaks as most meaningful endpoints. It can be concluded from the study results that predictive in vitro lung-focused toxicity screening of GRNP samples seems, in principle, to be possible when adequately choosing relevant cell models and endpoints, appropriate dispersion methods, incubation times and concentrations, and carefully adapting photometric and immunological detection methods in particular to the specific GRNP materials.

2. Materials and Methods

2.1. Graphene-Related 2D Nanomaterials (GRNP)

2.1.1. GRNP Species

For reasons of relevance, commercially available GRNP samples were chosen for the present study. Six GRNP samples, including three functionalized materials, were purchased from ACS Material, LLC (Medford, MA, USA). These samples comprised two single-layer graphene samples (SLG), graphene nanoplatelets (GNP), carboxyl graphene (CG), graphene oxide (GO) powder and graphite oxide. Additionally, an experimental GNP sample was provided by Avanzare (Logroño, Spain). Spherical carbon black (CB; Printex 90®; Evonik Industries AG, Essen, Germany), as a non-platelet reference, completed a total set of eight test samples, subsequently encoded P1-P8 (see Table 1).

Table 1. GRNP materials, codes, providers, identification, characteristics and synthesis.

Code	Material	Supplier/ Identification	Diameter [μm] *	Thickness [nm] *	BET Surface Area [m²/mg]	Dispersible in *	Preparation/Properties *
P1	Single-Layer Graphene (SLG)	ACS: GN1P0005	0.4–5	0.6–1.2	278 (400–1000) *	Water, ethanol and others	Thermal exfoliation + hydrogen reduction; sulfur impurities
P2	Single-Layer Graphene, (Graphene factory) (SLG)	ACS: GNP1F010	0.5–5	1–5 atomic layers	620 (650–750) *	No information	Thermal exfoliation + hydrogen reduction
P3	Carboxyl Graphene (CG)	ACS: GNCP0005	1–5	0.8–1.2	1.5	Polar solvents like DMF	(1) Modified Hummer's Method to make graphene oxide (2) Conversion of (–OH) and (C-O-C) into (-COOH); non-conductivity
P4	Graphene Nanoplatelets (GNP)	ACS: GNNP0051	2–7	2–10	15 (20–40) *	Easy to disperse	Stacks of multilayer graphene; high aspect ratio (width to thickness); D50 = 48.93 μm, phosphorous impurities
P5	Graphene Oxide (S Method) (GO)	ACS: GNOS0010	1–15	0.8–1.2	5.2 (5–10) *	No information	Staudenmaier Method; oxygen content: 35 wt %; single-layer ratio > 90%
P6	Graphite Oxide	ACS: GTOP0002	0.5–5	1–3	2.7	Polar solvents like water, ethanol, DMF	Modified Hummer's Method; oxygen content: 46 wt %; intercalated with either two ethanol or methanol monolayers
P7	Reference Graphene Nanoplatelets (GNP)	Avanzare: GR1	2	3	195 (70) *	No information	No XPS (low defects by Raman), all C1s; 8 ± 0.5 atomic graphene layers
P8	Spherical Carbon Black (CB)	Evonik: Printex 90®	14	n.a.	317 (337) *	No information	Spherical; specified as >99% pure carbon black, PAH = 0.039 ppm

* Values and information given by the supplier; BET = Brunauer–Emmett–Teller method; DMF = dimethylformamide; n.a. = not applicable.

2.1.2. Material Characterization

To supplement the PC information provided by the suppliers of the materials, the specific surface area was re-evaluated as a service by the Fraunhofer Institute for Ceramic Technologies and Systems IKTS (Dresden, Germany), using the Brunauer–Emmett–Teller (BET) method. Part of the resulting data did not match the BET values given by the material suppliers (Table 1). Sterility testing of GRNP was done by adding a defined amount of the GRNP dispersions to thioglycolate broth (supporting the growth of a broad panel of bacteria and fungi) and incubating duplicate samples at 34–35 °C for 14 days. Physiological saline (0.9% NaCl) served as negative and Staphylococcus aureus (ATCC 25923) as a positive control. Turbidity was finally determined by the naked eye. Additionally, endotoxin content was measured by a commercial service laboratory (Lonza Bioscience, Verviers, Belgium) to avoid unspecific pro-inflammatory effects due to bacterial contamination. Both sterility and endotoxin testing did not point to relevant contaminations. Additionally, scanning electron microscopic (SEM; Zeiss Supra 55, Carl Zeiss NTS GmbH, Oberkochen, Germany) pictures (Figure 1) were generated for all samples to look for material morphology and dispersion grade. Therefore, stock dispersions of 50 µg/mL were prepared by ultrasonication in cell culture medium (in vitro experiments, see Section 2.1.3) or in a dispersion medium (in vivo study), as described by [24]. Depending on the material, magnifications in the range of 1000–50,000× were used.

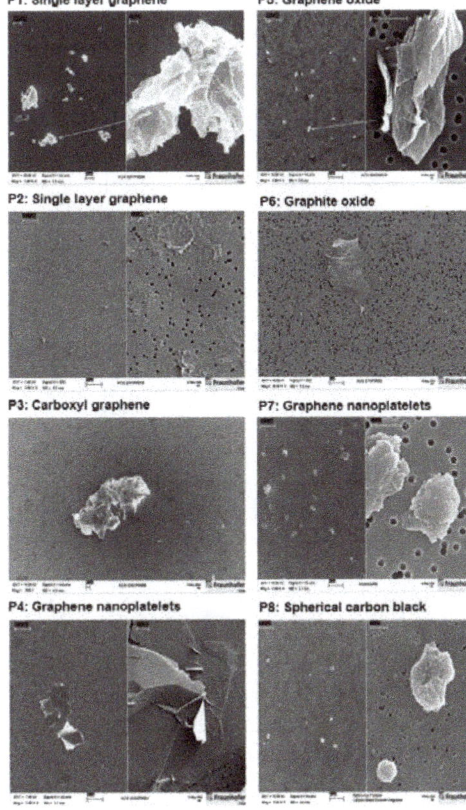

Figure 1. Scanning electron microscopic (SEM) pictures of the GRNP materials P1–P8, dispersed by sonication in cell culture medium with 10% FCS. Magnifications used for the split pictures were as follows, left: 1000–2000×; right: 20,000–50,000×. For the P3, a magnification of 500× was used, and for P6, a magnification of 10,000× was used.

2.1.3. Dispersion and In Vitro Treatment with GRNP Materials

For in vitro (geno)toxicity screening and respective SEM analyses, the various GRNP were accurately weighed into glass vials and stock dispersions were prepared by suspending them in the cell type-specific cell culture media with 10% fetal calf serum (FCS), stirring the dispersions for 10 min using a magnetic stirrer and finally applying ultrasonication. At Fraunhofer ITEM, ultrasonication was performed by placing the dispersion on ice and applying sonication using a Sonoplus HD2070 (70 W) ultrasonic homogenizer and a VS70T sonotrode (diameter: 13 nm), both from Bandelin (Berlin, Germany). Ultrasonication was applied three times for 5 min each with 1 min breaks in between (90% duty cycle amplitude of 80 μm/ss and 0.9 s working intervals, 0.1 s rest intervals). A similar procedure was followed at the University of Aveiro, except a Vibra-Cell ultrasonic processor (Sonics & Materials, Inc., Newtown, CT, USA) was used. Appropriate dispersion was estimated by light microscopy. Subsequent pH measurements of the GRNP dispersions indicated slightly basic conditions for all samples, ranging from pH 8.18 (P5) to pH 8.50 (P7). For cell exposure, the stock dispersions were finally diluted to the double-concentrated (AM and NR8383) or final (other cell types) mass-based concentrations with the respective FCS-containing cell culture media, and dispersions were vortexed for 5 sec directly before cell treatment. Submerse cell treatment with the GRNP materials was done for 4–48 h, depending on the cell type and endpoint used.

2.1.4. Sedimentation Kinetics

Sedimentation kinetics of the various GRNP species was estimated by determining the decay of turbidity using spectrophotometric OD600 measurements. Dispersions of the different GRNP (6.25 μg/mL, corresponding to the lowest concentration used for cell exposure, i.e., 3.125 μg/cm^2) were prepared, as described under Section 2.1.3. One milliliter of FCS-containing cell culture medium (blank) and the GRNP dispersions were placed into disposable semi-micro spectrophotometric cuvettes and turbidity (OD600) was measured directly after material dispersion (time point zero) and after 1, 18, 24 and 48 h with an Eppendorf BioPhotometer® (Eppendorf, Hamburg, Germany). In between measurements, cuvettes were stored vertically under cell culture conditions (37 °C, 5% CO$_2$, in a water-saturated atmosphere) and were handled very carefully during measurements to avoid unspecific swirling up of the sedimented GRNP fractions. The time point zero OD600 values were set to 100% and relative OD600 values were calculated to enable a direct comparison of relative decays in turbidity for all GRNP materials.

2.2. In Vitro Screening

2.2.1. Cell Models

For in vitro (geno)toxicity screening of the various GRNP and, in particular, the pristine GRNP, both lung-relevant primary/normal alveolar macrophages and lung fibroblasts and macrophage cell lines were used and evaluated for their ability to finally predict in vivo lung response.

Primary rat alveolar macrophages (AM) were chosen as the first primary lung-relevant cell model, as AM represent the first site of contact for particles in the lung and had previously been shown to represent a sensitive test system for the detection of membrane and direct DNA damage (DNA strand breaks, oxidative base modifications) as well as cytokine/chemokine release, when testing various particulate matter [25–27]. Cells were obtained from healthy female (nulliparous and non-pregnant) Wistar rats [strain Crl:WI (Han); Charles River Deutschland Co., Sulzfeld, Germany] by bronchoalveolar lavage (BAL) in compliance with the German Federal Act on the Protection of Animals [28], cultured and exposed to GRNP, as described previously [29]. After preparation, lungs were lavaged 3 times, each with 4 mL of 0.9% NaCl (brought to room temperature to avoid macrophage activation). Gentle massage enabled the isolation of sufficient cell numbers. After 10 min of centrifugation (300× g, 4 °C) of the lavages, cells were counted and plated at a density of 1.5 × 10^5 cells per well on 24-well plates with hydrophobic culture surface

(1.9 cm^2 cell culture surface per well; Nunc, Thermo Fisher Scientific, Brunswick, Germany). Before being exposed to GRNP, cells were pre-cultured for 24 h in 500 µL of Dulbecco's Modified Eagle Medium (DMEM) with high glucose (4.5 g/L), GlutaMax^{TM} and sodium pyruvate (110 mg/L), supplied by GIBCO, Thermo Fisher Scientific (Brunswick, Germany), and supplemented with 100 µg/mL streptomycin sulfate, 100 U penicillin G, sodium salt (Sigma-Aldrich, Taufkirchen, Germany) and 10% (*v/v*) FCS (Sigma-Aldrich, Taufkirchen, Germany) at 37 °C and 5% CO$_2$ in a humidified atmosphere using an incubator. The GRNP materials were added to the cultures two-fold concentrated in a volume of 500 µL without medium exchange to avoid unspecific cell activation. Cells were incubated for 4 or 24 h with the test materials prior to endpoint analysis.

As the second lung-relevant cell type, human MRC-5 fetal lung fibroblasts were used for in vitro GRNP (geno)toxicity screening. MRC-5 cells (pd19; ECACC catalogue number 05072101; lot: 13H007) were received from Sigma-Aldrich (Taufkirchen, Germany). MRC-5 cells exhibit fibroblast-like morphology, a predominantly normal 46, XY karyotype (evident in 56% of cells in the present study), and grow as an adherent monolayer. They are capable of approximately 45 population doublings before becoming senescent. Therefore, 19 population doublings in total were not exceeded. Under the culture conditions used, the mean population doubling time amounted to 24 h. Cells were propagated from the original pd19 aliquot, and the same working batch was used for validation purposes with a harmonized culture protocol by two different laboratories (Fraunhofer ITEM and the University of Aveiro). Cells were grown in Eagle's Minimal Essential Medium (EMEM), alpha modification, supplemented with 2 mM L-Glutamine, 1% non-essential amino acids, 100 µg/mL streptomycin sulfate, 100 U penicillin G, sodium salt (PAA Laboratories GmbH, Cölbe, Germany) and 10% (*v/v*) FCS (same FCS batch in the two laboratories; Biochrom, Berlin, Germany). For cell propagation, cells were grown in T75 flasks at 37 °C and 5% CO$_2$ in a humidified atmosphere using an incubator and were subcultured three times per week, when 70–80% confluent, using trypsin/EDTA (0.05%/0.02%). After thawing, cells were subcultured at least twice before the experiments, and then pre-cultured for 24 h before GRNP exposure. A complete medium exchange to the finally concentrated GRNP dispersions was performed at exposure start.

As an alternative to primary rat alveolar macrophages and to be independent of the use of animals for in vitro (geno)toxicity screening, the normal rat alveolar macrophage cell line NR8383 was included and evaluated regarding its predictivity for in vivo lung effects of GRNP. NR8383 cells were purchased from the American Type Culture Collection (ATTC, Manassas, VA, USA; AgC11x3A, NR8383.1]. BAL cells were originally isolated from a male Sprague-Dawley rat and subsequently transformed [30]. These cells grow as mixed cultures of adherent and suspension cells with a population doubling time of 32 h under the culture conditions used. NR8383 cells were grown in Kaighn's modification of Ham's F12 medium (F12-K; ATTC, Manassas, VA, USA) with 2 mM L-Glutamine, 1500 mg/L sodium hydrogen phosphate, 15% (*v/v*) FCS (Sigma-Aldrich, Germany), 100 µg/mL streptomycin sulfate, and 100 U penicillin G sodium salt (PAA Laboratories GmbH, Cölbe, Germany) on surface-activated culture vessels (TPP, Trasadingen, Switzerland) at 37 °C and 5% CO$_2$ in a humidified atmosphere using an incubator. Cells received new culture medium twice a week by respecting the two cell fractions (adherent and suspension cells) and were subcultured once a week by incubation at room temperature for about 20 min with subsequent tapping. For experiments, NR8383 cells were used during passages 10 to 30 and were pre-cultured for 24 h before treatment with GRNP without medium exchange using double-concentrated dispersions.

The murine macrophage-like cell line RAW 264.7 was purchased from the European Collection of Authenticated Cell Cultures (ECACC, Salisbury, UK). Cells were cultured in DMEM (Gibco, Life Technologies, Grand Island, NY, USA), supplemented with 10% (*v/v*) FCS (Gibco Life Technologies, Grand Island, NY, USA), 2 mM L-glutamine, 1% pen/strep (100 U/mL penicillin, 100 µg/mL streptomycin) and 2.5 µg/mL fungizone (Gibco, Life Technologies, Grand Island, NY, USA), in a humidified incubator at 37 °C and 5% CO$_2$.

Cell confluence and cell morphology were monitored frequently, and subculturing was performed when monolayers reached 75–80% of confluence.

2.2.2. Cellular Uptake

Light microscopy served as a screening tool for evaluation of both cell density, cellular uptake (NR8383 cells; Figure S2) and cell morphology, as well as for estimation of density and homogeneity of the GRNP suspensions. For documentation, pictures were taken using a camera-equipped Nikon ECLIPSE TS 100 infinity-corrected inverse microscope.

To look for cellular uptake of GRNP by AM and MRC-5 cells, fluorescence-coupled darkfield microscopy (CytoViva, Auburn, AL, USA) was used. Therefore, cells were plated in one-well Nunc^{TM} Lab-Tek^{TM} II glass chamber slides (Thermo Fisher Scientific, Brunswick, Germany) at a density of 6×10^5 and 2×10^5 cells in 2 mL or 4 mL of cell culture medium, respectively. A GRNP concentration of 6.25 μg/cm^2 was used for both cell types, as higher concentrations made it impossible to take meaningful pictures, because of the material overlay. For AM, the various materials were added as 2 mL of a double-concentrated dispersion without medium exchange. In contrast, for MRC-5 cells, GRNP were added after a medium exchange to the final mass-based concentration. Cells were subsequently incubated for 24 h, fixed with methanol/glacial acetic acid (3:1) and stained with 4$'$,6-diamidino-2-phenylindole (DAPI) to visualize cell nuclei. Finally, fluorescence-coupled darkfield microscopy was used to detect DAPI-stained cell nuclei via fluorescence and the GRNP materials via the darkfield microscopy unit.

2.2.3. Cytotoxicity Screening

For cytotoxicity screening of GRNP, different endpoints, i.e., membrane damage, metabolic activity and cell proliferation, were investigated.

The influence of GRNP exposure on membrane integrity was investigated in all cell types by measuring lactate dehydrogenase (LDH) activity in culture supernatants after 24 h (depending on cell type) of incubation at 3.125, 6.25, 12.5, 25 and 50 μg/cm^2. Concentrations were limited to a maximum of 50 μg/cm^2 to avoid artificial results due to in vivo irrelevant concentrations. AM, NR8383, RAW 264.7 cells and MRC-5 cells were plated at a density of 1.5 (in 0.5 mL of culture medium), 0.56 (in 0.5 mL), 0.1 (in 0.5 mL) or 0.5×10^5 cells (in 1 mL) per well of 24-well plates, respectively. Cells were pre-cultured for 24 h. For treatment with GRNP, double-concentrated stock dispersions (0.5 mL) were directly added to AM, NR8383 and RAW 264.7 cultures, whereas a medium exchange to 1 mL of the finally concentrated GRNP dispersions was performed for MRC-5 cells to eliminate persistent LDH activity resulting from subculturing. At the end of treatment, cell supernatants were carefully removed, transferred to 1.5-mL reaction tubes and centrifuged at $11,000 \times g$ for 30 min to eliminate residing particle material. LDH activity in culture supernatants was subsequently measured by transferring 100 μL of supernatant per well into optically clear 96-well flat-bottom microplates (3–4 technical replicates), adding 100 μL of the reaction mixture of the "Cytotoxicity Detection Kit" (Roche Diagnostics, Basel, Switzerland) and incubating of plates for 30 min at room temperature in the dark, before photometric measurement at 490 and 630 nm using an ELISA reader. Percent cytotoxicity was finally calculated using delta OD of the two wavelengths, subtracting the blank value and setting the result of Triton X-100-treated cells [1% (v/v) for 10–15 min] to 100% to finally calculate relative cytotoxicity.

For analysis of the relative increase in cell count (RICC) for estimating the impact of GRNP on cell proliferation, MRC-5 and NR8383 cells were seeded at a density of 0.56×10^5 cells in 0.5 mL (NR8383) or 0.5×10^5 cells in 1 mL (MRC-5) per well of 24-well plates and were pre-cultured for 24 h. Before treatment started, cell density was determined (starting cell count) using a Neubauer counting chamber. Cells were then treated with GRNP for 24 h. At the end of treatment, cells were detached, counted and RICC was calculated by subtracting the starting cell number from the cell count at the end of treatment (increase in cell count), dividing the results of the treated cells by the results of the negative controls and multiplying these by 100.

Metabolic activity was investigated using the AlamarBlue® test, based on the reduction of blue, oxidized resazurin to reduced pink resorufin. Cell culture medium without GRNP served as negative control and Triton X-100 [1% (v/v); 10–15 min] was used as a positive control (complete loss of viability). Additionally, cell culture medium without cells served as blank and GRNP background controls without cells (all materials and concentrations) were incubated in parallel to compensate for potential redox activity of GRNP or disturbance of photometric measurement by residual GRNP material. Incubation time amounted to 24 h. For AM and NR8383 cells, seeding and treatment of cells were performed as described above. After treatment, 10% (v/v) AlamarBlue® reagent (Bio-Rad AbD Serotec GmbH, Puchheim, Germany) or AlamarBlue™ Cell Viability Reagent (ThermoFisher Scientific, Brunswick, Germany) was added to both the cell cultures or cell-free background controls for 2 or 8 h, respectively. Supernatants were finally sampled, centrifuged for 11,000× g for 30 min at 4 °C, 4 × 100 µL per sample were transferred to wells of a 96-well plate and were subsequently measured spectrophotometrically using a measured wavelength of 570 nm and a reference wavelength of 600 nm. In the cases of MRC-5 and RAW 264.7 cells, 7000 cells/well were seeded in 96-well plates, pre-cultured for 24 h and then treated with the GRNP dispersions. At the end of treatment, the medium was replaced by 100 µL of 10% (v/v) AlamarBlue® reagent in cell culture medium, and cells were incubated for an additional 2 h before being directly measured. Like for the other cell types, GRNP background controls were used for background correction. After background correction, the percent reduction of AlamarBlue®, as an indicator of cell viability, was calculated according to the formula given by the providers.

2.2.4. Genotoxicity Screening

To look for induction of both DNA strand breaks and oxidative DNA damage, i.e., induction of the pre-mutagenic, oxidative DNA lesion 8 hydroxy-2′-deoxyguanosine (8-OHdG), the human 8-oxoguanine DNA N-glycosylase 1 (hOGG1)-modified alkaline comet assay [31] was performed with both AM and NR8383 cells. Therefore, AM (1.5×10^5 cells) and NR8383 cells (5.6×10^4) were plated in a volume of 0.5 mL in the respective 24-well plates and precultured for 24 h. Then, two cultures per treatment were incubated for 24 h with 25 µg/cm^2 of the various GRNP samples, added to the culture as two-fold concentrated dispersions. KBrO$_3$ (1 mM, 4 h) served as a methodological positive control, cell culture medium as vehicle control, and Al$_2$O$_3$ as a particle-like negative control. Al$_2$O$_3$ was dispersed in cell culture medium by ultrasonication for 15 min using a SONOREX Super RK 514 BH water bath (Bandelin electronics, Berlin, Germany). For gentle detachment, cells were placed on ice for at least 10 min at 4 °C.

The following steps were performed as described previously [32]. After centrifugation, cells were resuspended in 80 µL of 0.75% (w/v) pre-heated low melting point agarose (LMA, peqGold No. 35-2010, Sigma-Aldrich, Taufkirchen, Germany) and applied to purpose-made slides with one roughened surface (Menzel-Gläser, Brunswick, Germany), pre-coated with normal melting agarose. After gelation at 4 °C, one additional layer of 100 µL of 0.75% LMA was applied. Slides were finally immersed in lysis solution [2.5 M NaCl, 100 mM Na$_2$EDTA, 10 mM Tris-HCl, 8 g/L NaOH, 1% Triton-X100, 10% (v/v) dimethyl sulfoxide] and stored overnight at 4 °C. After cell lysis, slides were immersed in enzyme buffer (40 mM HEPES, 100 mM KCl, 0.5 mM Na$_2$EDTA, 0.2 mg/mL bovine serum albumin, pH 8.0) and one of the two slides per treatment was incubated for 12 min at 37 °C in a humidified atmosphere with human hOGG1 by adding 0.16 U/gel of hOGG1 enzyme (New England BioLabs, Ipswich, MA, USA) in 100 µL of enzyme buffer to detect oxidative DNA base modifications. The other slide was incubated in parallel with enzyme buffer only. Electrophoresis was subsequently done using a pre-cooled electrophoresis platform filled with pre-cooled electrophoresis buffer (300 mM NaOH, 1 mM Na$_2$EDTA, pH > 13). DNA was allowed to unwind for 20 min, before 24 V/300 mA were applied for 20 min. Finally, slides were neutralized with 0.4 M Tris-HCl (pH 7.4) and stained with ethidium bromide for at least 1 h. Slides were finally analyzed using an Axioskop fluorescence microscope (Carl

Zeiss, Göttingen, Germany) and the Comet Assay III software (Perceptive Instruments, Bury St Edmunds, UK). The hOGG1-modified comet assay was also used to look for genotoxicity in bronchoalveolar lavage (BAL) cells from 5 animals per treatment group of the inhalation study after 28 days of recovery. Therefore, the cell number in BAL fluid was determined using a Neubauer counting chamber and 1.5×10^5 BAL cells were then subjected to the hOGG1-modified alkaline comet assay as described above. As a follow-up approach, alkaline comet assays without enzyme modification were performed using the $0.5 \times$ BMD30 and BMD30 concentrations for the different materials (see Section 2.2.5) to equalize cytotoxicity.

BMD30 concentrations were also used for genotoxicity screening with RAW 264.7 cells. For RAW 264.7 cells, a slightly different comet assay protocol was used by the University of Aveiro. Briefly, cells were seeded in 6-well plates and incubated for 24 h at 37 °C with GFNs at BMD30 concentrations. Then, 20 µL of cell suspension (prepared in phosphate-buffered saline) were mixed with 70 µL of 1% low melting point agarose in distilled water. Eight drops with 6 µL of cell suspension were placed onto pre-coated slides (approximately 1000 cells). After solidification of agarose at 4 °C, slides were immersed in lysis solution (2.5 M NaCl, 0.1 M EDTA, 10 mM Tris, 1% Triton X-100, pH 10) for 1 h, at 4 °C. After lysis, slides were washed with cold buffer (0.1 M KCl, 0.5 mM EDTA, 40 mM HEPES, 0.2 mg/mL bovine serum albumin, pH 8) and incubated with enzyme buffer for 30 min at 37 °C in a humidified atmosphere. Slides were then immersed for 20 min in cold electrophoresis buffer (200 mM Na_2EDTA, 10 mM NaOH, pH 13) to unwind DNA strands and expose alkali-labile sites. Electrophoresis was then conducted at 0.7 V/cm for 30 min at 4 °C, adjusting the current to 300 mA by raising or lowering the buffer level. Slides were finally neutralized with cold 0.4 M Tris buffer (pH 7.5) for 5 min and left to dry in the dark until staining and analysis. Prior to analysis, slides were hydrated with chilled distilled water for 30 min and stained for 20 min with propidium iodide (10 µg/mL). Slides were subsequently rinsed with distilled water to remove excess stain. DNA strand break induction was analyzed using a fluorescence microscope with $400\times$ magnification and the Comet Assay IV software (Perceptive Instruments, Bury St Edmunds, UK).

Tail intensity (%) was selected by both laboratories as the main measure for DNA damage, as it represents the currently most accepted one. TI is, over a wide range, directly proportional to the number of DNA strand breaks induced. At least one hundred appropriately stained, non-overlapping nuclei were evaluated per treatment. Comets without heads were excluded. An increase in TI on the hOGG1-treated slides, as compared to the slides treated with enzyme buffer only, was indicative of the occurrence of the oxidative base lesion 8-OHdG. The mean of the single-cell data was calculated per slide. The means of three slides per treatment stemming from three independent experiments were finally subjected to statistical analysis.

2.2.5. Calculation of Benchmark Dose

The benchmark dose (BMD) is a dose level estimated from the fitted dose–response curve, associated with a specified change in response, the benchmark response (BMR) [31]. This approach is applicable to all toxicological effects and estimates the dose that causes a low but measurable effect. It can be used as an alternative or in parallel with other dose descriptors (e.g., NOAEL). The benchmark dose lower confidence limit (BMDL) is the lower confidence bound of the BMD, while BMDU represents the upper confidence bound. The BMD approach can be used to derive a point of departure for further risk assessment [33–35]. In this study, the dose–response in vitro data were processed in order to calculate the BMD30, which represents the estimated dose corresponding to 30% of the cell viability reduction, using the PROAST software (version 38.9) as an R package [36]. BMD30 values for the different cells were then used to compare and rank the GRNP based on their cytotoxic potential (the response of 30% was selected due to the generally low cytotoxicity of GRNP, a parameter that allowed comparison of the materials tested).

2.3. In Vivo Validation

2.3.1. Animals

Female Wistar rats [strain: Crl:WI (Han)] delivered by Charles River Deutschland Co. (Sulzfeld, Germany) were used for two animal experiments. The rats were acclimatized for 2 weeks in the animal facility and were aged 9 weeks at dosing. The animal studies had been approved by the competent authority (file # 33.19-42502-04-16/2286/LAVES, Oldenburg, Lower Saxony, Germany). For sacrifice, rats were anesthetized by intraperitoneal administration (0.1 mL per 100 g body weight) of sodium pentobarbital (Narcoren®) and exsanguinated by cutting the vena cava caudalis before preparation of the lungs.

2.3.2. Dose Range Finding (DRF) Study (Intratracheal Instillation)

A DRF test was conducted to define an adequate dosing scheme for the final 28-day inhalation study. The rats were anaesthetized by CO_2/O_2 at $67/33$ (v/v) for some seconds to perform intratracheal instillation of GRNP dispersions in physiological saline (0.9% NaCl). Concurrent controls were treated with vehicle only. Total particle doses were intratracheally instilled using two administrations on consecutive days (each day $\frac{1}{2}$ of the total dose). Total GRNP and CB doses amounted to: P2 low (0.02 mg/lung); P2 high (0.2 mg/lung); P4 low (0.02 mg/lung); P4 high (0.2 mg/lung); Printex 90® (0.2 mg/lung). Dose justification applied a space factor of 10 to mimic no lung overload in the low-dose groups and lung overload in the high-dose groups.

For BAL, prepared lungs were lavaged with 0.9% NaCl, using two lavages with 4 mL each. In the pooled BAL fluid, leukocyte concentration was determined using a manual counting chamber, and two cytoslides were subsequently prepared for differential cell counting (macrophages, neutrophils, eosinophils, lymphocytes). After centrifugation of the BAL fluid, biochemical indicators relevant for the diagnosis of lung tissue damage were determined by standard methodologies in the supernatant [LDH activity, β-Glucuronidase (ß-Glu), total protein (TP)] [37]. The differential cell counts and enzyme/protein analyses were performed as described previously [38].

2.3.3. 28-Day Nose-Only Inhalation Study

The two GRNP species P2 and P4 chosen for the 28-day nose-only inhalation study, based on in vitro screening results, were aerosolized using a dry dispersion system optimized for powdered substances and operated with pressurized air (high-pressure pneumatic disperser). For morphology of the aerosolized GRNP samples P2 and P4 see Figure 2. The disperser was fed under computerized control. The scattering light signal of an aerosol photometer was used to control the feed rate of the dispersion system in order to keep the aerosol concentration in the inhalation unit constant. Actual P2 and P4 concentrations were measured in the breathing zone of the animals. The aerosols were delivered to the rats in a flow-past nose-only inhalation exposure system.

A 28-day nose-only inhalation test was conducted in accordance with [39]. Based on the in vitro results with AM, the two pristine graphene nanoplates showing the highest and lowest toxicity impact were selected for in vivo validation of the in vitro screening results. The SLG graphene factory series sample P2 was chosen for the test material with high adverse activity. It was preferred to the SLG sample P1, because of the affordable price for the considerable amount of test items needed for inhalation exposure (in the gram range). The GNP sample P4 was included as pristine graphene material with the lowest biological activity in AM in vitro. The test design, as depicted in Table 2, allowed for the direct comparison of a pristine SLG vs. a pristine GNP sample and estimation, amongst others, of the impact of graphene layers on potential lung toxicity. As a CB nanoparticle reference, Printex 90® (=P8) completed the pattern of test items.

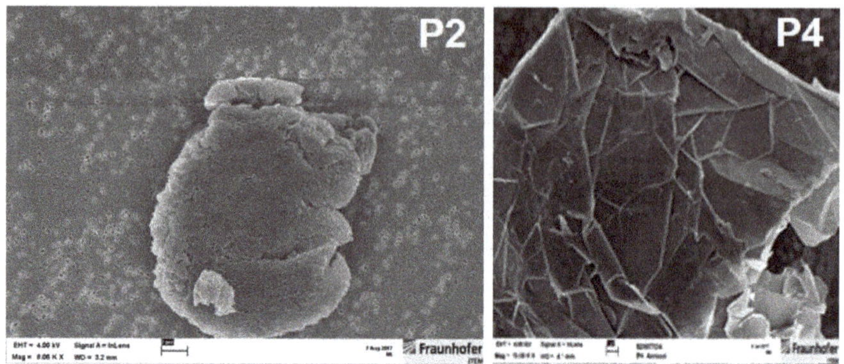

Figure 2. SEM photographs of aerosolized GRNP samples, i.e., P2 (left picture; SLG sample; magnification: 8000×) showing high and P4 (right picture; GNP sample; magnification: approx. 10,000×) showing low toxic potential in the in vitro screening with primary rat alveolar macrophages.

Table 2. Dosing scheme of the 28-day nose-only inhalation test with 28 days of recovery.

	Clean Air	P2 Low	P2 Mid	P2 High	P4 Low	P4 Mid	P4 High	Printex 90®
Aerosol conc. [mg/m^3]	0	0.2	0.8	3.2	0.2	0.8	3.2	3.2
MMAD [μm]	n.a.	1.92 B	2.52 B	3.11 M	-	2.72 B	2.63 B 3.87 B	0.92 M
GSD [–]	n.a.	(2.57)	(2.34)	(3.49)	-	(2.05)	(2.34) (3.12)	(3.53)

GSD: geometric standard deviation; n.a.: not applicable; M: Marple impactor; B: Berner impactor.

Because of the extremely light graphene materials ("mosquito behavior"), the precise measurement of the mass median aerodynamic diameter (MMAD) was challenging. In addition to the standard impactor-type Marple, an impactor of type Berner was also used. With the latter, the gravimetrical analysis is more reliable because higher absolute masses can be sampled on the filters. Measured MMAD values resulted in the ≤ 3 μm range, ensuring respirability of the aerosols.

Calculated doses (Multiple-Path Particle Dosimetry (MPPD) model) in the high-dose groups were similar to those administered in the intratracheal instillation DRF study: rat unisex: 0.2 L/min × 360 min/day × 20 days × 3.2 mg/m^3 × 2.4% (deposition fraction) = 0.11 mg/lung (for comparison: high dose in DRF: 0.2 mg/lung -> actually 0.14 mg/lung following rapid clearance) [40]. For the exact dosing scheme and MMAD values, see Table 2.

The endpoints investigated in BAL fluid were differential cell counts, LDH, ß-Glu and TP, all of which were determined at day 1 and day 29 post-exposure using standard methods. In addition, the cytokines CINC-1 (a marker for macrophage recruitment) and osteopontin (an indicator of inflammation) were analyzed using specific ELISA kits (MesoScale Discovery-MSD; Multiplex) and concentrations of the stable thromboxane A_2 (TXBA$_2$; marker for acute lung injury) metabolite thromboxane B2 (TXB2) were measured by Dr. Dirk Schäfer (TalkingCells c.o. dysantec, Wiesbaden, Germany) using a highly specific competitive ELISA kit (Cayman Chemical, Ann Arbor, MI, USA).

Histopathology of the respiratory tract was performed for the lungs, including all five lung lobes and their main bronchi, the lung-associated lymph nodes (LALN) from the hilar region of the lung (LALN, mediastinal and tracheobronchial), trachea, larynx, pharynx and the nasal cavities including the nasal-associated lymphoid tissue (NALT). For histopathological examination, tissues were fixed in buffered formalin (10%) for up to one day, embedded in paraffin, sectioned, and stained with hematoxylin and eosin (HE). In addition, Masson trichrome staining was used for the detection of connective tissue production within the lung. The nasal cavity was decalcified following formalin fixation prior to paraffin embedding.

For cell cycle analysis in BAL, cell suspensions were centrifuged, the cell pellet fixed in 1 mL of 80% (v/v) ethanol and stored at $-20\,^{\circ}$C until analysis. Cell suspensions were then centrifuged ($300\times g$, 5 min), resuspended in PBS and filtered through a nylon mesh into the test tubes before adding 50 µL of RNAse (Sigma-Aldrich, St. Louis, MO, USA) and 50 µL of propidium iodide (PI, $\geq$94%, Sigma-Aldrich, St. Louis, MO, USA). The samples were then incubated for 20 min at room temperature in the dark. PI-stained cells were finally analyzed on an Attune$^{\circledR}$ Acoustic Focusing Cytometer ((ThermoFisher Scientific, Brunswick, Germany)) and the percentages of cells at G0/G1, S and G2/M phases were determined using the FlowJo software (FlowJo LLC, Ashland, OR, USA).

2.3.4. Statistics

Differences between groups in both the in vitro and in vivo experiments were considered statistically significant at $p < 0.05$. For in vivo endpoints, data were analyzed using analysis of variance. If the group means differed significantly in the analysis of variance, the means of the treated groups were initially compared with the means of the control groups using Dunnett's test [41]. Finally, where indicated, the Student's t-test for unpaired values, two-tailed, combined with normality (Shapiro-Wilk) and equal variance testing (Brown-Forsythe), was used for statistical analysis of most of the in vitro and in vivo endpoints (e.g., differential cell count, the liberation of TXB$_2$, hOGG1-modified alkaline comet assay, LDH release), comparing the treatment groups with the respective vehicle/negative control group using the SigmaPlot 14.0 software (Inpixon GmbH, Düsseldorf, Germany). To estimate the occurrence of oxidative DNA lesions, TI values from hOGG1-treated slides were compared with enzyme buffer-treated slides using the Student's t-test for paired values, two-tailed. The statistical evaluation of the histopathological findings was done using the two-tailed Fisher test and statistical significance of flow cytometry data was calculated using the Dunnett's test [41].

3. Results

3.1. GRNP Dispersion and Sedimentation Kinetics

Material morphology and dispersion grade are critical factors for cell interference in nanomaterial toxicology. For example, MWCNT were previously shown to induce malignant mesothelioma in rats after intraperitoneal administration, with morphology (more straight or curved MWCNT) determining both tumor incidence and the earliest appearance of tumors [42]. Ma-Hock et al. [22] further demonstrated in a 5-day inhalation study that carbon-based materials with different morphology and agglomeration state, i.e., MWCNT, CB, graphene and graphite oxide, behave very differently with regard to pro-inflammatory potential, with no effect for CB and GO, but pro-inflammatory effects for both MWCNT (highest adverse potential) and GNP. Even within the same group of carbon-based nanomaterials, i.e., MWCNT, Reamon-Büttner et al. [43] demonstrated a variable extent of in vitro effects such as membrane damage, inhibition of cell proliferation and induction of gamma-H2A.X pan-stained nuclei in human peritoneal mesothelial LP9 cells. While strong effects were observed for long and straight MWCNTs, no or nearly no effects were seen for tangled-type and milled MWCNTs, thus underlining the importance of material morphology.

In the present study, the dry GRNP powders showed macroscopically different colors and granularity (Figure 3) and exhibited diverse morphologies after dispersion in FCS-containing cell culture medium (Figure 1) as determined by SEM. The microscopic appearance comprised thin and in part transparent morphology for P2 (SLG) and P6 (graphite oxide), more sheet-like morphology for P3 (CG) and P4 (GNP), intensely folded morphology in the case of P1 (SLG) and P5 (GO) and more particle-like morphology for P7 (GNP). When estimating the homogeneity of dispersions and agglomeration state by light microscopy, P1 (SLG), P2 (SLG), P7 (GNP) and P8 (CB) demonstrated more homogenous and denser dispersions than P3 (CG), P4 (GNP), P5 (GO) and P6 (graphite oxide), which showed, irrespective of ultrasonication, diverse and in part big agglomerate/aggregate sizes (Figure 3). For P4, the observed heterogeneity in agglomerate sizes was in line with the particle size distribution given by the supplier, with a D10 of 13.56 µm, a D50 of 48.93 µm and a D90 of 122.2 µm. In addition, the sedimentation kinetics demonstrated marked differences for the various GRNP samples. While no or nearly no sedimentation was observed for P8 (no reduction in turbidity ± 6.71%) and P2 (4.7 ± 4.45%) after 48 h of incubation, P1 showed intermediate (29.6 ± 6.00%) and P7 (75.7 ± 5.85%) and P4 (80.0 ± 6.22%) highest sedimentation rates (Figure 3). Notably, the lowest sedimentation rates correlated with the highest specific surface, higher dispersion homogeneity and lowest biocompatibility in AM.

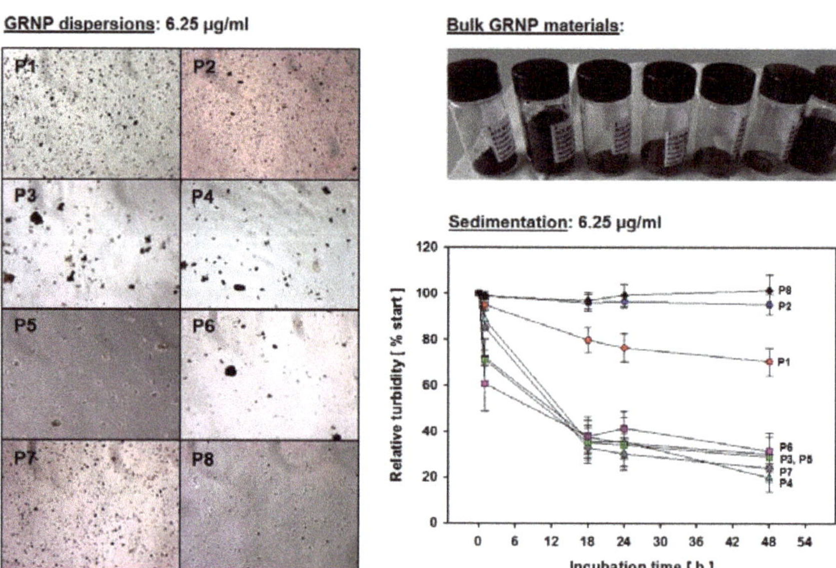

Figure 3. GRNP dispersions. Left: Light microscopic pictures of the different GRNP (P1–P8) dispersions at 6.25 µg/mL corresponding to the lowest concentration used for in vitro screening (3.125 µg/cm^2). Right: Sedimentation kinetics for the various GRNP (P1–P8) at 6.25 µg/mL. GRNP dispersions were placed in semi-micro cuvettes and turbidity (OD600) was measured directly after material suspension (time point zero) and after 1, 18, 24 and 48 h of incubation at 37 °C in a water saturation atmosphere using an incubator. OD600 at time point zero was set to 100% and relative turbidity was calculated for the other time points. Data represent the means ± SD of three independent experiments.

When evaluating the sedimentation kinetics, the question has to be raised whether differences in the extent of effects seen in the comparative in vitro screening experiments are in part a consequence of variable sedimentation rates and variable dispersion grades, leading to highly variable in vitro and intracellular material doses. Irrespective of the same mass-based concentrations, the different material dispersions might also considerably differ in nanoplate number, with, e.g., a higher nanoplate number for the two SLG species, leading to a completely black dispersion with the highest concentration used (50 µg/cm^2). A clear definition of in vitro doses for non-particle-like nanomaterials, i.e., nanoplates and fibers, thus poses a real challenge and should be an important topic for future research along with the quantification of intracellular material in the field of GRNP materials.

3.2. In Vitro Screening

3.2.1. Cytotoxicity Screening

After evaluating different dispersion methods for submerse GRNP exposure of cells, comparative cytotoxicity screening was performed using different lung-relevant cell models, which covered three different species (rat, mouse and human), primary cells and immortalized cell lines and two different cell types, i.e., rat alveolar macrophages/macrophage-like cells and fetal lung fibroblasts, to evaluate the predictivity of the cell models for the in vivo situation after lung exposure.

Cytotoxicity screening was started with the two lung-relevant, normal cell models, i.e., primary rat alveolar macrophages (AM) and human MRC-5 lung fibroblasts (investigated in two independent laboratories), to initially exclude an impact of cell immortalization. Cells were incubated for 24 h with the different GRNP samples at 3.125, 6.25, 12.5, 25 and 50 $\mu g/cm^2$ to enable the calculation of BMD30 values for the final ranking of material adversity. Membrane damage (LDH release), metabolic activity (AlamarBlue® assay), and RICC (cell counts; MRC-5 cells) were chosen as endpoints because of ease of measurement and, thus, good applicability for in vitro screening approaches. LDH activity, furthermore, represented a direct correlate to the in vivo investigations.

In MRC-5 cells, there was no GRNP sample that reproducibly mediated at least a 30% increase in membrane damage or reduction in metabolic activity in MRC-5 fibroblasts after 24 h of incubation in the concentration range tested (see BMD30 values in Figure 4). Despite the replacement of the GRNP dispersions before the final addition of the AlamarBlue® reagent and the use of cell-free background GRNP controls, however, an artificial reduction of the AlamarBlue® reagent seemed to occur, a phenomenon that was also observed in methylthiazoyldiphenyl-tetrazolium bromide (MTT) assays with GO and graphene sheets and human skin fibroblasts [44]. The resulting higher reduction values as compared to the negative control might mask cytotoxic effects. Notably, a proliferation-inhibiting effect of the GRNP materials was observed in MRC-5 cells using RICC and, thus, cell counts as an endpoint. The decrease in RICC in MRC-5 cells seemed to correlate with GRNP thickness, as P5, P3, P2 and P1 demonstrated the lowest BMD30 values (thickness between 0.6 and 1.2 nm), followed by P6 (1–3 nm), P4 (2–10 nm) and P7 (3 nm), and might indicate direct mechanical interference with the cytoskeleton and the mitotic apparatus of the cells, as discussed previously for induction of micronuclei by two types of pristine graphene nanoplatelets in THP-1 cells [45].

In contrast to the MRC-5 lung fibroblasts, both SLG materials (P1 and P2) with the highest specific surface and lowest sedimentation rates mediated strong, concentration-dependent membrane damage in primary rat alveolar macrophages (AM). BMD30 values amounted to as low as 3.22 and 2.47 $\mu g/cm^2$. Some increase in LDH activity was also observed for P5 (GO) and P7 (GNP), with remarkably higher BMD30 values of 39.25 and 45 $\mu g/cm^2$, respectively. All other GRNP samples, including P4 (later used for in vivo validation) as well as CB (P8) samples, exhibited no relevant membrane-damaging effect, as judged by BMD30 values equal to or below 50 $\mu g/cm^2$ (Figure 4). The final ranking was, thus, P2 (SLG) > P1 (SLG) > P5 (GO) > P7 (GNP).

Notably, AM seemed to be finally predictive of the in vivo situation when considering LDH release. The LDH data for AM were clearly in line with subsequent in vivo validation, as a strong effect was noted for P2, but no effect for P4. Disturbance of the cell membrane in AM by GRNP was, furthermore, shown to be time-dependent, with a statistically significant, small effect for P2 after 4 h at 50 $\mu g/cm^2$ (19.7 ± 6.98% cytotoxicity; 72.3 ± 8.32% cytotoxicity after 24 h; Figure S1) only, and there seemed to be some correlation with the specific surface area of the GRNP samples, as a ranking of P1 (BET: 278 m^2/g) > P7 (BET: 195 m^2/g) > P4 (BET: 15 m^2/g) was obvious (Figure 5). Interestingly, P3 (CG) seemed to quench LDH activity under certain conditions. Additional characteristics mediating the stronger membrane-disturbing effect of the SLG samples, compared to the other GRNP, should be further investigated but might comprise a higher particle number, better dispersion or different morphology, with potentially sharper edges, which could mediate membrane damage in the case of cell movements or cell overload or direct cutting of cell membranes.

Test system	Assay	P1	P2	P3	P4	P5	P6	P7	P8
Rat alveolar macrophages	Alamar Blue	50.00	50.00	50.00	50.00	50.00	50.00	50.00	50.00
	LDH	3.22	2.47	50.00	50.00	39.25	50.00	45.00	50.00
MRC-5	Alamar Blue	50.00	50.00	50.00	50.00	50.00	50.00	50.00	50.00
	LDH	50.00	50.00	50.00	50.00	50.00	50.00	50.00	50.00
	RICC	14.10	9.41	3.13	25.80	4.05	18.00	50.00	23.40
NR8383	Alamar Blue	41.40	32.7	38.80	50.00	34.40	50.00	50.00	50.00
	LDH	10.50	14.00	41.40	10.40	14.20	50.00	23.10	50.00
	RICC	20.80	50.00	50.00	12.80	44.20	50.00	50.00	50.00
RAW 264.7	Alamar Blue	46.80	50.00	22.00	25.30	29.40	23.60	50.00	50.00
	LDH	50.00	50.00	50.00	50.00	50.00	50.00	50.00	50.00

High toxicity   Low toxicity

Figure 4. Heatmap of in vitro cytotoxicity screening results. The cytotoxic potential of GRNP was screened in primary rat alveolar macrophages (AM), human MRC-5 lung fibroblasts, the rat alveolar macrophage cell line NR8383 and the murine macrophage-like cell line RAW 264.7. Cells were incubated for 24 h with the various GRNP samples and membrane damage (LDH release), metabolic activity (AlamarBlue® test) and cell proliferation (RICC; MRC-5 and NR8383 cells) were determined. Numbers given represent BMD30 values [$\mu g/cm^2$], calculated from three independent experiments with five concentrations (3.125, 6.25, 12.5, 25 and 50 $\mu g/cm^2$) each. A value of 50 is given where the final BMD30 value was calculated to be greater than 50 $\mu g/cm^2$.

To investigate the reason for the absence of LDH release in the MRC-5 model, GRNP uptake was determined after 24 h of incubation using fluorescence-coupled darkfield microscopy. Fluorescence-coupled darkfield microscopy showed that all GRNP materials were taken up by AM. The cells were more or less tightly packed with GRNP material, located in direct proximity to the cell nucleus (Figure 5). In contrast, when evaluating GRNP uptake in MRC-5 cells, the materials seemed predominantly attached to the cell surface but were not taken up to a marked extent. No obvious differences were observable between the different test materials, except for different cell morphologies and agglomerate/aggregate sizes.

As described for MRC-5 cells, GRNP samples mediated an unspecific reduction of the AlamarBlue® reagent in AM, however, to a lesser extent. This was perhaps based on a different assay protocol with the measurement of the supernatant after centrifugation and not a direct measurement of the cell culture plates. A very slight tendency towards a concentration-dependent decrease in reduced AlamarBlue® was only observed for the SLG sample P2, with 40.6 ± 6.03% at 3.125 $\mu g/cm^2$ as the lowest concentration and 29.4 ± 6.10% at 50 $\mu g/cm^2$ as the highest concentration. A very slight tendency was also observed for P1 and P7, but none of these reached statistical significance, compared to 36.2 ± 3.01% for the negative control. No effect was evident for P4 and all other GRNP.

In further experiments, cytotoxicity of the GRNP panel was investigated in NR8383 and RAW 264.7 cells as immortalized macrophage models to look for an animal-independent, generally available screening model. However, in vivo prediction seemed to be limited for the two cell types in the present study, based on cytotoxicity screening. RAW 264.7 cells showed no induction of membrane damage by the GRNP samples (Figures 4 and 5), and in the AlamarBlue® assay, the GNP sample P4, but not the in vivo active P2 (SLG) sample mediated a decrease in viability (Figure 4). However, this reciprocal ranking, compared to AM, was in line with the subsequent uptake experiment with RAW 264.7 cells, which demonstrated a significantly higher uptake for P4 than for P2. NR8383 cells, in fact, demonstrated a slightly lower BMD30 value for P2 than for P4 in the AlamarBlue® assay, but irrespective of a membrane-damaging potential of GRNP in NR8383 cells, the LDH assay demonstrated comparable BMD30 values for P2 and P4 and, additionally, a reciprocal activity ranking based on RICC results, making predictions difficult (Figure 4).

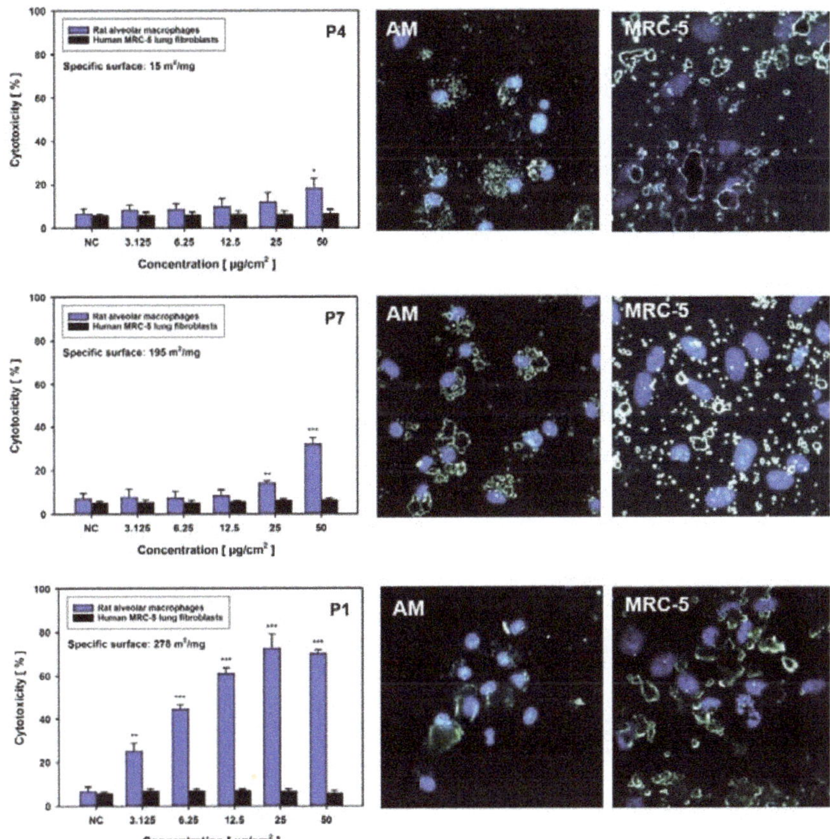

Figure 5. Material-dependent induction of membrane damage and cellular uptake in rat alveolar macrophages (AM; blue bars) and human MRC-5 lung fibroblasts (black bars). Cells were incubated without (NC) or with the given concentrations of P4, P7 or P1 for 24 h, with subsequent measurement of LDH activity in culture supernatants. Data represent arithmetic means $\pm$ SD of three independent experiments. */**/*** Statistically significant difference from NC: $p \leq 0.05$, $p \leq 0.01$ and $p \leq 0.001$, respectively. On the right-hand side, fluorescence-coupled darkfield microscopy pictures are given for both cell types. Cells were incubated for 24 h with P4, P7 or P1 at 6.25 µg/cm^2. Cell nuclei were subsequently stained with DAPI.

The results from AM were in line with previous reports on different cell types, pointing to higher cytotoxicity of pristine graphene and, in particular, SLG or FLG graphene, compared to GO (single-layer), graphite or surface-modified graphene nanoplates. In a previous study, Hinzmann et al. evaluated the impact of different GRNP on the viability of the glioblastoma cell line U87. Notably, as determined by trypan blue exclusion, the graphene sample demonstrated higher cytotoxicity than GO or graphite [46]. Using human bronchial epithelial Beas2B cells and 24 h of incubation, Chatterjee et al., furthermore, showed higher toxicity of pristine FLG (<4 layers) than of GO (single-layer) and CG using the endpoints of colony formation and metabolic activity [47]. Higher cytotoxicity of pristine graphene versus CG was thought to depend on hydrophobic interactions of pristine graphene with the cell membrane, with subsequent intracellular ROS generation and induction of apoptosis, whereas the more hydrophilic CG was taken up by cells without inducing marked cytotoxic effects [48]. The higher membrane-damaging activity of the SLG versus the GNP samples in AM might be based on the number of graphene layers, with higher stiffness and consequently lower activity of the GNP, as previously discussed by Muzi et al. to explain the very low cytotoxic potential of two multi-layer graphene samples in RAW 264.7 cells [49].

3.2.2. Genotoxicity Screening

In the initial cytotoxicity screening, the different macrophage models seemed to represent the more sensitive screening models when compared with human MRC-5 lung fibroblasts, most likely based on limited uptake of GRNP by fibroblasts (Figure 5 and Figure S2). Therefore, genotoxicity screening experiments were performed with the macrophage models only. There are only limited in vitro genotoxicity data, in particular for pristine graphene nanoplates in lung-relevant cell types, and potential modes of action are still under discussion, comprising direct DNA damage by entry into the cell nucleus and cutting or binding of DNA, amongst others, but also indirect DNA damage by the generation of reactive oxygen species by pristine graphene [13,48].

For genotoxicity screening, the hOGG1-modified comet assay with AM and NR8383 cells was used to look for both DNA strand break induction and oxidative DNA damage, i.e., 8-OHdG as a pre-mutagenic DNA lesion. The induction of reactive oxygen species with subsequent oxidative DNA damage has often been discussed as an indirect mode of action for nanomaterials. AM and NR8383 cells were incubated for 24 h with 25 μg/cm^2 with either the different GRNP, culture medium only (negative control), or Al_2O_3 (particle-like negative control). $KBrO_3$ served as technical positive control for both DNA strand break induction and oxidative DNA damage. Slight, but statistically significant induction of DNA strand breaks was observed in AM for all GRNP samples (Figure 6A) compared to the medium (mean TI: 0.29 $\pm$ 0.076%) and the particle-like negative control (Al_2O_3; mean TI: 0.43 $\pm$ 0.159%). The SLG samples P1 (mean TI: 7.93 $\pm$ 4.424%) and P2 (mean TI: 7.89 $\pm$ 5.351%), as the most active GRNP materials in cytotoxicity screening, mediated comparable, approximately 4.5-fold higher mean TI values than P8 (mean TI: 1.56 $\pm$ 0.590%) and P6 (1.79 $\pm$ 1.081%), as the GFN samples with the lowest activity (Figure 6A). In the present study, the arithmetic mean was used as summarizing measure for the cell nuclei analyzed per slide to consider the more bimodal than normal distribution of induced DNA damage in the case of particulate test items, and to enhance sensitivity. Like in AM, the SLG samples P1 and P2 mediated the highest effects in NR8383 cells (Figure 6B).

Higher genotoxic activity of pristine graphene (<4 layers) than of GO (single-layer) at 10 and 50 μg/mL was described previously in lung-relevant Beas2B cells after 24 h of incubation using the alkaline comet assay [3]. Additionally, Hinzmann et al. investigated the induction of DNA strand breaks in U87 cells after 24 h of incubation with subsequent activity ranking of pristine graphene > graphite > GO. Unfortunately, no nanoplate dimensions were given to estimate whether pristine SLG, FLG or GNP were used [46]. The results of the hOGG1-modified comet assay were finally in line with the comet assay data from the BAL cells of the 28-day inhalation study on day 29 post-exposure, with higher induction of DNA strand breaks for P2 than for P4.

None of the GRNP samples induced oxidative DNA lesions in both AM and NR8383 cells, as there were no higher mean TI values for the slides incubated with the DNA repair enzyme hOGG1 (specific for oxidative DNA lesions, i.e., 8-OHdG), compared to the slides incubated with enzyme buffer only. A false-negative result could be excluded based on the strong effect of $KBrO_3$ as a specific positive control for induction of 8-OHdG (Figure 6A,B). However, a trend toward oxidative DNA damage was noted in the BAL cells in the 28-day inhalation study.

As the comet assay is supposed to be sensitive to unspecific effects based on excessive cytotoxicity, follow-up experiments were performed using the BMD30 concentrations calculated from LDH release data to equalize for cytotoxicity. In these experiments, all GRNP samples mediated nearly the same effects, when compared to the respective negative controls in both AM and RAW 264.7 cells (Figure 6C,D). In AM, no GRNP material reached statistical significance, except for P8 (CB). In RAW 264.7 cells, higher TI values were observed for both the negative control and the GRNP-treated cultures. Statistically significant higher TI values were observed for P1–P4. P5 (GO) demonstrated the lowest TI values in both cell types, in RAW 264.7 cells, even comparable to the negative control. The outcome of the control experiments with AM and RAW 264.7 cells, thus, might indicate that GRNP-mediated DNA strand break induction might be of a more unspecific nature, based on cytotoxicity, and less likely due to direct genotoxicity. Alternatively, alignment of the genotoxic potential might indicate an adaptation of particle numbers.

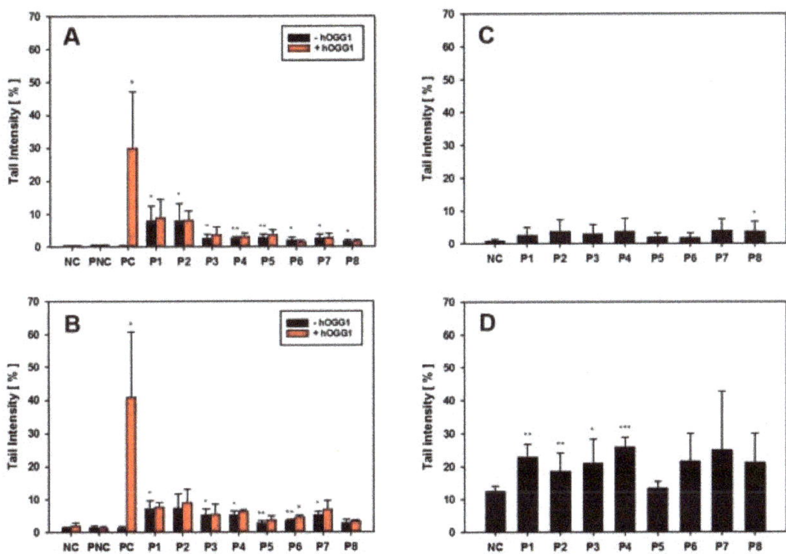

Figure 6. Genotoxicity screening of GRNP materials in AM, NR8383 cells and RAW 264.7 cells. (**A,B**) hOGG1-modified alkaline comet assay. AM (**A**) or NR8383 cells (**B**) were incubated for 24 h without (negative control, NC) or with 25 µg/cm^2 of P1–P8. Al$_2$O$_3$ served as particle-like negative control (PNC) and KBrO$_3$ was used as positive methodological control. Data represent the arithmetic mean tail intensities (TI) ± SD of three independent experiments. Increase in TI on slides with hOGG1 treatment, compared to the respective slides without hOGG1, is indicative of oxidative DNA lesions. (**B,C**) Alkaline comet assay. AM (**C**) or RAW 264.7 cells were incubated for 24 h with or without the calculated BMD30 concentrations of P1–P8. Data represent the arithmetic mean TI ± SD of three independent experiments. For all experiments, the arithmetic mean was used as summarizing measure for the single-cell data on slides. */**/*** Statistically significant difference from PNC (**A,B**) or the respective NC (**C,D**): $p \leq 0.05$, $p \leq 0.01$ and $p \leq 0.001$, respectively; Student's *t*-test for unpaired values, two-tailed. ° Statistically significant difference from slides without hOGG1 treatment: $p \leq 0.05$; Student's *t*-test for paired values, two-tailed.

3.3. In Vivo Validation

From the initial in vitro (geno)toxicity screening of various GRNP it was concluded that AM with the endpoints LDH release and DNA strand break induction might represent a fair and not hypersensitive screening model to predict in vivo lung toxicity of GRNP. To confirm this assumption, in vivo validation was performed with one of the two pristine graphene samples that were most active in vitro in AM, i.e., P2 (SLG; sufficient material available) and P4 (GNP) as the least biologically active pristine graphene sample. Two pristine graphene nanoplates were finally chosen for the 28-day inhalation study, not only due to their strikingly different cytotoxicity in AM in vitro, but also because in vivo lung toxicity data are still sparse for this subgroup of GRNP. This choice, furthermore, enabled comparison of graphene nanoplates with single- (P2) and multi-layer morphology (P4) and thus, different thicknesses (1–5 atomic layers versus 2–10 nm) and resulting stiffness, and investigation of GRNP materials with markedly varying specific surface areas (620 versus 15 m^2/mg).

3.3.1. Dose-Range Finding Test (DRF) with Instillation into Rat Lungs

In the DRF test with intratracheal instillation, P2 showed a very slight yet statistically significant increase in inflammatory lung effects, i.e., an increase in polymorphonuclear neutrophils (PMN; unspecific response) in the low-dose group 3 days after instillation, and a slightly higher and statistically significant increase in PMN values in the high-dose group. In contrast, P4 mediated no significant increase in PMN influx. Notably, in the differential cell counts, strong induction of eosinophils was observed in BAL fluid of the P2 high-dose group, which is normally not detected in the rat model after particle exposure. This pointed to a specific inflammatory response, whereas the PMN showed markedly lower values (Figure 7). In the P2 low-dose group, enzyme activities of LDH and ß-Glu, as well as TP, were not significantly increased, whereas in the P2 high-dose group, a statistically significant increase occurred. This was also observed in the P4 high-dose group, however, to a lower degree (Table S1).

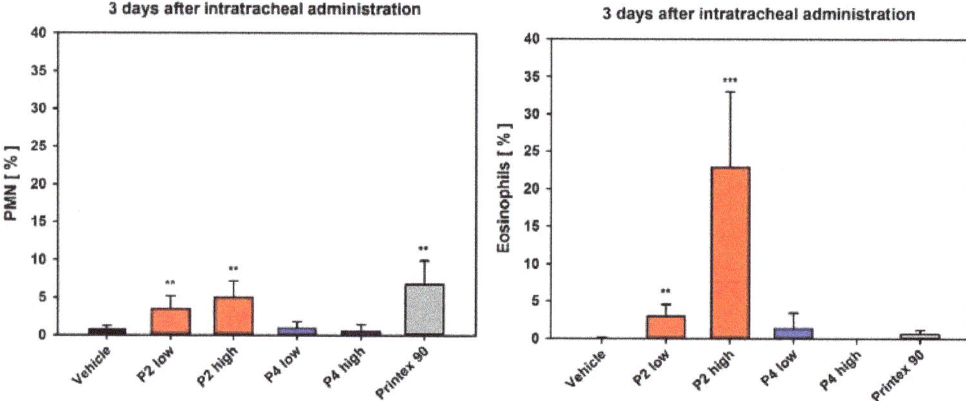

Figure 7. Differential cell counts 3 days after intratracheal instillation of P2 and P4. Note the high eosinophilic percentage, indicating a specific immunological response. Depicted are relative PMN and eosinophil numbers. Data represent means $\pm$ SD of four to five animals per group. **/*** Statistically significant difference from vehicle control (vehicle): $p \leq 0.01$ or $p \leq 0.001$, respectively; Student's *t*-test for unpaired values, two-tailed.

3.3.2. 28-Day Rat Inhalation Test

In the 28-day nose-only inhalation test, BAL fluid analysis in the P2 (SLG) low- and mid-dose groups revealed PMN values at the control level, whereas in the P2 high-dose group, an induction of moderate inflammation was noted. After recovery, the latter group still showed slight inflammation. All three P4 (GNP) groups, as well as the carbon black group, showed PMN values at clean air levels. Surprisingly, induction of eosinophil recruitment, as observed in the DRF study after intratracheal instillation, was not observed after inhalation. In the P2 high-dose group, but not in the P4 high-dose group, LDH levels were significantly increased, mirroring the results observed in the respective in vitro assay with AM (Figure 8).

In the P2 low- and mid-dose groups, BAL fluid analysis showed biochemical results close to clean air control levels. In the high-dose group, however, all three parameters (LDH, β-Glu and TP) were statistically significantly increased. All other groups, i.e., all P4 (GNP) dose groups as well as the Printex 90 (CB) group, remained at control levels. Following recovery, with the exception of the P2 high-dose group (SLG), all biochemical endpoints and PMN percentages, as well as absolute numbers in BAL fluid, had returned to clean air levels. In the P2 high-dose group, however, the PMN moiety still remained at approx. 10% and the biochemical endpoints β-Glu and TP, as well as the absolute PMN numbers, were still significantly increased (Figure 8 and Figure S4).

At day 1 post-exposure, CINC-1 showed clear dose-dependence. When comparing the high-dose groups of P2 and P4, the latter showed a 3.7-fold lower value. Osteopontin results mirrored this finding. At day 29 post-exposure, following a 4-week recovery period, the dose dependence of CINC-1 levels continued at reduced values in the P2 groups, whereas all P4 groups returned to control group values. For osteopontin, increased values were detected in the mid- and high-dose groups of P2 and P4, with P2 again exhibiting maximum values (Figure 8).

As an additional endpoint for the 28-day nose-only inhalation study, the content of TXB_2 as a stable metabolite of the eicosanoid TXA_2 was measured in BAL fluid. A concentration-dependent, highly statistically significant increase in TXB_2 was noted in BAL fluid of the P2 high-dose group at day 1 post-exposure, amounting to about 3-fold higher levels than in the vehicle controls. In contrast, P4 only mediated a very small increase. After 29 days of recovery, the P2 and P4 high-dose groups mediated a comparable, about 2-fold increase in TXB_2, as compared to the saline control (Figure 8, bottom graphs). TXB_2 seemed to represent a promising eicosanoid marker with regard to early detection and differentiation of inflammatory effects of different GRNP in the rat lung. Notably, concentration-dependent induction of TXB_2 liberation by GRNP treatment for 24 h was also observed in preliminary in vitro experiments with NR8383 cells (Figure S3). Increased TXB_2 levels were only seen in pristine graphene nanoplate samples (P1, P2, P4 and P7) and P5 (GO), but were nearly absent for P3 (CG), P6 (graphite oxide) and P8 (CB).

Cell cycle analysis with BAL cells demonstrated a decrease in the percentage of cells at G0/G1 with increasing P2 dose, which for the P2 mid- and P2 high-dose groups was followed by an increase in the percentage of cells at S-phase (Figure 9). This cell cycle profile for P2-exposed animals was mostly maintained after the 29-day recovery period, suggesting that P2 treatment can induce a prolonged S-phase arrest. Exposure to P4 at the high dose also induced a decrease in the percentage of cells at G0/G1 and an increase in the percentage of cells at the S-phase, which was recovered after 29 days of recovery. At this time point, however, a decrease in the percentage of cells at G2 for the P4 mid-dose group was detected. Previous studies have shown that graphene derivates may induce alterations in the cell cycle dynamics. For instance, Hashemi et al. [50] found that nano- and micron-sized GO triggered the cell cycle of fibroblasts and induced an arrest of cells at the S-phase. Furthermore, Wang et al. [51] showed an S-phase block in HEK293T, MCF-7, A549, and HepG2 cells exposed to GO-PEG-PEI. Proper progression through cell division is assured by cell cycle checkpoints that keep the cell from progressing to the next phase of the cell cycle before the prior phase has been completed. The S-phase checkpoint is activated under conditions of threatened DNA replication, such as DNA damage. This activation results in S-phase arrest, inhibiting DNA replication and promoting DNA repair mechanisms, checking the fidelity of DNA replication. Additionally, an increase in the coefficient of variation (CV) of the G0/G1 peak was observed in the P2 high-dose and P8 (CB) groups, which may reflect clastogenic effects (Figure S5).

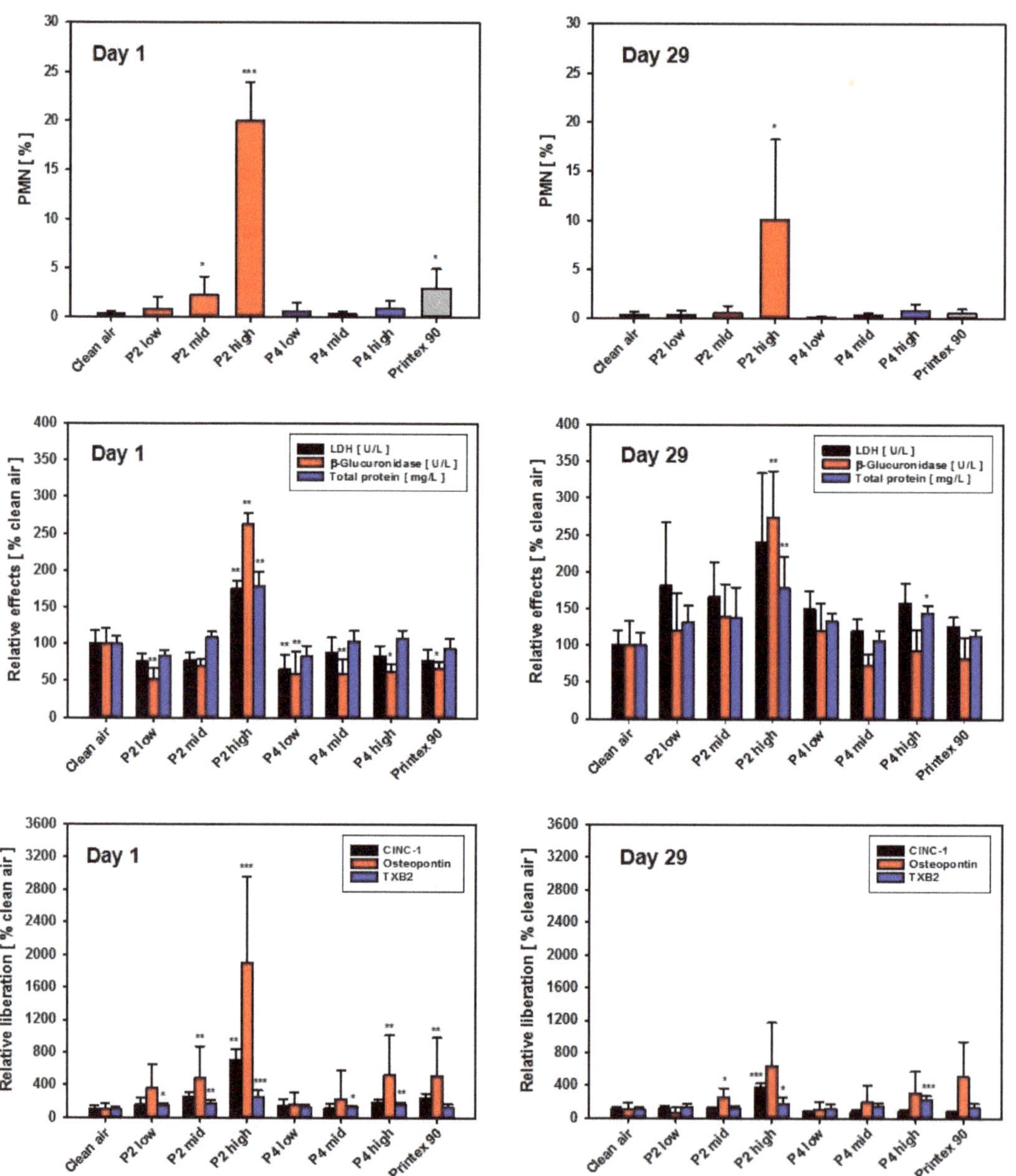

Figure 8. Results of BALF analysis at day 1 (left) and day 29 (right; 28 days of recovery) post-exposure. Depicted from differential cell counts are relative PMN numbers only. Top: PMN [% of total leukocytes]; middle: LDH activity, β-Glu activity, and total protein in BAL [relative to clean air controls set to 100%]; bottom: CINC-1, osteopontin and TXB_2 concentrations in BAL [relative to clean air controls set to 100%]. Data represent the means ± SD of five animals per group. */**/*** Statistically significant difference from negative control (clean air): $p \leq 0.05$ or $p \leq 0.01$ or $p \leq 0.001$, respectively; Student's t-test for unpaired values, two-tailed.

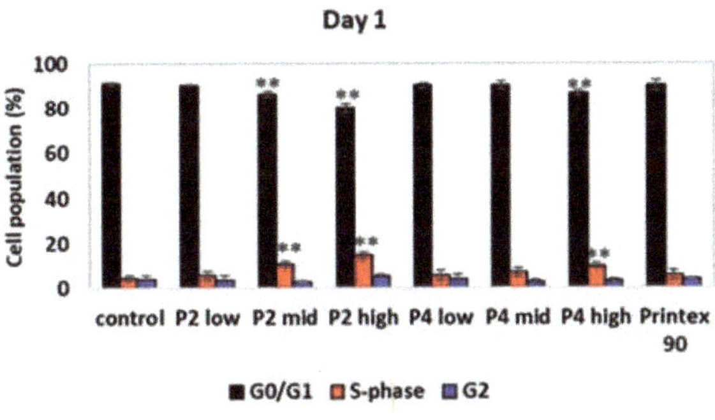

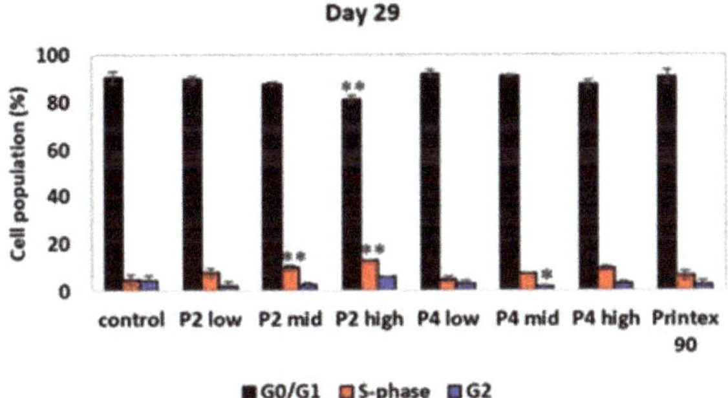

Figure 9. Cell cycle analysis of BAL cells on day 1 and day 29 post-exposure of the 28-day nose-only inhalation study. The distribution of cells in each of the cell cycle phases was based on cell DNA content by staining with propidium iodide. The percentage of cells at each of the cell cycle phases was determined using the FlowJo software. */** Statistically significant difference from control at $p \leq 0.05, p \leq 0.001$, respectively; Dunnett's test.

In the hOGG1-modified comet assay with BAL cells, a concentration-dependent induction of DNA strand breaks for P2, as well as a tendency towards oxidative DNA damage was observed. In contrast, P4- and P8-treated animals demonstrated markedly lower induction of DNA strand breaks after 29 days of recovery (Figure 10). This was in line with the results of the cell cycle analysis with BAL cells (Figure 9 and Figure S5) and the observation of still slight inflammatory effects in the P2 high-dose group after 29 days post-exposure, but the absence of marked inflammation in P4-treated animals and the P8-treated particle-like CB reference, as judged, in particular, by PMN, lymphocyte numbers, enzyme activities and TP in BAL fluid (Figure 8 and Figure S4). A tendency toward the 8-OHdG formation and, thus, an increase in oxidative DNA damage was most likely linked to P2-induced secondary genotoxicity by the inflammation-dependent generation of reactive oxygen species.

Histopathological examination of the respiratory tract revealed findings related to the inhalation of the GRNP samples P2 (SLG), P4 (GNP) and P8 (CB particle-like reference). Findings were observed in the lung, the lung-associated lymph nodes and the nasal cavity. In the nasal cavity, eosinophilic globules were seen in a multifocal pattern in the olfactory epithelium, most prominently in the P2 high-dose group, with a slight increase during the 29-day recovery period (Tables 3 and 4).

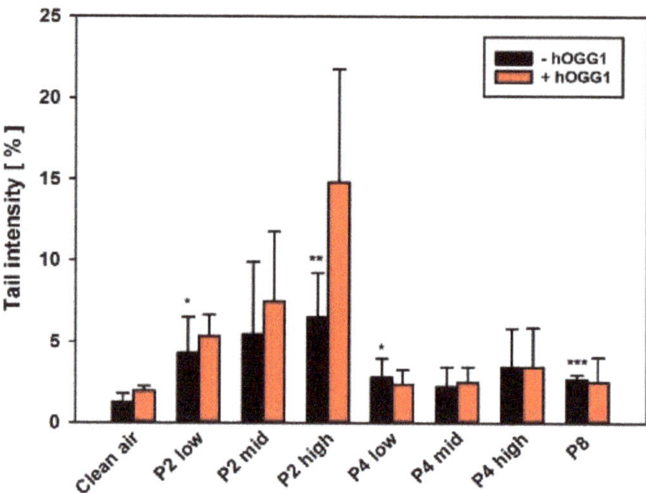

Figure 10. The hOGG1-modified comet assay with BAL cells on day 29 after the last inhalation. AM were isolated by alveolar lavage and subsequently analyzed regarding induction of DNS strand breaks and oxidative DNA lesions. Data represent the arithmetic mean tail intensities (TI) ± SD of 5 animals per treatment group, using the arithmetic mean as summarizing measure for the single-cell data on slides. An increase in TI on slides with hOGG1 treatment, as compared to the respective slides without hOGG1 incubation, is indicative of induction of oxidative DNA lesions, i.e., 8-OHdG. */**/*** Statistically significant difference from clean air control: $p \leq 0.05$, $p \leq 0.001$ and $p \leq 0.001$, respectively; Student's *t*-test for unpaired values, two-sided.

Table 3. GRNP-related histopathological findings at day 1 post-exposure.

Test Material Group				Clean Air	P2 Low	P2 Mid	P2 High	P4 Low	P4 Mid	P4 High	P8
	Number of Animals			5	5	5	5	5	5	5	5
Lung	Accumulation of particle-laden macrophages	Alveolar	Occ.	0	5	5	5	2	4	5	5
			Grade	0	1	1.2	2	0.4	0.8	1.2	1.8
		Interstitial	Occ.	0	0	0	1	0	0	0	1
			Grade	0	0	0	0.2	0	0	0	0.2
		BALT	Occ.	0	0	0	2	0	0	0	4
			Grade	0	0	0	0.4	0	0	0	0.8
	Particle-laden giant cells (syncytia)		Occ.	0	0	0	5	0	0	1	0
			Grade	0	0	0	1	0	0	0.2	0
	Infiltration granulocytes	Alveolar	Occ.	0	0	0	4	0	0	0	0
			Grade	0	0	0	0.8	0	0	0	0
LALN	Accumulation of particle-laden macrophages		Occ.	0	0	0	0	0	0	0	3
			Grade	0	0	0	0	0	0	0	0.6
Nasal cavity	Eosinophilic globules	Olfactory epithelium	Occ.	0	1	1	3	0	0	1	0
			Grade	0	0.2	0.2	0.6	0	0	0.2	0

Occ. = Occurrence: number of animals showing the findings in the respective group; Grade: mean group grade of this lesion. Grades can range from 0 (normal) up to 5 (very severe change):

Table 4. GRNP-related histopathological findings at day 29 post-exposure.

Test Material Group				Clean Air	P2 Low	P2 Mid	P2 High	P4 Low	P4 Mid	P4 High	P8
Number of Animals				5	5	5	5	5	5	5	5
Lung	Accumulation of particle-laden macrophages	Alveolar	Occ.	0	3	5	5	1	3	5	5
			Grade	0	0.6	1	1.6	0.2	0.6	1	1.4
		Interstitial	Occ.	0	0	0	3	0	0	0	1
			Grade	0	0	0	0.6	0	0	0	0.2
		BALT	Occ.	0	0	1	0	0	0	0	5
			Grade	0	0	0.2	0	0	0	0	1
	Particle-laden giant cells (syncytia)		Occ.	0	0	1	5	0	0	0	0
			Grade	0	0	0.2	1.6	0	0	0	0
	Infiltration granulocytes	Alveolar	Occ.	0	0	0	1	0	0	0	0
			Grade	0	0	0	0.2	0	0	0	0
LALN	Accumulation of particle-laden macrophages		Occ.	0	0	0	4	0	0	0	2
			Grade	0	0	0	0.8	0	0	0	0.4
Nasal cavity	Eosinophilic globules	Olfactory epithelium	Occ.	0	0	1	5	1	0	3	1
			Grade	0	0	0.2	1.4	0.2	0	0.6	0.2

Occ. = Occurrence: number of animals showing the findings in the respective group; Grade: mean group grade of this lesion. Grades can range from 0 (normal) up to 5 (very severe change):

0	1	2	3	4	5

The lung showed the dose-dependent occurrence of particle-laden macrophages in the alveoli, lung interstitium (Figures 11 and 12), and the bronchus-associated lymphoid tissue (BALT). Notably, some macrophages fused time-dependently to giant cells (syncytia), which were visible mainly in the P4 high-dose group (Figures 11 and 12 and Tables 3 and 4). Additionally, alveolar infiltration of granulocytes occurred in the P2 high-dose group on day 1 post-exposure after 28 days of inhalation, declining during the recovery period (Tables 3 and 4). In the lung-associated lymph nodes, accumulation of particle-laden macrophages was found in the P8 carbon black-treated group at day 1 post-exposure after 28 days of inhalation, as well as in the P2 high-dose and the P8 groups after 29 days of recovery (Tables 3 and 4). A histopathological finding that was interpreted as adverse was alveolar infiltration of granulocytes, which was observed only in the P2 high-dose group, but not in any of the P4-treated groups and, thus, only for the pristine graphene SLG sample, but not for the pristine graphene GNP sample. The other histopathological findings observed in the present study were interpreted to represent adaptive non-adverse findings.

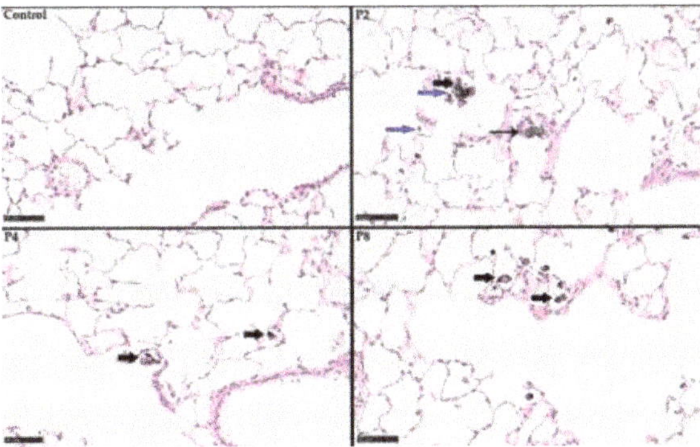

Figure 11. Histopathological findings at day 1 post-exposure following 28 days of inhalation. Control without any histopathological findings. P2 (SLG) high-dose group with particle-laden macrophages (bold black arrows), particle-laden giant cells (syncytia; black arrow) and infiltration of granulocytes (blue arrow). P4 (GNP) high-dose group with particle-laden macrophages (bold black arrows). P8 (CB) also with particle-laden macrophages (bold black arrows).

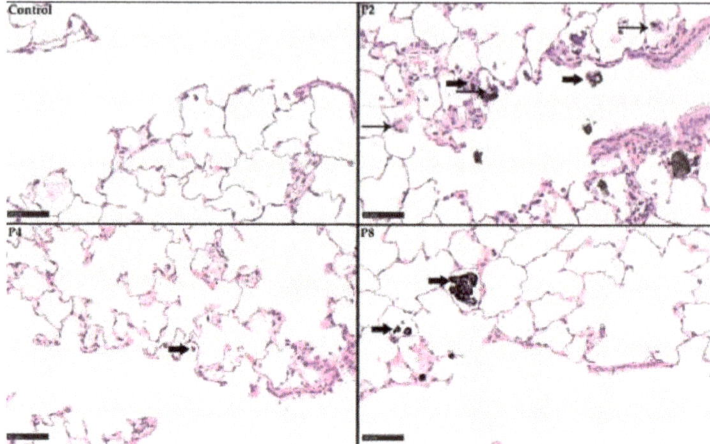

Figure 12. Histopathological findings at day 29 post-exposure following 28 days of inhalation. Control without any histopathological findings. P2 (SLG) high-dose group with particle-laden macrophages (bold black arrows) and particle-laden giant cells (syncytia; black arrow). P4 (GNP) high-dose group with particle-laden macrophages (bold black arrows). P8 (CB) also with particle-laden macrophages (bold black arrows).

4. Discussion

In the present study, we aimed to define a combined in vitro/in vivo approach to predict adverse lung effects of graphene-related nanoplates (GRNP), in particular, pristine single-layer graphene (SLG) and pristine graphene nanoplatelets (GNP), some of which have previously been shown to exhibit at least in vitro cytotoxic and genotoxic potential, depending, amongst others, on the specific material characteristics, the cell model used, the endpoints, concentrations and incubation times. For the prediction of adverse lung effects of graphene materials, it would, nevertheless, be highly desirable to have meaningful and differentiating in vitro screening approaches in place to enable prescreening of new materials before starting in vivo evaluations.

Up to now, a variety of in vitro assays have been established to predict the in vivo lung toxicity of engineered nanoparticles in subacute intratracheal instillation or inhalation studies in rats. Often, commercially available permanent lung epithelial or alveolar macrophage cell lines are used as model systems. Cells are then exposed in a submerse manner to nanomaterial dispersions, or the test items are dosed as aerosols using air-liquid interface approaches. Acute lung inflammogenicity in a rat instillation model had previously been compared with certain in vitro toxicity endpoints (comprising cytotoxicity, pro-inflammatory cytokine expression or hemolytic potential) by [52]. Nanoparticles acting via soluble toxic ions (e.g., ZnO) showed positive results in most of the assays and were consistent with the lung inflammation data, whereas low soluble dust showed a good correlation only in the hemolysis assay. Wiemann et al. [53] selected the permanent rat alveolar macrophage cell line NR8383, which was also used in the present study, to investigate up to 18 inorganic nanoparticles. The authors were able to sort the test item panel into toxic and non-toxic subgroups and recommend only the biologically active nano-dusts for subsequent in vivo testing. Gorki et al. [54] described mouse ex vivo cultured alveolar macrophages (AM) as a suitable model, maintaining typical morphological features and expressing AM surface markers.

A combined in vitro/in vivo approach was reported by Sayes et al. [55], who compared three different cell culture systems, i.e., ((i) immortalized rat L2 lung epithelial cells; (ii) primary rat alveolar macrophages; (iii) co-cultures of (i) and (ii) regarding their capability to predict the cytotoxic potential and inflammogenicity of various samples of dust in vivo (here using intratracheal instillation in the rat model). The authors only found an unsatisfactory correlation. However, they were still convinced that the approach could be successful in principle but concluded that there was a need for better culture systems.

In the PLATOX project presented here, the aim was to establish an unpretentious yet robust and well reproducible in vitro cell model able to deliver a good correlation of results as compared to short-term inhalation tests in rats. Additional considerations were to focus on a stable, easily accessible cell system, with preferably primary, non-transformed cells. In this context, AM, representing primary cells of the lung, were chosen and harvested from non-treated rats. These cells were finally shown to exhibit better predictivity than the immortalized macrophage models. The combination of in vitro screening assays included both cytotoxicity (membrane damage, cell proliferation and metabolic activity) and DNA damage (DNA strand break induction) testing, thus covering decisive entry-level tests to allow the commercialization of new materials. Notably, in contrast to cytotoxicity screening, both AM and NR8383 cells were able to predict induction of DNA strand breaks in BAL cells, observed in vivo in the 28-day inhalation study on day 29 post-exposure. However, due to the monoculture type of the macrophage models, the in vitro models were not able to predict the trend toward oxidative DNA damage, most likely representing secondary genotoxicity due to ongoing inflammation. Irrespective of the chemical nature of pristine graphene, which is rather inert in principle, genotoxicity is nevertheless a topic to be considered. However, the exact mechanisms, besides secondary genotoxicity based on inflammation and production of reactive oxygen species, still have to be determined [14,56]. Notably, in a study by Ursini et al., some genotoxicity was detected in workers with occupational exposure to graphene by using the FPG-modified comet assay and analysis of micronucleus frequencies [57].

The induction of micronuclei by pristine graphene nanoplates and, thus, of fixed DNA damage was furthermore demonstrated in THP-1 cells, underlining the need to include genotoxicity testing in the screening of pristine graphene nanoplates [45]. The same robust endpoints could be mirrored and validated in the in vivo inhalation testing using cytotoxicity, differential cell count in lung lavage fluid as well as histopathological equivalents.

In the DRF study with intratracheal instillation, the single-layer graphene P2 induced a statistically significant increase in eosinophils at day 3 post-treatment, whereas carbon black did not (Figure 3). Interestingly, this effect was not observed in the subsequent 28-day inhalation study. An increase in eosinophils in BAL fluid is not regularly found in powder studies with rats; however, a paper by Lee et al. [58] reported similar effects after intratracheal instillation of nickel oxide nanoparticles. Inflammatory cells were evaluated at days 1, 2, 3 and 4 post-treatment. The authors concluded that NiO nanoparticles in rats induced a unique mixed type of neutrophilic and eosinophilic inflammation 3 and 4 days after instillation, which was consistent with the inflammation by $NiCl_2$ at day 1 after instillation. The mechanism of eosinophilia recruitment by nano-NiO seemed to be based on the direct rupture of cells, releasing a significant level of intracellular eotaxin. Notably, GRNP, in particular those with a low layer number, are thought to exhibit sharp edges, potentially enabling membrane damage. Perhaps a bolus application by intratracheal instillation might have triggered a NiO-comparable effect for the P2 SLG sample. A strong, concentration-dependent membrane-damaging effect was also observed for P2 in the AM in vitro model.

Moghimian and Nazarpour [59] reported an acute inhalation test performed under Good Laboratory Practice (GLP) at Charles River Laboratories Montreal using graphene powder with 6–10 graphene layers. The maximally feasible aerosol concentration of $1900 \, mg/m^3$ was found as the NOAEL value, thus showing the same low in vivo toxicity as found for the graphene nanoplatelet sample P4 in the present study. Kim et al. [21] investigated a commercial multilayer graphene (Cabot Corp, USA; GPX-205; 20–30 layers) in a 28-day inhalation test at aerosol concentrations of 0.12, 0.47 and $1.88 \, mg/m^3$. No dose-dependent effects were recorded for body weights, organ weights, BAL fluid inflammatory markers and blood biochemical parameters at days 1 and 28 post-exposure. No distinct lung pathology was observed at days 1, 28 and 90 post-exposure, suggesting low toxicity and a NOAEL of no less than 1.88 mg/m^3. These results are also in agreement with the outcome of P4 in the 28-day inhalation study in the present PLATOX project that used a similar dosing scheme.

In another study, mice were exposed to graphene nanoplates by pharyngeal aspiration at doses of 4–40 µg/mouse [60]. Three types with lateral sizes of 1, 5 or 20 µm were analyzed. At the high dose, increased lung inflammation was induced in lavage fluid to a higher degree by the 5- and 20-µm-sized platelets (G5, G20) than observed for the 1-µm-sized (G1) and the carbon black reference. G5 and G20 showed no or minimal lung epithelial hypertrophy and hyperplasia and no development of fibrosis at 2 months post-exposure. Interestingly, with regard to fibrosis, the MRC-5 lung fibroblasts, used as one in vitro screening model in the present study, were shown to lack significant uptake of GRNP and induction of membrane damage.

For the new material class of GRNP, appropriate risk characterization is presently missing due to the limited database. In the PLATOX project, the highest biological response both in vitro and in vivo was detected for the single-layer graphene. All other GRNP showed a weaker or irrelevant toxic impact. According to literature searches, the modified GRNP (e.g., CG, GO) of the in vitro test item set are relatively well toxicologically characterized, with, e.g., CG exhibiting, in contrast to pristine graphene platelets, a higher hydrophilicity, supposed to result in a lower disturbance of cells and subsequent cytotoxicity [44], and to enable better clearance. Therefore, it was decided to use another non-functionalized, i.e., the multi-layer graphene P4 (GNP), as a counterpart for the final validation test in vivo.

Thus, the main output of this study is a justified step forward toward a validated in vitro (geno)toxicity screening tool for unmodified/non-functionalized graphene nanoplate species based on AM as a cell model. The screening tool, however, should be further supplemented by endpoints that more concretely mirror the liberation of pro-inflammatory mediators, with eicosanoids, TXB2 in particular, representing a promising candidate. Additionally, with regard to the dispersion of nanomaterials in protein-containing media for in vitro screening, protein corona should be considered to potentially have an impact on screening outcomes. Any pristine particle interacting with a biological medium forms a protein corona on its surface. However, its potential influence on particle toxicity is still strongly under debate, as the protein corona is not equivalent to a covalent technical coating that could hide particle surface-mediated toxicity excessively in the respiratory tract or in cell line assays [61].

Finally, the screening approach used predicted conclusively the adverse findings observed in the in vivo inhalation test. Based on these findings, the following toxicological ranking of single-layer (SLG) vs. multilayer graphene (GNP) was derived, as compared to the carbon black reference (Printex 90): SLG > MLG > P90. Based on BAL fluid analysis (PMN percentage) and histopathological examination (granulocytic infiltration), a NOAEC of 0.8 mg/m^3 was finally derived for the investigated SLG. The evaluated GNP sample and CB reference, however, were weaker in the toxic response and the calculated NOAEC amounted to >3.2 mg/m^3.

5. Conclusions

The present study has shown that the graphene nanoplatelet sample P4 (GNP; multi-layer graphene) had almost no lung toxicity/pro-inflammatory potential following a 28-day inhalation study, in contrast to a moderate effect found with the single-layer graphene nanoplates P2 (SLG). When considering these in vivo results with a clear cytotoxic, genotoxic and pro-inflammatory potential of SLG, but less activity of GNP after 28 days of inhalation, alveolar macrophages, as primary rat cells derived from the respiratory tract, showed the highest reliability in predicting the in vivo adverse lung effects of the tested pristine graphene nanoplates, as members of the graphene-related two-dimensional (2D) nanomaterials group (GRNP). This includes lactate dehydrogenase (LDH) release and the induction of DNA strand breaks as the most meaningful endpoints. It can be concluded from the study results that, in principle, a predictive in vitro lung-focused toxicity screening of GRNP seems possible. However, relevant cell types and endpoints, as well as appropriate culture conditions, incubation times, and concentrations, should be chosen. Furthermore, the photometric and immunological detection methods should always be carefully adapted to GRNP properties.

Supplementary Materials: The following supporting information can be downloaded at: https: //www.mdpi.com/article/10.3390/nano12081254/s1, Figure S1. Time-dependent induction of membrane damage in primary rat alveolar macrophages by incubation with P2 (SLG); Figure S2. Uptake of GRNP by NR8383 and RAW 264.7 cells, as assessed by light microscopy or flow cytometry, respectively; Figure S3: TXB2 release from NR8383 cells, preliminary data; Figure S4: Differential cell counts in BAL fluid at day 1 and day 29 post-exposure of the 28-day nose-only inhalation study with the P2 (SLG), P4 (GNP) and P8 (CB) materials, in absolute cell numbers; Figure S5: Coefficient of variation (CV) of G0/G1 cell cycle peak of BALF cells from the 28-day nose-only inhalation study estimated by the FlowJo software. Table S1. Biochemical parameters analyzed in the BAL fluid of the DRF in vivo study.

Author Contributions: Conceptualization: O.C., C.Z., H.O. and L.F.; methodology: D.S., A.M., A.C.M., T.T. and S.B.; investigation: D.S., A.M., A.C.M., T.T. and S.B.; resources: O.C., C.Z., H.O. and L.F.; writing—original draft preparation: O.C., C.Z., D.S., H.O. and L.F.; writing—review and editing: O.C., C.Z., D.S., H.O. and L.F.; visualization: L.F., C.Z., D.S. and H.O.; project administration: O.C., C.Z., H.O. and L.F.; funding acquisition: O.C., C.Z., H.O. and L.F. All authors have read and agreed to the published version of the manuscript.

Funding: This joint research was funded within the European FP7 SIINN ERA-NET program on Nanosafety under the acronym PLATOX by the respective national public funding bodies, i.e., the German Federal Ministry of Education and Research (project number: 03XP0007), the Portuguese Fundação para a Ciência e a Tecnologia for the funding of the ERA-SIINN/0003/2013 project, and the Romanian Executive Unit for the Financing of Higher Education, Research, Development and Innovation (UEFISCDI) (contract number: 14/2015). H.O. thanks FCT for the research contract under Scientific Employment Stimulus (CEECIND/04050/2017).

Institutional Review Board Statement: The animal study protocol was approved by the Institutional Review Board (or Ethics Committee), the competent authority (file # 33.19-42502-04-16/2286/LAVES, Oldenburg, Lower Saxony, Germany) for Fraunhofer ITEM; date of approval: 25-Oct-2016).

Data Availability Statement: Not applicable.

Acknowledgments: We thank A. Westendorf, R. Korolewitz, G. Kühne, H.C. Brockmeyer, J. Wallus and H. Rahmer for their excellent technical assistance and Karin Schlemminger for proofreading (all Fraunhofer ITEM).

Conflicts of Interest: The authors declare no conflict of interest.

Abbreviations

AM	Primary rat alveolar macrophages
BAL	Bronchoalveolar lavage
β-Glu	β-Glucuronidase
BMD	Benchmark dose
BMD30	Benchmark dose 30%
BMR	Benchmark response
BMDL	Benchmark dose lower confidence limit
BMDU	Upper confidence limit of BMD
CB	Carbon black
CG	Carboxyl graphene
CNT	Carbon nanotubes
1D	One-dimensional
2D	Two-dimensional
3D	Three-dimensional
DAPI	4′,6-diamidino-2-phenylindole
DMEM	Dulbecco's Modified Eagle Medium
DRF	Dose range finding
EMEM	Eagle's Minimal Essential Medium
FCS	Fetal calf serum
FLG	Few-layer graphene
GFN	Family of graphene-based materials
GNP	Graphene nanoplatelets
GRNP	Graphene-related nanoplates
GO	Graphene oxide
GSD	Geometric standard deviation
HE	Hematoxylin and eosin
hOGG1	Human 8-oxoguanine DNA N-glycosylase 1
LALN	Lung-associated lymph nodes
LDH	Lactate dehydrogenase
LMA	Low melting point agarose
LPS	Lipopolysaccharide
MLG	Multilayer graphene
MMAD	Mass median aerodynamic diameter
MPPD	Multiple-Path Particle Dosimetry
MWCNT	Multiwalled carbon nanotubes
n.a.	Not applicable
NALT	Nasal-associated lymphoid tissue
NC	Negative control
NOAEL	No observed adverse effect level
8-OHdG	8-hydroxy-2′-deoxyguanosine
PI	Propidium iodide
PMN	Polymorphonuclear neutrophils
rGO	Reduced graphene oxide
RICC	Relative increase in cell count
SD	Standard deviation
SEM	Scanning electron microscopy
SLG	Single-layer graphene
TI	Tail intensity
TP	Total protein
TXA_2	Thromboxane A_2
TXB2	Thromboxane B_2

References

1. Kiew, S.F.; Kiew, L.V.; Lee, H.B.; Imae, T.; Chung, L.Y. Assessing biocompatibility of graphene oxide-based nanocarriers: A review. *J. Control. Release* **2016**, *226*, 217–228. [CrossRef] [PubMed]
2. Jastrzębska, A.M.; Kurtycz, P.; Olszyna, A.R. Recent advances in graphene family materials toxicity investigations. *J. Nanopart. Res.* **2012**, *14*, 1320. [CrossRef] [PubMed]
3. Chatterjee, N.; Yang, J.; Choi, J. Differential genotoxic and epigenotoxic effects of graphene family nanomaterials (GFNs) in human bronchial epithelial cells. *Mutat. Res. Genet. Toxicol. Environ. Mutagen.* **2016**, *798*, 1–10. [CrossRef] [PubMed]
4. European Union Observatory for Nanomaterials (EUON); European Chemicals Agency (ECHA). Nanopinions. 18 January 2022. Available online: https://euon.echa.europa.eu/nanopinion/-/blogs/what-does-graphene-really-look-like-and-why-is-it-not-carbon-nanotubes-%2021.%20Januar%202022.%2013:25 (accessed on 21 January 2022).
5. Singh, Z.S. Applications and toxicity of graphene family nanomaterials and their composites. *Nanotechnol. Sci. Appl.* **2016**, *15*, 1. [CrossRef] [PubMed]
6. *ISO/TS 80004-13:2017*; A Nanotechnologies–Vocabulary–Part 13: Graphene and Related Two-Dimensional (2D) Materials. Article No.: 225120; VDE-Verlag: Berlin, Germany, 2017.
7. Tang, F.; Gao, J.; Ruan, Q.; Wu, X.; Wu, X.; Zhang, T.; Liu, Z.; Xiang, Y.; He, Z.; Wu, X. Graphene-Wrapped MnO/C Composites by MOFs-Derived as Cathode Material for Aqueous Zinc ion Batteries. *Electrochim. Acta* **2020**, *353*, 136570. [CrossRef]
8. Zhang, Z.; Cai, R.; Long, F.; Wang, J. Development and application of tetrabromobisphenol A imprinted electrochemical sensor based on graphene/carbon nanotubes three-dimensional nanocomposites modified carbon electrode. *Talanta* **2015**, *134*, 435–442. [CrossRef]
9. Sanchez, V.C.; Jachak, A.; Hurt, R.H.; Kane, A.B. Biological Interactions of Graphene-Family Nanomaterials: An Interdisciplinary Review. *Chem. Res. Toxicol.* **2012**, *25*, 15–34. [CrossRef]
10. European Chemicals Agency (ECHA). Substance InfoCard. 2022. Available online: https://echa.europa.eu/substance-information/-/substanceinfo/100.227.924 (accessed on 21 January 2022).
11. European Chemicals Agency (ECHA). Graphene. Available online: https://echa.europa.eu/registration-dossier/-/registered-dossier/24678/1/1 (accessed on 21 January 2022).
12. Guo, X.; Mei, N. Assessment of the toxic potential of graphene family nanomaterials. *J. Food Drug Anal.* **2014**, *22*, 105–115. [CrossRef]
13. Ruiz, A.; Lucherelli, A.; Murera, D.; Lamon, D.; Ménard-Moyon, C.; Bianco, A. Toxicological evaluation of highly water dispersible few-layer graphene in vivo. *Carbon* **2020**, *170*, 347–360. [CrossRef]
14. Wu, K.; Zhou, Q.; Ouyang, S. Direct and Indirect Genotoxicity of Graphene Family Nanomaterials on DNA—A Review. *Nanomaterials* **2021**, *11*, 2889. [CrossRef]
15. Spinazzè, A.; Cattaneo, A.; Borghi, F.; Del Buono, L.; Campagnolo, D.; Rovelli, S.; Cavallo, D.M. Exposure to airborne particles associated with the handling of graphene nanoplatelets. *Med. Lav.* **2018**, *109*, 285–296. [PubMed]
16. Syama, S.; Mohanan, P.V. Safety and biocompatibility of graphene: A new generation nanomaterial for biomedical application. *Int. J. Biol. Macromol.* **2016**, *86*, 546–555. [CrossRef] [PubMed]
17. Schinwald, A.; Murphy, F.A.; Jones, A.; MacNee, W.; Donaldson, K. Graphene-based nanoplatelets: A new risk to the respiratory system as a consequence of their unusual aerodynamic properties. *ACS Nano* **2012**, *6*, 736–746. [CrossRef] [PubMed]
18. Bellagamba, I.; Boccuni, F.; Ferrante, R.; Tombolini, F.; Marra, F.; Sarto, M.S.; Iavicoli, S. Workers' exposure assessment during the production of graphene nanoplatelets in R&D laboratory. *Nanomaterials* **2020**, *10*, 1520.
19. Pelin, M.; Sosa, S.; Prato, M.; Tubaro, A. Occupational exposure to graphene based nanomaterials: Risk assessment. *Nanoscale* **2018**, *10*, 15894–15903. [CrossRef]
20. Fadeel, B.; Bussy, C.; Merino, S.; Vasquez, E.; Flahaut, E.; Mouchet, F.; Evariste, L.; Gauthier, L.; Koivisto, A.J.; Vogel, U.; et al. Safety assessment of graphene-based materials: Focus on human health and the environment. *ACS Nano* **2018**, *12*, 10582–10620. [CrossRef]
21. Kim, J.K.; Shin, J.H.; Lee, J.S.; Hwang, J.H.; Lee, J.H.; Baek, J.E.; Kim, T.G.; Kim, B.W.; Kim, J.S.; Lee, G.H.; et al. 28-day inhalation toxicity study of graphene nanoplatelets in Sprague-Dawley rats. *Nanotoxicology* **2016**, *10*, 891–901. [CrossRef]
22. Ma-Hock, L.; Strauss, V.; Treumann, S.; Küttler, K.; Wohlleben, W.; Hofmann, T.; Gröters, S.; Wiench, K.; van Ravenzwaay, B.; Landsiedel, R. Comparative inhalation toxicity of multi-wall carbon nanotubes, graphene, graphite nanoplatelets and low surface carbon black. *Part. Fibre Tox.* **2013**, *10*, 23. [CrossRef]
23. Vision Statement: SRA [Internet]. 2017. Available online: https://www.sra.org/vision-statement (accessed on 20 February 2017).
24. Porter, D.; Sriram, K.; Wolfarth, M.; Jefferson, A.; Schwegler-Berry, D.; Andrew, E.; Castranova, V. A biocompatible medium for nanoparticle dispersion. *Nanotox* **2008**, *2*, 144–154. [CrossRef]
25. Ziemann, C.; Jackson, P.; Brown, R.; Attik, G.; Rihn, B.H.; Creutzenberg, O. Quartz-containing ceramic dusts: In vitro screening of the cytotoxic, genotoxic and pro-inflammatory potential of 5 factory samples. *J. Phys. Conf. Ser.* **2009**, *151*, 1–6. [CrossRef]
26. Ziemann, C.; Harrison, P.T.C.; Bellmann, B.; Brown, R.C.; Zoitos, B.K.; Class, P. Lack of marked cyto- and genotoxicity of cristobalite in devitrified (heated) alkaline earth silicate wools in short-term assays with cultured primary rat alveolar macrophages. *Inhal. Toxicol.* **2014**, *26*, 113–127. [CrossRef] [PubMed]
27. Monfort, E.; López-Lilao, A.; Ibáñeza, M.J.; Ziemann, C.; Creutzenberg, O.; Bonvicinic, G. Feasibility of using organosilane dry-coated detoxified quartzes as raw material in different industrial sectors. *Clean. Eng. Technol.* **2021**, *5*, 100331. [CrossRef]

28. German Animal Protection Law 2006 (Tierschutzgesetz of 18 May 2006, with update on 13 July 2013.). Available online: https://www.globalanimallaw.org/database/national/germany/ (accessed on 27 February 2022).

29. Ziemann, C.; Escrig, A.; Bonvicini, G.; Ibáñez, M.J.; Monfort, E.; Salomoni, A.; Creutzenberg, O. Organosilane-Based Coating of Quartz Species from the Traditional Ceramics Industry: Evidence of Hazard Reduction Using In Vitro and In Vivo Tests. *Ann. Work. Expo. Health* **2017**, *61*, 468–480. [CrossRef] [PubMed]

30. Helmke, R.J.; Boyd, R.L.; German, V.F.; Mangos, J.A. From growth factor dependence to growth factor responsiveness. The genesis of an alveolar macrophage cell line. *In Vitro Cell. Dev. Biol.* **1987**, *23*, 567–574. [CrossRef]

31. Smith, C.C.; O'Donovan, M.R.; Martin, E.A. hOGG1 recognizes oxidative damage using the comet assay with greater specificity than FPG or ENDOIII. *Mutagenesis* **2006**, *21*, 185–190. [CrossRef]

32. Kirkland, D.; Brock, T.; Haddouk, H.; Hargeaves, V.; Lloyd, M.; Mc Garry, S.; Proudlock, R.; Sarlang, S.; Sewald, K.; Sire, G.; et al. New investigations into the genotoxicity of cobalt compounds and their impact on overall assessment of genotoxic risk. *Regul. Toxicol. Pharmacol.* **2015**, *73*, 311–338. [CrossRef]

33. European Food Safety Authority (EFSA). Update: Use of the benchmark dose approach in risk assessment. *EFSA J.* **2017**, *15*, e04658. Available online: https://efsa.onlinelibrary.wiley.com/doi/full/10.2903/j.efsa.2017.4658 (accessed on 29 January 2022).

34. European Chemicals Agency (ECHA). *Guidance on Information Requirements and Chemical Safety Assessment: Chapter R.8: Characterisation of Dose [Concentration]-Response for Human Health*; ECHA: Helsinki, Finland, 2012. Available online: https://echa.europa.eu/documents/10162/13632/information_requirements_r8_en.pdf/e153243a-03f0-44c5-8808-88af66223258 (accessed on 29 January 2022).

35. World Health Organization. *Environmental Health Criteria 239. Principles for Modelling Dose–Response for The Risk Assessment of Chemicals*; World Health Organization: Geneva, Switzerland, 2009. Available online: http://apps.who.int/iris/bitstream/10665/43940/1/9789241572392_eng.pdf (accessed on 29 January 2022).

36. PROAST Software. 2022. Available online: https://www.rivm.nl/en/proast (accessed on 29 January 2022).

37. Henderson, R.F.; Mauderly, J.L.; Pickrell, J.A.; Hahn, R.F.; Muhle, H.; Rebar, A.H. Comparative study of bronchoalveolar lavage fluid: Effect of species, age and method of lavage. *Exp. Lung Res.* **1987**, *13*, 329–342. [CrossRef]

38. Creutzenberg, O.; Hansen, T.; Ernst, H.; Muhle, H.; Oberdörster, G.; Hamilton, R. Toxicity of a quartz with occluded surfaces in a 90-day intratracheal instillation study in rats. *Inhalation Tox.* **2008**, *20*, 995–1008. [CrossRef]

39. Organisation for Economic Co-operation and Development (OECD). *Guidelines for the Testing of Chemicals, Section 4*; Test No. 412: Subacute Inhalation Toxicity: 28-Day Study; OECD Publishing: Paris, France, 2018. Available online: https://doi.org/10.1787/9789264070783-en (accessed on 27 February 2022).

40. Anjilvel, S.; Asgharian, B. A multiple-path model of particle deposition in the rat lung. *Fund Appl Tox* **1995**, *28*, 41–50. [CrossRef]

41. Dunnett, C.W. New tables for multiple comparisons with a control. *Biometrics* **1964**, *20*, 482–491. [CrossRef]

42. Rittinghausen, S.; Hackbarth, A.; Creutzenberg, O.; Ernst, H.; Heinrich, U.; Leonhardt, A.; Schaudien, D. The carcinogenic effect of various multi-walled carbon nanotubes (MWCNTs) after intraperitoneal injection in rats. *Part. Fibre Toxicol.* **2014**, *11*, 59. [CrossRef] [PubMed]

43. Reamon-Buettner, S.M.; Hackbarth, A.; Leonhardt, A.; Braun, A.; Ziemann, C. Cellular senescence as a response to multiwalled carbon nanotube (MWCNT) exposure in human mesothelial cells. *Mech. Ageing Dev.* **2021**, *193*, 111412. [CrossRef] [PubMed]

44. Liao, K.-H.; Lin, Y.-S.; Macosko, C.W.; Haynes, C.L. Cytotoxicity of graphene oxide and graphene in human erythrocytes and skin fibroblasts. *ACS Appl. Mater. Interfaces* **2011**, *3*, 2607–2615. [CrossRef]

45. Malkova, A.; Svadlakova, T.; Singh, A.; Kolackova, M.; Vankova, R.; Borsky, P.; Holmannova, D.; Karas, A.; Borska, L.; Fiala, Z. In Vitro Assessment of the Genotoxic Potential of Pristine Graphene Platelets. *Nanomaterials* **2021**, *11*, 2210. [CrossRef]

46. Hinzmann, M.; Jaworski, S.; Kutwin, M.; Jagiello, J.; Kozinski, R.; Wierzbicki, M.; Grodzik, M.; Lipinska, L.; Sawosz, E.; Chwalibog, A. Nanoparticles containing allotropes of carbon have genotoxic effects on glioblastoma multiforme cells. *Int. J. Nanomed.* **2014**, *9*, 2409–2417.

47. Chatterjee, N.; Yang, J.S.; Park, K.; Oh, S.M.; Park, J.; Choi, J. Screening of toxic potential of graphene family nanomaterials using in vitro and alternative in vivo toxicity testing systems. *Environ. Health Tox.* **2015**, *30*, e2015007. [CrossRef]

48. Sasidharan, A.; Panchakarla, L.S.; Chandran, P.; Menon, D.; Nair, S.; Rao, C.N.R.; Koyakutty, M. Differential nano-bio interactions and toxicity effects of pristine versus functionalized graphene. *Nanoscale* **2011**, *3*, 2461. [CrossRef]

49. Muzi, L.; Mouchet, F.; Cadarsi, S.; Janowska, I.; Russier, J.; Menard-Moyon, C.; Risuleo, G.; Soula, B.; Galibert, A.M.; Flahaut, E.; et al. Examining the impact of multi-layer graphene using cellular and amphibian models. *2D Mater* **2016**, *3*, 025009. [CrossRef]

50. Hashemi, E.; Akhavan, O.; Shamsara, M.; Majd, S.A.; Sanati, M.H.; Joupari, M.D.; Farmany, A. Graphene Oxide Negatively Regulates Cell Cycle in Embryonic Fibroblast Cells. *Int. J. Nanomed.* **2020**, *15*, 6201–6209. [CrossRef]

51. Wang, Y.; Xu, J.; Xu, L.; Tan, X.; Feng, L.; Luo, Y.; Liu, J.; Liu, Z.; Peng, R. Functionalized graphene oxide triggers cell cycle checkpoint control through both the ATM and the ATR signaling pathways. *Carbon* **2018**, *129*, 495–503. [CrossRef]

52. Cho, W.S.; Duffin, R.; Bradley, M.; Megson, I.L.; MacNee, W.; Lee, J.K.; Jeong, J.; Donaldson, K. Predictive value of in vitro assays depends on the mechanism of toxicity of metal oxide nanoparticles. *Part. Fibre Tox.* **2013**, *10*, 55. [CrossRef] [PubMed]

53. Wiemann, M.; Vennemann, A.; Sauer, U.G.; Wiench, K.; Ma-Hock, L.; Landsiedel, R. An in vitro alveolar macrophage assay for predicting the short-term inhalation toxicity of nanomaterials. *J. Nanobiotech.* **2016**, *14*, 16. [CrossRef] [PubMed]

54. Gorki, A.-D.; Symmank, D.; Zahalka, S.; Lakovits, K.; Hladik, A. Murine ex vivo cultured alveolar macrophages provide a novel tool to study tissue-resident macrophage behavior and function. *Am. J. Resp. Cell Mol. Biol.* **2021**, *66*, 64–75. Available online: https://doi.org/10.1165/rcmb.2021-0190OC (accessed on 28 February 2022). [CrossRef] [PubMed]
55. Sayes, C.M.; Reed, C.L.; Warheit, D.B. Assessing toxicity of fine and nanoparticles: Comparing in vitro measurements to in vivo pulmonary toxicity profile. *Tox. Sci.* **2007**, *97*, 163–180. [CrossRef] [PubMed]
56. Gurcan, C.; Taheri, H.; Bianco, A.; Delogu, L.G.; Yilmazer, A. A closer look at the genotoxicity of graphene based materials. *J. Phys. Mater.* **2019**, *3*, 014007. [CrossRef]
57. Ursini, C.L.; Fresegna, A.M.; Ciervo, A.; Maiello, R.; Del Frate, V.; Folesani, G.; Galetti, M.; Poli, D.; Buresti, G.; Di Cristo, L.; et al. Occupational exposure to graphene and silica nanoparticles. Part II: Pilot study to identify a panel of sensitive biomarkers of genotoxic, oxidative and inflammatory effects on suitable biological matrices. *Nanotoxicology* **2021**, *15*, 223–237. [CrossRef]
58. Lee, S.; Hwang, S.H.; Jeong, J.; Han, Y.; Kim, S.H.; Lee, D.-K.; Lee, H.-S.; Chung, S.-T.; Jeong, J.; Roh, C.; et al. Nickel oxide nanoparticles can recruit eosinophils in the lungs of rats by the direct release of intracellular eotaxin. *Part. Fibre Tox.* **2016**, *13*, 30. [CrossRef]
59. Moghimian, N.; Nazarpour, S. The future of carbon: An update on graphene's dermal, inhalation and gene toxicity. *Crystals* **2020**, *10*, 718. [CrossRef]
60. Roberts, J.R.; Mercer, R.R.; Stefaniak, A.B.; Seehra, M.S.; Geddam, U.K.; Chaudhuri, I.S.; Kyrlidis, A.; Kodali, V.K.; Sager, T.; Kenyon, A.; et al. Evaluation of pulmonary and systemic toxicity following lung exposure to graphite nanoplates: A member of the graphene-based nanomaterial family. *Part. Fibre Tox.* **2016**, *13*, 34. [CrossRef]
61. Corbo, C.; Molinaro, R.; Parodi, A.; Toledano Furman, N.E.; Salvatore, F.; Tasciotti, E. The impact of nanoparticle protein corona on cytotoxicity, immunotoxicity and target drug delivery. *Nanomedicine* **2016**, *11*, 81–100. [CrossRef] [PubMed]

nanomaterials

Article

How Structured Metadata Acquisition Contributes to the Reproducibility of Nanosafety Studies: Evaluation by a Round-Robin Test

Linda Elberskirch [1], Adriana Sofranko [2], Julia Liebing [3], Norbert Riefler [4], Kunigunde Binder [5], Christian Bonatto Minella [5], Matthias Razum [5], Lutz Mädler [4], Klaus Unfried [2], Roel P. F. Schins [2], Annette Kraegeloh [1,*,†] and Christoph van Thriel [3,*,†]

1   INM—Leibniz Institute for New Materials, Campus D2 2, 66123 Saarbrücken, Germany; linda.elberskirch@leibniz-inm.de
2   IUF—Leibniz Research Institute for Environmental Medicine, Auf'm Hennekamp 50, 40225 Düsseldorf, Germany; adriana.sofranko@iuf-duesseldorf.de (A.S.); klaus.unfried@iuf-duesseldorf.de (K.U.); roel.schins@iuf-duesseldorf.de (R.P.F.S.)
3   IfADo—Leibniz Research Centre for Working Environment and Human Factors, Ardeystraße 67, 44139 Dortmund, Germany; liebing@ifado.de
4   IWT—Leibniz-Institut für Werkstofforientierte Technologien, Badgasteiner Str. 3, 28359 Bremen, Germany; riefler@iwt.uni-bremen.de (N.R.); lmaedler@iwt.uni-bremen.de (L.M.)
5   FIZ Karlsruhe—Leibniz Institute for Information Infrastructure, Hermann-von-Helmholtz-Platz 1, 76133 Eggenstein-Leopoldshafen, Germany; kunigunde.binder@fiz-karlsruhe.de (K.B.); christian.bonatto-minella@fiz-karlsruhe.de (C.B.M.); matthias.razum@fiz-karlsruhe.de (M.R.)
*   Correspondence: annette.kraegeloh@leibniz-inm.de (A.K.); thriel@ifado.de (C.v.T.)
†   These authors contributed equally to this work.

Citation: Elberskirch, L.; Sofranko, A.; Liebing, J.; Riefler, N.; Binder, K.; Bonatto Minella, C.; Razum, M.; Mädler, L.; Unfried, K.; Schins, R.P.F.; et al. How Structured Metadata Acquisition Contributes to the Reproducibility of Nanosafety Studies: Evaluation by a Round-Robin Test. *Nanomaterials* **2022**, *12*, 1053. https://doi.org/10.3390/nano12071053

Academic Editor: Laura Canesi

Received: 21 February 2022
Accepted: 16 March 2022
Published: 24 March 2022

Publisher's Note: MDPI stays neutral with regard to jurisdictional claims in published maps and institutional affiliations.

Abstract: It has been widely recognized that nanosafety studies are limited in reproducibility, caused by missing or inadequate information and data gaps. Reliable and comprehensive studies should be performed supported by standards or guidelines, which need to be harmonized and usable for the multidisciplinary field of nanosafety research. The previously described minimal information table (MIT), based on existing standards or guidelines, represents one approach towards harmonization. Here, we demonstrate the applicability and advantages of the MIT by a round-robin test. Its modular structure enables describing individual studies comprehensively by a combination of various relevant aspects. Three laboratories conducted a WST-1 cell viability assay using A549 cells to analyze the effects of the reference nanomaterials NM101 and NM110 according to predefined (S)OPs. The MIT contains relevant and defined descriptive information and quality criteria and thus supported the implementation of the round-robin test from planning, investigation to analysis and data interpretation. As a result, we could identify sources of variability and justify deviating results attributed to differences in specific procedures. Consequently, the use of the MIT contributes to the acquisition of reliable and comprehensive datasets and therefore improves the significance and reusability of nanosafety studies.

Keywords: interlaboratory comparison; minimal information; quality criteria; description standards

1. Introduction

Nanosafety studies are necessary for the assessment of the potential human health hazards of engineered nanomaterials (ENMs) and play a key role in addressing safety issues already during the development and design phase of new ENMs [1,2]. In vitro assays are used to assess the cellular effects of nanomaterials within decision-making frameworks [3]. At this level, mechanism-linked bioactivity assays, e.g., focusing on inflammatory responses, in combination with traditional cytotoxicity assays are thought to be important tools as long as they provide reliable and reproducible data [4,5]. These assays are easy to handle, cost-efficient, fast, and enable high-throughput screening [6,7]. Moreover, cells and in vitro systems derived from relevant target organs are available, according to the predominant exposure routes. Their use in regulatory processes is twofold as they can (1) be used to prioritize the generation of in vivo data and (2) facilitate the use of read-across approaches to avoid more animal experiments [8].

According to the advantages described above, the generated results are expected to provide comparability among laboratories and high confidence, which are required for safety assessment. In reality, established in vitro screening assays often lead to conflicting results, even if similar assays are used for the analysis of identical ENMs. Such contradictory and inconclusive research results have generated a wide discussion on data quality in nanosafety assessment [9,10]. This is clearly demonstrated by round-robin tests or interlaboratory comparisons, which are implemented in order to evaluate the comparability of results from multiple laboratories. To enhance the consistency in terms of quality and comprehensibility, standard operating procedures (SOPs) providing detailed instructions on how to perform different steps of the experiment are recommended. As an outcome of previous interlaboratory comparisons, the use of identical laboratory equipment, materials, and preceding training for the experimenters to minimize potential variability have been proposed [5,11]. This strategy improves the comprehensibility of results within single projects or collaborating laboratories [12]. However, considering the number of research groups in the field of nanosafety assessment, this time- and labor-intense approach cannot be realized throughout the community. Furthermore, the development of new approach methodologies (NAMs) as non-animal alternative models to mimic more realistic exposure scenarios and physiological conditions poses additional requirements due to their complexity [13,14]. These could only be partially described by already established toxicological assays, which is challenging for interlaboratory comparisons. Therefore, description standards and quality criteria for nanosafety studies are a substantial issue in academic and industrial research.

Various scientific projects and international organizations are developing standards or guidelines and promote harmonization throughout the nanosafety community [5,9,12]. Widely known examples of scientific guidelines directly related to nanosafety research are the Minimum Information Reporting in Bio–nano Experimental Literature (MIRIBEL) [15] and the documents of the Organization for Economic Cooperation and Development (OECD) Working party on Manufactured Nanomaterials (WPMNs) [16]. MIRIBEL focuses on minimum information that should be reported with study results of experiments investigating bio–nano interactions. The OECD WPMN works on the development of methods and strategies to identify and manage the potential health and environmental risks of nanomaterials and provide Test Guidelines (TGs) and Guidance Documents (GDs) [16]. In regulatory toxicology, which also involves industry and contract research, good laboratory practice (GLP) has been introduced as a comprehensive tool to ensure quality standards [17]. Even though in academia, this standard is not adhered to, it was recently shown that in nanoparticle-induced cytotoxicity testing, GLP can improve the quality of results from in vitro toxicity assays [5]. Following the recommendations of minimum information standards or guidelines will enhance the quality, reliability, and comprehensibility of published research data. Furthermore, the resulting published research data are limited in their findability, which complicates re-use. In this context, the collection of information about, e.g., experimenter-dependent handling or deviations from SOPs could support machine-readable data and metadata and therefore enhance findability and accessibility. To implement this information and achieve FAIRification of nanosafety research data, metadata schemas could be used [18]. Metadata schemas represent a common consensus on the hierarchical and relational structure of the descriptive information and specify designations of parameters and required content. A starting point for the development of metadata schemas is the previously introduced minimum information table (MIT), which is a collection of descriptive information and quality criteria based on existing standards or guidelines [19]. In order to facilitate multidisciplinary use, the

MIT is divided into six modules: general information, material information, biological model information, exposure information, endpoint readout information, an, analysis and statistics. The relevance of containing descriptive information and quality criteria depend on the types of studies (e.g., in vitro or in vivo) or methods (e.g., dynamic light scattering or fluorescence microscopy) that have been used and therefore can be chosen in accordance with the individual requirements. Furthermore, the modular structure enables the addition of study- and method-specific parameters, e.g., there are multiple methods for ENM production or sample preparation that influence the behavior and interaction of ENMs in biological systems. The use of the MIT could support data collection throughout the data life cycle, bring improvement of data quality including completeness and reproducibility, and facilitate the data comprehensibility and re-usability of future research in accordance with the FAIR data principles.

This study was designed to evaluate the parameters and criteria collected by the MIT to identify similar outcomes and explain differences in assay results based on the descriptive data and implementation of quality criteria. Descriptive information, provided by subject-specific metadata and quality criteria, plays a key role in data reliability and comprehensibility. Furthermore, descriptive information helps to understand the study purpose and design. The corresponding quality criteria are used to verify and evaluate the test system and study conditions. For example, the conduction of a cytotoxicity assay by using high ENM doses may not correlate with human exposure; however, high doses are required in mechanistic studies, which could be justified by the study purpose. In this context, quality criteria help to identify the interferences of ENMs with cytotoxicity assays, e.g., caused by the optical properties of the ENMs, and the adsorption or reactions with assay reagents or biomolecules and could improve the reusability and significance of the resulting data [20,21] (Figure 1).

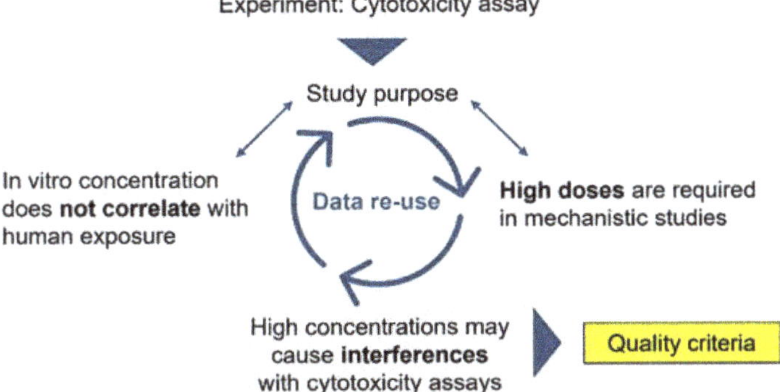

Figure 1. Role of descriptive information and quality criteria. Bold tags are used for highlighting keywords.

The objective was to show how the content of the MIT could help to elucidate the variation in results to systematically understand the potential sources of variability in the study. Therefore, we demonstrate the study-dependent implementation and application of the previously defined quality criteria and minimum information by the implementation of a three-laboratory comparison of a nanotoxicity assay. The JRC reference nanomaterials NM101 and NM110 were chosen as ENM models for analyzing their effects on the human cell line A549. The study was conducted based on a round-robin approach that was performed in the framework of the EU FP7 nanosafety project ENPRA [22]. To be able to compare the experiments among laboratories, SOPs with detailed information and defined quality criteria on ENM handling and dispersion, cell cultivation and exposure, and the implementation of the WST-1 water-soluble tetrazolium salt-based assay were exchanged. The corresponding results were analyzed, and important aspects, deviations, and their effects were identified based on the MIT. Consequently, we were able to identify and justify reasons for the differences of the study results, which were mainly attributed to the MIT modules' biological model information and exposure information.

2. Materials and Methods

2.1. Study Design and SOP Development

The study design phase started with the planning of the experiments and the acquisition of information on the used materials and methods and the roles of the various partners in the round-robin test. These steps were supported by the MIT groups "General Information", "Material Information", "Biological Model Information", and "Exposure Information". The experimental procedure was based on protocols previously used as a part of the ENPRA project [22,23]. Based on these protocols, (S)OPs for cell cultivation, probe sonicator calibration, sample preparation, exposure of the cells, and the WST-1 assay were defined. The SOPs were reviewed and discussed by one experimenter of each participating partner, whereby a device- and experience-independent description should be achieved. Additionally, the definition of the acceptance criteria (Table 1) served to ensure the comparability of the results despite the use of different materials and equipment. The detailed SOPs are available in the Supplement (Supplementary SOPs: SOP S1: Culturing A549 cells; SOP S2: Sample preparation; SOP S3: Sonicator calibration; SOP S4: Cellular viability—WST-1 assay in A549 cells). The procedures are briefly described below. Some materials and methods, e.g., sources of chemicals and instruments, instrument settings, cell culture conditions, etc., varied depending on the partner, but were recorded (Table S1: Minimal information table (MIT)).

Table 1. Acceptance criteria to be fulfilled within the modules' biological model and endpoint read out information.

Acceptance Criteria	Brief Description
Source cells	Microscopy observation of cell morphology and viability during cultivation: adherent cell growth and cuboidal cell morphology.
Biological test system	Healthy culture should contain at least 80% viable cells and exhibit a confluence of >70%. Microscopy check of cell morphology and confluence prior to cell cultivation, ENM treatment, and performance of the WST-1 assay.
Viability assay	Corrected absorption of controls representing viable cells between 0.5 and 2, standard deviation of 4 replicates < 0.3.Corrected absorption of cytotoxicity controls should be lower than the viability controls.

2.2. Preparation of Test Materials

Test materials: Two test materials (JRC reference nanomaterials) were used that were also analyzed in the previously conducted ENPRA project [22]: NM 101 (Hombikat UV100; titanium dioxide (TiO_2), rutile with minor anatase; mean diameter 29.9 nm $\pm$ 13.1 nm) and NM 110 (BASF Z-Cote; nonfunctionalized zinc oxide (ZnO), mean diameter 62.7 nm $\pm$ 32.6 nm) [24,25]. To exclude the effects of different material batches, one batch of each of these ENMs was split among the partners.

Sample preparation: The ENMs were supplied as dry powders. In order to obtain a concentration of 2.56 mg ENM/mL, ENMs were weighed into a 50 mL Falcon tube, followed by the addition of an appropriate amount of ultrapure water supplemented with 2% (*v/v*) fetal calf serum (2% FCS). Dispersion was achieved by ultrasonic treatment according to the ENPRA protocol: After application of a total energy of 7056 $\pm$ 103 J using a probe sonicator, samples were immediately transferred to ice (see Figure S1) and used within 1 h. The ENM stock dispersion was further diluted 1:10 in full-cell culture media without phenol red to obtain a concentration of 256 µg/mL (corresponding to 80 µg/cm^2). Starting from this concentration, twofold serial dilutions were prepared in a mixture containing assay medium (cell culture medium with 10% FCS, v/v) and ENM dispersion medium (2% FCS in ultrapure water, v/v) at a ratio of 9:1 (v/v).

Probe sonicator calibration: The sonicators available at the three partner institutes were calibrated to obtain comparable ENM dispersions. The calibration procedure (see Supplementary SOPs) used was based on a calorimetric protocol described by Taurozzi et al. [26], which was further improved in the frame of the NANoREG project (NANoREG D4.12 SOP Probe Sonicator Calibration for ecotoxicological testing [27]).

According to this protocol, the sonicator was operated in continuous mode by starting with the lowest output setting and increasing it to 20% of the maximum amplitude. The temperature increase was recorded with a time resolution of no more than 30 s. The recorded data were plotted, and the temperature vs. time values were fit using a least-squares regression. The effective delivered power was calculated by the equation:

$$P_{ac}\ (Watt) = \frac{\Delta T}{\Delta t}\ MC_P \tag{1}$$

where P_{ac} is the delivered acoustic power (W), $\Delta T / \Delta t$ the slope of the regression curve with temperature T (K) and time t (s), C_P the specific heat of the liquid (4.18 J/g × K for water), and M the mass of the liquid (g). To deliver a total energy of 7056 ± 103 J, the amplitude settings and the sonication time t were adjusted according to the equation [28]:

$$t\ (s) = \frac{E\ (7056\ J)}{P_{ac}(W)} \tag{2}$$

2.3. Sedimentation Analysis

Cell responses to ENM exposure are considered to be dependent on the ENM dose delivered to the cells rather than due to the administered concentration during an in vitro experiment [18]. ENM transport in liquid media is determined by diffusion and gravitational settling and therefore dependent on ENM size (including aggregates), density, and agglomeration behavior [29]. The resulting particle sedimentation was expected to influence the particle delivery during the exposure phase. In addition, sedimentation might affect the quantitative handling of ENMs already during sample preparation, thereby influencing the effective administered ENM concentration. In order to determine the relevance of particle sedimentation for both instances, the time-dependent sedimentation of the used ENMs dispersed in cell culture medium was examined by use of UV-Vis spectroscopy and compared to theoretical sedimentation curves, described in the Supplement (Method S1). Measurements were performed with a UV-Vis spectrometer (UV-2600, Shimadzu Deutschland GmbH, Duisburg, Germany) at room temperature (20 °C within a laboratory with air conditioning), using a cuvette holder. In a first step, spectra of the cell culture medium were recorded as a reference. After cleaning and drying, the cuvette was refilled with the ENM dispersions prepared as described in Section 2.2. By adjusting the time interval between measurements, long-term studies over up to 45 h were performed. At high particle concentrations (c_0 = 2.56 mg/mL), a 2 mm (layer thickness) quartz cuvette was used, while all measurements at lower particle concentrations were performed with standard (10 mm) polystyrene cuvettes.

2.4. Cell Culture

The cell line A549 (DSMZ No.: ACC 107) [30] was used as a model for human alveolar epithelial type II cells as it is widely used for nanotoxicity studies [31]. Cells of the same passage were distributed among the three partners. A549 cells were cultivated in Dulbecco's Modified Eagle's Medium (DMEM) supplemented with 10% FCS in 75 cm^2 cell culture flasks in a humidified atmosphere at 37 °C and 5–10% CO_2 (constant value, depending on the sodium bicarbonate concentration of the used medium). At a confluence of 70–90%, cells were detached by the addition of 2 mL 0.05% Trypsin/0.02% EDTA, incubation at

37 °C for 3–5 min, and centrifugation at $200\times g$ for 5 min. For subcultivation, cells were replated at a ratio of 1:5–1:10 into 75 cm^2 cell culture flasks.

2.5. Exposure of Cells to ENMs

For cytotoxicity experiments, A549 cells were seeded into 96-well plates at a density of 1×10^4 cells per well and allowed to attach for 24 h. Cells were then washed using serum containing medium without phenol red. Then, 100 µL of freshly prepared ENM dispersions with administered concentrations of 0.3125 µg/cm^2, 0.625 µg/cm^2, 1.25 µg/cm^2, 2.5 µg/cm^2, 5 µg/cm^2, 10 µg/cm^2, 20 µg/cm^2, 40 µg/cm^2, and 80 µg/cm^2 (see Section 2.2) or medium (as the control representing viable cells) was added and the cells incubated for 24 h. Afterwards, two of all replicates were exposed to 0.5% Triton-X 100 for additional 15 min at 37 °C (as the cytotoxicity control).

2.6. Viability (WST-1 Assay)

To determine viability (metabolic activity), the WST-1 assay (Roche Diagnostics) was used according to a protocol initially adapted by Vietti et al. [32] and further adjusted in the frame of the ENPRA project [22]. After incubation with 10 µL of WST-1 solution for 1 h, absorption was measured at 450 nm and at 630 nm as the reference wavelength.

2.7. Analysis and Statistics

The WST-1 assay was performed by each experimenter (all in all, 6) within the three partner institutes (nested design) three times independently ($n = 3$, biological replicates) on different days, using different cell passages, freshly dispersed ENMs, and freshly diluted Triton-X-100. Four wells per treatment per setup were used and served as technical replicates. All these different "factors" were used as possible sources of variance in the statistical analysis. For data analysis, the raw absorption data were entered into a shared MS Excel calculation template. Corrected absorption values and normalized values were calculated as follows:

1. Corrected absorption: To correct for unspecific medium and ENM absorption, the absorption of the medium and the ENM-containing (at the corresponding concentrations) medium at 450 nm was subtracted from the corresponding absorption values obtained in the presence of cells. The values at 450 nm were further corrected by subtraction of the reference absorption values obtained at 630 nm. To further correct for cell absorption (see SOP), the absorption of cells treated with the positive control (0.5% Triton X-100) was subtracted from the absorption values of the ENM-treated cells;
2. Normalized values: The corrected quadruple ENM-treated cell sample values from each treatment group were converted to normalized values by the following equation and used for further statistical analyses:

$$normalized\ value\ x\ (\%) = \frac{A\ sample}{A\ untreated\ control} \times 100 \qquad (3)$$

The statistical analyses were performed using IBM SPSS Statistics for Macintosh, Version 28.0 (IBM Corp. Released 2021. IBM SPSS Statistics for Macintosh, Version 28.0. Armonk, NY, USA: IBM Corp) and GraphPad Prism version 9.2.0 for Mac OS X (GraphPad Software, San Diego, CA, USA: www.graphpad.com, accessed on 20 February 2022). In general, two statistical approaches were used to explore the data, one approach with and another without normalization to the respective control conditions. The normalized data were expressed as the percentage of the controls (range: 0% to 100%), and these values were used to calculate sigmoidal functions describing the inhibitory potency of the tested concentrations. The half maximal inhibitory concentrations (IC$_{50}$) for the separate datasets generated by each experimenter were calculated based on normalized values of the three biological replicates. Before calculating these values, the data were inspected for outliers (see Krebs et al., 2019 [33]) using the algorithm implemented in the box and whiskers plot procedure of SPSS. Here, outliers are defined as values above the 3rd quartile + 1.5 ×

interquartile range (IQR) or below the 1st quartile—1.5 × IQR. This procedure does not assume normally distributed values and is therefore suitable for cytotoxicity tests where this assumption is difficult to achieve due to the often skewed distribution of the dependent variable in the range of the high or low concentrations. To avoid non-linear transformation introduced by the normalization to the control wells, we used the corrected absorption as provided by the plate readers available at the partners labs for the second analysis. Again, these data were inspected for outliers as described before. To identify factors contributing to the interlaboratory variation as nested (hierarchical), an analysis of variance (ANOVA) model was used. The model (UNIANOVA in SPSS 28.0) consisted of four factors that were (1) concentration (fixed factor), (2) biological replicate (random factor), (3) partner (random factor), and (4) experimenter (nested factor within partner). All possible main effects and interactions were included in the model, and post hoc tests and paired comparisons were used to explore the significant effects in detail.

The IC50 values and their respective 95% confidence intervals (95%-CI) were calculated in Prism using the nonlinear fitting algorithm "[Inhibitor] vs. normalized response—Variable slope" that uses the following equation to describe the dose–response relationship obtained by the six experimenters on the three biological replicates with the concentration (Y) and the normalized response in the WST-1 assay (X).

$$Y = \frac{100}{1 + \left(\frac{IC_{50}}{X}\right)^{HillSlope}} \tag{4}$$

3. Results

3.1. Cytotoxicity

The WST-1 assay, indicating metabolic activity, was selected to analyze the effects of NM101 and NM110 on cell viability after 24 h of exposure. The round-robin testing was carried out by two experimenters each at the three involved institutes.

As shown in Figure 2, all experimenters demonstrated a concentration-dependent decrease in the viability of A549 cells down to 0% after exposure to NM110 (ZnO). For NM101 (TiO$_2$), no concentration-dependent decrease of the viability was observed ($F_{(9,710)} = 1.22$; $p = 0.28$; Supplementary Figure S1). Therefore, only the results of NM110 were used here to evaluate whether the MIT can be used to identify sources of variance.

As described in Section 2.7, outliers were excluded before estimating the IC$_{50}$ values mathematically. The percentage of outliers in the normalized dataset for the six experimenters ranged from 0% to 4.1%. Thus, a maximum of 5 of the 120 values (4 [measures/well] × 3 [biological replicates] × 10 [concentrations]) that were obtained in the experiments of one experimenter were recognized as outliers. The results of the nonlinear fitting algorithm for the outlier-corrected, normalized data are shown in Figure 2. The three partners and the two experimenters within these institutes are given separately. The normalized data clearly showed variations in the IC$_{50}$ values obtained by the six experimenters (Figure 2). The IC$_{50}$ values varied in the range of 5.91–22.78 µg/cm^2. Moreover, the 95% confidence intervals (CIs) indicated differences in the accuracy of the IC$_{50}$ estimates across the experimenters. Experimenter B at Partner 1 and Experimenter A at Partner 3 obtained very narrow 95%-CIs, indicating high reproducibility across the technical and biological replicates, while other sets spanned a broader range of uncertainty. Figure 2 also illustrates that the dose–response curves of the two experimenters at Partners 1 and 2 ran parallel to each other, while at Partner 3, the slopes of the curves differed, and they crossed each other at 10 µg/cm^2.

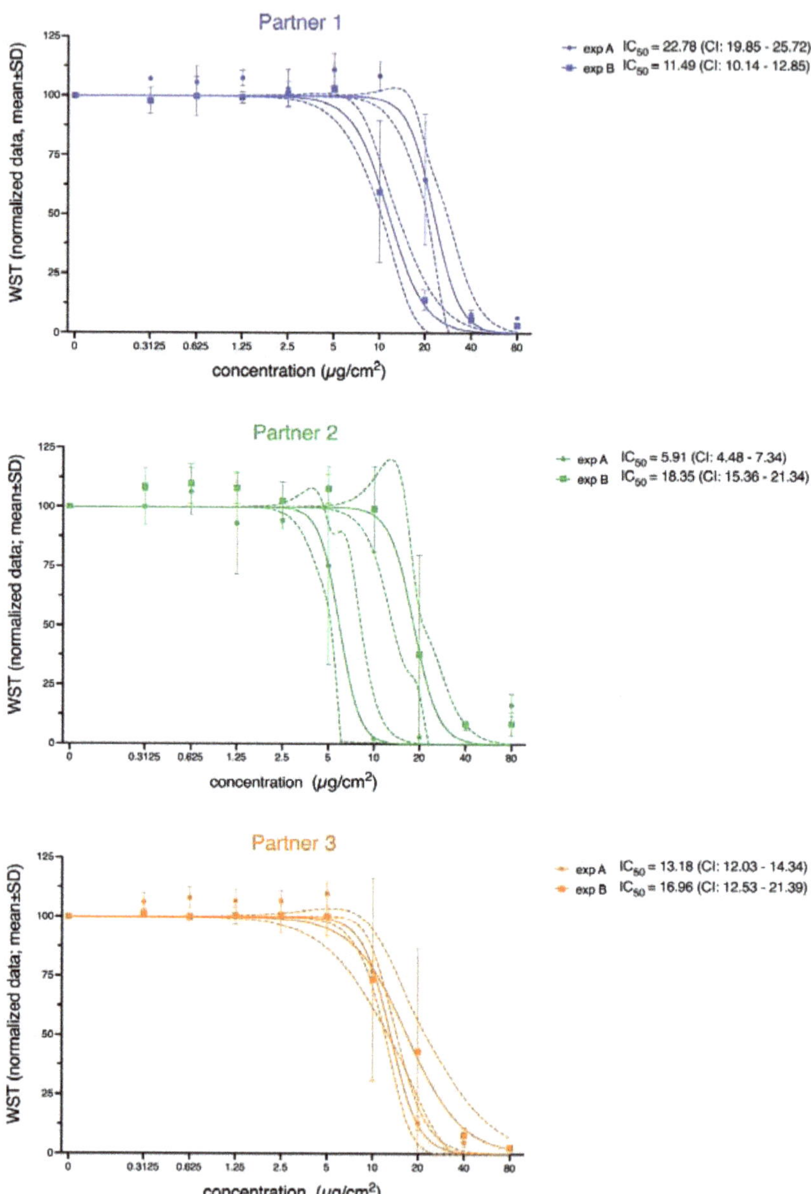

Figure 2. Viability of A549 cells, as indicated by the WST-1 assay after 24 h of exposure to NM110 (ZnO). Normalized data are presented corresponding to measurements performed by the various experimenters (exp A/B) at the participating institutes (Partners 1–3). Solid lines represent the nonlinear regression based on the calculated mean values; dashed lines represent the corresponding 95% confidence bands. The calculated IC_{50} values are given in the figure legends along with the corresponding confidence intervals (CIs). The concentration is the administered concentration of the ENMs based on the initial dispersion.

To explore these differences in more detail, as the next step of the analyses, the sources of these variations should be identified by using the nested (hierarchical) ANOVA model.

Explorative data analyses and box and whiskers plots were used to identify outliers per concentration (see the Supplement) within the single datasets of the six experimenters. In the datasets of three experimenters, no outliers were detected. In the other three sets, only a few outliers were identified (0.8%, 1.7%, and 2.5% of the 120 data points/experimenter). For these sets, mean values were calculated omitting the outlier values and subsequently used to substitute the outlier values. Thereby, the calculation of the sum of squares (total variation observed in the sample/factor of the model) and the mean square (total variation divided by the degrees of freedom) given in the ANOVA table were based on an identical number of observations. The results of the ANOVA are given in Table 2.

Table 2. Results of the nested (hierarchical) ANOVA showing type III sum of squares as an estimate of the variation caused by the different factors and their interactions (sources, interactions of two or more factors indicated by *) in comparison to the variation caused by the respective error terms. The degree of freedom (df) values are used to calculate the mean squares of the respective sources, and their ratio provides (hypothesis/error) the F-value.

Source		Type III Sum of Squares	df	Mean Square	F	Sig.
Intercept	Hypothesis	964.142	1	964.142	12.790	0.067
	Error	155.074	2.057	75.382		
Concentration	Hypothesis	286.274	9	31.808	11.748	0.000
	Error	45.014	16.625	2.708		
Replicate	Hypothesis	3.932	2	1.966	2.508	0.181
	Error	3.700	4.720	0.784		
Partner	Hypothesis	148.599	2	74.299	8.988	0.017
	Error	47.589	5.757	8.267		
Experimenter (partner)	Hypothesis	17.565	3	5.855	5.744	0.023
	Error	7.871	7.722	1019		
Concentration * partner	Hypothesis	50.524	18	2.807	6.956	0.000
	Error	7.449	18.459	0.404		
Concentration * experimenter (partner)	Hypothesis	2.879	18	0.160	0.617	0.862
	Error	9.331	36	0.259		
Experimenter * replicate (partner)	Hypothesis	10.674	27	0.395	1.575	0.078
	Error	13.555	54	0.251		
Concentration * partner * replicate	Hypothesis	5.250	6	0.875	3.486	0.006
	Error	13.555	54	0.251		
Concentration * experimenter * replicate (partner)	Hypothesis	9.331	36	0.259	1.033	0.450
	Error	13.555	54	0.251		

First, the main effect of the fixed factor concentration was shown to be highly significant (F = 11.75, $p < 0.001$; see the black squares in Figure 3). This strong effect of the applied concentrations on the corresponding corrected absorption values of the three partners is illustrated by the roughly sigmoidal courses (see Figure 2 for plots using normalized data) of the curves, when plotted against concentration. The curves approach y = 0 only at the highest concentrations. The significant main effect of the random factor partner (F = 8.99, $p = 0.02$) is also visible in Figure 3 as Partner 3 yielded an approximately two-fold higher absorption value than the other two partners. However, as indicated by the significant interaction of the factors partners and concentration (F = 6.96, $p < 0.001$), these absorption differences were more pronounced at the lower concentrations. At the two highest concentrations, the absorption values of all partners approached zero. For all partners, post hoc tests revealed that the absorption values obtained at 10 µg/cm^2 differed significantly from the values obtained at the next lower concentration (Partner 1: mean difference: −0.26, $p < 0.001$; Partner 2: mean difference: −0.36, $p < 0.001$; Partner 3: mean difference: −0.70, $p < 0.001$). This is also the concentration where the Dunnett t post hoc test of the random factor concentration indicated the first significant difference from the control condition (mean difference: −0.39, $p < 0.001$). However, from concentration step 20 µg/cm^2 on, no significant differences in the measured absorption (see overlapping CIs) among the partners could be found. The functions plotted in Figure 3 also illustrate the significant interaction of the two factors: partner and concentration. The slopes of the sigmoidal curves were slightly different, while Partner 2 yielded a flatter slope, while the slope of Partner 3 was steeper.

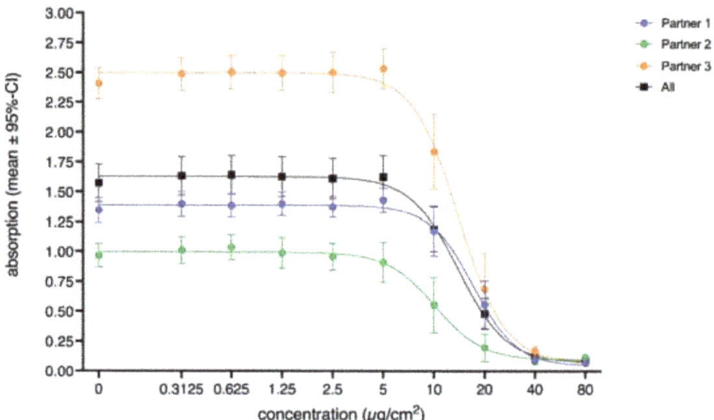

Figure 3. Concentration-dependent effects of the random factor partner on the absorption. Values represent mean values (corrected absorption) of the three partners, as well as the average (black square). Error bars indicate the corresponding 95% confidence intervals. The concentration is the administered concentration of the ENMs based on the initial dispersion.

By evaluating the records of the experimenters in the MIT, we could identify deviations in the seeded cell numbers (see MIT parameter "cell seeding details") as a possible reason for the high absorption values of Partner 3 at low concentrations. According to the SOP, 10,000 cells per well in a 96-well plate format should be seeded for the WST-1 assay. The deviation of Partner 3 was caused by the attempt to meet the quality criteria of treating the cells at 70% confluence. Partner 3 noted that this was not achieved with the given cell number of 10,000 cells. Therefore, the information documented according to the MIT revealed that the cultivation conditions differed between the individual partners, resulting in variable cell growth.

Interestingly, the random factor replicates yielded neither a significant main effect, nor a significant interaction with the fixed factor concentration (Table 2 and Figure 4) Thus, the replication of the experiments yielded comparable results (Figure 4) across the entire concentration range. Overall, Replicate 2 revealed slightly lower absorption values; however, in general, the results were in good agreement, and overall, the replication of the experiments showed comparable results.

However, the ANOVA results (Table 2) revealed that within the three partners (nested effect), the values of the three replicate sets markedly differed for Experimenters A and B, especially at lower concentrations. Thus, a huge amount of the variation can be explained by differences between the three replicates performed by the two experimenters of the three partners. This highly significant interaction of all factors (F = 25.52, $p < 0.001$) involved in the model is depicted in Figure 5.

Within the three partners, the WST-1 results were differently affected by the concentration and/or the replicates of the respective experimenters. For all partners, the two involved experimenters produced significantly different results for most of the concentration steps. These differences were systematic, as one experimenter always yielded lower absorption values than the other at least up to the concentrations 2.5 μg/cm² and 5 μg/cm². Only at the lab of Partner 3, the two experimenters yielded comparable values for the 5 μg/cm² concentration, a generally non-cytotoxic concentration (see the Dunnett t post hoc test). For Partners 1 and 2, the two experimenters yielded highly different results for the concentration steps 10 μg/cm^2 and 20 μg/cm^2. These differences were to some extent less pronounced for Partner 3. However, all paired comparisons were significant, and here, Experimenter B yielded highly significant differences among the biological replicates. Surprisingly, the cell viability increased with concentrations higher than 20 μg/cm^2 for Experimenter A at Partner 2 with no strong variation across the replicates. The concentration-dependent differences of the replicates were less pronounced for Partner 3, but here, Experimenter B showed huge differences among the replicates. In general, the variation across the replicates was highest in the concentration steps around the IC$_{50}$ values (see Figure 1). However, even in this dose range, some experimenters showed almost no (see Partner 2) or small differences (see Partner 3, Experimenter A).

Again, we matched these results to the entries in the MIT. Some explanations could be found in the group "Endpoint Read Out Information". As an acceptance criterion for assessing the test system at its start, 70% confluence of the cells is defined in the SOP. It has to be mentioned that this value was estimated by the experimenters and no values were given by Partner 3. Nevertheless, the differences in absorption between the two experimenters of Partner 2 may be in part due to the fact that Experimenter A had only 40–50% cell confluence and Experimenter B between 70% and 80%. Lower cell numbers may result in decreased absorption in the WST-1 assay because of less reagent reduction. Likewise, Experimenter A of Partner 1 estimated 80–85% confluence and Experimenter B estimated values of 70–90%. This is reflected by the absorption values in Figure 5, revealing that the absorption values of Experimenter A were usually significantly higher compared to Experimenter B.

Additionally, if a lesser quantity of cells were treated, toxicity may also be increased at lower ENM concentrations, which may explain the difference in IC$_{50}$ values between Experimenter A with an IC$_{50}$ of 5.3 and B with an IC$_{50}$ of 16.7. Kim et al. investigated the effect of cell density and nanoparticle uptake, and they showed that lower cell density results in higher nanoparticle uptake [34].

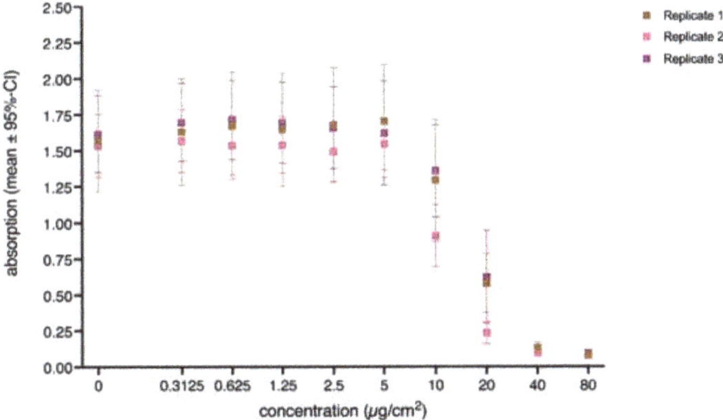

Figure 4. Concentration-dependent effects of the random factor replicate on the absorption. Values represent mean values (corrected absorption) of the three biological replicates. Error bars indicate the corresponding 95% confidence intervals. The concentration is the administered concentration of the ENMs based on the initial dispersion.

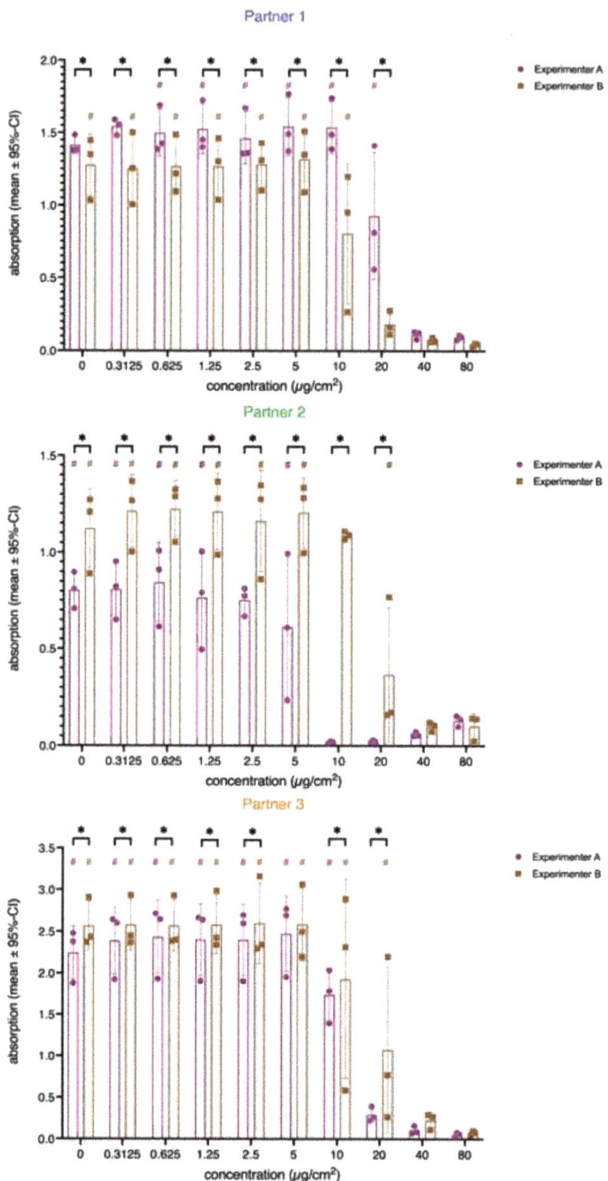

Figure 5. Bar chart showing the interaction of the factors concentration (x-axis), partner (panels Partner 1–3), experimenter within the partners (color-coded bars), and biological replicate (purple and brown dots and squares). Bars represent mean values (corrected absorption), and error bars indicate the corresponding standard deviation. Individual measures of the three biological replicates are given as color-coded dots and squares. # above the bars indicate significant differences among the biological replicates of the two experimenters at a certain concentration step ($p < 0.05$ according to pairwise comparisons, adjusted for multiple tests). Adjusted significant comparisons between the two experimenters within in the partners are given above the bars (*: $p < 0.05$ according to pairwise comparisons, adjusted for multiple tests). The concentration is the administered concentration of the ENMs based on the initial dispersion.

A further acceptance criterion was defined in the SOP for the test system at the end of compound exposure. At this time point, the absorption should be between 0.5 and 2 with a standard deviation of <0.3. The standard deviation criterion was met by all experimenters except for one concentration (2.5 µg/cm^2) in the sets of Experimenter B of Partner 3, showing a standard deviation of 1.03. The criteria for raw absorption values were met for most of the experiments, except for Partner 3. Here, Experimenter A obtained absorption values between 2.2 and 2.5 in Replicate 1 and 2.4 and 2.6 in Replicate 3. Experimenter B measured absorption values between 2.8 and 3.0, 2.3 and 2.6, and 2.2 and 2.4 in Replicates 1–3, respectively. This could be due to the increased cell numbers seeded into the well plates. Although the measured values were outside the acceptance criteria, the resulting curves and the IC$_{50}$ of Partner 3 were within a suitable range. Overall, the acceptance criteria for the test system were not achieved by all experimenters.

Finally, the overall impact of normalization should be analyzed. Figure 6 shows the fitted sigmoidal curves for normalized and raw absorption data of the WST-1 assay. While the estimated IC$_{50}$ values and fitted curves were almost identical, the confidence intervals, as well as the SDs of the raw data were wider. This is also expressed in the goodness-of-fit values with an R2 of 0.50 for the raw data and an R2 of 0.80 for the normalized data. Thus, by normalizing the data to the respective control condition, the impact of some sources of variance identified and described in the previous sections could be reduced.

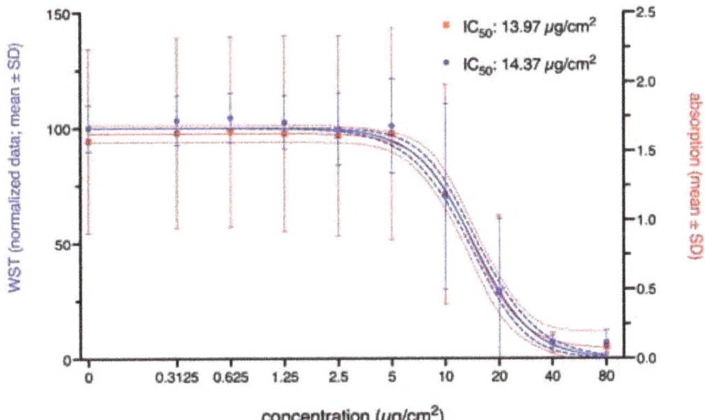

Figure 6. Viability of A549 cells, as indicated by the WST-1 assay, after 24 h of exposure to NM110 (ZnO). Both normalized data (blue) and corrected absorption (red) are given. Solid lines represent the nonlinear regression based on the calculated mean values; dashed and dotted lines represent the corresponding 95% confidence bands. The calculated IC$_{50}$ values are given in the figure legend. The concentration is the administered concentration of the ENMs based on the initial dispersion.

3.2. ENM Sedimentation

Deviations in the absorption values at IC$_{50}$ not only between partners, but also between the replicates of single experimenters could be due to the non-uniform dosage of the ENMs. Such inconsistent dosage might be caused by particle agglomeration and/or resulting sedimentation during sample preparation. In order to estimate the influence of these effects on the reproducibility of the assay, the stability of the ENM dispersion was tested by sedimentation analyses. Due to the unknown agglomeration state of the nanoparticles in complex media, e.g., cell culture media, theoretical considerations about the sedimentation (see the Supplement) cannot be used to predict particle sedimentation times. However, qualitative comparisons are possible. An example illustrating qualitative particle sedimentation analysis by recording UV-Vis spectra is shown in Figure 7 for the used ZnO (NM110) nanoparticles. A gradient of nanoparticles from the top to the ground of the cuvette is clearly visible.

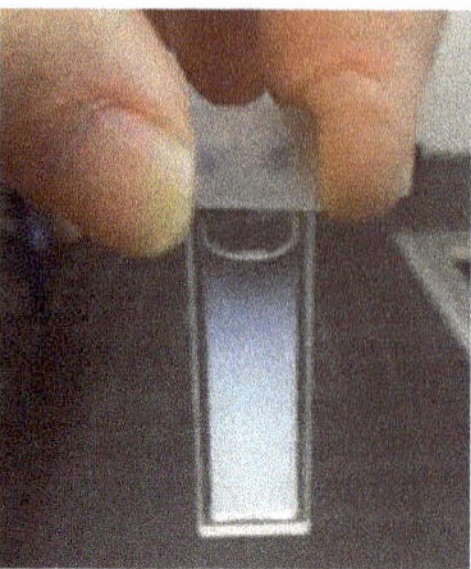

Figure 7. Sedimentation of ZnO (NM110) nanoparticles at c = 2.56 mg/mL and t = 30 h after dispersion in cell culture medium. In the UV-Vis spectrophotometer used for qualitative analysis, the light passes through the suspension as a flat vertical light sheet in the lower third of the cuvette.

In the UV range, absorption was very strong and noisy and clipped at units higher than 10. Over the course of time, within a timescale of several hours, the absorption decreased significantly in the visible and infrared wavelength range (Figure 8). This can be explained by a reduction in the number of particles due to sedimentation. At particle concentrations used for cytotoxicity assays (e.g., c = 256 µg/mL, the highest concentration used), a comparable decrease in the absorption over time was observed (Figure 8, right). Only at the lower concentration range used for the cytotoxicity assays (e.g., c = 8 µg/mL, corresponding to 2.5 µg/cm^2), hardly any decline in the absorption over time was observed. However, under the conditions used, the observed particle sedimentation and the concomitant increase in the delivered dose did not result in a decrease of cell viability.

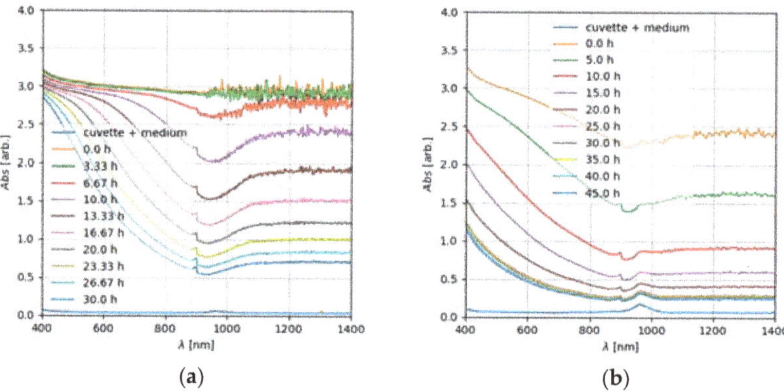

(a)   (b)

Figure 8. Time series of the UV-Vis spectra of NM110 TiO$_2$ dispersed in cell culture medium (treatment; see Section 2.2) at a concentration of c_0 = 2.56 mg/mL (a) and 0.256 mg/mL (b); the lowermost spectrum is from the cuvette and the medium without particles.

In comparison, the dispersion of NM 110 (ZnO) particles (Figure 9a for c_0 = 2.56 mg/mL) exhibited a sloping spectrum in the visible wavelength range and a roughly constant, low absorption in the infrared range. Over time, only a minor decrease in the overall absorption was observed, indicating a limited alteration of the particle dispersion, which in the case of ZnO might be due to sedimentation, but also to particle dissolution [35]. However, apparently not all ZnO particles transformed during the time course of the experiment. In contrast, a prompt transformation of ZnO nanoparticles in the cell culture medium was reported by Ivask et al. [35].

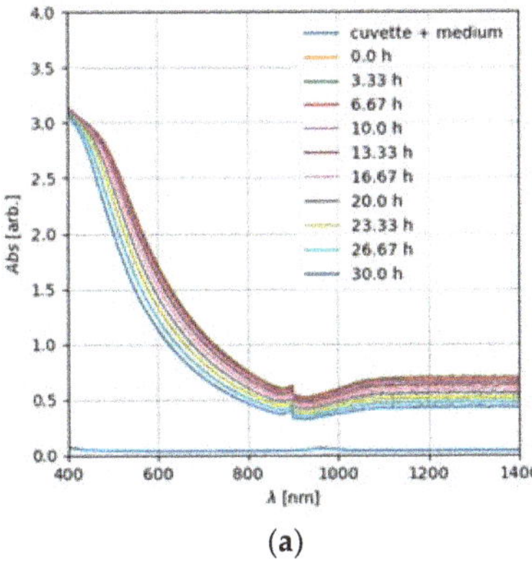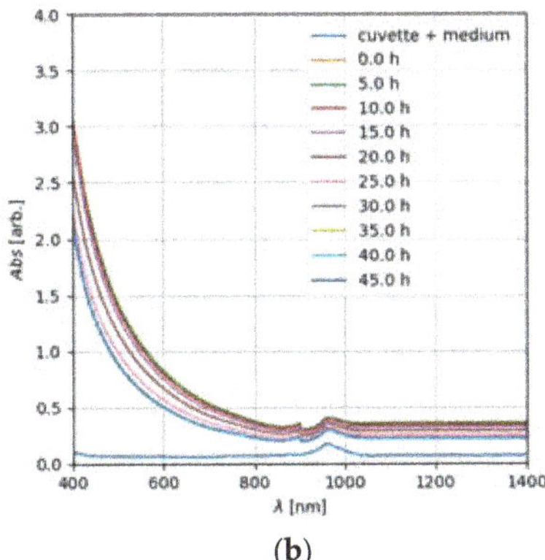

(a)   (b)

Figure 9. Time series of UV-Vis spectra of ZnO nanoparticles (NM 110) dispersed in cell culture medium (treatment; see Section 2.2) at a concentration of c = 2.56 mg/mL (**a**) and 0.256 mg/mL (**b**); the lowermost spectrum represents the medium without particles.

The results indicated that due to the slowness of the sedimentation process (timescale of hours), a strong influence on quantitative and reproducible handling of the dispersion during sample preparation can be neglected. Following the SOP, the ENM dispersions should be utilized within 60 min.

In comparison, particle sedimentation during the exposure of submerged, static cell cultures increased the dose delivered to the cells over the time course of the experiment. Particle sedimentation might be increased by agglomeration but, on the other hand, decreased by dissolution. Dissolution is regarded to be relevant for the impact of ZnO nanoparticles. Data considering the sedimentation and dissolution behavior of ENMs should be provided (see [20], Supplementary File) A recent review on mathematical models including in vitro dosimetry was provided by Lamon et al., 2019 [36].

4. Discussion

Data from nanosafety studies frequently show discrepant outcomes and low reproducibility even within single studies, which leads to contradictory assessments and limits their use for regulatory purposes. In this round-robin test, we aimed to improve the comprehensibility of variable results by using the MIT containing descriptive information and quality criteria [19]. The round-robin test was composed by a WST-1 assay for toxicity measurement of ENMs to evaluate various sources of variance such as the laboratory (partners), the experimenters performing the assay, and the independent replicates of these experiments. Initially, SOPs for cell cultivation, ENM preparation, and WST-1 performance were developed. Furthermore, the SOPs included acceptance criteria to monitor the biological model system. Since the chosen test procedure is a common working process in a toxicology laboratory, we assumed that the implementation could be carried out by the experimenters without further training or explanation. In addition, it should be assumed that the use of different materials and devices, provided they are approved for the described application, should only have a limited influence on the results. In fact, this is not confirmed by various published interlaboratory comparisons [11,12,22]. In previous round-robin tests in the field of nanosafety research, the exchange of materials and the training of experimenters played a key role in achieving less variability in the study results [5]. In contrast, we used the MIT to identify reasons and sources of variances in the results. The main reasons for variances could be found in the MIT groups General Information, Biological Model Information, Exposure information, and Endpoint and Readout Information. Deviations from the SOP based on the experimenter's decision were a main factor.

Two experimenters at each partner laboratory performed three rounds (replicates) of the test procedure. This setup allowed comparison of variances within one laboratory and between the laboratories concerning the WST-1 assay results. Variances could indeed be detected between the three partners, but also between the experimenters within the same laboratory. The results appeared to be influenced by the so-called human factor. Specific laboratory practices such as pipetting techniques, which are not explicitly described in the SOP, or experience and routine regarding the test procedure can be mentioned as a source of variance. Therefore, critical steps in the protocol need to be identified and defined in the SOP to ensure the overall robustness and reproducibility of assay results within and between different laboratories. Additionally, appropriate and standardized ENMs, including their reliable and curated characterization, are needed to ensure the test performance.

Based on the concentration-dependent results obtained for NM-110, the IC_{50} values were calculated. The values varied in the range of 5.91–22.78 $\mu g/cm^2$. In comparison, a multilaboratory toxicological assessment by Kermanizadeh et al. [22] obtained IC_{50} values in the range of 5.39 $\pm$ 6.03 $\mu g/cm^2$. Differences in these values were attributed to potential differences in the handling of ENMs or a variation in cell stocks. ENMs tend to age at different conditions during storage, such as, e.g., silver ENM [37]. A single-laboratory study by Thongkam et al. obtained a higher IC_{50} value of 44 $\mu g/cm^2$ while using the same ENMs. Moreover, in the study of Thongkam et al., cells were seeded at lower densities in order to align the conditions to mutagenicity studies performed in this context. Overall, literature data on IC_{50} values for NM110 or ZnO varied in the range of 2.53–44 $\mu g/cm^2$, which is comparable to the results of the presented round-robin test (Table 3).

Table 3. Literature data on NM110 or ZnO ENM IC_{50} values.

Cell Type	Assay	ENM	Incubation Time	Source	IC_{50} ($\mu g/cm^2$)
THP-1	WST-1	NM110	24 h	Safar et al. 2019 [38]	2.53
A549	WST-1	NM110	24 h	Ding et al. 2020 [39]	3.3
A549	WST-1	ZnO (17 nm)	24 h	Remzova et al. 2019 [40]	9.6
A549	WST-1	NM110	24 h	Thongkam et al. 2017 [23]	44

4.1. Statistics

In risk assessment procedures, in vitro cytotoxicity assays are often used in the context of read-across approaches [41] as they provide, first, quantitative information about adverse effects. Thus, these estimations should be reliable and valid, and the data handling and statistical analyses should be appropriate and described in detail. By analyzing and describing the proportion of outliers using standardized statistical methods, first, information about the quality of the test results could be given and bias due to extreme values could be excluded. Our analysis showed that only a few values of the WST assay were identified as outliers, and they were either excluded from the analysis or substituted by mean values of the remaining measures. Our results also showed that some interlaboratory variation could be reduced by using normalized data (see Figure 6). However, these mathematical transformations mainly affect the confidence interval of the cytotoxicity estimations without having a strong impact on the absolute values. When providing in vitro data in repositories following the FAIR principles, information about outliers and normalization should be given to enable quality assessment of the provided information.

4.2. Metadata Acquisition

The round-robin test was performed to evaluate the applicability and contribution of the MIT to the analysis and identification of sources of variance. The process of the round-robin test comprised the organization of the study design, including the development of the SOPs, the measurements performed by the experimenters, up to the analysis of the results (Figure 10). Nevertheless, these are only major steps that need to be considered within a round-robin test. With regard to data comprehensibility and reuse, specifically, the interfaces of information and data transfer play a key role with regard to data comprehensibility and reuse. Usually, data are stored on private servers, and the data are published in the end. However, the researcher decides which data and information will be shared. In contrast, digital data transfer and non-publication outputs are demanding for collaboration, curation, and (data) management activities. This process includes recording important information about the data and their generation, sources, analysis methods, and changes to the data, which can be fulfilled by the defined requirements for descriptive information and quality criteria given in the MIT. Its use could result in an ongoing data curation throughout the research data life cycle, contributing to data preserving in, e.g., repositories and enabling access and reuse in the future.

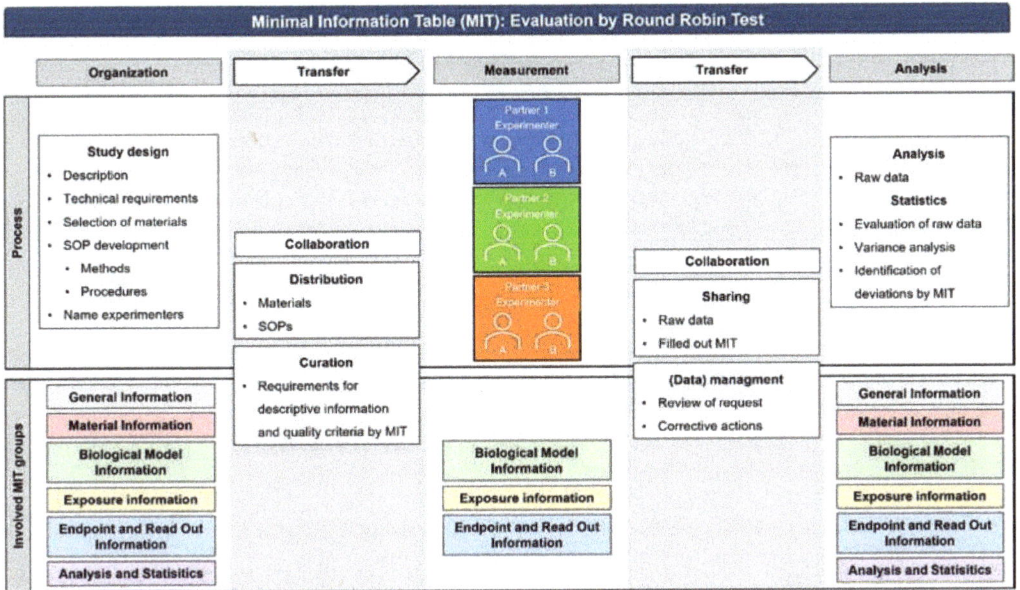

Figure 10. Evaluation process of the MIT by round-robin test. The planning and implementation of the round-robin test could be separated into organization, measurement, and analysis. The connective interfaces are transfer areas with collaboration processes including distribution, curation, sharing, and (data) management.

However, there are aspects and actions that cannot be achieved by using the MIT as a single tool. Therefore, further developments regarding comprehensibility and re-use are necessary. For the first transfer point, it should be considered that, e.g., SOPs need to be available and access conditions should be regulated; furthermore, they should be provided with an identifier (e.g., DOI). Related descriptions should be generally comprehensive and complete, to ensure that the SOP is reproducible by third parties. If necessary, appropriate metadata should be made available, and criteria to evaluate the SOP quality should be given. At the point of data transfer, the generated data are made accessible with the assumption that sensitive data may only be accessible to a restricted circle. Additionally, a persistent identifier is needed that is relevant for subsequent publication. To improve data quality, the persistent identifier should be linked to corresponding SOPs and metadata, e.g., in a standardized format, which could be provided and defined by the MIT. Further developments are necessary to assess the completeness and quality of the data. With regard to ensuring interoperability, SOPs and metadata need to be in a machine-readable format to enable interchangeability across various systems and generate findable metadata (e.g., in a database).

It will be a future challenge to further optimize curation boundaries and develop repositories for data sharing and units for (data) management in line with tasks such as review of requests or corrective actions in the international and multidisciplinary field of nanosafety research.

Supplementary Materials: The following Supporting Information can be downloaded at: https://www.mdpi.com/article/10.3390/nano12071053/s1, Table S1: Minimal information table (MIT); SOP S1: Culturing A549 cells; SOP S2: Sample preparation; SOP S3: Sonicator calibration; SOP S4: Cellular viability—WST-1 assay in A549 cells; Method S1: Background information for the sedimentation analysis; Figure S1: Viability of A549 cells in NM101 (TiO_2). References [42–48] are cited in the supplementary materials.

Author Contributions: Conceptualization: M.R., C.v.T., K.U., R.P.F.S., L.M., N.R., L.E., A.S., J.L. and A.K. methodology: C.v.T., K.U., R.P.F.S., L.M., N.R., L.E., A.S., J.L. and A.K. formal analysis: C.v.T., N.R., L.E., A.S., J.L. and A.K. investigation: C.v.T., N.R., L.E., A.S., J.L. and A.K. writing—review and editing: L.E., A.S., J.L., N.R., K.B., C.B.M., M.R., L.M., K.U., C.v.T., R.P.F.S. and A.K., visualization: L.E., N.R., J.L. and C.v.T.; funding acquisition: M.R., C.v.T., K.U., R.P.F.S., L.M., N.R. and A.K. All authors have read and agreed to the published version of the manuscript.

Funding: This work is part of the NanoS-QM project and received funding by the German Federal Ministry of Education and Research (BMBF) under FKZ 16QK07.

Institutional Review Board Statement: Not applicable.

Informed Consent Statement: Not applicable.

Data Availability Statement: See Supplementary Materials.

Acknowledgments: The authors thank Silke Kiefer, Andrea Boltendahl, Karolina Zajac, and Marwa Bahri for their excellent technical assistance.

Conflicts of Interest: The authors declare no conflict of interest.

References

1. Warheit, D.B. Hazard and risk assessment strategies for nanoparticle exposures: How far have we come in the past 10 years? *F1000Research* **2018**, *7*, 376. [CrossRef] [PubMed]
2. Johnston, L.J.; Gonzalez-Rojano, N.; Wilkinson, K.J.; Xing, B. Key challenges for evaluation of the safety of engineered nanomaterials. *NanoImpact* **2020**, *18*, 100219. [CrossRef]
3. Arts, J.H.E.; Hadi, M.; Irfan, M.-A.; Keene, A.M.; Kreiling, R.; Lyon, D.; Maier, M.; Michel, K.; Petry, T.; Sauer, U.G.; et al. A decision-making framework for the grouping and testing of nanomaterials (DF4nanoGrouping). *Regul. Toxicol. Pharmacol.* **2015**, *71*, S1–S27. [CrossRef] [PubMed]
4. Xia, T.; Hamilton, R.F.; Bonner, J.C.; Crandall, E.D.; Elder, A.; Fazlollahi, F.; Girtsman, T.A.; Kim, K.; Mitra, S.; Ntim, S.A.; et al. Interlaboratory Evaluation of in Vitro Cytotoxicity and Inflammatory Responses to Engineered Nanomaterials: The NIEHS Nano GO Consortium. *Environ. Health Perspect.* **2013**, *121*, 683–690. [CrossRef] [PubMed]
5. Nelissen, I.; Haase, A.; Anguissola, S.; Rocks, L.; Jacobs, A.; Willems, H.; Riebeling, C.; Luch, A.; Piret, J.-P.; Toussaint, O.; et al. Improving Quality in Nanoparticle-Induced Cytotoxicity Testing by a Tiered Inter-Laboratory Comparison Study. *Nanomaterials* **2020**, *10*, 1430. [CrossRef]
6. Haase; Klaessig. EU US Roadmap Nanoinformatics 2030. EU Nanosafety Cluster. 2018. Available online: https://zenodo.org/record/1486012#.YjrNfTYzaUk (accessed on 20 February 2022).
7. Halappanavar, S.; Brule, S.V.D.; Nymark, P.; Gaté, L.; Seidel, C.; Valentino, S.; Zhernovkov, V.; Danielsen, P.H.; De Vizcaya-Ruiz, A.; Wolff, H.; et al. Adverse outcome pathways as a tool for the design of testing strategies to support the safety assessment of emerging advanced materials at the nanoscale. *Part. Fibre Toxicol.* **2020**, *17*, 16. [CrossRef]
8. Landsiedel, R.; Ma-Hock, L.; Wiench, K.; Wohlleben, W.; Sauer, U.G. Safety assessment of nanomaterials using an advanced decision-making framework, the DF4nanoGrouping. *J. Nanopart. Res.* **2017**, *19*, 171. [CrossRef]
9. Hirsch, C.; Roesslein, M.; Krug, H.F.; Wick, P. Nanomaterial cell interactions: Are current in vitro tests reliable? *Nanomedicine* **2011**, *6*, 837–847. [CrossRef]
10. Krug, H.F. Nanosafety research-are we on the right track? *Angew. Chem. Int. Ed.* **2014**, *53*, 12304–12319. [CrossRef]
11. Barosova, H.; Meldrum, K.; Karakocak, B.B.; Balog, S.; Doak, S.H.; Petri-Fink, A.; Clift, M.J.D.; Rothen-Rutishauser, B. Inter-laboratory variability of A549 epithelial cells grown under submerged and air-liquid interface conditions. *Toxicol. Vitr.* **2021**, *75*, 105178. [CrossRef]

12. Elliott, J.T.; Rösslein, M.; Song, N.W.; Toman, B.; Kinsner-Ovaskainen, A.; Maniratanachote, R.; Salit, M.L.; Petersen, E.J.; Sequeira, F.; Romsos, E.L.; et al. Toward achieving harmonization in a nano-cytotoxicity assay measurement through an interlaboratory comparison study. *ALTEX* **2017**, *34*, 201–218. [CrossRef] [PubMed]

13. Nymark, P.; Bakker, M.; Dekkers, S.; Franken, R.; Fransman, W.; García-Bilbao, A.; Greco, D.; Gulumian, M.; Hadrup, N.; Halappanavar, S.; et al. Toward Rigorous Materials Production: New Approach Methodologies Have Extensive Potential to Improve Current Safety Assessment Practices. *Small* **2020**, *16*, 1904749. [CrossRef] [PubMed]

14. Llewellyn, S.V.; Conway, G.E.; Zanoni, I.; Jørgensen, A.K.; Shah, U.-K.; Seleci, D.A.; Keller, J.G.; Kim, J.W.; Wohlleben, W.; Jensen, K.A.; et al. Understanding the impact of more realistic low-dose, prolonged engineered nanomaterial exposure on genotoxicity using 3D models of the human liver. *J. Nanobiotechnol.* **2021**, *19*, 193. [CrossRef] [PubMed]

15. Faria, M.; Björnmalm, M.; Thurecht, K.J.; Kent, S.J.; Parton, R.G.; Kavallaris, M.; Johnston, A.P.R.; Gooding, J.J.; Corrie, S.R.; Boyd, B.J.; et al. Minimum information reporting in bio–nano experimental literature. *Nat. Nanotechnol.* **2018**, *13*, 777–785. [CrossRef] [PubMed]

16. Rasmussen, K.; Rauscher, H.; Kearns, P.; González, M.; Sintes, J.R. Developing OECD test guidelines for regulatory testing of nanomaterials to ensure mutual acceptance of test data. *Regul. Toxicol. Pharmacol.* **2019**, *104*, 74–83. [CrossRef]

17. Seiler, J.P. *Good Laboratory Practice—The Why and the How*; Springer: Berlin/Heidelberg, Germany, 2005; ISBN 9783540282341.

18. Papadiamantis, A.G.; Klaessig, F.C.; Exner, T.E.; Hofer, S.; Hofstaetter, N.; Himly, M.; Williams, M.A.; Doganis, P.; Hoover, M.D.; Afantitis, A.; et al. Metadata Stewardship in Nanosafety Research: Community-Driven Organisation of Metadata Schemas to Support FAIR Nanoscience Data. *Nanomaterials* **2020**, *10*, 2033. [CrossRef]

19. Elberskirch, L.; Binder, K.; Riefler, N.; Sofranko, A.; Liebing, J.; Minella, C.B.; Mädler, L.; Razum, M.; van Thriel, C.; Unfried, K.; et al. Digital research data: From analysis of existing standards to a scientific foundation for a modular metadata schema in nanosafety. *Part. Fibre Toxicol.* **2022**, *19*, 1–19. [CrossRef]

20. Ong, K.J.; MacCormack, T.J.; Clark, R.J.; Ede, J.D.; Ortega, V.A.; Felix, L.C.; Dang, M.K.M.; Ma, G.; Fenniri, H.; Veinot, J.G.C.; et al. Widespread Nanoparticle-Assay Interference: Implications for Nanotoxicity Testing. *PLoS ONE* **2014**, *9*, e90650. [CrossRef]

21. Guadagnini, R.; Halamoda Kenzaoui, B.; Walker, L.; Pojana, G.; Magdolenova, Z.; Bilanicova, D.; Saunders, M.; Juillerat-Jeanneret, L.; Marcomini, A.; Huk, A.; et al. Toxicity screenings of nanomaterials: Challenges due to interference with assay processes and components of classic in vitrotests. *Nanotoxicology* **2015**, *9*, 13–24. [CrossRef]

22. Kermanizadeh, A.; Gosens, I.; MacCalman, L.; Johnston, H.; Danielsen, P.H.; Jacobsen, N.R.; Lenz, A.G.; Fernandes, T.; Schins, R.P.F.; Cassee, F.R.; et al. A Multilaboratory Toxicological Assessment of a Panel of 10 Engineered Nanomaterials to Human Health—ENPRA Project—The Highlights, Limitations, and Current and Future Challenges. *J. Toxicol. Environ. Health—Part B Crit. Rev.* **2016**, *19*, 1–28. [CrossRef]

23. Thongkam, W.; Gerloff, K.; Van Berlo, D.; Albrecht, C.; Schins, R.P. Oxidant generation, DNA damage and cytotoxicity by a panel of engineered nanomaterials in three different human epithelial cell lines. *Mutagenesis* **2017**, *32*, 105–115. [CrossRef] [PubMed]

24. Rasmussen, K.; Mast, J.; De Temmerman, P.-J.; Verleysen, E.; Waegeneers, N.; Van Steen, F.; Pizzolon, J.C.; De Temmerman, L.; Van Doren, E.; Jensen, K.A.; et al. *Titanium Dioxide, NM-100, NM-101, NM-102, NM-103, NM-104, NM-105: Characterisation and Physico-Chemical Properties*; European Comission: Brussels, Belgium, 2014; ISBN 9789279381881.

25. Singh, C.; Friedrichs, S.; Levin, M.; Birkedal, R.; Jensen, K.A.; Pojana, G.; Wohlleben, W.; Schulte, S.; Wiench, K.; Turney, T.; et al. *NM-Series of Representative Manufactured Nanomaterials—Zinc Oxide NM-110, NM-111, NM-112, NM-113: Characterisation and Test Item Preparation*; European Comission: Brussels, Belgium, 2011. [CrossRef]

26. Taurozzi, J.S.; Hackley, V.A. Protocol for Preparation of Nanoparticle Dispersions From Powdered Material Using Ultrasonic Disruption. *CEINT Natl. Inst. Standars Technol.* **2010**, *1200*, 1200–1202. [CrossRef]

27. Booth, A.; Jensen, K.A. NANoREG D4.12 SOP Probe Sonicator Calibration for Ecotoxicological Testing. Available online: https://www.rivm.nl/en/documenten/nanoreg-d412-sop-probe-sonicator-calibration-for-ecotoxicological-testing (accessed on 11 February 2022).

28. Jensen, K.A.; Kembouche, Y.; Christiansen, E.; Jacobsen, N.R.; Wallin, H.; Guiot, C.; Spalla, O.; Witschger, O. *Physicochemical Characterisation of Manufactured Nanomaterials (MNs) and Exposure Media (EMs) Deliverable 3: Final Protocol for Producing Suitable MN Exposure Media*; The National Research Centre for the Working Environment: Copenhagen, Denmark, 2009.

29. Teeguarden, J.G.; Hinderliter, P.M.; Orr, G.; Thrall, B.D.; Pounds, J.G. Particokinetics In Vitro: Dosimetry Considerations for In Vitro Nanoparticle Toxicity Assessments. *Toxicol. Sci.* **2007**, *95*, 300–312. [CrossRef]

30. Lieber, M.; Todaro, G.; Smith, B.; Szakal, A.; Nelson-Rees, W. A continuous tumor-cell line from a human lung carcinoma with properties of type II alveolar epithelial cells. *Int. J. Cancer* **1976**, *17*, 62–70. [CrossRef] [PubMed]

31. Leibrock, L.; Wagener, S.; Singh, A.V.; Laux, P.; Luch, A. Nanoparticle induced barrier function assessment at liquid–liquid and air–liquid interface in novel human lung epithelia cell lines. *Toxicol. Res.* **2019**, *8*, 1016–1027. [CrossRef] [PubMed]

32. Vietti, G.; Ibouraadaten, S.; Palmai-Pallag, M.; Yakoub, Y.; Bailly, C.; Fenoglio, I.; Marbaix, E.; Lison, D.; van den Brule, S. Towards predicting the lung fibrogenic activity of nanomaterials: Experimental validation of an in vitro fibroblast proliferation assay. *Part. Fibre Toxicol.* **2013**, *10*, 52. [CrossRef]

33. Krebs, A.; Waldmann, T.; Wilks, M.F.; van Vugt-Lussenburg, B.M.A.; van der Burg, B.; Terron, A.; Steger-Hartmann, T.; Ruegg, J.; Rovida, C.; Pedersen, E.; et al. Template for the description of cell-based toxicological test methods to allow evaluation and regulatory use of the data. *ALTEX* **2019**, *36*, 682–699. [CrossRef]

34. Kim, J.A.; Åberg, C.; Salvati, A.; Dawson, K.A. Role of cell cycle on the cellular uptake and dilution of nanoparticles in a cell population. *Nat. Nanotechnol.* **2012**, *7*, 62–68. [CrossRef]

35. Ivask, A.; Scheckel, K.G.; Kapruwan, P.; Stone, V.; Yin, H.; Voelcker, N.H.; Lombi, E. Complete transformation of ZnO and CuO nanoparticles in culture medium and lymphocyte cells during toxicity testing. *Nanotoxicology* **2017**, *11*, 150–156. [CrossRef]

36. Lamon, L.; Asturiol, D.; Vilchez, A.; Cabellos, J.; Damásio, J.; Janer, G.; Richarz, A.; Worth, A. Physiologically based mathematical models of nanomaterials for regulatory toxicology: A review. *Comput. Toxicol.* **2019**, *9*, 133–142. [CrossRef]

37. Izak-Nau, E.; Huk, A.; Reidy, B.; Uggerud, H.; Vadset, M.; Eiden, S.; Voetz, M.; Himly, M.; Duschl, A.; Dusinska, M.; et al. Impact of storage conditions and storage time on silver nanoparticles' physicochemical properties and implications for their biological effects. *RSC Adv.* **2015**, *5*, 84172–84185. [CrossRef]

38. Safar, R.; Doumandji, Z.; Saidou, T.; Ferrari, L.; Nahle, S.; Rihn, B.H.; Joubert, O. Cytotoxicity and global transcriptional responses induced by zinc oxide nanoparticles NM 110 in PMA-differentiated THP-1 cells. *Toxicol. Lett.* **2019**, *308*, 65–73. [CrossRef]

39. Ding, Y.; Weindl, P.; Lenz, A.-G.; Mayer, P.; Krebs, T.; Schmid, O. Quartz crystal microbalances (QCM) are suitable for real-time dosimetry in nanotoxicological studies using VITROCELL®Cloud cell exposure systems. *Part. Fibre Toxicol.* **2020**, *17*, 44. [CrossRef] [PubMed]

40. Remzova, M.; Zouzelka, R.; Brzicova, T.; Vrbova, K.; Pinkas, D.; Rőssner, P.; Topinka, J.; Rathousky, J. Toxicity of TiO$_2$, ZnO, and SiO$_2$ Nanoparticles in Human Lung Cells: Safe-by-Design Development of Construction Materials. *Nanomaterials* **2019**, *9*, 968. [CrossRef] [PubMed]

41. Gajewicz, A.; Schaeublin, N.; Rasulev, B.; Hussain, S.; Leszczynska, D.; Puzyn, T.; Leszczynski, J. Towards understanding mechanisms governing cytotoxicity of metal oxides nanoparticles: Hints from nano-QSAR studies. *Nanotoxicology* **2015**, *9*, 313–325. [CrossRef]

42. DSMZ. Available online: https://www.dsmz.de/collection/catalogue/details/culture/ACC-107 (accessed on 11 February 2022).

43. Available online: https://www.nanopartikel.info/files/methodik/VIGO/cell_culture_A549.pdf (accessed on 11 February 2022).

44. Foster, K.A.; Oster, C.G.; Mayer, M.M.; Avery, M.L.; Audus, K.L. Characterization of the A549 cell line as a type II pulmonary epithelial cell model for drug metabolism. *Exp. Cell Res.* **1998**, *243*, 359–366. [CrossRef]

45. Giard, D.J.; Aaronson, S.A.; Todaro, G.J.; Arnstein, P.; Kersey, J.H.; Dosik, H.; Parks, W.P. In vitro cultivation of human tumors: establishment of cell lines derived from a series of solid tumors. *J. Natl. Cancer Inst.* **1973**, *51*, 1417–1423. [CrossRef]

46. Available online: https://www.ols-bio.de/wp-content/uploads/2020/02/OLS_-CASY_TTT-OperatorsGuide_2018-8.pdf (accessed on 11 February 2022).

47. Jensen, K.A. *The ENPRA Dispersion Protocol for NANoREG*; National Research Centre for the Working Environment: Copenhagen, Denmark, 2014.

48. Booth, A.; Jensen, K.A. SOP for Probe Sonicator Calibration of Delivered Acoustic Power and Deagglomeration Efficiency for Ecotoxicological Testing. 2015. Available online: http://creativecommons.org/licenses/by-nc-sa/4.0/ (accessed on 11 February 2022).

 nanomaterials

Article

Cellular Uptake of Silica and Gold Nanoparticles Induces Early Activation of Nuclear Receptor NR4A1

Mauro Sousa de Almeida [1], Patricia Taladriz-Blanco [1,2], Barbara Drasler [1], Sandor Balog [1], Phattadon Yajan [1], Alke Petri-Fink [1,3] and Barbara Rothen-Rutishauser [1,*]

[1] Adolphe Merkle Institute, University of Fribourg, Chemin des Verdiers 4, 1700 Fribourg, Switzerland; mauro.sousadealmeida@unifr.ch (M.S.d.A.); patricia.taladriz@inl.int (P.T.-B.); barbara.drasler@unifr.ch (B.D.); sandor.balog@unifr.ch (S.B.); phattadon.yajan@unifr.ch (P.Y.); alke.fink@unifr.ch (A.P.-F.)
[2] Water Quality Group, International Iberian Nanotechnology Laboratory (INL), Av. Mestre José Veiga s/n, 4715-330 Braga, Portugal
[3] Department of Chemistry, University of Fribourg, Chemin du Musée 9, 1700 Fribourg, Switzerland
* Correspondence: barbara.rothen@unifr.ch; Tel.: +41-26-300-9502

Abstract: The approval of new nanomedicines requires a deeper understanding of the interaction between cells and nanoparticles (NPs). Silica (SiO_2) and gold (Au) NPs have shown great potential in biomedical applications, such as the delivery of therapeutic agents, diagnostics, and biosensors. NP-cell interaction and internalization can trigger several cellular responses, including gene expression regulation. The identification of differentially expressed genes in response to NP uptake contributes to a better understanding of the cellular processes involved, including potential side effects. We investigated gene regulation in human macrophages and lung epithelial cells after acute exposure to spherical 60 nm SiO_2 NPs. SiO_2 NPs uptake did not considerably affect gene expression in epithelial cells, whereas five genes were up-regulated in macrophages. These genes are principally related to inflammation, chemotaxis, and cell adhesion. Nuclear receptor NR4A1, an important modulator of inflammation in macrophages, was found to be up-regulated. The expression of this gene was also increased upon 1 h of macrophage exposure to spherical 50 nm AuNPs and 200 nm spherical SiO_2 NPs. NR4A1 can thus be an important immediate regulator of inflammation provoked by NP uptake in macrophages.

Keywords: nanoparticles; gene regulation; endocytosis; inflammation; NR4A1

Citation: Sousa de Almeida, M.; Taladriz-Blanco, P.; Drasler, B.; Balog, S.; Yajan, P.; Petri-Fink, A.; Rothen-Rutishauser, B. Cellular Uptake of Silica and Gold Nanoparticles Induces Early Activation of Nuclear Receptor NR4A1. *Nanomaterials* **2022**, *12*, 690. https://doi.org/10.3390/nano12040690

Academic Editors: Andrea Hartwig and Christoph Van Thriel

Received: 28 December 2021
Accepted: 16 February 2022
Published: 18 February 2022

Publisher's Note: MDPI stays neutral with regard to jurisdictional claims in published maps and institutional affiliations.

1. Introduction

Administration of clinically relevant nanoparticles (NPs) to humans can occur in various ways, including inhalation, oral ingestion, injection (intravenous, intramuscular, and subcutaneous), and dermal and ocular penetration [1,2]. Once inside the human body, the NPs can overcome organs and tissue barriers and then come into contact with single cells. Strong evidence has indicated that cellular responses to NPs are cell-type- and NP-dependent [1,3,4]. This means that each type of NP, with its intrinsic properties (e.g., size, shape, stiffness, surface chemistry, etc.), may lead to different cellular responses in different cell types [1,3–5]. For the design and optimization of biomedically relevant NPs, it is important to understand the mechanisms induced at the single-cell level. Cell-NP interaction may activate signaling cascades, leading to structural modifications inside cells and at the cell surface, interfering with normal cell function [6].

When NPs are deposited on the outer cellular membrane, they may interact and be internalized, mainly via endocytosis [7]. Endocytosis occurs via multiple mechanisms, including phagocytosis and pinocytosis (macropinocytosis, clathrin-mediated endocytosis (CME), caveolae-mediated endocytosis, and clathrin- and caveolae-independent endocytosis) [8]. All of the aforementioned mechanisms are complex and involve a wide range of molecules (e.g., surface receptors, lipids, and adaptor proteins) that work together to ensure an efficient process of endocytosis [9]. For example, non-porous silica (SiO_2) and gold (Au) NPs, which have been extensively studied in biomedical context thanks to their controllable and large-scale syntheses, facile surface modification and biocompatibility, revealed different uptake mechanisms in different cell types [10,11]. Shapero et al. reported that spherical 50, 100, and 300 nm SiO_2 NPs do not enter human lung epithelial cells(A549) via clathrin- or caveolae-mediated endocytosis. However, independent of their size, all NPs were internalized via an energy-dependent mechanism and ended up in lysosomes [12]. On the other hand, a similar study led by Hsiao et al. concluded that spherical 15, 60, and 200 nm SiO_2 NPs are internalized via clathrin-mediated endocytosis in A549 cells, but also in macrophage-derived THP-1 cells [13]. The authors also proved that caveolae-mediated endocytosis contributes to the uptake of the 200 nm SiO_2 NPs in A549 cells and the uptake of 60 and 200 nm NPs in macrophage-like THP-1 cells. In this study, the exposure of cells to NPs was conducted in the absence of serum, which might explain the different findings. Clearly, both studies prove that the presence of proteins in the cell-culture medium influence NP uptake and support the idea that observations in one type of cell should not be extrapolated to another. Spherical-shaped SiO_2 NPs have received the most attention in nanomedicine; however it is known that NPs' shape can also influence the cellular internalization mechanism [7]. Similarly, while spherical AuNPs are the main investigated type of NPs, several other studies looked into the effect of other shapes, such as rods, stars, and triangles, on the cellular uptake mechanisms. Ding et al. conducted a study to evaluate the effect of the cellular internalization of AuNPs in the form of spheres, rods, and stars in mouse breast cancer (4T1), human hepatoma (SMCC-7721), and human gastric mucosal (GES-1) cells [11]. The results showed that spherical AuNPs (average diameter of 15 and 45 nm), rod-shaped AuNPs (33 × 10 nm), and star-shaped AuNPs (average diameter 15 nm) were principally internalized through clathrin-mediated endocytosis in the different cell types. In addition, authors revealed that the uptake of Au nanostars also occurred via caveolin-mediated endocytosis, and that macropinocytosis was involved in the internalization of larger spherical AuNPs (average diameter 80 nm).

Aside from the internalization mechanism, several researchers have focused their investigations on different biological effects of SiO_2 and Au NPs, such as cell viability, oxidative stress, and pro-inflammation [14–17]. Lin et al. reported an increase in intracellular reactive oxygen species (ROS), leading to oxidative stress and apoptosis after 12 h exposure of A549 cells to spherical 37 nm SiO_2 NPs at 50 µg/mL [18]. Zhao et al. concluded that spherical SiO_2 NPs of 27 nm can block the autophagic flux and impair lysosomal acidification in A549 cells after 24 h exposure to 50 µg/mL [19]. Kusaka et al. found that exposure of bone marrow-derived macrophages to 100 µg/mL of spherical SiO_2 NPs with 30, 100, and 300 nm for 4 h, trigger inflammation, lysosomal destabilization, and cell death [20]. In summary, the recent in vitro experiments revealed that SiO_2 NPs can impair normal cell function, inducing autophagic dysfunction, oxidative stress, and inflammatory response in different cell types in a dose and size-dependent manner. The exposure of different cells to AuNPs also revealed potential harmful effects. Uboldi et al. showed that 72 h exposure of spherical AuNPs with 9.5, 11.2, and 25 nm to A549 cells, at ~138 µg/mL, induced cytotoxicity [17]. Contrarily, Zhang et al. did not observe any sign of cytotoxicity and inflammation in murine macrophage cells (RAW 264.7) after 48 h exposure to spherical 60 nm AuNPs (100 µg/mL) [21]. Another study from D. Bachand et al. concluded that the exposure of spherical 20 and 60 nm AuNPs at concentrations of ~350 pg/mL did not cause any significant change in neither oxidative stress, nor cytotoxicity after 24 and 48 h in A549 cells [22]. Nevertheless, an increase in IL-8 secretion was observed for both AuNPs, revealing that these NPs can trigger inflammation. A higher IL-8 release from A549 cells was observed for 20 nm AuNPs, confirming that NP size affects the cellular responses. In short, cellular responses are influenced not only by the physicochemical properties of the NPs and the cell type, but also by other factors such as administered and delivered dose, and exposure time. Omics-based research has been applied in some of the aforementioned studies as it allows a thorough and systematic investigation of the changes that occur at the gene/transcript/protein level. This is crucial for a deeper understanding of the potential molecular mechanisms associated with NP uptake and potential adverse effects. Thus far, most of the research in this area has focused on the analysis of the transcriptome profiling after extended exposure to NPs (i.e., 24, 48, and 72 h). Nevertheless, we presume that cellular responses to NPs start a few minutes/hours after exposure, as cellular internalization occurs within this timeframe [12,23].

In this study, we evaluated the overall impact of spherical SiO_2 NPs in human macrophages and lung epithelial cells, by using RNA-Seq to examine genome-wide transcriptional changes with a focus on endocytic and early-response genes. Based on previous findings [12,24] and on the in vitro sedimentation, diffusion, and dosimetry model (ISDD)[25], we showed that a small fraction of administered NPs can reach the cells after 1 h and be internalized. In this sense, we have decided to evaluate the cellular effects of SiO_2 NPs exposure at three different time points: 1, 6, and 24 h. The current study is one of the first to investigate whether NP uptake influences the expression of endocytic genes at early time points. Our results show that the uptake of 60 nm SiO_2 NPs did not affect the expression of endocytosis-related genes in macrophages and lung epithelial cells, but did increase the expression of five genes involved in inflammation, chemotaxis, and cell adhesion in macrophages. In particular, the most relevant gene involved in inflammation, nuclear receptor 4A1 (*NR4A1*), was up-regulated in macrophages upon 1 h exposure to 60 nm and 200 SiO_2 NPs. Furthermore, to study the effect of material composition, macrophages were exposed for 1 h to spherical 50 nm Au NPs, which resulted in the up-regulation of *NR4A1*, revealing *NR4A1* as an early response gene and possibly an immediate regulator of inflammation.

2. Materials and Methods

2.1. Synthesis of Nonporous SiO₂-Rhodamine B NPs

SiO_2 NPs measuring 60 nm were synthesized following a modified-Stöber method [26,27]. Briefly, a mixture of absolute ethanol (144 mL, EtOH, $\geq$99.8%VWR, Dietikon, Switzerland), Milli-Q water (6.75 mL), and ammonium hydroxide (3.9 mL, NH_4OH $\geq$25% NH_3 in water, Merck, Zug, Switzerland was heated in a 500 mL rounded-bottom flask provided with a reflux system at 60°C under magnetic stirring. After 30 min, tetraethyl orthosilicate (11 mL, TEOS, >99%, Sigma Aldrich, Buchs, Switzerland) was added to the mixture and stirred for 2 min. Then, 100 μL of a mixture containing 100 μL Rhodamine B isothiocyanate (10 mg/mL RhoB in EtOH, Dye content ~95%, Sigma Aldrich, Buchs, Switzerland) and 1.5 μL of (3-aminopropyl) triethoxysilane (APTES, 99%, Sigma Aldrich, Buchs, Switzerland) were added to the flask using a syringe. After 4 h at 60 °C, the fluorescently labeled NPs were dialyzed against water for three days and stored at 4 °C in the darkness. The 200 nm SiO_2 NPs were prepared using a similar synthetic approach. A mixture of 11 mL of TEOS, 180 mL of EtOH, 36 mL of Milli-Q water, and 24 mL of NH_4OH was stirred at room temperature for 2 min before adding 300 μL of an APTES-RhodB mixture (10 mg/mL RhodB in EtOH) and 7.5 μL of APTES. The mixture was stirred overnight, and the particles were cleaned twice by centrifugation at 100 g and redispersed in EtOH. Three additional washes were carried out by centrifugation at 988 g to finally redisperse the NPs in autoclaved Milli-Q water. The concentration of the particles was determined by measuring the weighted average of dried 1 mL particle suspension in three different Eppendorf tubes.

2.2. Synthesis of Au NPs

Au NPs measuring 50 nm were synthesized by the Brown method [28,29]. Briefly, 1.34 mL of 0.22 M of hydroxylamine hydrochloride ($NH_2OH \cdot HCl$, ACS reagent, 98%, Sigma-Aldrich, Buchs, Switzerland) was added to a solution containing 144 mL of gold (III) chloride trihydrate (0.25 mM, $HAuCl_4 \cdot 3H_2O$, $\geq$99.9%, Sigma-Aldrich, Buchs, Switzerland), as-prepared 15 nm gold seeds ([Au] = 0.0125 mM) and sodium citrate tribasic dihydrate (0.5 mM NaCit, $C_6H_5Na_3O_7 \cdot 2H_2O$, $\geq$98%, Sigma-Aldrich, Buchs, Switzerland) under vigorous magnetic stirring. After 15 min under magnetic stirring, the NPs were cleaned by centrifugation at 3500 rpm for 20 min and redispersed in 0.5 mM NaCit. Au seeds measuring 15 nm were prepared by the well-known Turkevich method [30]. Briefly, 0.5 mM HAuCl$_4$ was boiled in the presence of 1.7 mM NaCit for 15 min. Au NPs measuring 15 nm were cooled down to room temperature and stored at 4°C overnight before using them to synthesize the larger particles.

2.3. Fluoresbrite® Yellow-Green Polystyrene (PS)-Based Latex NPs

Yellow-green PS microspheres with ~50 nm in diameter were purchased from Chemie Brunschwig AG (Basel, Switzerland). The company states that PS NPs are stable and dye leaching is not expected, making them suitable for use in cell experiments.

2.4. Physicochemical Characterization

NPs were drop-cast onto a 300-mesh carbon-membrane-coated copper grid and imaged using a Tecnai Spirit transmission electron microscope (TEM) (FEI, Hillsboro, OR, USA) operating at 120 kV equipped with a CCD camera (Eagle, Thermo Fischer, Waltham, MA, USA). The core diameter and size distribution were calculated using an open-source image processing program (ImageJ). UV−Vis extinction spectrum of Au NPs was recorded in a Jasco V-670 spectrophotometer using 10 mm path length quartz Suprasil-grade cuvettes (Hellma Analytics, Plainview, NY, USA) at 25 °C. The stability of the NPs in the cell culture media was tested at 0 and 24 h by DLS at 25 °C and one scattering angle (90°) using a commercial goniometer instrument (3D LS Spectrometer, LS Instruments AG, Fribourg, Switzerland) equipped with a linearly polarized and a collimated laser beam (Cobolt 05-01 diode-pumped solid-state laser, λ = 660 nm, P max. = 500 mW). Two APD detectors, assembled for pseudo-cross-correlation, were used to improve the signal-to-noise ratio. The scattering signal of complete RPMI 1640 (cRPMI 1640) and serum-free RPMI 1640 (i.e., without Fetal Bovine Serum (FBS) but supplemented with L-glutamine and penicillin/streptomycin) obtained by DLS was subtracted from the signal of the NPs suspended in the media, as presented elsewhere [31], to obtain only the contribution of the particles. NPs dispersed in Milli-Q water were used as a control. The mean and standard deviation were calculated from five independent measurements. The surface charge was determined by phase-amplitude light scattering (ZetaPALS, Brookhaven Instruments Corp., Holtsville, NY, USA) in Milli-Q water.

2.5. Cell Culture

Cell culture reagents were purchased from Gibco, Thermo Fisher Scientific (Zug, Switzerland), unless otherwise specified. Human alveolar epithelial type II cells (A549 cell line) from American Type Culture Collection (ATCC, Rockville, MD, USA) were cultured in Roswell Park Memorial Institute (RPMI)-1640 cell culture medium supplemented with 10 vol.% FBS, 2 mM L-Glutamine (100 Units/mL), and Penicillin-Streptomycin (100 µg/mL). The final solution is referred to as cRPMI 1640. Cell cultures were kept in a humidified incubator (37 °C, 5% CO$_2$, and 95% humidity). A549 epithelial cells were sub-cultured twice per week using a mixture of 0.25% Trypsin-Ethylenediaminetetraacetic acid (EDTA) according to the ATCC recommendations. Prior to seeding, cell concentration was determined using the trypan blue exclusion assay (0.4%vol. in phosphate-buffered saline solution (PBS, pH 7.2, Gibco, Life Technologies Europe B.V., Zug, Switzerland)) and an automated cell counter (EVE, NanoEnTek Inc., Seoul, South Korea). A549 cells (5.26×10^4 cells/cm^2), in the passage range of 4–15, were seeded for 24 h in cRPMI 1640 followed by serum starvation for 24 h before NP exposure. Serum starvation was performed to synchronize all cells to the same cell cycle phase. Primary human monocyte-derived macrophages (MDMs) were obtained by isolating and further differentiating human peripheral blood monocytes from human blood buffy coats (Blood Donation Service, Bern University Hospital, Bern, Switzerland), as previously described [32,33]. The work involving primary monocytes isolation from human blood was approved by the committee of the Federal Office for Public Health Switzerland (reference number: 611-1, Meldung A110635/2). Briefly, human blood was separated using density gradient filtration (Lymphoprep, Grogg Chemie, Stettlen, Switzerland) and the monocyte fraction was extracted from the mixture and purified using CD14+ magnetic MicroBeads (Miltenyi Biotec GmbH, Bergisch Gladbach, Germany) according to the manufacturer's protocol. For the monocyte differentiation, the isolated blood monocytes were cultured in 6-well plates (Corning®Falcon, Reinach, Switzerland) containing 3 mL of cRPMI 1640 supplemented with the macrophage colony-stimulating factor (M-CSF) (10 ng/mL, Milteny Biotech, Bergisch Gladbach, Germany) for seven days at a density of 10^6 cells/mL. After this period, cRPMI containing the M-CSF was removed and MDMs (1.05×10^5 cells/cm^2) were seeded for 24 h before NP exposure.

2.6. NPs Exposure

A549 and MDMs grown in 6-well plates and 35 mm glass-bottom dish (MatTek Inc., Ashland, MA, USA) were exposed to 3 mL of 60 nm SiO$_2$-RhodB, 200 nm SiO$_2$-RhodB, 60 nm PS particles ([NP] = 20 µg/mL), or Au NPs ([Au] = 20 µg/mL) previously suspended in cRPMI 1640. For experiments where µ-Slide 8 Wells (Ibidi, Graefelfing, Germany) were used (i.e., sections "Fluorescence imaging" and "Co-localization analysis"), cells were exposed to 316 µL of previously suspended NPs. ISDD model was used to estimate the particle deposition [25]. The relative densities and the diameter of each NP, based on TEM analysis, were taken in consideration. Amorphous silica, 2.2 g/cm^3; gold, 19.32 g/cm^3; polystyrene, 1.05 g/cm^3. After exposure, cells were washed 3 times with PBS to remove the non-cell adhered NPs.

2.7. Cytotoxicity Assay

Lactate dehydrogenase (LDH, Roche Applied Science, Mannheim, Germany) assay was performed on the cell supernatants after NPs exposure in 6-well plates. Triton X-100 at 0.2 vol.% was added to the cell culture medium as a positive control for 6 h prior to collecting the supernatant. LDH levels were measured in triplicate by following the manufacturer's protocol. The absorbance of the colorimetric product was determined by spectrophotometry (Benchmark Microplate reader, BioRad, Cressier, Switzerland) at 490 nm with a reference wavelength of 630 nm. Interference analysis was performed for SiO$_2$ NPs as recommended by Petersen et al. [34], at the administered dose (20 µg/mL), in one independent experiment with 3 technical replicates. There was no evidence of quenching or auto-absorption.

2.8. Flow Cytometry

After cell growth and NPs exposure for 1, 6, and 24 h in 6-well plates, MDMs were scraped off in 1 mL of cRPMI, using a cell scraper (Sarstedt, Sevelen, Switzerland) and collected in a flow cytometry tube (5 mL Polystyrene Round-Bottom Tube, Corning® Falcon, Reinach, Switzerland). A549 cells were detached with Trypsin-EDTA (300 µL) for 6 min followed by the addition of 700 µL of cRPMI 1640. Cells were centrifuged at 4 °C for 5 min at $300\times g$, washed 2 times in PBS and then resuspended and fixed with 2 vol.% paraformaldehyde (PFA, Sigma-Aldrich, Buchs, Switzerland) in PBS for 15 min at 4 °C. Two additional washing steps were performed in PBS, before resuspension in cold FC buffer (PBS with 1 w/v.% bovine serum albumin (BSA, Sigma-Aldrich, Buchs, Switzerland), 0.1 vol.% sodium azide (Sigma-Aldrich, Buchs, Switzerland), and 1 mM EDTA (Sigma-Aldrich, Buchs, Switzerland) at pH 7.4. Data acquisition was performed on a BD LSR FORTESSA (BD Biosciences, San Jose, CA, USA) equipped with a yellow-green laser and PE filter where 30,000 events were recorded. Flow cytometry data was analyzed using the FlowJo software v.10.

2.9. Fluorescence Imaging

After NPs exposure for 1, 6, and 24 h, cells were fixed with 4 vol.% PFA in PBS for 15 min at room temperature and permeabilized for 10 min in 0.2 vol.% Triton X-100 in PBS. Samples were washed thrice with PBS between steps. F-actin was stained with Alexa Fluor

488 Phalloidin (0.66 µM in PBS, Cat. # A12379, Invitrogen, Thermo Fisher Scientific Inc., Zug, Switzerland) for 1 h, and cell nuclei counterstained using 4′,6-diamidino-2-phenylindole (DAPI, 1 µg/mL in PBS, Sigma-Aldrich, Buchs, Switzerland) for 5 min in PBS. Samples were washed 3 times using PBS and kept in PBS until further analysis. During fixation and staining, samples were kept at room temperature and dark conditions between steps. Images were acquired in an inverted Zeiss LSM 710 Meta apparatus (Axio Observer.Z1, Zeiss, Oberkochen, Germany) using an excitation laser of 405 nm (DAPI), 488 nm (Alexa Fluor 488), and 561 nm (rhodamine B) equipped with a Plan-Apochromat 63x/1.4 Oil M27 objective (Zeiss GmbH, Oberkochen, Germany).

2.10. Co-Localization Analysis

Co-localization of the exposed NPs with early endosomes was evaluated after 1 and 6 h. NPs co-localization with lysosomes was studied at 6 and 24 h. After NPs exposure, cells were washed 3 times with PBS and incubated with fresh cRPMI supplemented with 75 nM LysoTracker Green DND-26 (Invitrogen, Thermo Fisher Scientific Inc., Zug, Switzerland) for 15 min to stain the lysosomes. Then, the cells were washed twice with PBS and immediately imaged after the addition of cRPMI. For early endosomes labeling, immunostaining with early endosome antigen 1 (EEA1) was performed. Cells were fixed and permeabilized, as mentioned in the previous section. After, 20 µg/mL of EEA1 (Cat. # ab109110, Abcam, Cambridge, UK) in antibody solution (1 w/v.% bovine serum albumin (BSA, Cat. # A7030, Sigma-Aldrich, Zug, Switzerland) and 0.1 vol.% Triton-X in PBS) was added for 2 h. A secondary antibody goat anti-rabbit Alexa Fluor 647 (2 µg/mL, Cat. # A21244, Invitrogen, Thermo Fisher Scientific Inc., Zug, Switzerland) in antibody solution was added for 1 h. Finally, cells were counterstained with DAPI. All steps were conducted at room temperature and under dark conditions. The NPs co-localization was calculated using the Pearson's correlation coefficient (PCC) and the open source plugin for ImageJ, EzColocalization [35].

2.11. Dye Leaching from SiO_2-Rho B NPs

The potential release of dye from SiO_2 NPs in cell culture medium and lysosomal milieu was investigated by incubating the NPs in cRPMI (without phenol red) and artificial lysosomal fluid (ALF). ALF was prepared as previously reported [36]. Briefly, sodium chloride (3.210 g), sodium hydroxide (6.000 g), citric acid (20.800 g), calcium chloride (0.097 g), sodium phosphate heptahydrate (0.179 g), sodium sulfate (0.039 g), magnesium chloride hexahydrate (0.106 g), glycerin (0.059 g), sodium citrate dihydrate (0.077 g), sodium tartrate dihydrate (0.090 g), sodium lactate (0.085 g), sodium pyruvate (0.086 g), and formaldehyde (1.000 mL, added fresh before use) were dissolved in 200 mL of MilliQ water to obtain a 5× stock solution. The stock solution was later diluted with MilliQ water and NPs to a final concentration of 20 µg/mL. NPs were incubated in cRPMI and ALF for 24 h at 37 °C. Then, the NPs were centrifuged at high speed (16,000× g) for 1 h and the supernatants were collected. The supernatants were centrifuged again at the same speed, to guarantee that a minimum number of particles remains in suspension. A control containing the NPs in water, at the administered dose (20 µg/mL), was included in the experiments. Fluorescence emission intensity was measured on a Fluorolog TCSPC spectrofluorometer (Horiba, Northampton, UK) with the FluorEssence software (v3.8). For each sample, emission spectrum with a λem between 560 to 700 nm and a fixed λex of 550 nm, was recorded. The excitation and emission slits were fixed to 4 nm.

2.12. Focused Ion Beam-Scanning Electron Microscopy (FIB-SEM)

Cells were seeded in a 35 mm glass-bottom dish (MatTek Inc., Ashland, MA, USA), exposed to SIO_2 NPs for 6 h and fixed with 2 vol.% PFA and 2.5 vol.% glutaraldehyde (25%, Electron Microscopy Sciences, Hatfield, PA, USA) in PBS for 3 h on ice. Samples were then treated with a mixture of 3 w/v.% potassium hexacyanoferrate(II) trihydrate (≥99.95%, Sigma Aldrich, Buchs, Switzerland) and 2 vol.% osmium tetroxide (4% in H_2O, Sigma Aldrich, Buchs, Switzerland) in Milli-Q water, together with 0.2 M cacodylate buffer

(Polysciences, Eppelheim, Germany) for 1 h at room temperature. After, a treatment with 1 w/v% thiocarbohydrazide (98%, Sigma Aldrich, Buchs, Switzerland) in Milli-Q water was performed for 20 min at 60 °C. Finally, samples were incubated with 2 vol.% osmium tetroxide in Milli-Q water for 30 min. Samples were dehydrated using increasing graded ethanol series, followed by embedding and polymerization in Epon resin (Epoxy embedding kit, Sigma Aldrich, Buchs, Switzerland) for 48 h at 60 °C. Polymerized Epon blocks were then attached to an aluminum stub with carbon tape, and a thin layer of ~2 nm Au was sputtered onto the sample surface to render them conductive. FIB-SEM milling and imaging was performed using a Thermo Scientific Scios 2 Dual Beam microscope (Thermo Fisher Scientific, Waltham, MA, USA). A FIB operated at 30 kV was used to localize the cells of interest underneath the resin block. Once the region of interest was chosen, a trench was created using an ion (Ga^+) beam (30 kV and current of 1 nA). A final polishing step was performed 1 nA at 30 kV. The freshly milled cross-section was imaged using an electron beam with an acceleration voltage of 2 kV and a current of 50 pA and the backscattered electron detector. This investigation was complemented by energy-dispersive x-ray spectroscopy (Thermo Fisher Scientific, Waltham, MA, USA) for the chemical analysis of silicon (Si).

2.13. Total RNA Isolation and Illumina Sequencing (RNA-Seq)

After NPs exposure, total RNAs were isolated from cells growing in 6-well plates. Cell lysis was performed directly in the well, using 250 µL of BL + TG buffer (Promega Madison, WI, USA), and total RNA was extracted using ReliaPrep™ RNA Cell Miniprep System (Promega, Z6012, Madison, WI, USA) following the manufacturer's protocol. The quantity and quality of RNA were examined by Thermo Scientific™ NanoDrop™ 2000 Spectrophotometer and Agilent 2100 Bioanalyzer (Agilent Technologies, Santa Clara, CA, USA). Only RNA with OD 260/280 $\geq$ 1.8 and RNA integrity number ≥ 7 were selected for the subsequent experiments. Equal quantities of high-quality RNA, i.e., that met the above-stated criteria, from each sample were pooled together for mRNA library preparation (TruSeq Stranded RNA) and sequencing (HiSeq400 SR 150) at the Genomic Technologies Facility, Lausanne, Switzerland. Statistical analysis was performed in R (R version 4.0.2). Genes with low counts were filtered out according to the rule of one count per million (1 cpm). Library sizes were scaled using TMM normalization. The normalized counts were transformed to cpm values, and a log2 transformation was applied by means of the function cpm with the parameter setting prior.count = 1 (EdgeR v 3.30.3) [37]. After data normalization, a quality control analysis was performed through hierarchical clustering and sample PCA. Differential expression was computed with the R Bioconductor package limma [38] by fitting data to a linear model. The approach limma-trend was used. Sample pairings were taken into account by including a factor in the model. Fold changes were computed, and a moderated *t*-test was applied. *p*-values were adjusted using the Benjamini-Hochberg (BH) method.

2.14. Real-Time qRT-PCR

The reverse transcriptase reaction was performed with the Omniscript RT system (Qiagen, Hilden, Germany), OligodT (Microsynth, Balgach, Switzerland), and RNasin Plus RNase Inhibitor (Promega, Madison, WI, USA, Switzerland). The synthesis of complementary DNA (cDNA) was performed by using 6.5 µL of isolated RNA (250 ng), 1 µL oligo-dT primer (10 µM), 0.25 µL RNase inhibitor, 1 µL dNTP Mix (5 mM), 0.25 µL Omniscript reverse transcriptase (1 Unit), and 1 µL buffer RT. The real-time PCR was performed on the 7500 fast real-time PCR system (Applied Biosystems, Thermo Fisher Scientific, Waltham, MA, USA) by mixing 2 µL 5-fold diluted cDNA with 5 µL SYBR-green master mix (Fast SYBR Green master mix, Applied Biosystems, Thermo Fisher Scientific, Waltham, MA, USA), 2 µL nuclease-free water (Promega, Madison, WI, USA), and 2 µL primer mix (91 nM). Relative expression levels were calculated using the Pfall method [39] with glyceraldehyde-3-phosphate dehydrogenase (GAPDH) and tyrosine

3-monooxygenase/tryptophan 5-monooxygenase activation protein zeta (YWHAZ) as internal standard genes. Primers were purchased from Thermo Fisher Scientific (Zug, Switzerland). Details about the primers are included in the Supplementary Information (Table S1).

2.15. Western Blot

Total protein was isolated from cells growing in 6-well plates. Cell lysis was performed directly in the wells by adding 50 µL of ice-cold T-PER buffer (Cat. # 78510, Thermo Fisher Scientific, Zug, Switzerland) supplemented with Halt^{TM} Protease Inhibitor Cocktail, EDTA-free (Cat. # 78425, Thermo Fisher Scientific, Zug, Switzerland) and sodium fluoride (Cat. # 27860, 20 mM, VWR, Dietikon, Switzerland). Plates were kept at 4 °C for 20 min. Protein lysates were pipetted up and down, transferred to 1.5 mL Eppendorf tubes, kept on ice for 10 min and centrifuged at $10,000\times g$ for 5 min. The protein in the supernatant was collected and quantified via Bradford assay. The samples were boiled in a reducing Laemmli buffer for 5 min, and the same amount of protein was loaded in a 7.5% SDS-PAGE (Bio-Rad, Hercules, CA, USA). Proteins were electrophoretically transferred onto polyvinylidene difluoride (PVDF) membranes at 150 mA for 75 min under wet conditions. A solution of 0.1 w/v.% Ponceau S (Cat. # 141194, Sigma-Aldrich, Buchs, Switzerland) was used to confirm the transfer of proteins. A solution containing 3 w/v.% BSA and 0.1 vol.%Tween 20 (Cat. # P9416, Sigma-Aldrich, Buchs, Switzerland) in Tris-buffered saline (TBS) solution was used to block the nonspecific sites for 1 h. The same solution was used for immunostaining with primary and secondary antibodies. Three rounds of washing with TTBS (0.1 vol.% Tween 20 in TBS) were performed between steps. Primary antibody was added to the blots overnight at 4 °C. The following concentrations of antibodies were used: α-tubulin (1 µg/mL, Cat. # sc-32293, Santa Cruz Biotechnology, Heidelberg, Germany) GAPDH (1 µg/mL, sc-47724, Santa Cruz Biotechnology, Heidelberg, Germany), EEA1 (2 µg/mL, Cat. # ab109110, Abcam, Cambridge, UK), and Nur77 (2 µg/mL, Cat. # sc-365113, Santa Cruz Technology, Heidelberg, Germany). The blots were then incubated with a goat anti-mouse HRP conjugated secondary antibody (Cat. # HAF007, R&D, Abingdon, UK) for 1 h at 1:2000 (α-tubulin), 1:4000 (GAPDH), 1:1000 (EEA1), and 1:1000 (Nur77). A molecular weight marker mPAGE® Color Protein Standard (Cat. # MPSTD4, Sigma-Aldrich, Buchs, Switzerland) was used to identify the corresponding detected bands. Protein bands were visualized using the chemiluminescent HRP detection reagent Immobilon Forte Western HRP substrate (Cat. # WBLUF0020, Sigma-Aldrich, Buchs, Switzerland). The optical density of the bands was estimated using ImageJ. The housekeeping proteins α-tubulin and GAPDH were used as normalization controls.

2.16. Statistical Analyses

Comparisons between two related groups were made by paired *t*-test. Two-way ANOVA (Dunnett's post-hoc test for multiple comparisons) was used to compare more than two groups with more than one variable. Statistical analyses were performed with GraphPad Prism 9.2 software (GraphPad Software, San Diego, CA, USA).

3. Results and Discussion

3.1. Interaction and Localization of SiO$_2$ NPs in MDMs and A549 Cells

Non-porous SiO$_2$ NPs, measuring 60 and 200 nm in diameter, were synthesized and functionalized with the fluorescent dye RhoB, to enable their detection within cells using flow cytometry and confocal laser scanning microscopy (CLSM). Representative TEM micrographs of the individual particles and physicochemical properties are represented in Figures 1A and S1. It has been demonstrated that cellular responses differ significantly after interacting with NPs in an aggregated form or as individual particles [40]. Therefore, the stability of both 60 and 200 nm SiO$_2$ NPs in serum-free RPMI 1640 and complete RPMI 1640 (cRPMI) was evaluated by dynamic light scattering (DLS) at 24 h. An increase in size, a consequence of the aggregation of the NPs, was observed for both NPs in serum-free

RPMI 1640, whereas in cRPMI the particles remain stable (Figure 1A inset table). The latter is explained by the ability of proteins in the serum to adsorb on the NPs surface, creating the so-called protein corona [41] and assisting NP stabilization via steric and/or hydration interactions [3]. Due to the aggregation of both SiO_2 NPs in serum-free RPMI 1640, as demonstrated by DLS and CLSM (Figure S2), the exposure of cells to NPs was performed in cRPMI. In addition, the stability of the fluorescent probes on the SiO_2 NPs in cRPMI and lysosomal fluid was evaluated (Figure S3). The performed fluorescence measurements revealed that there is a minor signal from the supernatants after SiO_2 NPs incubation in cRPMI and ALF for 24 h at 37 °C. The findings confirm the stability of the fluorophore RhoB on SiO_2 NPs.

MDM and A549 cells were selected for this study as MDMs are phagocytic cells and one of the first cell types to interact with NPs in the body, contributing to rapid NP clearance from the tissue where these cells reside [42]. The non-phagocytic A549 cell line is frequently used to mimic an alveolar type II epithelial barrier [43] and is one the most widely used cell lines in human research in a wide range of applications, including in the testing of novel drugs [44] and in particle uptake mechanism studies [14,45,46]. Initially, cellular cytotoxicity was evaluated using a lactate dehydrogenase assay after 6 and 24 h exposure (Figure S4). No significant alterations in cell membrane permeability after exposure to 60 nm and 200 nm SiO_2 NPs were observed for either cell type, confirming their non-cytotoxicity.

The interaction and association (i.e., both plasma membrane-bound and internalized) of 60 and 200 nm SiO_2 NPs with A549 and MDM cells were evaluated after 1, 6, and 24 h of NP exposure using CLSM, focused ion beam-scanning electron microscopy (FIB-SEM) and flow cytometry (Figure 1, Figures S5 and S6). CLSM and flow cytometry showed that the interaction between both cell types and NPs occurs within the first hour. However, in CLSM images, it is possible to conclude that the majority of NPs interact with the A549 cell membrane in the first hour but are not internalized. CLSM and FIB-SEM micrographs reveal the internalization of both SiO_2 NPs by A549 cells after 6 h of exposure. In MDM cells, the association of SiO_2 NPs begins within the first hour and increases over time. A high number of 60 and 200 nm SiO_2 NPs can be seen inside MDMs at the 6 h timepoint. The association rate of NPs with MDM cells was higher than with A549 cells, which is to be expected based on the higher clearance capability of the latter cell type. In addition, cell division in dividing cells (A549) takes place and the NP load might thus be lower than in non-dividing cells (MDM) [47]. Macrophages are a type of cell of the reticuloendothelial system (RES) [1]. RES is the biggest limitation in NP drug delivery because it is one of the main factors responsible for the sequestration and clearance of NPs [48].

After uptake, it is expected that NPs will be localized within endocytic vesicles that fuse together with the early endosomes/phagosomes and, later, with the lysosomes [1]. With this in mind, the co-localization of SiO_2 NPs with early endosomes and lysosomes was investigated. Our results (Figure 2) show a very weak co-localization (Pearson correlation coefficient (PCC) < 0.3) with early endosome antigen 1 (EEA1) for both 60 and 200 nm NPs in MDMs or A549 cells, confirming that SiO_2 NPs only stay in the early endosomes for a short period of time. As described in the literature, early endosomes rapidly fuse with late endosomes, over an 8–15 min period [45], which is consistent with the obtained results. In contrast, higher PCCs were obtained for co-localization between lysosomes (Lysotracker) and SiO_2 NPs. Higher co-localization values were obtained for 60 nm NPs in both MDMs and A549 cells, which can be rationalized by a different phagosomal/lysosomal transport mechanism of the NPs. The endosomes/phagosomes formed during NP uptake have varying sizes, which can strongly affect endosomal/phagosomal transport [46] and, consequently, maturation and fusion with lysosomes. After 24 h, there was an increase in co-localization of both particles with lysosomes in MDM cells, possibly due to accumulation of these particles in this compartment. Furthermore, the PCC is higher for both particle types at all investigated time points in MDMs, which can be related to the higher number of intracellular NPs.

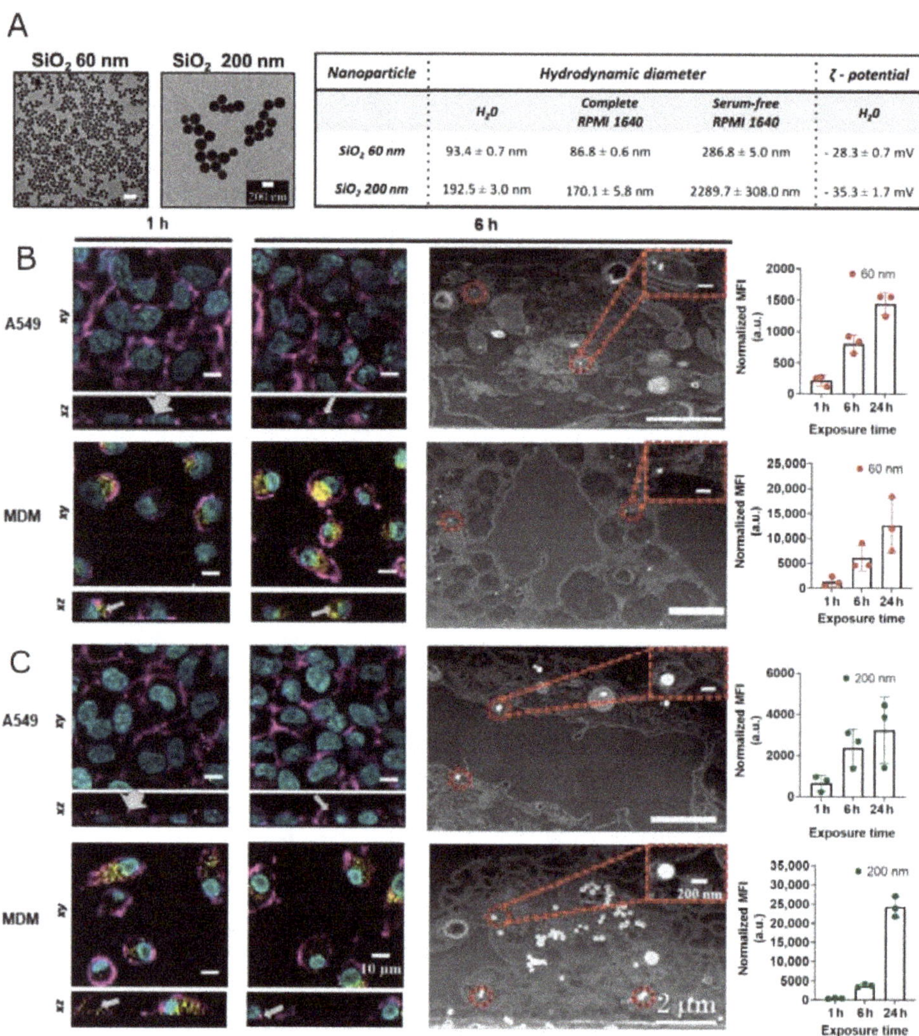

Figure 1. SiO$_2$ NP characterization and cellular uptake. (**A**) Representative transmission electron microscopy (TEM) micrographs and physicochemical characterization of SiO$_2$ NPs used in this study. Scale bar = 200 nm. Hydrodynamic diameter measured by dynamic light scattering in H$_2$O, complete RPMI 1640, and serum-free RPMI 1640, revealing the aggregation of both NPs in serum-free RPMI 1640. Confocal laser scanning microscopy (CLSM), focused ion beam-scanning electron microscopy (FIB-SEM) micrographs and flow cytometry data (bar graphs on the left), revealing the association of 60 (**B**) and 200 nm (**C**) SiO$_2$ with lung epithelial cells (A549) and primary human monocyte-derived macrophages (MDMs) after 1, 6, and 24 h of exposure to 20 µg/mL. Bar graphs represent the median fluorescence intensity (MFI), normalized to untreated cells. Data is presented as mean ± standard deviation (n = 3). Cell nuclei (cyan), cytoskeleton (magenta), and NPs (yellow). Thicker and thinner grey arrows indicate extracellular (surface bound) and intracellular localization of NPs, respectively. The red dashed circles indicate the intracellular localization of NPs. Scale bar = 10 µm for CLSM pictures. Scale bar = 2 µm for FIB-SEM pictures and scale bar = 200 nm for zoomed-in images.

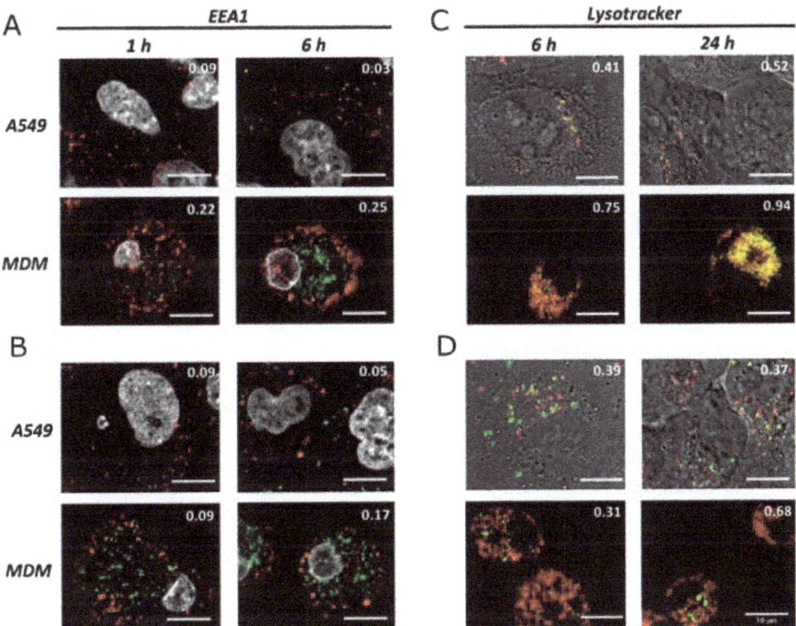

Figure 2. Co-localization of SiO$_2$ NPs with early endosomes (EEA1, **A,B**) and lysosomes (Lysotracker, **C,D**) at different time points. SiO$_2$ of 60 nm are represented in (**A,C**), and 200 nm in (**B,D**). Cell nuclei (grey), EEA1 and lysotracker (red), and NPs (green). The co-localization was determined by the Pearson correlation coefficient and values are represented at the top right of each image. Phase contrast images for NP-lysosome co-localization, in lung epithelial cells (A549), were included to facilitate cell structure visualization. Scale bar = 10 μm.

3.2. Regulation of Gene and Protein Expression upon Uptake of SiO$_2$ NPs

To investigate the early changes that occur at the genetic level upon cell-NP uptake, A549 and MDMs were exposed to 60 nm SiO$_2$ NPs for 6 h, followed by a genome-wide transcriptome analysis via RNA-seq (Figure 3). Only 60 nm SiO$_2$ were included in this first screening. The results revealed that no gene was differentially expressed in A549 cells. On the other hand, 117 genes (adjusted *p*-value < 0.2) were found to be differentially expressed in MDM cells, but only five (*NR4A1, NR4A2, FOSB, MIF, and ASIC3*) showed a change greater than 1.5-fold. Surprisingly, none of these five genes are related to endocytic mechanisms. The most significantly changed gene, *NR4A1*, revealed a 2.06-fold change. *NR4A1* is an orphan nuclear receptor and is part of the nuclear receptor group 4A (NR4A) subfamily of nuclear hormone receptors [49]. It modulates the inflammatory response of macrophages through a number of mechanisms, including transcriptional reprogramming of mitochondrial metabolism [50]. The up-regulation of this gene can be triggered via physical stimulation and by inflammatory and growth factors [51]. Waters et al. also revealed the up-regulation of *NR4A1* in macrophages after 2 h of exposure to amorphous SiO$_2$ NPs [52]. A different analytical technique, gene set enrichment analysis (GSEA), was performed for MDM cells, including all the differentially expressed genes, and the principal gene ontology (GO) biological processes (BPs) are shown in Figure S7. The uptake of 60 nm SiO$_2$ NPs led to significant changes in the group of genes involved in cell–cell adhesion, cell chemotaxis, immune response, and inflammation. The changes in these processes have also been observed in previous studies [53–55]. These findings reveal that NP internalization does not lead to major transcriptional changes at early time points in the genes related to endocytosis, and lets us suggest that regulation might occur at the protein level (i.e., post-translation modifications). Furthermore, it demonstrates that NPs can cause inflammation even when they are not cytotoxic. The main reason for NP uptake in macrophages is their capability to recognize the opsonins present at NPs surface. The process of opsonization occurs upon NPs interaction with physiological fluids, containing different biomolecules including opsonins that promote cellular recognition and clearance by macrophages [7]. The opsonins at NPs surface can also dictate the extent of uptake and toxicity [42]. Fedeli et al. produced 26 nm spherical SiO_2 NPs and proved that high amounts of histidine rich glycoprotein adsorbed at NPs surface avoided human macrophage recognition [56].

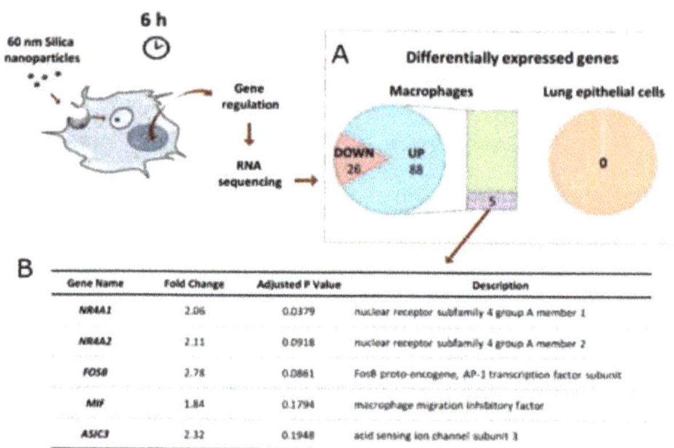

Gene Name	Fold Change	Adjusted P Value	Description
NR4A1	2.06	0.0379	nuclear receptor subfamily 4 group A member 1
NR4A2	2.11	0.0918	nuclear receptor subfamily 4 group A member 2
FOSB	2.78	0.0861	FosB proto-oncogene, AP-1 transcription factor subunit
MIF	1.84	0.1794	macrophage migration inhibitory factor
ASIC3	2.32	0.1948	acid sensing ion channel subunit 3

Figure 3. (**A**) Whole transcriptome screening via RNA sequencing upon exposure of A549 lung epithelial cells and primary human monocyte-derived macrophages (MDMs) to 60 nm SiO_2 NPs for 6 h. Number of differentially expressed genes in lung epithelial cells and macrophages with an adjusted *p*-value of less than 0.2. No changes in the gene expression were observed for A549 cells. In MDM cells, 117 genes were found to be differentially regulated (26 down-regulated and 88 up-regulated), but only five revealed a greater than 1.5 fold change. (**B**) List with the five up-regulated genes in MDM cells.

In order to confirm the previous findings and to validate the RNA-seq results, the endocytosis related-genes clathrin light chain (*CLTC*), caveolin-1 (*CAV1*), early endosome antigen 1 (*EEA1*), lysosomal-associated membrane protein 1 (*LAMP1*), Rac family small GTPase 1 (*RAC1*), and dynamin 2 (*DNM2*) were evaluated at 1, 6, and 24 h after exposure to 60 and 200 nm SiO_2 NPs by real-time qRT-PCR (Figure 4A). SiO_2 NPs measuring 200 nm were included to investigate if particle size impacts cellular response at the gene level. No gene was found to be differentially expressed in A549 cells after exposure to either NPs. In MDM cells, *LAMP1* was down-regulated (Fold change = −1.7) upon 24 h exposure to 200 nm SiO_2 NPs. As previously stated, lower co-localization values between 200 nm SiO_2 NPs and lysosomes were observed after 24 h. In this regard, down-regulation of *LAMP1* might be associated with an impaired process of autophagic lysosome reformation [21,22,57]. Since *CAV1* is only weakly expressed in MDM cells, it was not included in the analysis. Nevertheless, and to confirm the results of RNA-Seq, the nuclear receptor subfamily 4 group A member 1 (*NR4A1*), one of the genes that was found to be up-regulated in MDM cells, was included in the real-time qRT-PCR analysis. The results confirmed the up-regulation of *NR4A1* with 60 nm SiO_2 NPs after exposure for 6 h. In addition, we observed the up-regulation of this gene after 1 h, for 60 nm (Fold change = 2.3) and 200 nm SiO_2 NPs (Fold change = 3.1), but not after 24 h, which suggests that *NR4A1* is an immediate-early response gene.

The comparison between protein and gene expression was evaluated by Western blot (Figure 4B). The results showed an increase of the expression for the NR4A1 protein in MDM cells at 6 h but not after 1 h. This can be explained by the fact that other regulation events occur between transcript and protein products [58]. The expression of the protein EEA1 in A549 cells was investigated and, as expected, its expression did not change considerably.

A computation model [25] was used to estimate the dose of SiO_2 NPs that reaches the cells (i.e., delivered dose) (Figure S8). ISDD is a useful tool for calculating the delivered dose of NPs and performing more accurate analysis of cellular responses. A higher number of 60 nm SiO_2 NPs reaches the cells in comparison with 200 nm SiO_2 NPs after 1, 6, and 24 h. However, the obtained results do not directly correlate with the number of delivered NPs, as a higher *NR4A1* expression effect was observed after 1 h for the bigger NPs. It has been proved that the biologically most relevant dose metric for the evaluation of NP effects is the particle surface area [59]. When particle surface area was used as a metric, the simulation showed similar results, demonstrating a greater delivered dose for 60 nm SiO_2 NPs. In brief, there is not a good correlation between cellular response and delivered dose in our study. This is due to the fact that the delivered dose is not the only factor influencing the cellular response. Other factors, such as the uptake mechanism and intracellular fate, can have an effect on biological responses [60].

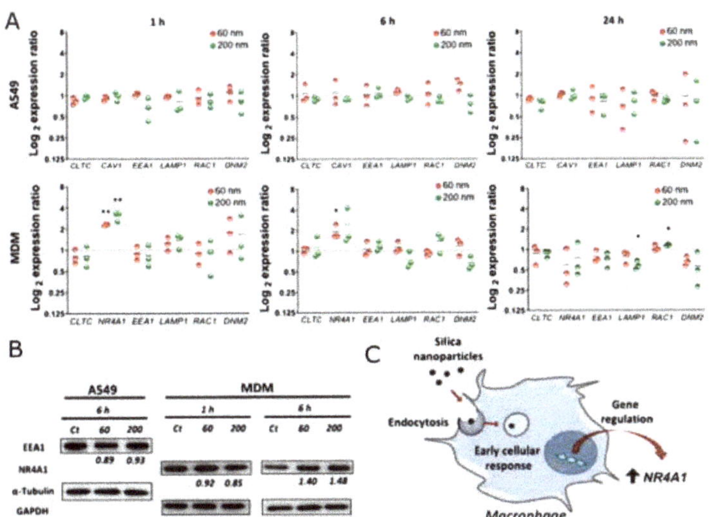

Figure 4. Gene and protein expression upon exposure to SiO_2 NPs. (**A**) Real-time qRT-PCR results representing the expression of several genes upon exposure to 60 and 200 nm SiO_2 NPs at different time points (1, 6, and 24 h) in lung epithelial cells (A549) and primary human monocyte-derived macrophages (MDMs). Statistically significant differences among the groups (Two-way ANOVA Dunnett's post-hoc test for multiple comparisons): * $p \leq 0.05$; ** $p \leq 0.01$. (**B**) Expression of proteins NR4A1 and EEA1 were analyzed by Western blot. The representative images are shown. The mean expression ratios of the indicated protein, determined via densitometry from three independent experiments, are shown at the bottom of each blot. (**C**) Scheme representing the early activation of *NR4A1* upon SiO_2 NP uptake.

3.3. Expression of NR4A1 in Macrophages upon Exposure to Au and PS NPs

To evaluate the effect of NP material composition on the expression of NR4A1, MDMs were exposed to Au and PS NPs of similar size and shape for 1 and 6 h. Both NPs were characterized by TEM, DLS, and UV-Vis spectroscopy (Figures 5A and S1). Similarly to SiO_2 NPs, Au NPs of ca. 50 nm diameter are colloidally stable in complete RPMI 1640, but tend to aggregate in serum-free RPMI 1640. In contrast, the presence or absence of serum does not affect the stability of PS NPs of ca. 60 nm diameter. This can be explained by the fact that PS NPs possess a different surface chemistry than Au and SiO_2 NPs, and the ionic strength of the cell-culture medium appears not to affect their stability.

Internalization of both NPs was confirmed at 1 and 6 h (Figure 5). Similarly to SiO_2 NPs, *NR4A1* was up-regulated at early time points after exposure to Au NPs. The gene up-regulation was observed after 1 h but not after 6 h. A slight increase at the protein level was detected at 6 h, as also observed for SiO_2 NPs. In contrast, exposure to PS NPs did not affect *NR4A1* expression. This can be explained by the fact that NPs possess different properties (e.g., material and surface chemistry), which per se can affect NP uptake. In addition, due to the different NP surface properties, distinct protein corona can be formed [57,61], which might also influence the internalization of NPs and, consequently, the signaling pathways. In vivo studies revealed that AuNPs tend to primarily accumulate in the liver and spleen [62]. Smaller NPs (<8 nm) are cleared through renal clearance and larger ones via hepatobiliary excretion [48]. As shown, AuNPs can trigger inflammation, even at earlier stages, which requires a deep understanding of the cellular processes that are initiating this process. This knowledge is critical for the development of new NPs with optimized properties for optimal clearance and the ability to modulate and control inflammation.

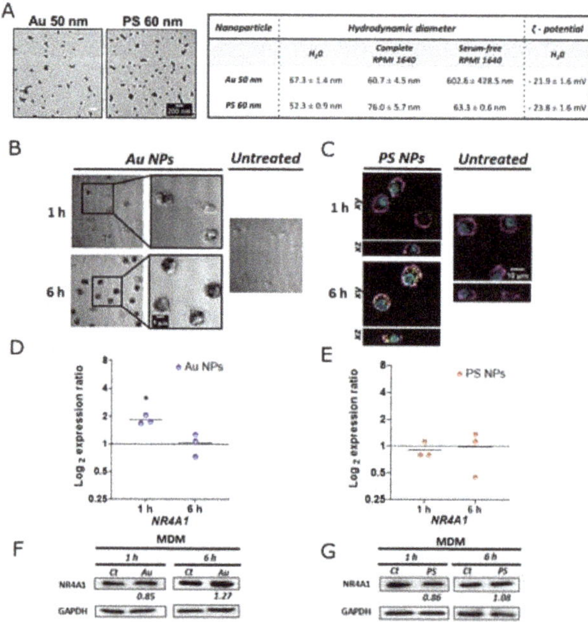

Figure 5. Effects of the cellular uptake of 50 nm Au and 60 nm PS in the expression of NR4A1 in primary human monocyte-derived macrophages (MDMs). (**A**) Representative transmission electron microscopy micrographs and physicochemical characterization of Au and PS NPs. Scale bar = 200 nm. Hydrodynamic diameter measured by dynamic light scattering in H_2O, complete RPMI 1640 and serum-free RPMI 1640. (**B**) Differential interference contrast images showing the internalization of Au NPs in MDMs upon 1 and 6 h exposure. Scale bar = 5 μm. (**C**) Confocal laser scanning microscopy images revealing the uptake of PS NPs in MDM cells after exposure to PS NPs for 1 and 6 h. Scale bar = 10 μm. *NR4A1* gene expression upon exposure to Au NPs (**D**) and PS NPs. (**E**) NR4A1 protein expression after exposure to Au NPs (**F**) and PS NPs (**G**). In (**D,E**) comparisons between groups were performed with a paired *t*-test: * $p \leq 0.05$. The representative Western blot images are shown in (**F,G**). The mean expression ratios of the indicated protein, determined via densitometry from three independent experiments, are shown at the bottom of each blot.

4. Conclusions

In the present study, we confirm that the type of NP and the type of cell both influence NP uptake. Macrophages revealed a higher uptake rate for SiO_2 NPs than with lung epithelial cells, which is attributed to the strong clearance, i.e., phagocytic, capability of macrophage cell types. Internalization of SiO_2 NPs by lung epithelial cells was much slower and occurred to a lesser extent. Furthermore, cellular internalization of 60 nm SiO_2 NPs did not lead to significant transcriptional changes after 6 h exposure to lung epithelial cells. In macrophages, despite our observation that genes related to endocytosis were not differentially expressed, we were able to identify the significant modification of the expression of gene *NR4A1*. The early up-regulation of *NR4A1* also occurred when macrophages were exposed for 1 h to 200 nm SiO_2 and 50 nm Au NPs, which suggests that *NR4A1* is an immediate-early response gene for this type of NPs. NR4A1 is an important modulator of inflammatory response and it would be useful to investigate whether NR4A1 could be a potential therapeutic target to avoid exacerbated inflammation caused by NP uptake.

Supplementary Materials: The following supporting information can be downloaded at: https://www.mdpi.com/article/10.3390/nano12040690/s1, Figure S1: Characterization of the different NPs; Figure S2. Stability of SiO_2 NPs in cell culture medium; Figure S3. Dye leaching study of SiO_2-rhodamine B NPs; Figure S4. Phase-contrast images and cytotoxicity assay of lung epithelial cells (A549) and primary human monocyte-derived macrophages (MDMs) after exposure to SiO_2 NPs; Figure S5. Cellular uptake of SiO_2 NPs; Figure S6. Energy Dispersive X-ray (EDX) analysis of a cross-section of primary human monocyte-derived macrophages (MDMs) at three different spots (red cross), which confirms the intracellular presence of Silicon (Si); Figure S7. Gene set enrichment analysis (GSEA) with gene ontology (GO) biological processes (BP) of primary human monocyte-derived macrophages (MDMs) after 6 h exposure to 60 nm SiO_2 NPs in comparison with untreated cells; Figure S8. Prediction of the effective cellular dose based on In vitro Sedimentation, Diffusion and Dosimetry (ISDD) model; Table S1. Information about the primers used for Real-time qRT-PCR.

Author Contributions: M.S.d.A. performed the experiments, analyzed the results, and wrote the manuscript; P.T.-B. performed DLS measurements and contributed significantly to the manuscript writing; B.D. revised the manuscript and provided scientific suggestions; S.B. performed DLS data analysis; P.Y. performed FIB-SEM experiments; A.P.-F. supervised the project; B.R.-R. thoroughly reviewed the manuscript, supervised, and designed the project. All authors contributed to the scientific discussions, manuscript revisions. All authors have read and agreed to the published version of the manuscript.

Funding: This research was supported by the Swiss National Science Foundation through the National Center of Competence in Research Bio-Inspired Materials (Grant number 51NF40-182881), and the Adolphe Merkle Foundation.

Data Availability Statement: The datasets underlying the results discussed in this article can be found in the online repository Zenodo under the doi 10.5281/zenodo.6125809 (Accessed on 18 February 2022).

Acknowledgments: The authors are grateful to Laetitia Haeni and Liliane Ackermann Hirschi for their kind support with particle synthesis, cell cultures, and biological assays.

References

1. Bourquin, J.; Milosevic, A.; Hauser, D.; Lehner, R.; Blank, F.; Petri-Fink, A.; Rothen-Rutishauser, B. Biodistribution, Clearance, and Long-Term Fate of Clinically Relevant Nanomaterials. *Adv. Mater.* **2018**, *30*, 1704307. [CrossRef] [PubMed]
2. Yildirimer, L.; Thanh, N.T.K.; Loizidou, M.; Seifalian, A.M. Toxicology and clinical potential of nanoparticles. *Nano Today* **2011**, *6*, 585–607. [CrossRef] [PubMed]
3. Moore, T.L.; Rodriguez-Lorenzo, L.; Hirsch, V.; Balog, S.; Urban, D.; Jud, C.; Rothen-Rutishauser, B.; Lattuada, M.; Petri-Fink, A. Nanoparticle colloidal stability in cell culture media and impact on cellular interactions. *Chem. Soc. Rev.* **2015**, *44*, 6287–6305. [CrossRef] [PubMed]

4. Shen, Z.; Ye, H.; Yi, X.; Li, Y. Membrane Wrapping Efficiency of Elastic Nanoparticles during Endocytosis: Size and Shape Matter. *ACS Nano* **2019**, *13*, 215–228. [CrossRef]

5. Inoue, M.; Sakamoto, K.; Suzuki, A.; Nakai, S.; Ando, A.; Shiraki, Y.; Nakahara, Y.; Omura, M.; Enomoto, A.; Nakase, I.; et al. Size and surface modification of silica nanoparticles affect the severity of lung toxicity by modulating endosomal ROS generation in macrophages. *Part. Fibre Toxicol.* **2021**, *18*, 21. [CrossRef]

6. Jiang, W.; Kim, B.Y.S.; Rutka, J.T.; Chan, W.C.W. Nanoparticle-mediated cellular response is size-dependent. *Nat. Nanotechnol.* **2008**, *3*, 145–150. [CrossRef]

7. Sousa de Almeida, M.; Susnik, E.; Drasler, B.; Taladriz-Blanco, P.; Petri-Fink, A.; Rothen-Rutishauser, B. Understanding nanoparticle endocytosis to improve targeting strategies in nanomedicine. *Chem. Soc. Rev.* **2021**. [CrossRef]

8. Iversen, T.-G.; Skotland, T.; Sandvig, K. Endocytosis and intracellular transport of nanoparticles: Present knowledge and need for future studies. *Nano Today* **2011**, *6*, 176–185. [CrossRef]

9. Park, J.H.; Oh, N. Endocytosis and exocytosis of nanoparticles in mammalian cells. *Int. J. Nanomed.* **2014**, *9* (Suppl. 1), 51. [CrossRef]

10. Yang, S.-A.; Choi, S.; Jeon, S.M.; Yu, J. Silica nanoparticle stability in biological media revisited. *Sci. Rep.* **2018**, *8*, 185. [CrossRef]

11. Ding, L.; Yao, C.; Yin, X.; Li, C.; Huang, Y.; Wu, M.; Wang, B.; Guo, X.; Wang, Y.; Wu, M. Size, Shape, and Protein Corona Determine Cellular Uptake and Removal Mechanisms of Gold Nanoparticles. *Small* **2018**, *14*, 1801451. [CrossRef]

12. Shapero, K.; Fenaroli, F.; Lynch, I.; Cottell, D.C.; Salvati, A.; Dawson, K.A. Time and space resolved uptake study of silicananoparticles by human cells. *Mol. BioSyst.* **2011**, *7*, 371–378. [CrossRef]

13. Annika Mareike Gramatke, I.-L.H. Size and Cell Type Dependent Uptake of Silica Nanoparticles. *J. Nanomed. Nanotechnol.* **2014**, *5*. [CrossRef]

14. Sun, D.; Gong, L.; Xie, J.; He, X.; Chen, S.; Luodan, A.; Li, Q.; Gu, Z.; Xu, H. Evaluating the toxicity of silicon dioxide nanoparticles on neural stem cells using RNA-Seq. *RSC Adv.* **2017**, *7*, 47552–47564. [CrossRef]

15. Yazdimamaghani, M.; Moos, P.J.; Ghandehari, H. Global gene expression analysis of macrophage response induced by nonporous and porous silica nanoparticles. *Nanomed. Nanotechnol. Biol. Med.* **2018**, *14*, 533–545. [CrossRef]

16. Murugadoss, S.; Lison, D.; Godderis, L.; Van Den Brule, S.; Mast, J.; Brassinne, F.; Sebaihi, N.; Hoet, P.H. Toxicology of silica nanoparticles: An update. *Arch. Toxicol.* **2017**, *91*, 2967–3010. [CrossRef]

17. Uboldi, C.; Bonacchi, D.; Lorenzi, G.; Hermanns, M.I.; Pohl, C.; Baldi, G.; Unger, R.E.; Kirkpatrick, C.J. Gold nanoparticles induce cytotoxicity in the alveolar type-II cell lines A549 and NCIH441. *Part. Fibre Toxicol.* **2009**, *6*, 18. [CrossRef]

18. Lin, C.; Zhao, X.; Sun, D.; Zhang, L.; Fang, W.; Zhu, T.; Wang, Q.; Liu, B.; Wei, S.; Chen, G.; et al. Transcriptional activation of follistatin by Nrf2 protects pulmonary epithelial cells against silica nanoparticle-induced oxidative stress. *Sci. Rep.* **2016**, *6*, 21133. [CrossRef]

19. Zhao, X.; Wei, S.; Li, Z.; Lin, C.; Zhu, Z.; Sun, D.; Bai, R.; Qian, J.; Gao, X.; Chen, G.; et al. Autophagic flux blockage in alveolar epithelial cells is essential in silica nanoparticle-induced pulmonary fibrosis. *Cell Death Dis.* **2019**, *10*, 127. [CrossRef]

20. Kusaka, T.; Nakayama, M.; Nakamura, K.; Ishimiya, M.; Furusawa, E.; Ogasawara, K. Effect of Silica Particle Size on Macrophage Inflammatory Responses. *PLoS ONE* **2014**, *9*, e92634. [CrossRef]

21. Zhang, Q.; Hitchins, V.M.; Schrand, A.M.; Hussain, S.M.; Goering, P.L. Uptake of gold nanoparticles in murine macrophage cells without cytotoxicity or production of pro-inflammatory mediators. *Nanotoxicology* **2011**, *5*, 284–295. [CrossRef] [PubMed]

22. Bachand, G.D.; Allen, A.; Bachand, M.; Achyuthan, K.E.; Seagrave, J.C.; Brozik, S.M. Cytotoxicity and inflammation in human alveolar epithelial cells following exposure to occupational levels of gold and silver nanoparticles. *J. Nanoparticle Res.* **2012**, *14*, 1212. [CrossRef]

23. Vanhecke, D.; Kuhn, D.A.; de Aberasturi, D.J.; Balog, S.; Milosevic, A.; Urban, D.; Peckys, D.; de Jonge, N.; Parak, W.J.; Petri-Fink, A.; et al. Involvement of two uptake mechanisms of gold and iron oxide nanoparticles in a co-exposure scenario using mouse macrophages. *Beilstein J. Nanotechnol.* **2017**, *8*, 2396–2409. [CrossRef] [PubMed]

24. Hoppstädter, J.; Seif, M.; Dembek, A.; Cavelius, C.; Huwer, H.; Kraegeloh, A.; Kiemer, A.K. M2 polarization enhances silica nanoparticle uptake by macrophages. *Front. Pharmacol.* **2015**, *6*, 1–12. [CrossRef]

25. Hinderliter, P.M.; Minard, K.R.; Orr, G.; Chrisler, W.B.; Thrall, B.D.; Pounds, J.G.; Teeguarden, J.G. ISDD: A computational model of particle sedimentation, diffusion and target cell dosimetry for in vitro toxicity studies. *Part. Fibre Toxicol.* **2010**, *7*, 36. [CrossRef]

26. Stöber, W.; Fink, A.; Bohn, E. Controlled growth of monodisperse silica spheres in the micron size range. *J. Colloid Interface Sci.* **1968**, *26*, 62–69. [CrossRef]

27. Larson, D.R.; Ow, H.; Vishwasrao, H.D.; Heikal, A.A.; Wiesner, U.; Webb, W.W. Silica Nanoparticle Architecture Determines Radiative Properties of Encapsulated Fluorophores. *Chem. Mater.* **2008**, *20*, 2677–2684. [CrossRef]

28. Brown, K.R.; Natan, M.J. Hydroxylamine Seeding of Colloidal Au Nanoparticles in Solution and on Surfaces. *Langmuir* **1998**, *14*, 726–728. [CrossRef]

29. Brown, K.R.; Walter, D.G.; Natan, M.J. Seeding of Colloidal Au Nanoparticle Solutions. 2. Improved Control of Particle Size and Shape. *Chem. Mater.* **2000**, *12*, 306–313. [CrossRef]

30. Enustun, B.V.; Turkevich, J. Coagulation of Colloidal Gold. *J. Am. Chem. Soc.* **1963**, *85*, 3317–3328. [CrossRef]

31. Crippa, F.; Rodriguez-Lorenzo, L.; Hua, X.; Goris, B.; Bals, S.; Garitaonandia, J.S.; Balog, S.; Burnand, D.; Hirt, A.M.; Haeni, L.; et al. Phase Transformation of Superparamagnetic Iron Oxide Nanoparticles via Thermal Annealing: Implications for Hyperthermia Applications. *ACS Appl. Nano Mater.* **2019**, *2*, 4462–4470. [CrossRef]

32. Lehmann, A.D.; Brandenberger, C.; Blank, F.; Gehr, P.; Rothen-Rutishauser, B. A 3D Model of the Human Epithelial Airway Barrier. *Altern. Technol. Anim. Test.* **2010**, *14*, 239–260.
33. Barosova, H.; Drasler, B.; Petri-Fink, A.; Rothen-Rutishauser, B. Multicellular Human Alveolar Model Composed of Epithelial Cells and Primary Immune Cells for Hazard Assessment. *J. Vis. Exp.* **2020**. [CrossRef]
34. Petersen, E.J.; Hirsch, C.; Elliott, J.T.; Krug, H.F.; Aengenheister, L.; Arif, A.T.; Bogni, A.; Kinsner-Ovaskainen, A.; May, S.; Walser, T.; et al. Cause-and-Effect Analysis as a Tool To Improve the Reproducibility of Nanobioassays: Four Case Studies. *Chem. Res. Toxicol.* **2020**, *33*, 1039–1054. [CrossRef]
35. Stauffer, W.; Sheng, H.; Lim, H.N. EzColocalization: An ImageJ plugin for visualizing and measuring colocalization in cells and organisms. *Sci. Rep.* **2018**, *8*, 15764. [CrossRef]
36. Stopford, W.; Turner, J.; Cappellini, D.; Brock, T. Bioaccessibility testing of cobalt compounds. *J. Environ. Monit.* **2003**, *5*, 675. [CrossRef]
37. Robinson, M.D.; Oshlack, A. A scaling normalization method for differential expression analysis of RNA-seq data. *Genome Biol.* **2010**. [CrossRef]
38. Ritchie, M.E.; Phipson, B.; Wu, D.; Hu, Y.; Law, C.W.; Shi, W.; Smyth, G.K. limma powers differential expression analyses for RNA-sequencing and microarray studies. *Nucleic Acids Res.* **2015**, *43*, e47. [CrossRef]
39. Pfaffl, M.W. A new mathematical model for relative quantification in real-time RT-PCR. *Nucleic Acids Res.* **2001**, *29*, e45. [CrossRef]
40. Zook, J.M.; MacCuspie, R.I.; Locascio, L.E.; Halter, M.D.; Elliott, J.T. Stable nanoparticle aggregates/agglomerates of different sizes and the effect of their size on hemolytic cytotoxicity. *Nanotoxicology* **2011**, *5*, 517–530. [CrossRef]
41. Gunawan, C.; Lim, M.; Marquis, C.P.; Amal, R. Nanoparticle–protein corona complexes govern the biological fates and functions of nanoparticles. *J. Mater. Chem. B* **2014**, *2*, 2060. [CrossRef] [PubMed]
42. Gustafson, H.H.; Holt-Casper, D.; Grainger, D.W.; Ghandehari, H. Nanoparticle uptake: The phagocyte problem. *Nano Today* **2015**, *10*, 487–510. [CrossRef] [PubMed]
43. Rothen-Rutishauser, B.M.; Kiama, S.G.; Gehr, P. A Three-Dimensional Cellular Model of the Human Respiratory Tract to Study the Interaction with Particles. *Am. J. Respir. Cell Mol. Biol.* **2005**, *32*, 281–289. [CrossRef] [PubMed]
44. Wang, X.-J.; Sun, Z.; Villeneuve, N.F.; Zhang, S.; Zhao, F.; Li, Y.; Chen, W.; Yi, X.; Zheng, W.; Wondrak, G.T.; et al. Nrf2 enhances resistance of cancer cells to chemotherapeutic drugs, the dark side of Nrf2. *Carcinogenesis* **2008**, *29*, 1235–1243. [CrossRef] [PubMed]
45. Huotari, J.; Helenius, A. Endosome maturation. *EMBO J.* **2011**, *30*, 3481–3500. [CrossRef] [PubMed]
46. Keller, S.; Berghoff, K.; Kress, H. Phagosomal transport depends strongly on phagosome size. *Sci. Rep.* **2017**, *7*, 17068. [CrossRef]
47. Bourquin, J.; Septiadi, D.; Vanhecke, D.; Balog, S.; Steinmetz, L.; Spuch-Calvar, M.; Taladriz-Blanco, P.; Petri-Fink, A.; Rothen-Rutishauser, B. Reduction of Nanoparticle Load in Cells by Mitosis but Not Exocytosis. *ACS Nano* **2019**, *13*, 7759–7770. [CrossRef]
48. Longmire, M.; Choyke, P.L.; Kobayashi, H. Clearance properties of nano-sized particles and molecules as imaging agents: Considerations and caveats. *Nanomedicine* **2008**, *3*, 703–717. [CrossRef]
49. Kim, S.O.; Ono, K.; Tobias, P.S.; Han, J. Orphan Nuclear Receptor Nur77 Is Involved in Caspase-independent Macrophage Cell Death. *J. Exp. Med.* **2003**, *197*, 1441–1452. [CrossRef]
50. Koenis, D.S.; Medzikovic, L.; van Loenen, P.B.; van Weeghel, M.; Huveneers, S.; Vos, M.; Evers-van Gogh, I.J.; Van den Bossche, J.; Speijer, D.; Kim, Y.; et al. Nuclear Receptor Nur77 Limits the Macrophage Inflammatory Response through Transcriptional Reprogramming of Mitochondrial Metabolism. *Cell Rep.* **2018**, *24*, 2127–2140.e7. [CrossRef]
51. Hamers, A.A.J.; Argmann, C.; Moerland, P.D.; Koenis, D.S.; Marinković, G.; Sokolović, M.; de Vos, A.F.; de Vries, C.J.M.; van Tiel, C.M. Nur77-deficiency in bone marrow-derived macrophages modulates inflammatory responses, extracellular matrix homeostasis, phagocytosis and tolerance. *BMC Genom.* **2016**, *17*, 162. [CrossRef]
52. Waters, K.M.; Masiello, L.M.; Zangar, R.C.; Tarasevich, B.J.; Karin, N.J.; Quesenberry, R.D.; Bandyopadhyay, S.; Teeguarden, J.G.; Pounds, J.G.; Thrall, B.D. Macrophage Responses to Silica Nanoparticles are Highly Conserved across Particle Sizes. *Toxicol. Sci.* **2009**, *107*, 553–569. [CrossRef]
53. Kersting, M.; Olejnik, M.; Rosenkranz, N.; Loza, K.; Breisch, M.; Rostek, A.; Westphal, G.; Bünger, J.; Ziegler, N.; Ludwig, A.; et al. Subtoxic cell responses to silica particles with different size and shape. *Sci. Rep.* **2020**, *10*, 21591. [CrossRef]
54. Dalzon, B.; Aude-Garcia, C.; Collin-Faure, V.; Diemer, H.; Béal, D.; Dussert, F.; Fenel, D.; Schoehn, G.; Cianférani, S.; Carrière, M.; et al. Differential proteomics highlights macrophage-specific responses to amorphous silica nanoparticles. *Nanoscale* **2017**, *9*, 9641–9658. [CrossRef]
55. Chen, L.; Liu, J.; Zhang, Y.; Zhang, G.; Kang, Y.; Chen, A.; Feng, X.; Shao, L. The toxicity of silica nanoparticles to the immune system. *Nanomedicine* **2018**, *13*, 1939–1962. [CrossRef]
56. Fedeli, C.; Segat, D.; Tavano, R.; Bubacco, L.; De Franceschi, G.; de Laureto, P.P.; Lubian, E.; Selvestrel, F.; Mancin, F.; Papini, E. The functional dissection of the plasma corona of SiO 2 -NPs spots histidine rich glycoprotein as a major player able to hamper nanoparticle capture by macrophages. *Nanoscale* **2015**, *7*, 17710–17728. [CrossRef]
57. Francia, V.; Yang, K.; Deville, S.; Reker-Smit, C.; Nelissen, I.; Salvati, A. Corona Composition Can Affect the Mechanisms Cells Use to Internalize Nanoparticles. *ACS Nano* **2019**, *13*, 11107–11121. [CrossRef]
58. Maier, T.; Güell, M.; Serrano, L. Correlation of mRNA and protein in complex biological samples. *FEBS Lett.* **2009**, *583*, 3966–3973. [CrossRef]

59. Schmid, O.; Stoeger, T. Surface area is the biologically most effective dose metric for acute nanoparticle toxicity in the lung. *J. Aerosol Sci.* **2016**, *99*, 133–143. [CrossRef]
60. Kou, L.; Sun, J.; Zhai, Y.; He, Z. The endocytosis and intracellular fate of nanomedicines: Implication for rational design. *Asian J. Pharm. Sci.* **2013**, *8*, 1–10. [CrossRef]
61. Konduru, N.V.; Molina, R.M.; Swami, A.; Damiani, F.; Pyrgiotakis, G.; Lin, P.; Andreozzi, P.; Donaghey, T.C.; Demokritou, P.; Krol, S.; et al. Protein corona: Implications for nanoparticle interactions with pulmonary cells. *Part. Fibre Toxicol.* **2017**, *14*, 42. [CrossRef] [PubMed]
62. Tabish, T.A.; Dey, P.; Mosca, S.; Salimi, M.; Palombo, F.; Matousek, P.; Stone, N. Smart Gold Nanostructures for Light Mediated Cancer Theranostics: Combining Optical Diagnostics with Photothermal Therapy. *Adv. Sci.* **2020**, *7*, 1903441. [CrossRef] [PubMed]

 *nanomaterials*

Article

Collection of Controlled Nanosafety Data—The CoCoN-Database, a Tool to Assess Nanomaterial Hazard

Harald F. Krug

NanoCASE GmbH, St. Gallerstr. 58, CH-9032 Engelburg, Switzerland; hfk@nanocase.ch; Tel.: +41-793565188

Abstract: Hazard assessment is the first step in nanomaterial risk assessment. The overall number of studies on the biological effects of nanomaterials or innovative materials is steadily increasing and is above 40,000. Several databases have been established to make the amount of data manageable, but these are often highly specialized or can be used only by experts. This paper describes a new database which uses an already existing data collection of about 35,000 publications. The collection from the first phase between the years 2000 and 2013 contains about 11,000 articles and this number has been reduced by specific selection criteria. The resulting publications have been evaluated for their quality regarding the toxicological content and the experimental data have been extracted. In addition to material properties, the most important value to be extracted is the no-observed-adverse-effect-level (NOAEL) for in vivo and the no-observed-effect-concentration (NOEC) for in vitro studies. The correlation of the NOAEL/NOEC values with the nanomaterial properties and the investigated endpoints has been tested in projects such as the OECD-AOP project, where the available data for inflammatory responses have been analysed. In addition, special attention was paid to titanium dioxide particles and this example is used to show with searches for in vitro and in vivo experiments on possible lung toxicity what a typical result of a database query can look like. In this review, an emerging database is described that contains valuable information for nanomaterial hazard estimation and should aid in the progress of nanosafety research.

Keywords: nanomaterials; hazard assessment; database; lung toxicity; titanium dioxide; study quality

Citation: Krug, H.F. Collection of Controlled Nanosafety Data—The CoCoN-Database, a Tool to Assess Nanomaterial Hazard. *Nanomaterials* **2022**, *12*, 441. https://doi.org/10.3390/nano12030441

Academic Editors: Christoph Van Thriel and Andrea Hartwig

Received: 9 December 2021
Accepted: 24 January 2022
Published: 28 January 2022

Publisher's Note: MDPI stays neutral with regard to jurisdictional claims in published maps and institutional affiliations.

1. Introduction

The technical and chemical developments in nanotechnology have clearly demonstrated that the schedule from invention to market release has become much shorter compared to the last century. Additionally, the protection of the environment and human health against toxic chemicals or materials and regulatory demands have also been intensified. The combination of these two areas has conflict potential, thus, information about new technological developments is needed to establish a comprehensive risk assessment for new chemicals and innovative materials [1,2]. To support research and development activities in industry and science institutes, many funding programs support the accompanying risk research. In the case of nanotechnology, national and international action plans coordinate the funding of research on hazard and exposure of nanomaterials, e.g., the Nano Safety Cluster of the European Commission (https://www.nanosafetycluster.eu/, last access 8 December 2021). These activities have led to an enormous increase in published studies since the 2000s, as can be seen in Figure 1. In the PubMed literature database, there exists more than 40,000 published studies on a huge variety of nanomaterials investigated in different biological models, such as animals, cell and tissue cultures or plants. Keeping track of this tremendous number of published data is very difficult, both for scientists but even more for representatives of the industry or regulatory authorities [3]. In addition to the sheer number of publications, the quality issue is of great importance [4]. Not all published data are produced under good laboratory practice, or even under good scientific practice conditions. Many labs do not use standard operating procedures (SOPs) or harmonized protocols for toxicological studies, although many projects funded in Europe for example produced such standardized methods and published the protocols on different websites (e.g., the national DaNa project https://nanopartikel.info/en/knowledge/operating-instructions accessed on 8 December 2021 and the European PATROLS-GRACIOUS project consortium—https://www.nanosafetycluster.eu/joint-patrols-gracious-nanosafety-cluster-event-on-harmonization-of-standard-operating-procedures-sops/ accessed on 8 December 2021).

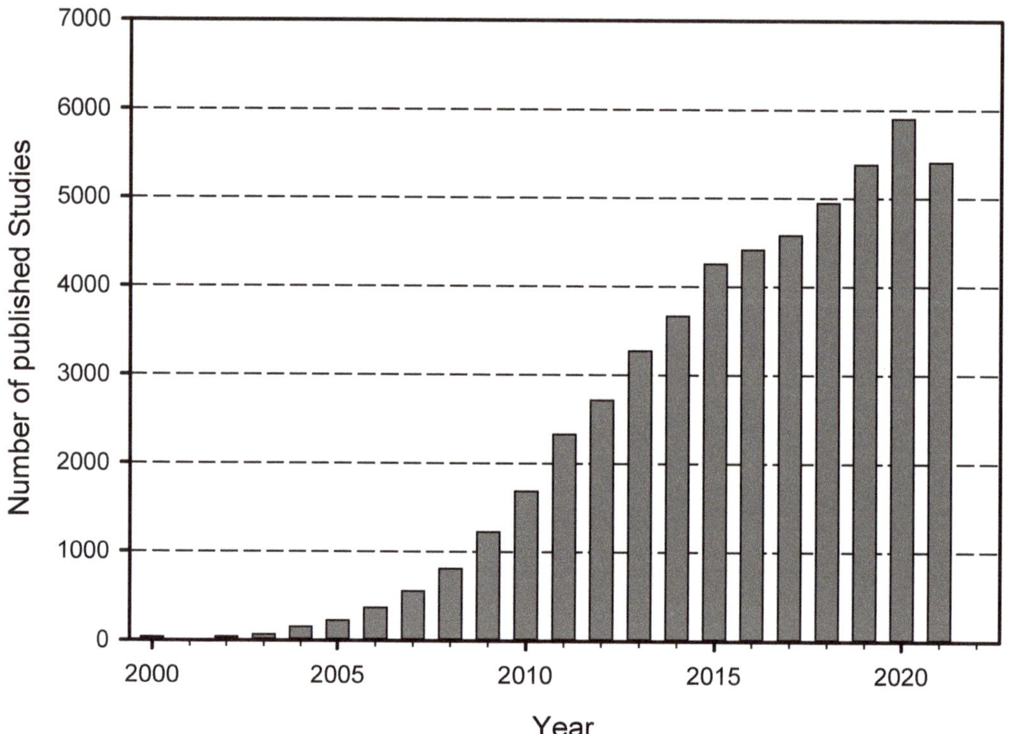

Figure 1. Publications per year found with a specific search profile in PubMed for the topic "Nanotoxicology". The bar for the year 2021 represents all publications until the 4th of November of this year.

The constant growth of data and the diversity in quality requires a way to use the existing knowledge effectively and properly. Therefore, several projects and activities in recent years have established many databases on nanomaterials with different contents (Table 1). It has already been discussed in detail that it is not easy to decide which database fits best to the needs of a specific user [5]. Besides the fact that most databases contain information about products containing nanomaterials or show relationships between nanomaterial properties and possible applications, good toxicological data are often missing. Only three of the databases available online contain toxicological data, these are eNanoMapper, NanoE-Tox and NBIK. Whereas NBIK is exclusively contains data from zebrafish embryo exposure experiments [6], NanoE-Tox contains ecotoxicological data on a huge variety of animal and plant species [7]. eNanoMapper delivers further additional toxicological relevant data an many species and information about ontology and material characteristics [8]. The two databases on ecotoxicological data are very special on the one hand but, on the other hand will no longer be updated. The only living database with toxicological relevant data is eNanoMapper but it is relatively complex and can mainly only be used by experts. Furthermore, there are no comparative values as result of a query, that could indicate whether a nanomaterial poses a hazard or not.

The data collection which will be used for the establishing of the database has been generated in several phases. The first phase reflects the literature between 2000 and 2013. During this phase the focus was on all materials without exception and some of the results have already been published [10]. In the year 2019 in a second phase the search was extended for the years 2014 to 2018. From the literature found in the second phase, only material specific evaluation has been carried out until now. Thus, the data presented here mainly concern the first phase if not mentioned otherwise.

Table 1. Popular databases of nanomaterials (all the web links were accessed on 8 December 2021).

Databases	Website	Remark
caNanoLab	https://cananolab.nci.nih.gov/	Nanotechnology in biomedical research on cancer
eNanoMapper	https://data.enanomapper.net/	Ontology and safety assessment of nanomaterials
NR	https://nanomaterialregistry.net/	Physicochemical properties of selected nanomaterials
NanoData	https://nanodata.echa.europa.eu/	Nanotechnology in products in 8 different sectors
Nanodatabase	https://nanodb.dk/en/	Nanomaterials in products
NanoE-Tox	http://www.beilstein-journals.org/bjnano/content/supplementary/2190-4286-6-183-S2.xls	Ecotoxicological data available as an excel sheet which will not be updated anymore
NanoNature	https://nano.nature.com/	Literature database on nanomaterials; will be retired in June 2022
Nanowatch	https://www.bund.net/themen/chemie/nanotechnologie/nanoprodukte-im-alltag/nanoproduktdatenbank/	Commercially available products containing nanomaterials; available in German only
Nanowerk	https://www.nanowerk.com/	Database on suppliers of commercially available nanomaterials
NBIK	http://nbi.oregonstate.edu/ no safe connection on access day	Database on study results of nanomaterial exposure effects in embryo zebrafish
NECID	https://necid.ifa.dguv.de/Login.aspx?ReturnUrl=%2fUser%2fFirstpage.aspx	Data on occupational nanomaterial exposure during various exposure scenarios
NIL	http://nanoparticlelibrary.net/	Physicochemical characteristics of very specifically produced nanomaterials
NKB	https://ssl.biomax.de/nanocommons/	Nano-safety knowledge infrastructure
PubVINAS	http://www.pubvinas.com/ no safe connection on access day	Virtual nanostructure simulation tool
PaFTox	https://publica.fraunhofer.de/dokumente/N-277711.html	Data on genotoxicity of nanomaterials; the database is not available anymore although funded by government money
StatNano	https://statnano.com/	Applications and properties of nanomaterials

All these reasons have forced the consideration of the establishment of a new database, as the relationship between the material properties and the biological effects that may be induced can be very informative. To make scientific results especially available for users outside the scientific community without excluding scientists, the CoCoN® database is to be created based on the "Collection of Controlled Nanosafety Data". Searching for toxicological data related to the properties of specific nanomaterials should be possible and the result would ideally have to be a comparable value that has good validity for different biological models. One such value could be the NOAEL in animal studies and the NOEC in cell or tissue culture experiments [9]. During the last 8 years, a very large data collection has been developing, containing several thousands of datasets on many different nanomaterials, tested in various biological models for their toxicological potential. This data collection is the basis for the programming of the CoCoN® database, which aims to deliver a material property-related outcome. The CoCoN® database, which is an active development, will deliver quality-assessed data on a huge variety of biological endpoints affected by nanomaterials which will be taken from the available published literature. This paper describes the idea behind CoCoN® and provides use case examples. The database should help to develop a better understanding of the material property to hazard relationship and its query tools may produce new knowledge for predicting the toxic potential of advanced materials.

2. Materials and Methods

2.1. Literature Search and Study Selection

In the first phase, the literature was taken only from the PubMed database and imported into an Endnote Library. The search profile is shown in Table 2. The number of found citations with this search profile per year can be seen in Figure 1. For the first phase it sums up to 11,058 and for the second phase to 24,303 publications.

Table 2. Search profile in Endnote for PubMed database for publications on nanotoxicological studies.

Field Delimiter	Where	How	What
	All Fields	Contains	nanotox *
Or	All Fields	Contains	fulleren * AND toxic *
Or	All Fields	Contains	carbo nanotube * AND toxic *
Or	All Fields	Contains	bucky ball * AND toxic
Or	All Fields	Contains	nanotube * AND toxic
Or	All Fields	Contains	nanoparticle * AND toxic *
Or	All Fields	Contains	nanomat * AND toxic *
Or	All Fields	Contains	Nano * AND toxic
Or	Year	Contains	2021 [1]

[1] for 2021 5515 records have been found on the 11th of November. *: wildcard for the search words.

As mentioned above in the first phase all investigated nanomaterials have been recorded. The number of studies was reduced by a defined selection process focused on toxicological investigations on nanomaterials. The exclusion criteria were defined as follows:

- The study deals with medical treatment or therapeutic application of the nanomaterial only, e.g., drug delivery or therapeutic application via injection.
- No real toxicological experiments have been shown; this is the case if a material- or engineering-oriented publication uses the buzz word "toxicity" or similar within the introduction or discussion, which led to the inclusion of the publication at the beginning.
- Neither animal studies nor cell- or tissue culture experiments have been described, which excludes reviews as well.
- The content deals exclusively with environmental issues, e.g., air pollution with diesel exhaust particles or ultrafine dust, or describes solely ecotoxicological topics such as plant research or distribution in environmental compartments such as water, soil or air.
- The publication is not available as pdf file and/or not written in English or German language.

Based on this selection process the number of studies was reduced from 11,058 to 6626 for the first project phase [11]. To better handle the huge number of citations the excluded non-relevant publications have been removed from the library.

In the second phase more than 24,000 publications have been found with the same search profile and included into a second Endnote library. The number of publications of the second phase has not been reduced with the same selection process as described above. Until now only a rough selection (e.g., wrong language, etc.) has been carried out and the number of citations in this library is actually 19,732. These newer studies are the basis for specific searches on defined materials such as silica, graphene, or titanium dioxide. As the second library is not at the same stage as the first one, the following examples and results concentrate only on publications from the first phase but represent examples for the CoCoN® database as it is planned to be established.

2.2. Data Extraction and Collection

The publications from the first phase have been implemented into a huge table with four different sections. The first section contains bibliographic data such as author names, publication year, and the digital object identifier (DOI) as a specific identifier. Moreover, information about the study type (in vitro or in vivo), the intended application of the nanomaterial (food sector, cosmetics, technical applications, etc.), and a first indication of whether this study is toxicology oriented or more of a mechanistic study is collected here. The second part presents all the available information about the used material. Name, size of primary particles, source, shape, surface charge and specific surface area, crystal phase, coatings, aggregation or agglomerations state, the used concentrations within the experiments and some more information are given in this section. Within the third part, the information about the biological model is presented. Animal species for in vivo studies or cell type for in vitro studies, the type of culture, the treatment design with information on duration, repeats, exposure pathways, dose concentrations and OECD guideline number, if available, and more. All these data are included as the authors have mentioned them in the publication. Following this fundamental information about the respective study, the biological endpoints are then recorded with their associated assays.

Each publication may result in several datasets within this table because each investigated material examined, and even each variation of it, e.g., different sizes or different coatings, are recorded as a separate dataset. It may happen that one publication generates a huge number of datasets when many materials have been tested in a large number of cell cultures (e.g., [12]; with more than 200 datasets) but on average there are 2 to 4 datasets per publication.

2.3. Quality Evaluation—Scoring

During the collection of the data, (readout) the quality of the study was evaluated. During the first phase of the project, a rough estimation was made about the quality just by looking for the obligatory characterization of the nanomaterial and the minimum information about the biological model. Later, after several national and international projects established criteria catalogues for the data quality of published studies (literature criteria catalogue from the German DaNa-project: https://nanopartikel.info/en/knowledge/literature-criteria-checklist/ accessed on 8 December 2021 [13]; the European GuideNano project established an Excel-Sheet for the quantitative evaluation of studies on nanomaterials [14]; the ToxRTool, established by a European project to refine the Klimisch scores for toxicological studies [15]). The studies were also evaluated for quality when the datasets were implemented into the existing table, which led to a disappointing outcome, in that more than 70% of the toxicological and mechanistical studies on nanomaterials were not reliable [16]. Taken together, the most important criteria from the different criteria catalogues and tools for the CoCoN® database a variation of the GuideNano-evaluation scheme have been used to automatically generate a quality score which is attached to the publication for the traceability of the process. The quality evaluation was performed in four different sections and for each tox category (in vivo, in vitro, and ecotox) there was a value calculated separately in combination with the material characterization. The rating can take the following values (Figure 2), where 0 and 0.5 are low quality studies, 0.8 are good studies and 1 is the value for very reliable studies.

The score is an important value which is included in the table (and later in the established database as well). This gives the user of such a database the chance to select "all" available data or "only the good" studies for their queries.

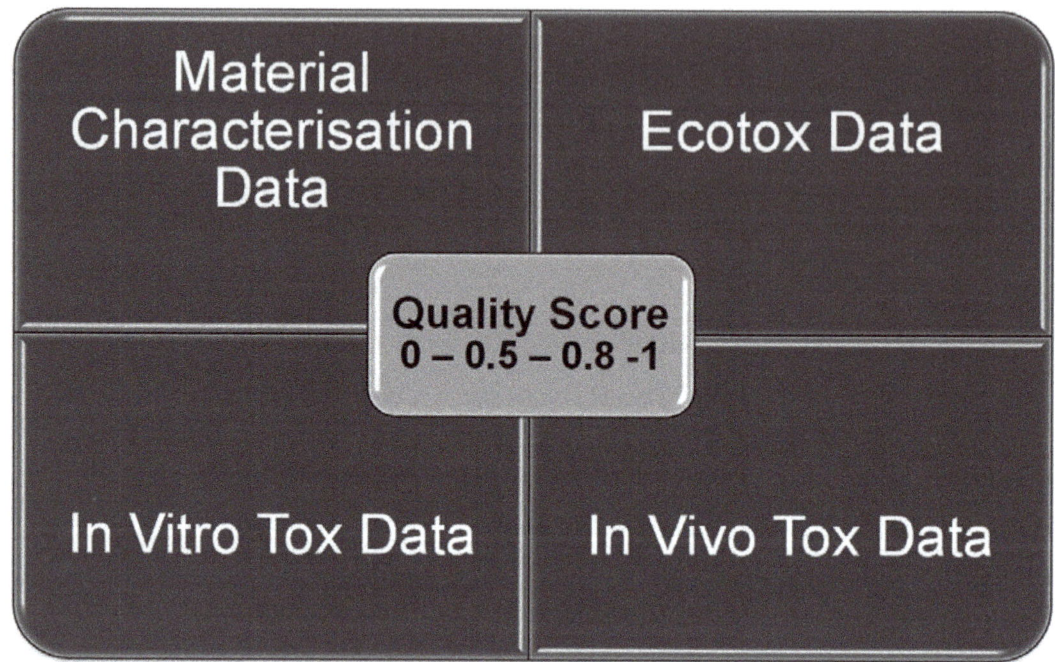

Figure 2. A scheme of the four sections for which quality scores are generated. Each tox-section calculated together with the material characterisation reveals a quality score for the specific kind of study.

2.4. NOAEL/NOEC—Readout from Published Studies

The heart of the database and thus also the most important result is the possibility to readout a real comparative value between the studies. For toxicological assessments, the dose or concentration at which a biological effect is just not triggered is of importance. This no-observed adverse-effect-level (NOAEL—in vivo) or no-observed-effect-concentration (NOEC—in vitro) is included from each study for the observed biological effect or key event which is described by the authors. One difficulty for the readout of this value from published data is the fact that not all studies cover the concentration or dose range to directly determine the NOAEL/NOEC. As shown in Figure 3, there exist mainly three variations. In the best case, the concentration range is chosen correctly by the authors and the NOAEL/NOEC can be read off from the data directly (Figure 3a). The second case is that an effect was observed even at the lowest concentration used in the experiment. In this situation, the NOAEL/NOEC is lower than the lowest concentration, which in this case represents the lowest-observed-adverse-effect-level (LOAEL) or lowest-observed-effect-concentration (LOEC). In such a case the value in the table is given with "<" to demonstrate that there might be a concentration less than the number extracted from the publication (Figure 3b). The last case consists of the fact that none of the used concentrations changes the studied biological activity, that is, not even the highest one. Thus, the highest concentration used is the upper-no-observed-adverse-effect-level (UNOAEL) or upper-no-observed-effect-concentration (UNOEC), which means that there is no toxic effect at all (Figure 3c). To show this in the table, the respective value is preceded by the character ">".

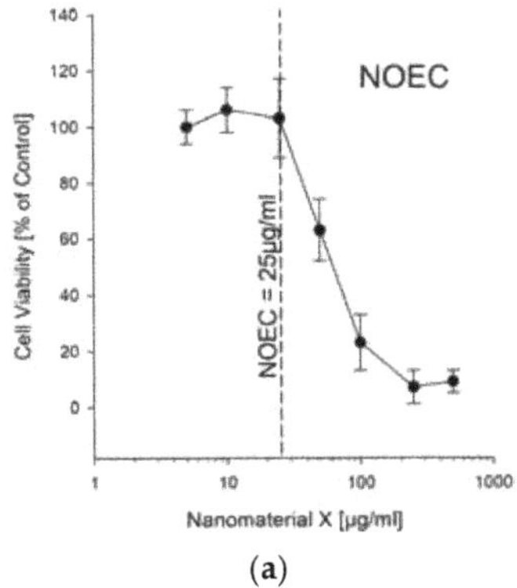

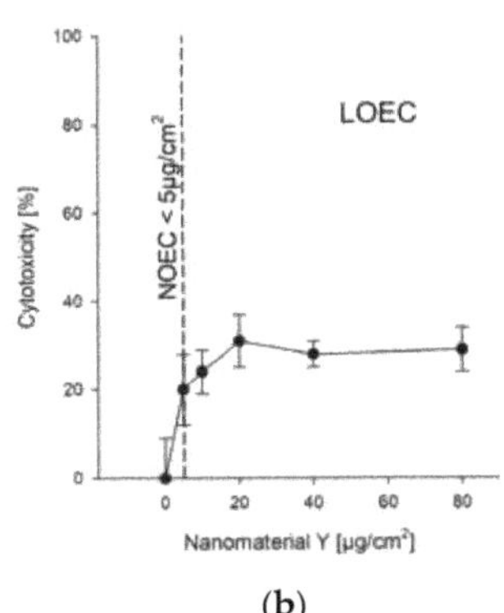

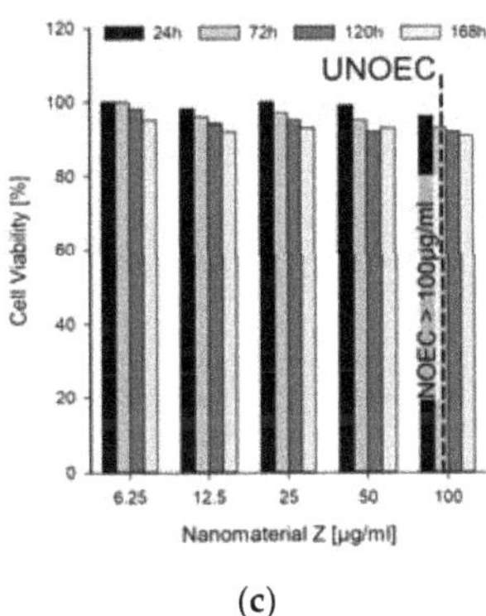

(a) (b) (c)

Figure 3. An explanation of how the NOAEL/NOEC were readout from the data presented within the published studies. The three examples shown here are based on real data taken from publications but reproduced without any relation to the materials and the publications, as these are just to demonstrate the process. All three examples are in vitro studies and measured cell viability or cytotoxicity. On the left (**a**) the NOEC can be determined directly from the graph as 25 µg/mL. In the middle (**b**) the lowest concentration used is still affecting the cells, thus it represents the LOEC. On the right (**c**) even the highest concentration does not induce any toxic effect on the cells and the value of 100 µg/mL is the UNOEC.

During the extraction of the data from the publications, a major hurdle to the usability of the data became apparent, namely the information on the dose or concentration units used. These are generally not standardized in the studies and therefore differ greatly Table 3). Moreover, the exposure pathway is different too, especially for the exposure pathway via the air, as there exists a huge variety of application methods. In addition to inhalation (nose-only or whole-body exposure) and instillation (intranasal, intratracheal, or oropharyngeal), the authors of the studies have used aspiration (pharyngeal or oropharyngeal), and intrapulmonary spraying. Wherever possible, units have been recalculated to finally result in a greater number of matching dose units.

Table 3. Quantities in the dosing of nanomaterials in the in vitro and in vivo experiments as given by the authors.

Given Concentration Units for In Vitro Studies	Given Dose Units for In Vivo Studies	
	oral exposure	inhalation/instillation
$\mu g/mL$		
$\mu g/cm^2$	$\mu g/animal$	$\mu g/m^3$
cm^2/mL	$\mu g/kg$ BW and	mg/m^3
m^2/cm^2	mg/kg BW	cm^2/m^3
$mM/\mu M/nM$	dermal or intradermal	$/cm^3$
ppm	$\mu g/\mu L$ or mL	$\mu g/animal$
/mL	g or mg/ear	$\mu g/g$ lung tissue
$/cm^2$	mg/animal	$\mu g/lung$
/cell	mg/kg	

3. Results

Currently, more than 1500 studies of the first phase have been evaluated and included in the data collection, which represent more than 4000 datasets. Additionally, more than 2000 studies from the second phase have been evaluated for specific materials and 888 have been included in the data collection, representing 1318 datasets for these specific nanomaterials chosen. During the next phase of programming the new CoCoN® database, more studies will be included taken from the pre-evaluated selection of both phases. In the following, only evaluations of studies from the first phase are presented, as they were evaluated with the same approach and provide a consistent picture.

3.1. Key Figures of the Data Collection

The more than 4000 datasets of the first phase regarding the literature between 2000 and 2013 show results for more than 100 different nanomaterials which have been investigated. To really put this number in perspective, it is important to know that for this calculation, e.g., all multi-walled carbon nanotubes are taken as "One" material. The same applies for graphene, titanium dioxide, silica, and all other materials. Discriminating between different crystallinities, fiber lengths, oxidation states and other properties of the nanomaterials would result in a much larger number of materials. Additionally, more than 150 different coatings have been found, but most of them have been used only once or less than five times. A comparable high number of cell cultures have been used for the in vitro studies, as around 260 different cell lines are treated with nanomaterials. The variability is much less for animal studies, where mainly rats, mice, fishes, nematodes, daphnia, mussels, or drosophila have been investigated. The biological endpoints, which are described within the studies, sum up currently to 107 in the table with the results, but are still increasing. For the future database, some of the endpoints will have more than one choice concerning gene expression, for example, other omics technologies which result in several alterations of genes or proteins, which are probably up or down regulated. At the end of the table a selection can be made during input of data which pathways of toxicity may be involved and which adverse outcome pathway could be the result of the biological effects in the investigated system. The input is mainly based on the statements of the authors. If there is no clear statement or another possibility to derive this information from the data, then these fields are left blank.

Not all the information provided by the authors is unambiguous. The most obvious indication with high potential for misunderstandings is to present the applied amount of nanomaterial to cell cultures in molarities. It is very difficult to re-calculate the given concentration to $\mu g/mL$ or similar for TiO_2 nanoparticles [17,18] but the question remains as to how to re-calculate the molarity of multi-walled carbon nanotubes into a comparable dimension [19]. Such concentration units are not meaningful and are irrelevant to hazard or risk assessment. Perhaps these papers should be rejected during the review process because the most important part of the experiments is not comprehensible. In addition, the concentration data in $\mu g/mL$ for in vitro experiments are not always unambiguous and traceable. Without the knowledge of which Petri dishes or multi-well plates have been used, how much volume has been applied per dish, and how many cells have been plated into these dishes, the exact design of the experiment cannot be reproduced. Without the above-mentioned details, the real concentrations the cells are treated with may vary by more than 100%. Such details become obvious when the studies are re-evaluated and analyzed during data extraction for the database.

3.2. Readout Results from the Data Collection

A first look at the distribution of the studies in relation to the endpoints gives a clear picture of the focus of the studies. Specifically, the in vitro studies have clear priorities in the effects studied (Figure 4). The overall data collection has been the basis for several projects and activities in recent years. Therefore, an excerpt will be given here concerning which possibilities exist to be able to use the data of the collection.

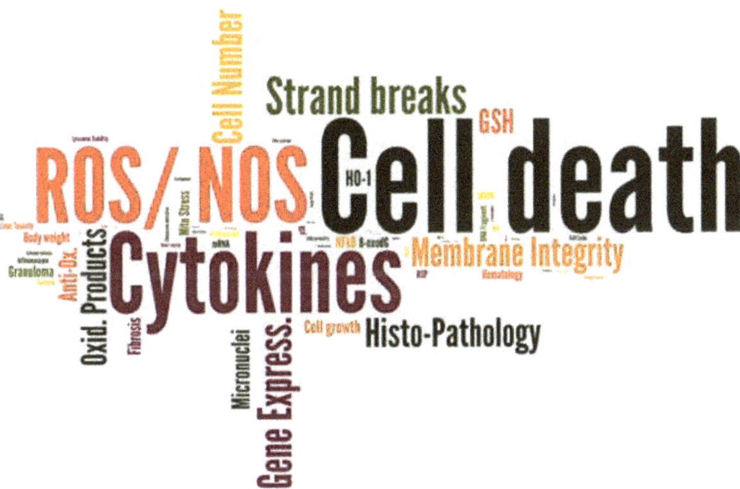

Figure 4. Word Cloud results after computing of all in vitro experiments within the data collection. Letter sizes are directly proportional to the number of data sets for the respective analyzed endpoint.

3.2.1. OECD-Project on Adverse Outcome Pathways for Nanomaterial Risk Assessment

Besides the already published data [10,16,20], one of the first intense queries has been made for the international OECD WPMN Pilot Project on Advancing Adverse Outcome Pathway (AOP) Development for Nanomaterial Risk Assessment and Categorization [11]. To follow the selection process for such specific queries, the PRISMA system has been applied [21]. From the 6626 Studies at the end, 191 were selected for the adverse outcome "inflammation", which resulted in 447 datasets. The main goal of this activity was developing criteria and methods to identify and prioritize potential key events (KEs) from the nanotoxicity literature, which may be inflammation-associated. Within this subset of data around 60 different endpoints have been reported for 45 different nanomaterials and the KEs have been prioritized for their contribution to the inflammation process [11]. Additionally, for a selection of seven nanomaterials specific criteria have been discussed. Which endpoints are relevant for this AOP and if these are induced or not have been some of the key questions. The results on these questions have been published in 2020 [22]. Moreover, the relationship between the literature data extracted from the data collection and the possibility to assess the nanomaterials toxicity via non-animal strategies is another idea to use these datasets [23]. There exists another example in using data from dossiers to assess possible toxic activities of chemicals with the tools of machine learning and data mining. The group of Thomas Hartung at the Johns Hopkins University in Baltimore evaluated ECHA-dossiers, extracted the toxicity data, and calculated based on structure similarities the possible toxic potential of new chemicals and outperformed by this process the results of animal testing [24].

3.2.2. Evaluation of Titanium Dioxide Lung Toxicity

Another example at this point concerns the evaluation of data on titanium dioxide. This material, which has been on the market for more than five decades, has repeatedly been the subject of discussion, especially since it has also been used as a nanomaterial directly in cosmetics and food. To specifically compare in vitro and in vivo results in this chapter, some possibilities will be demonstrated concerning how the display of the results of a query with appropriate filters could be presented.

The filter for the in vitro data was as follows:

- Material = titanium dioxide;
- Type of study = in vitro studies;

- Cell type = mammalian lung cells (exclusively macrophages or immune cells);
- Concentration units = μg/mL (or μg/cm^2, if recalculation is possible);
- Biological endpoints = acute cytotoxicity, formation of reactive oxygen species (ROS), cytokine production;

The filter for the in vivo data was as follows:

- Material = titanium dioxide;
- Type of study = in vivo studies;
- Species = rats or mice;
- Exposure = inhalation (mg/m^3) or;
- Exposure = instillation or aspiration (μg/kg or μg/animal);
- Biological endpoint = immune cell migration (neutrophil or macrophage influx).

The query for the in vitro studies revealed 56, 51, and 32 datasets for the three different endpoints: cytotoxicity, ROS formation and cytokine production, respectively. The database offers the possibility to distinguish between different size groups of the investigated nanomaterials. In this case, the primary particle sizes of the applied titanium dioxide variants were assigned to the following five size groups:

0–10 nm
11–20 nm
21–50 nm
51–100 nm
101–500 nm

The data are shown in Figure 5 and demonstrate a typical result for a corresponding query from the data collection. Although the number of data points for cytokine production is relatively low, all three endpoints tell the same story: in most cases, an effect is triggered only at very high doses. This is particularly well illustrated if the work of Klaus Wittmaack is taken into account [25,26], who was able to show impressively that at concentrations higher than 27 μg/mL of titanium dioxide the cells in the Petri dish are "buried" under a layer of sediment particles, which prevents the normal supply of nutrients and oxygen to the cells. This is what the author calls the "landslide effect", as the cells under such a layer of solid material will show signs of deficiency and start to die after some period of treatment. Nevertheless, most cell lines are robust and survive for the first 24 h, as can be seen by most data points beyond this limit shown in Figure 5A–C. Therefore, the statement from the in vitro studies is that titanium dioxide is not critical to lung cells and belongs to the nanomaterials of low toxicity.

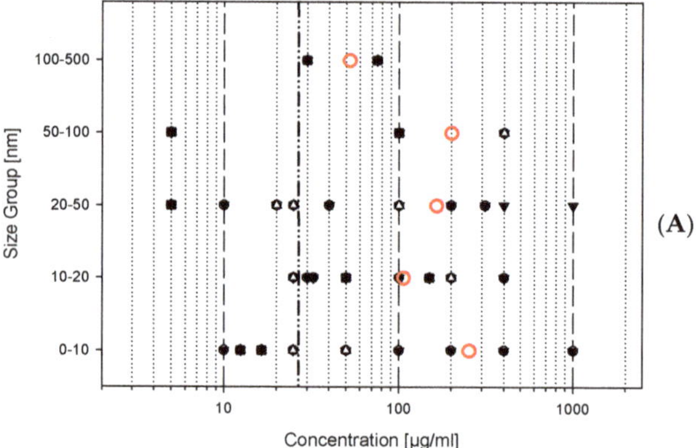

Figure 5. *Cont.*

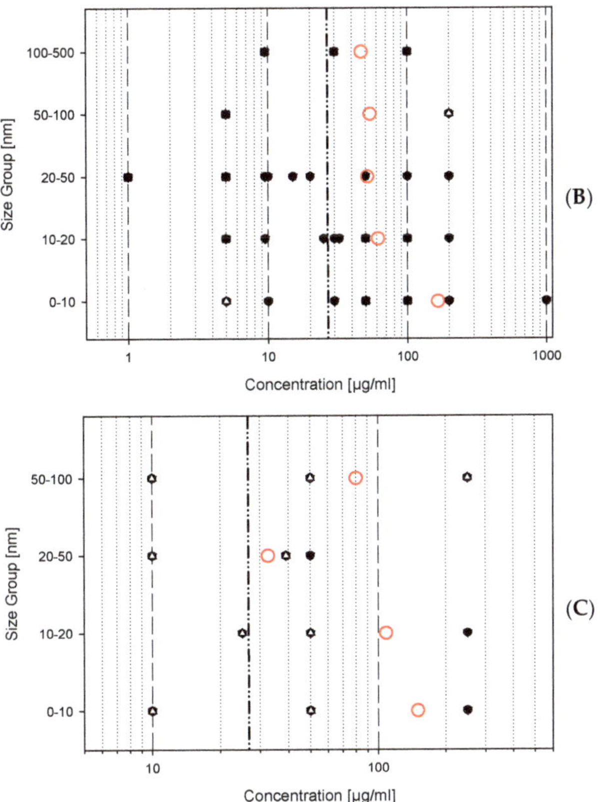

Figure 5. NOEC for titanium dioxide treatment of mammalian lung cells. Shown are 56 datapoints for cytotoxicity (**A**), 51 datapoints for ROS formation (**B**) and 32 for cytokine production as extracted from the data collection with the specific filter given in the text. Abscissa gives the concentrations at which no effect could be observed in µg/mL (NOEC). The ordinate shows the result for five (**A,B**) or four different size groups (**C**) of the primary particle sizes as given by the authors. Often, multiple data points overlie each other. The red open circles represent the calculated mean for each size class. No data available for the size class 100–500 (**C**) The dash-dotted-vertical line (— •• —) represents the limit for the "landslide" situation above 27 µg/mL (explanation see text).

The query for the in vivo studies revealed 25 datasets, 17 for instillation or aspiration and 8 for inhalation experiments from 14 studies in total. Ten Studies have been published in this period between 2000 and 2013 on the exposure pathway instillation or aspiration. The applied doses spread over a wide range from 400 to 5000 µg/kg for rats and 100 to 30,000 µg/kg for mice. The method of exposure via instillation or aspiration is paralleled by a very fast dose rate [27] and thus, it can be expected that the material provokes in most of the treatments a transient inflammatory response because of the delivery of a high amount of particles in a very short time. This usually manifests in immune cell migration into the lung and cytokine production by these cells. Two studies which have been carried out with mice used overload concentrations ($\geq$500 µg/mouse lung), whereas no overload conditions have been applied in the experiments with rats. The results in detail can be looked up in the Supplementary Materials (Tables S1 and S2), but here the most important facts will be listed:

- The overall number of studies with a comparable experimental study design is low; most toxicological studies do not yet use standardised protocols, e.g., OECD testing guidelines as suggested recently [28].
- The variety of investigated titanium dioxide is high; in this period, not even two studies used the same material.
- In rats, NOAEL range from 200 µg/kg to 5000 µg/kg, which might represent the huge variety in the materials.
- Some studies found a response for the only investigated dose; in many cases the observation time was only 1 d and in this treatment period the influx of neutrophils can be expected after exposure with dust particles as a "normal defence response".
- For mice, the NOAEL is between 30 µg/animal to 500 µg/animal, which represents the overload dose.
- In mice and rats, the inflammatory response is strongly dependent on the material tested.
- For the more realistic inhalation studies, 7 datasets out of 8 do not describe any cell migration into the lungs; the only study which found an inflammatory response used 100 mg/m^3, 6 h per day for 5 days for the inhalation exposure.

There are some more animal studies in the first phase of this data collection regarding lung exposure with TiO$_2$, but these described other endpoints or used slightly different material shapes (e.g., rods or short fibres). A total of 29 publications deal with instillation/aspiration experiments and 26 publications are on inhalation experiments (reference list is given in the Supplementary Materials). Nevertheless, it is very interesting to evaluate these data and consider the authors' assessment when it comes to the overall impression of the effect of TiO$_2$. Based on the described effects, a categorization was attempted to evaluate the effects in terms of an adverse outcome. The following five categories were defined, and the results are shown in Figure 6:

1. Overload: this reflects the situation that for mice a higher dose than 500 µg/lung and for rats more than 2500 µg/lung were applied.
2. No effect: describes the situation that even the highest dose used for the experiments did not induced any adverse effect in the animals during the observation time.
3. GBP-like: the behaviour as a granular biopersistent particle (GBP) is described to induce a transient inflammatory response, which usually subsides after about 7 days and the animal recovers completely without permanent sequelae.
4. Quartz-like: the relationship to quartz describes a more intense inflammation accompanied by oxidative stress, a formation of granuloma with signs of fibrosis.
5. Asbestos-like: the effects which indicate quartz-like mechanisms and additionally genotoxicity and the induction of a higher tumour incidence.

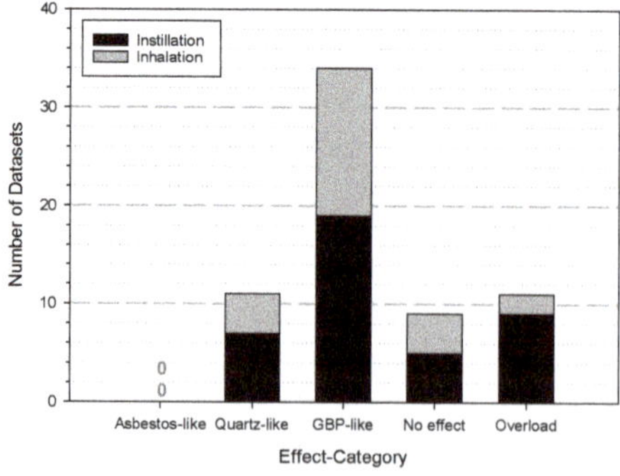

Figure 6. The attempt to categorize the effects of TiO$_2$ in lung exposure studies. The distribution of the datasets from the data collection over the five different categories is shown. Black bars show instillation experiments and grey bars the data for inhalation studies. For definition of the categories see text.

The use of the data collection enables clear presentations of the results as shown above. The example of titanium dioxide and its effects to lungs and lung cells makes it clear that the relatively large number of data sets allows a good statement to be made about the effects of a specific material. Significant is the value for the NOAEL/NOEC which allows considerations for the classification in the Globally Harmonised System for Hazard Classification. A complete list of the references for all studies, for the in vitro as well as the in vivo studies, are given within the Supplementary Materials.

4. Discussion

The tremendous development in the number of publications overall, but also for nanotoxicological topics, gives rise to considerations on how this flood of data could be harnessed. Moreover, the ongoing discussion on the safety of nanomaterials and innovative materials is often driven by single investigative studies with questionable toxicological backgrounds [29]. Based on these facts, there exists several activities in establishing databases for a variety of applications and different user groups (compare Table 1). Keeping in mind the registration of new products and materials during the REACH process (Registration, Evaluation, Authorisation and Restriction of Chemicals) it is of utmost importance that industry on the one hand and the registration authorities on the other have access to already prepared data [3]. This was the reason to establish the data collection under the aspect that a quick access can take place and important information can be queried in simple representation. So far, the existing tables containing most of the datasets is much too complex for such queries by clients themselves. Thus, the idea was born to transfer the datasets to a database in order to fulfil the criteria of easy access for clients and present an informative representation of the results.

In order to keep the complexity of the data in the database in check, certain details of the studies are not recorded (e.g., the in-depth details of the methods used in the studies), but the emphasis is on the information which is of great importance (e.g., the test design such as treatment of animals, duration, repeats, etc.). The scheme of the data collection follows the already published workflow for "nanodata" curation (Figure 7) as described a few years ago [30]. The existing data collection cannot and does not claim to be complete but gives an impression of the existing knowledge in its entirety. As soon as the database has been established and data entry has been simplified, the information from the clients should also be taken up (feedback system) when it comes to updating the content of the database and incorporating new studies.

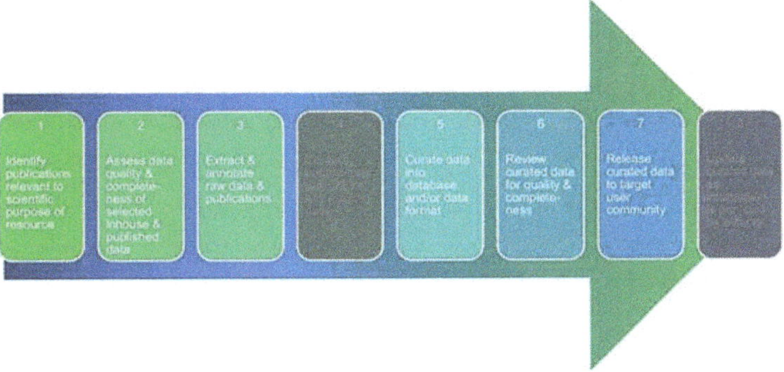

Figure 7. The workflow during data collection and implementation into the CoCoN® database. Steps 4 and 8 in this process were cancelled as it takes too much time to contact the authors to complete their data for so many publications. Figure 7 was adapted from [30] ("Nanocuration workflows: Establishing best practices for identifying, inputting, and sharing data to inform decisions on nanomaterials", ©2015 C. M. Powers et al., distributed under the terms of the Creative Commons Attribution 2.0 International License, https://creativecommons.org/licenses/by/2.0).

For the incorporation of new study results into CoCoN® in the future, it is of utmost importance that toxicological publications respect the FAIR principles. This would make the data needed for the database input findable, accessible, interoperable, and reusable [31]. The introduction of electronic lab books and accompanying Supplementary Materials with publications in the form of toxicological dossiers with the most important information about the material, the biological model and test design would make it additionally much easier to implement such data into databases [32,33]. It was impressively demonstrated that the data from machine-readable REACH dossiers with the aid of machine learning processes provided a better estimate of the toxicological potential of chemicals than the corresponding animal tests [24]. This example illustrates the opportunity that lies in digitizing existing data. Of course, this will not replace all cell culture or animal experiments, but the evaluation and calculation based on the existing studies will help to significantly advance read-across [34,35] and other possibilities for risk assessment, following the 3R principles [36]. However, some efforts are still needed to achieve this goal, particularly as the range of variation in the studies on nanotoxicology published to date is very large. Test models, test design, materials used, and the exposure conditions mean that hardly any study is comparable to another, making it difficult to combine the results. In addition, studies often described as toxicological studies do not follow the principles of toxicology as demanded a long time ago [37]. This includes missing concentration dependencies or dose–response relationships, as well as the complete lack of standardized procedures [28]. Thus, mechanistic studies are very often overlooked in the literature as toxicological ones, which leads to a misjudgement from a toxicological point of view. This explains why the assessment of a substance could be understood based exclusively on high-dose experiments or, in the case of nanomaterials, so-called "overload conditions".

In addition to these very conspicuous errors, however, there are other, less easily recognizable flaws that lead to the disqualification of published studies [38]. Very often, for example, the analysis of the properties of the nanomaterials used for the investigations is weak, right down to the size information, which can also be missing. However, without a reference of the results to the properties of the material, a study is of little value from a toxicologist point of view. Therefore, it is of importance to also look at and evaluate the quality of the studies when collecting data for a data base.

There exist many suggestions on how to increase the quality of studies, and the most important prerequisites for good quality papers on the toxicology of nanomaterials are listed here and have been published elsewhere [4,14,39–41]:

1. A rigorous and adequate physicochemical characterization of the test materials is needed;
2. Adequate particle controls must be included;
3. Possible contaminations, such as endotoxins, should be analysed;
4. Interferences of the tested material with the assay should be investigated;
5. High-dose experiments designed to produce toxicological effects—which are publishable (and sensational)—should be avoided
6. As far as possible, standardized protocols should be used to better compare results

These and more recommendations are also called for by project consortia and experts in the field. To face the problems of costly development of products and the pace regulators are confronted with, new approach methodologies are a possible solution for the improvement of risk assessment [1–3,42]. Additionally, here the implementation of searchable databases plays a central role. Such databases, as the herewith introduced CoCoN® database, can provide answers to classify the biological effect of a material, but also allow read-across or grouping. As shown above for the example material titanium dioxide, the data can be filtered for a specific material, specific cell types and special biological endpoints. Figure 5 demonstrates clearly that the majority of datapoints show low toxicity and the mean value of the individual size classes is well above the maximum reasonable concentration of 27 µg/mL, as given by Wittmaack [25].

The presented results of the selected in vivo studies from 2000 up to 2013 (Tables S1 and S2) show typical characteristics for lung exposure experiments. First, the study design of instillation or aspiration induced a high dose rate, which usually induces an inflammatory response to particle exposure, which decreases during the first week after treatment [27]. This response is not observed within inhalation experiments that reflect more realistic dose rates, except for only one dataset (#1, Table S2). Here, an extremely high dose of 100 mg/m^3 for 6 h per day and 5 consecutive exposure days was used for this study. Secondly, many of the investigated materials are self-made materials or materials not allowed to be used on the market. Thirdly, the collection of in vivo data also shows the weaknesses of some studies which even do not announce the doses applied for the experiments shown in the figures of the publication (datasets 10 and 11, Table S1, taken from [43]).

Nevertheless, the following period from 2013 to now is of importance as well, and these studies are under evaluation. It should soon be possible to compare the data from the first phase with those from more actual studies and to draw further conclusions on the biological effects of titanium dioxide and other materials.

5. Conclusions

Safety concerns in the use of nanomaterials must be considered and their potential risks to human health as well as their undesirable effects on the environment should be avoided. To assess such risks, the totality of published data is a good source, but a compilation of all existing data is lacking. Therefore, the CoCoN® database is an attempt to process the already existing but very complex data collection in such a way that an easy and user-friendly retrieval of the study results is possible. This allows users of the database to enter their own queries and search for results for specific materials. This will allow the assessment of a potential hazard for nanomaterials or innovative materials and make the regulatory processes smoother and faster. Moreover, producers of new materials may get a first picture of the biological effects of their material and can assess how useful the further development of new material will be for specific applications. In the best case, the use of the metadata from the database can enable a cross-comparison of important details, both in terms of material properties and hazard potentials [44]. Scientists can use the data to make comparisons with their own work or between different materials, which may support grouping and read-across. Regulators can draw conclusions about material safety from their queries and manufacturers can identify and close existing knowledge gaps at an early stage of product development.

Supplementary Materials: The following are available online at https://www.mdpi.com/article/10.3390/nano12030441/s1, Table S1: Read out from the tabular form of the data collection on the effects of TiO$_2$ after lung exposure by instillation or aspiration, Table S2: Read out from the tabular form of the data collection on the endpoint "inflammation/cell migration" induced by TiO$_2$ after lung exposure via inhalation. Full reference lists for the in vitro data used for Figure 5A–C and full reference list for the in vivo data used for Figure 6.

Funding: This project was made possible with financial support by the Swiss Federal Office of Public Health (FOPH) and the German VCI (Association of the Chemical Industry).

Data Availability Statement: The data points in all figures have been taken directly from original publications which are included in the CoCoN data collection. The selected contributions used for this study are listed in the Supplementary Materials. Further inquiries can be directed to the corresponding author.

Conflicts of Interest: The author declares no conflict of interest. The funders had no role in the design of the study; in the collection, analyses, or interpretation of data; in the writing of the manuscript, or in the decision to publish the results.

References

1. Gottardo, S.; Mech, A.; Drbohlavova, J.; Malyska, A.; Bowadt, S.; Riego Sintes, J.; Rauscher, H. Towards safe and sustainable innovation in nanotechnology: State-of-play for smart nanomaterials. *NanoImpact* **2021**, *21*, 100297. [CrossRef] [PubMed]
2. Tavernaro, I.; Dekkers, S.; Soeteman-Hernandez, L.G.; Herbeck-Engel, P.; Noorlander, C.; Kraegeloh, A. Safe-by-design part ii: A strategy for balancing safety and functionality in the different stages of the innovation process. *Nanoimpact* **2021**, *24*, 100354. [CrossRef]
3. Soeteman-Hernández, L.G.; Bekker, C.; Groenewold, M.; Jantunen, P.; Mech, A.; Rasmussen, K.; Sintes, J.R.; Sips, A.J.A.M.; Noorlander, C.W. Perspective on how regulators can keep pace with innovation: Outcomes of a european regulatory preparedness workshop on nanomaterials and nano-enabled products. *NanoImpact* **2019**, *14*, 100166. [CrossRef]
4. Hristozov, D.R.; Gottardo, S.; Critto, A.; Marcomini, A. Risk assessment of engineered nanomaterials: A review of available data and approaches from a regulatory perspective. *Nanotoxicology* **2012**, *6*, 880–898. [CrossRef] [PubMed]
5. Ji, Z.; Guo, W.; Sakkiah, S.; Liu, J.; Patterson, T.A.; Hong, H. Nanomaterial databases: Data sources for promoting design and risk assessment of nanomaterials. *Nanomaterials* **2021**, *11*, 1599. [CrossRef]
6. Karcher, S.C.; Harper, B.J.; Harper, S.L.; Hendren, C.O.; Wiesner, M.R.; Lowry, G.V. Visualization tool for correlating nanomaterial properties and biological responses in zebrafish. *Environ. Sci. Nano* **2016**, *3*, 1280–1292. [CrossRef]
7. Juganson, K.; Ivask, A.; Blinova, I.; Mortimer, M.; Kahru, A. Nanoe-tox: New and in-depth database concerning ecotoxicity of nanomaterials. *Beilstein J. Nanotechnol.* **2015**, *6*, 1788–1804. [CrossRef]
8. Comandella, D.; Gottardo, S.; Rio-Echevarria, I.M.; Rauscher, H. Quality of physicochemical data on nanomaterials: An assessment of data completeness and variability. *Nanoscale* **2020**, *12*, 4695–4708. [CrossRef]
9. Romeo, D.; Salieri, B.; Hischier, R.; Nowack, B.; Wick, P. An integrated pathway based on in vitro data for the human hazard assessment of nanomaterials. *Environ. Int.* **2020**, *137*, 105505. [CrossRef]
10. Krug, H.F. Nanosafety research—Are we on the right track? *Angew Chem. Int. Ed. Engl.* **2014**, *53*, 12304–12319. [CrossRef]
11. Halappanavar, S.; Ede, J.D.; Shatkin, J.A.; Krug, H.F. A systematic process for identifying key events for advancing the development of nanomaterial relevant adverse outcome pathways. *NanoImpact* **2019**, *15*, 100178. [CrossRef]
12. Kroll, A.; Dierker, C.; Rommel, C.; Hahn, D.; Wohlleben, W.; Schulze-Isfort, C.; Gobbert, C.; Voetz, M.; Hardinghaus, F.; Schnekenburger, J. Cytotoxicity screening of 23 engineered nanomaterials using a test matrix of ten cell lines and three different assays. *Part Fibre Toxicol.* **2011**, *8*, 9. [CrossRef]
13. Krug, H.F.; Bohmer, N.; Kühnel, D.; Marquardt, C.; Nau, K.; Steinbach, C. The dana(2.0) knowledge base nanomaterials-an important measure accompanying nanomaterials development. *Nanomaterials* **2018**, *8*, 204. [CrossRef]
14. Fernández-Cruz, M.L.; Hernández-Moreno, D.; Catalán, J.; Cross, R.K.; Stockmann-Juvala, H.; Cabellos, J.; Lopes, V.R.; Matzke, M.; Ferraz, N.; Izquierdo, J.J.; et al. Quality evaluation of human and environmental toxicity studies performed with nanomaterials— The guidenano approach. *Environ. Sci. Nano* **2018**, *5*, 381–397. [CrossRef]
15. Schneider, K.; Schwarz, M.; Burkholder, I.; Kopp-Schneider, A.; Edler, L.; Kinsner-Ovaskainen, A.; Hartung, T.; Hoffmann, S. "Toxrtool", a new tool to assess the reliability of toxicological data. *Toxicol. Lett.* **2009**, *189*, 138–144. [CrossRef]
16. Krug, H.F. The uncertainty with nanosafety: Validity and reliability of published data. *Colloids Surf. B Biointerfaces* **2018**, *172*, 113–117. [CrossRef]
17. Ghosh, M.; Bandyopadhyay, M.; Mukherjee, A. Genotoxicity of titanium dioxide (tio2) nanoparticles at two trophic levels: Plant and human lymphocytes. *Chemosphere* **2010**, *81*, 1253–1262. [CrossRef]
18. Schanen, B.C.; Karakoti, A.S.; Seal, S.; Drake, D.R., 3rd; Warren, W.L.; Self, W.T. Exposure to titanium dioxide nanomaterials provokes inflammation of an in vitro human immune construct. *ACS Nano* **2009**, *3*, 2523–2532. [CrossRef]
19. Panessa-Warren, B.J.; Warren, J.B.; Wong, B.A.; Misewich, J.A. Biological cellular response to carbon nanoparticle toxicity. *J. Phys. Condens. Matter* **2006**, *18*, S2185–S2201. [CrossRef]
20. Krug, H.F.; Nau, K. Reliability for nanosafety research—Considerations on the basis of a comprehensive literature review. *ChemBioEng Rev.* **2017**, *4*, 331–338. [CrossRef]
21. Moher, D.; Liberati, A.; Tetzlaff, J.; Altman, D.G.; Group, P. Preferred reporting items for systematic reviews and meta-analyses: The prisma statement. *J. Clin. Epidemiol.* **2009**, *62*, 1006–1012. [CrossRef]
22. Halappanavar, S.; Ede, J.D.; Mahapatra, I.; Krug, H.F.; Kuempel, E.D.; Lynch, I.; Vandebriel, R.J.; Shatkin, J.A. A methodology for developing key events to advance nanomaterial-relevant adverse outcome pathways to inform risk assessment. *Nanotoxicology* **2021**, *15*, 289–310. [CrossRef]
23. Halappanavar, S.; Nymark, P.; Krug, H.F.; Clift, M.J.D.; Rothen-Rutishauser, B.; Vogel, U. Non-animal strategies for toxicity assessment of nanoscale materials: Role of adverse outcome pathways in the selection of endpoints. *Small* **2021**, *17*, e2007628. [CrossRef]
24. Luechtefeld, T.; Marsh, D.; Rowlands, C.; Hartung, T. Machine learning of toxicological big data enables read-across structure activity relationships (rasar) outperforming animal test reproducibility. *Toxicol. Sci.* **2018**, *165*, 198–212. [CrossRef]
25. Wittmaack, K. Excessive delivery of nanostructured matter to submersed cells caused by rapid gravitational settling. *ACS Nano* **2011**, *5*, 3766–3778. [CrossRef]
26. Wittmaack, K. Novel dose metric for apparent cytotoxicity effects generated by in vitro cell exposure to silica nanoparticles. *Chem. Res. Toxicol.* **2011**, *24*, 150–158. [CrossRef]

27. Oberdorster, G. Safety assessment for nanotechnology and nanomedicine: Concepts of nanotoxicology. *J. Intern. Med.* **2010**, *267*, 89–105. [CrossRef]
28. Rasmussen, K.; Rauscher, H.; Kearns, P.; Gonzalez, M.; Riego Sintes, J. Developing oecd test guidelines for regulatory testing of nanomaterials to ensure mutual acceptance of test data. *Regul. Toxicol. Pharmacol.* **2019**, *104*, 74–83. [CrossRef]
29. Warheit, D.B.; Donner, E.M. How meaningful are risk determinations in the absence of a complete dataset? Making the case for publishing standardized test guideline and 'no effect' studies for evaluating the safety of nanoparticulates versus spurious 'high effect' results from single investigative studies. *Sci. Technol. Adv. Mater.* **2015**, *16*, 034603. [CrossRef]
30. Powers, C.M.; Mills, K.A.; Morris, S.A.; Klaessig, F.; Gaheen, S.; Lewinski, N.; Ogilvie Hendren, C. Nanocuration workflows: Establishing best practices for identifying, inputting, and sharing data to inform decisions on nanomaterials. *Beilstein J. Nanotechnol.* **2015**, *6*, 1860–1871. [CrossRef]
31. Jeliazkova, N.; Apostolova, M.D.; Andreoli, C.; Barone, F.; Barrick, A.; Battistelli, C.; Bossa, C.; Botea-Petcu, A.; Chatel, A.; De Angelis, I.; et al. Towards fair nanosafety data. *Nat. Nanotechnol.* **2021**, *16*, 644–654. [CrossRef]
32. Higgins, S.G.; Nogiwa-Valdez, A.A.; Stevens, M.M. Considerations for implementing electronic laboratory notebooks in an academic research environment. *Nat. Protoc.* **2022**. *Epub ahead of print.* [CrossRef]
33. Solle, D. Be fair to your data. *Anal. Bioanal. Chem.* **2020**, *412*, 3961–3965. [CrossRef]
34. Rovida, C.; Barton-Maclaren, T.; Benfenati, E.; Caloni, F.; Chandrasekera, P.C.; Chesne, C.; Cronin, M.T.D.; De Knecht, J.; Dietrich, D.R.; Escher, S.E.; et al. Internationalization of read-across as a validated new approach method (nam) for regulatory toxicology. *Altern. Anim. Exp. ALTEX* **2020**, *37*, 579–606. [CrossRef]
35. Wigger, H.; Nowack, B. Material-specific properties applied to an environmental risk assessment of engineered nanomaterials—Implications on grouping and read-across concepts. *Nanotoxicology* **2019**, *13*, 623–643. [CrossRef]
36. Hristozov, D.; Gottardo, S.; Semenzin, E.; Oomen, A.; Bos, P.; Peijnenburg, W.; van Tongeren, M.; Nowack, B.; Hunt, N.; Brunelli, A.; et al. Frameworks and tools for risk assessment of manufactured nanomaterials. *Environ. Int.* **2016**, *95*, 36–53. [CrossRef]
37. Henschler, D. Toxicological problems relating to changes in the environment. *Angew. Chem. Int. Ed. Engl.* **1973**, *12*, 274–282. [CrossRef]
38. Wörle-Knirsch, J.M.; Pulskamp, K.; Krug, H.F. Oops they did it again! Carbon nanotubes hoax scientists in viability assays. *Nano Lett.* **2006**, *6*, 1261–1268. [CrossRef]
39. Faria, M.; Bjornmalm, M.; Thurecht, K.J.; Kent, S.J.; Parton, R.G.; Kavallaris, M.; Johnston, A.P.R.; Gooding, J.J.; Corrie, S.R.; Boyd, B.J.; et al. Minimum information reporting in bio-nano experimental literature. *Nat. Nanotechnol.* **2018**, *13*, 777–785. [CrossRef]
40. Hirsch, C.; Roesslein, M.; Krug, H.F.; Wick, P. Nanomaterial cell interactions: Are current in vitro tests reliable? *Nanomedicine* **2011**, *6*, 837–847. [CrossRef]
41. Schulze, C.; Kroll, A.; Lehr, C.M.; Sch„fer, U.F.; Becker, K.; Schnekenburger, J.; Schulze-Isfort, C.; Landsiedel, R.; Wohlleben, W. Not ready to use—Overcoming pitfalls when dispersing nanoparticles in physiological media. *Nanotoxicology* **2008**, *2*, 51–61. [CrossRef]
42. Nymark, P.; Bakker, M.; Dekkers, S.; Franken, R.; Fransman, W.; Garcia-Bilbao, A.; Greco, D.; Gulumian, M.; Hadrup, N.; Halappanavar, S.; et al. Toward rigorous materials production: New approach methodologies have extensive potential to improve current safety assessment practices. *Small* **2020**, *16*, e1904749. [CrossRef]
43. Yazdi, A.S.; Guarda, G.; Riteau, N.; Drexler, S.K.; Tardivel, A.; Couillin, I.; Tschopp, J. Nanoparticles activate the nlr pyrin domain containing 3 (nlrp3) inflammasome and cause pulmonary inflammation through release of il-1alpha and il-1beta. *Proc. Natl. Acad. Sci. USA* **2010**, *107*, 19449–19454. [CrossRef] [PubMed]
44. Zielinska, A.; Costa, B.; Ferreira, M.V.; Migueis, D.; Louros, J.M.S.; Durazzo, A.; Lucarini, M.; Eder, P.; Chaud, M.V.; Morsink, M.; et al. Nanotoxicology and nanosafety: Safety-by-design and testing at a glance. *Int. J. Environ. Res. Public Health* **2020**, *17*, 4657. [CrossRef]

Article

Assessing Genotoxicity of Ten Different Engineered Nanomaterials by the Novel Semi-Automated FADU Assay and the Alkaline Comet Assay

Sarah May [1,2], Cordula Hirsch [1], Alexandra Rippl [1], Alexander Bürkle [2] and Peter Wick [1,*]

[1] Particles-Biology Interactions Lab, Swiss Federal Laboratories for Materials Science and Technology (EMPA), Lerchenfeldstrasse 5, 9014 St. Gallen, Switzerland; sf.may@hotmail.de (S.M.); cordula.hirsch@empa.ch (C.H.); alexandra.rippl@empa.ch (A.R.)

[2] Molecular Toxicology Group, University of Konstanz, Universitätsstrasse 10, 78464 Konstanz, Germany; alexander.buerkle@uni-konstanz.de

* Correspondence: peter.wick@empa.ch

Abstract: Increased engineered nanomaterial (ENM) production and incorporation in consumer and biomedical products has raised concerns about the potential adverse effects. The DNA damaging capacity is of particular importance since damaged genetic material can lead to carcinogenesis. Consequently, reliable and robust in vitro studies assessing ENM genotoxicity are of great value. We utilized two complementary assays based on different measurement principles: (1) comet assay and (2) FADU (fluorimetric detection of alkaline DNA unwinding) assay. Assessing cell viability ruled out false-positive results due to DNA fragmentation during cell death. Potential structure–activity relationships of 10 ENMs were investigated: three silica nanoparticles (SiO_2-NP) with varying degrees of porosity, titanium dioxide (TiO_2-NP), polystyrene (PS-NP), zinc oxide (ZnO-NP), gold (Au-NP), graphene oxide (GO) and two multi-walled carbon nanotubes (MWNT). SiO_2-NPs, TiO_2-NP and GO were neither cytotoxic nor genotoxic to Jurkat E6-I cells. Quantitative interference corrections derived from GO results can make the FADU assay a promising screening tool for a variety of ENMs. MWNT merely induced cytotoxicity, while dose- and time-dependent cytotoxicity of PS-NP was accompanied by DNA fragmentation. Hence, PS-NP served to benchmark threshold levels of cytotoxicity at which DNA fragmentation was expected. Considering all controls revealed the true genotoxicity for Au-NP and ZnO-NP at early time points.

Keywords: comet assay; FADU assay; engineered nanomaterials; DNA strand breaks; genotoxicity; ENM interference

Citation: May, S.; Hirsch, C.; Rippl, A.; Bürkle, A.; Wick, P. Assessing Genotoxicity of Ten Different Engineered Nanomaterials by the Novel Semi-Automated FADU Assay and the Alkaline Comet Assay. *Nanomaterials* **2022**, *12*, 220. https://doi.org/10.3390/nano12020220

Academic Editors: Saura Sahu and David M. Brown

Received: 28 October 2021
Accepted: 7 January 2022
Published: 10 January 2022

Publisher's Note: MDPI stays neutral with regard to jurisdictional claims in published maps and institutional affiliations.

1. Introduction

Due to their novel and unique physico-chemical properties, engineered nanomaterials (ENMs) are remarkable for their rapid technological development accompanied by increasing production in the last two decades. Because of their small size, particulate shape, increased surface area and surface reactivity, some of these new nanosized materials might cause different biological effects compared to the bulk, composite or ionic form. Nevertheless, many ENMs find application in diverse areas, such as energy, information technology, electronics, cosmetics and healthcare, e.g., as diagnostics or therapeutics (for an overview see [1] and references therein). Despite the undeniable benefits of ENMs for many applications, there is a growing concern that these novel properties may cause harm to human beings and the environment [1–6]. The conflictive literature on ENM toxicity impedes their evaluation by regulatory bodies and hinders the efficient transfer of nano-based applications into the clinic. For these reasons, the toxicological impact of ENMs needs to be thoroughly investigated by in vitro and in vivo studies. In particular, the evaluation of ENM genotoxicity in human cells is of great importance since genotoxic insults independent of their origin are clearly linked to adverse health effects, most notably to cancer development [7]. Over the past decades, the literature on ENM-induced genotoxicity has grown but the results remain inconsistent, inconclusive or even contradictory. Numerous reviews, surveying the literature suggests diverse factors influencing the obtained results [5,8–12]. Accordingly, contradictory findings of different studies can be explained by the variability in physico-chemical characteristics of ENMs, for instance in size, shape, structure and coating and the type of biological model systems used. Furthermore, the unique characteristics of ENMs increase their likelihood of interference with analytical methods, detection systems and assay components, which may lead to data artefacts and inconsistent results of genotoxicological assays [13–15]. In some cases, the variability of experimental results between different laboratories might originate from the diversity of methods for handling ENM and the differences in dispersion protocols [16]. The most important factor, however, is the lack of standardized assay conditions and protocols [17]. Therefore, extensive investigations on ENM genotoxicity with appropriate assay systems in human cells are still required and significant [5,18].

Genotoxicity is a complex research field comprising not only the detection of specific DNA lesions (DNA strand breaks, alkali-labile sites [ALS] or chromosomal aberrations to name only a few) but also including modes of DNA repair. Here, we focus particularly on the detection of DNA strand breaks as one well-known type of DNA damage. In this context, two assays shall be highlighted: the comet (or single-cell gel electrophoresis) assay which is one of the most used assays for the detection of DNA strand breaks and the FADU (fluorimetric detection of alkaline DNA unwinding) assay as a new emerging semi-automated technology [19,20]. The comet assay was originally developed in 1984 by Ostling and Johanson and further modified in 1988 by Singh and colleagues [21,22] and is commonly used to determine the induction of DNA breakage by genotoxic agents and ionizing radiation at the level of individual cells. The assay principle is relatively straightforward. An OECD guideline (TG 489) [23] for an in vivo version exists and detailed publications on technical aspects for the in vivo and in vitro assay are available [24]. Therefore, this method has become the most frequently used method for in vitro ENM genotoxicity [25]. However, the assay itself and the evaluation of the results is quite tedious and time-consuming. Furthermore, several interlaboratory comparison studies have demonstrated the poor reproducibility of this assay [26–29]. The FADU assay was first described by Birnboim and Jevcak in 1981 [30] and is based on the progressive unwinding of DNA under the controlled conditions of time, temperature and alkaline pH. Besides replication forks and chromosome ends, DNA strand breaks are the origin of the unwinding process. For the evaluation of DNA damage, a fluorescent dye intercalating preferentially into double-stranded DNA (e.g., SybrGreen®) is used. A decrease in SybrGreen® fluorescence intensity in the cell lysates indicates an increase in progressive DNA unwinding and, thus, represents a greater number of DNA strand breaks. The applicability of this method has been demonstrated for the detection of mutagen-induced DNA strand breaks [20,31], the repair of UV-induced DNA strand breaks [32], environmental genotoxic effects [33] and ZnO nanoparticles [34]. The novel semi-automated version of this assay enables the fast assessment of DNA breakage in a 96-well-plate format and the integration of different control samples for ENM interference detection. Very recent research has extended its field of application to the detection of oxidative and methylation-induced DNA damage [19].

The goal of this case study was to assess the potential of a broad range of nanomaterials to induce DNA strand breaks in vitro in order to provide the first data set on potential structure–activity relationships (SAR). Physico-chemical properties of interest included differences in ENM chemistry, size, shape and porosity. Therefore, the following set of ENM was chosen: two metal oxide nanoparticles, non-soluble and soluble (TiO_2-NP and ZnO-NP) that are produced in high quantities. Three carbon-based materials, i.e., graphene oxide (GO) as a novel and highly interesting 2-D material, and two different multi-walled carbon nanotubes (MWNTs) as well-known benchmark materials. A medically relevant gold NP (Au-NP I) that had been shown to induce DNA strand breaks in a previous study [35]. Amine-modified polystyrene NPs (PS-NP) as a known cytotoxic material as well as three types of silica NPs (SiO_2-NPs) of low, middle and high porosity.

Upon ENM exposure, their interaction with cells of the immune system was very likely [36]. Therefore, we chose the T-lymphocytic cell line Jurkat E6-I as one potential immune system candidate to make contact with ENMs early on. The sublethal working concentrations of all ENM were determined using the MTT viability assay. DNA fragmentation occurring during cell death is the main cause of false-positive genotoxicity results. Therefore, ruling out cytotoxicity was indispensable for a reliable interpretation of gentoxicity data. Furthermore, two complementary assays to assess DNA strand breaks that are based on different read-out systems were used, namely the well-established comet assay and the semi-automated FADU assay. Thus, assay intrinsic errors such as ENM interferences could be detected and potentially even avoided which would increase the reliability and robustness of the acquired data sets [37]. Consolidating results from both methods finally facilitated a better fundamental understanding of the DNA-damaging potential of ENM.

2. Materials and Methods

Cell Culture: The Jurkat E6-I lymphoblastoid cell line (ATCC: TIB-152) was purchased from the American Type Culture Collection (Manassas, VA, USA). Cells were maintained in a complete medium (CM) consisting of RPMI (Roswell Park Memorial Institute) 1640 medium (Gibco) supplemented with 10% fetal bovine serum (FBS, Invitrogen) and 1% penicillin/streptomycin/neomycin (Gibco) at standard growth conditions of 37 °C and 5% CO_2 in a humidified atmosphere. Jurkat E6-I cells were cultured in suspension and cell titers were not allowed to exceed 3×10^6 cells/mL.

ENM handling and Dilutions: ENMs, delivered as powder, were prepared as 1 mg/mL stock suspensions either in ddH_2O or 160 ppm Pluronic F-127 (Sigma), as summarized in Table 1. ZnO, TiO_2, MWNT A and MWNT C suspensions were ultrasonicated in an ultrasound bath (Bandelin electronic, Berlin, Germany) for 10 min. GO suspension was ultrasonicated for 2 min. ENM stock suspensions were serially pre-diluted in the respective solvent. The same volume from different pre-dilutions and solvents was then used to prepare the final concentrations in CM. All suspensions were prepared and diluted directly before application to cells. To reduce formation of aggregates, tubes containing the solvent were placed on a continuously shaking Vortex® (Witeg, Wertheim am Main, Germany) as described by Zook and co-workers in 2011 [38]. The stock suspensions or respective sub-dilutions were added to the shaking solvent in a drop-wise manner. The resulting suspension remained on the Vortex® for an additional 3 s. Directly before application to cells, all final suspensions were vortexed again.

Table 1. ENM suspension preparation.

	Delivered as	Prepared Stock Concentration	Solvent	Ultrasonication [1]
PS-NP	100 mg/mL in ddH_2O		ddH_2O	-
TiO$_2$-NP	powder	1 mg/mL	ddH_2O	10 min
ZnO-NP	powder	1 mg/mL	ddH_2O	10 min
Au-NP I	4.7 mg/mL in ddH_2O		ddH_2O	-
MWNT A	powder	0.5 mg/mL	Pluronic F-127	10 min
MWNT C	powder	0.5 mg/mL	Pluronic F-127	10 min
GO	powder	1 mg/mL	ddH_2O	2 min
SiO$_2$-160	11.2 mg/mL in ddH_2O		ddH_2O	-
MS-SiO$_2$-140	4.7 mg/mL in ddH_2O		ddH_2O	-
MSHT-SiO$_2$-300	18.6 mg/mL in ddH_2O		ddH_2O	-

[1] Sonication using an ultrasound bath.

MTT Assay: The cell viability of Jurkat E6-I cells was determined by measuring the reduction of water-soluble MTT (3-(4,5-dimethylthiazol-2-yl)-2,5-diphenyltetrazolium bromide, Sigma) to water-insoluble formazan by metabolically active cells. Briefly, 7×10^4

Jurkat E6-I cells were seeded in 96-well plates in a volume of 50 µL directly before treatment with 50 µL of double-concentrated ENM suspensions. ZnO, TiO_2, PS and GO stock suspensions were serially diluted 1:2 as described above and reached final concentrations in a medium of 100, 50, 25, 12.5, 6.25, 3.125, 1.56, 0 µg/mL. The MWNTs suspensions were serially diluted in Pluronic F-127 and reached final assay concentrations in a medium of 50, 25, 12.5, 6.25, 3.125, 1.56, 0.78, 0 µg/mL. "0 µg/mL" samples received the corresponding amount of the respective solvent. Following incubation for 30 min or 21 h, 10 µL of MTT solution (5 mg/mL in PBS) were added and cells were incubated for 3 h under standard growth conditions. Sodium dodecyl sulfate (SDS, Sigma, 200 µM) served as positive control. After incubation, 100 µL of solubilizing solution (10% SDS in 0.01 M HCl) was added and the samples were incubated at 37 °C overnight. Absorbance was measured at 590 nm and 750 nm as references (Mithras2 LB943, Berthold Technologies, Bad Wildbad, Germany). Wells without cells were used as blanks and subtracted from the corresponding sample values. The mean and the corresponding standard deviation of at least three independent experiments with six technical replicates each are shown.

Comet Assay: The alkaline comet assay was performed as previously described by Singh and co-workers in 1988 [22] with the following modifications. Jurkat E6-I cells were seeded in 6-well plates at a density of 5×10^5 cells per well in 1.25 mL CM directly before ENM application. For treatment, a volume of 1.25 mL double-concentrated ENM suspension was added to reach the following final concentrations for ZnO, TiO_2, amine-modified PS, GO, SiO_2 and Au: 100, 6.25, 3.125, 1.56 µg/mL and 50, 3.125, 1.56, 0.78 µg/mL for MWNTs. The application of 20 mM EMS (Sigma) 30 min before the end of the incubation time served as the positive control. Following 3 h or 24 h of incubation, cells were collected and centrifuged for 6 min at $125 \times g$. The pellets obtained were resuspended in 300 µL CM. All subsequent steps were identical to the protocol described by May and co-workers in 2018 [35]. If not otherwise stated, samples were blinded and 100 randomly chosen comets per sample were analyzed for each experiment. Tail intensities in percentages (=tail intensity (%)) are expressed as the mean of at least three independent experiments and their corresponding standard deviations.

FADU Assay: The automated FADU assay was performed according to the protocol published with minor modifications [20,34]. Directly prior to experimentation 3.6×10^4 Jurkat E6-I cells were seeded into deep-well 96-well plates (Greiner) in a volume of 80 µL CM. Five-times-concentrated ENM pre-dilutions were prepared in CM. 20 µL of these suspensions were added to the 80 µL of cells to reach final ENM concentrations of 100, 50, 25, 12.5, 6.25, 3.125, 1.56, 0 µg/mL in a total volume of 100 µL CM. The final concentrations of MWNTs differed from those of all other ENM and equaled 50, 25, 12.5, 6.25, 3.125, 1.56, 0.78, 0 µg/mL. Etoposide (20 µM, 30 min) served as the chemical positive control. Untreated samples received CM only. After 3 h and 24 h, DNA strand break analysis using the AUREA gTOXXs Anlayzer (3T analytik; www.aurea.solutions (accessed on 27 October 2021) was carried out. Fluorescence measurements were performed using a multi-well plate reader (Mithras^2, Berthold Technologies) with an excitation wavelength of 492 nm and an emission of 530 nm. The overall fluorescence intensity of the *p*-values was expressed as the percentage of the fluorescence of the control cells, i.e., cells that were not been exposed to ENM. A decrease in the fluorescence intensity indicated an increase in DNA strand breaks. Two different conditions were applied for each sample, one in which the DNA was not unwound and, therefore, represents the total amount of double-stranded DNA (T-value), and one in which the DNA was unwound by the addition of unwinding solution prior to the neutralization solution (*p*-value). Only for the assessment of ENM-induced interference were the T-values and *p*-values used. To take the potential ENM-derived influence (fluorescence quenching or enhancement) to the resulting *p*-values into account, each *p*-value was expressed as the percentage of its respective T-value before being normalized to the solvent control as described above. Results processed like this are declared as "corrected". In all cases, data shown represent the mean and corresponding standard deviation of at least three independent experiments with four technical replicates each.

Statistical Analysis: Microsoft Excel (2016) was used for figures and statistical calculations. Statistical differences were assessed by the Student's *t*-test. *p* values ≤ 0.05 were considered as statistically significant. For FADU and viability (MTT) assessments, an additional criterion for "biological significance" was used, i.e., only reductions in cell viability and intact DNA below 80% were considered biologically relevant. Since variability in the comet assay is known to be high [39], only 2-fold differences compared to the untreated control were considered relevant.

3. Results

3.1. ENM Characterization

For this study, a panel of ten different ENMs was investigated regarding their ability to induce DNA damage in Jurkat E6-I cells. The characterization of these materials was published earlier and is summarized in Tables 2 and 3.

Table 2. Characteristics of cytotoxic ENM.

Description	Au-NP	MWNT A	MWNT C	ZnO-NP	PS-NP
Source	collaboration partners of the CCMX NanoScreen consortium [a]	Bayer Technologies Service, Baytubes, Leverkusen, Germany	Cheap Tubes Inc., Grafton, Vermon, USA	IBUtec, Weimar, Germany	Bangs Laboratories, Inc., Fishers, IN, USA
Delivered as	suspension (4.7 mg Au/mL in ddH$_2$O)	powder	powder	powder	suspension (100 mg/mL in ddH$_2$O)
Manufacturing process	see Bohmer et al., 2018			pulsation reactor technique	
Size/Size distribution (diameter)	TEM: 3.1 ± 1.3 nm DLS [b]: 147 nm	inner diameter: 1–9 nm outer diameter: 4–24 nm	inner diameter: 2–13 nm outer diameter: 6–34 nm	TEM: 15.5 ± 3.9 nm	57 nm [c] SEM: 51 ± 9 nm DLS [b]: 56 nm
Lateral dimensions		1–5 μm	1–16 μm		
Surface area				60 ± 5 m^2/g [c]	99 m^2/g [c]
Density	19.3 g/cm^3 [d]				1.05 g/cm^3 [c]
Zeta potential [e]	24.5 mV	−5 mV in Pluronic F-127	−15 mV in Pluronic F-127	−24.3 mV	48.8 mV
Surface modification	[AL]$_{21}$[α-gal]$_{23}$				NH$_2$ (amine)
Publication on characterization details	Bohmer et al., 2018 Rademacher et al., 2013 patent [a]	Thurnherr et al., 2009	Thurnherr et al., 2009	Buerki-Turnherr et al., 2013	Elliott et al., 2017

[a] for details see patent US 8,568,781 B2, 2013. [b] DLS values are given as Z-average from measurements in ddH$_2$O. [c] Manufacturer's information. [d] Density of Au, ratio of NP core to ligands unknown. [e] If not otherwise specified zeta potential was measured in water. abbreviations: DLS: dynamic light scattering; MWNT: multi-walled carbon nanotubes; NP: nanoparticle; SEM: scanning electron microscopy; TEM: transmission electron microscopy.

Table 3. Characteristics of cytotoxic ENM.

Description	SiO$_2$-160	MS-SiO$_2$-160	MSHT-SiO$_2$-300	TiO$_2$-NP	GO
Source	collaboration partners of the CCMX NanoScreen consortium [a]	collaboration partners of the CCMX NanoScreen consortium [a]	collaboration partners of the CCMX NanoScreen consortium [a]	Sigma-Aldrich	Cheap Tubes, Inc.
Delivered as	suspension (11.2 mg/mL in ddH$_2$O)	suspension (4.7 mg/mL in ddH$_2$O)	suspension (18.6 mg/mL in ddH$_2$O)	powder	powder
Manufacturing process	Stöber synthesis	CTAB-method	CTAB-method with additional hydrothermal treatment		modified Hummers method
Size/Size distribution (diameter)	TEM: 161 ± 15 nm DLS [b]: 204 ± 2 nm	TEM: 128 nm DLS [b]: 209 nm	TEM: 288 nm DLS [b]: 270 nm	<25 nm [c] DLS [b]: 279 ± 51 nm	thickness: 0.7–1.2 nm [d]
Lateral dimensions					SEM: 1–40 μm AFM: 300–800 nm
Surface area	23 m^2/g [e]	1092 m^2/g [e]	462 m^2/g [e]	200–220 m^2/g [c]	
Density				3.9 g/cm^3 [c]	
Zeta potential [f]	−49 ± 3 mV	−35.2 mV	−47.7 mV	−36.1 ± 1 mV	−39.4 ± 1.3 mV
Publication on characterization details	Bohmer et al., 2018	unpublished	unpublished	unpublished	Kucki et al., 2016

[a] Powder Technology Laboratory, EPFL, Lausanne, Switzerland. [b] DLS values are given as Z-average from measurements in ddH$_2$O. [c] Manufacturer's information. [d] Corresponds to few- or even single-layer graphene. [e] Assessed by N2-BET. [f] If not otherwise specified zeta potential was measured in water.

3.2. ENM Influence on Cell Viability and DNA Damage

The cell viability of Jurkat E6-I cells was investigated after 3 and 24 h of treatment with different ENMs by using the MTT assay. Upon viability assessment, and for the ease of reading, data are grouped throughout the manuscript according to ENM cytotoxicity. Firstly, results from non-cytotoxic ENMs, including SiO$_2$-160, MS-SiO$_2$-140, MSHT-SiO$_2$-300, TiO$_2$-NP and GO are presented, followed by results from the cytotoxic panel consisting of Au-NP I, MWNT A, MWNT C, ZnO-NP and PS-NP.

No decrease in cell viability could be observed after 3 h of incubation with the first group of ENMs (Figure 1a). The alkaline comet assay revealed the absence of DNA damage after 3 h of exposure to the same set of ENMs (Figure 1b). Similarly, in the FADU assay, no significant reduction of intact DNA was observed for the three types of silica particles (SiO$_2$-160, MS-SiO$_2$-160 and MSHT-SiO$_2$-300) and TiO$_2$-NP (Figure 1c). A significantly strong dose-dependent decrease in the percentage of intact DNA was detected for GO, starting at a concentration of 3.13 μg/mL; however, for the highest concentration of 100 μg/mL, a slight increase was observed (Figure 1c).

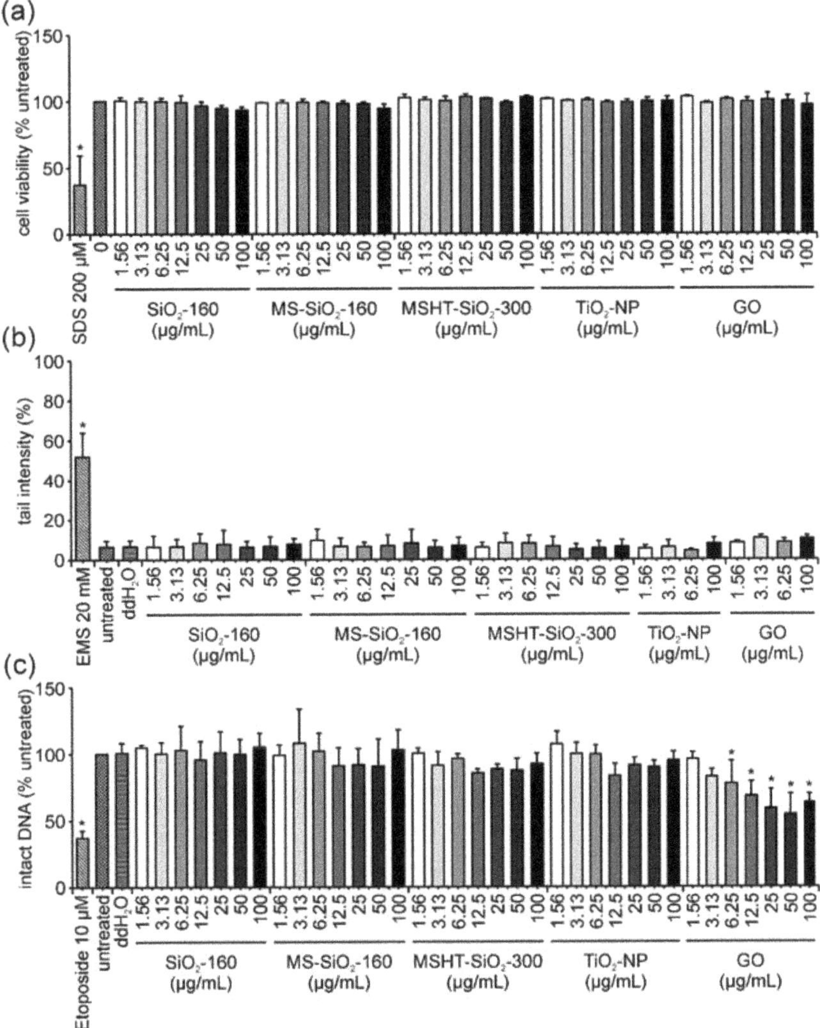

Figure 1. Influence of SiO$_2$-160, MS-SiO$_2$-160, MSHT-SiO$_2$-300, TiO$_2$ and GO on Jurkat E6-I cell viability and DNA damage induction after 3 h of incubation. Following the incubation of Jurkat E6-I cells with different concentrations of SiO$_2$-160, MS-SiO$_2$-160, MSHT-SiO$_2$-300, TiO$_2$-NP and GO for 3 h, cell viability was determined by MTT assay (**a**). As a positive control, cells were incubated with 200 µM sodium dodecylsulfate (SDS, 3 h). DNA damage expressed as tail intensity percentage was assessed by alkaline comet assay (**b**). Ethyl methanesulfonate (EMS, 30 min) served as the positive control. The FADU assay was performed as a second method for genotoxicity assessment (**c**). Treatment with etoposide (30 min) served as the positive control. Results represent the mean and corresponding standard deviations from at least three independent experiments. (* $p \leq 0.05$).

In the case of GO-treated cells, the interference correction of FADU results proved to be necessary because this material quenched the fluorescence signal. This could be observed by a dose-dependent reduction in T-values, which is a strong indication for fluorescence quenching (Figure 2a). T-values represent the total amount of DNA, which is obtained by preventing DNA unwinding through the neutralization of an alkaline unwinding buffer. Therefore, these T-values should have stayed equally high for all samples, even genotoxic ones, as double-stranded DNA remains unwound. Only *p*-values, where DNA unwinding is allowed to take place, decrease following treatment with genotoxic stimuli. Consequently, a reduction in T-values indicates either ENM interference in the form of fluorescence quenching or variations in cell density per well. Since cell viability was not influenced by GO treatment, fluorescence quenching seemed to be the reason for the observed decrease in T-values. Following interference correction of the results obtained for GO, according to calculations summarized in materials and methods, no reduction in intact DNA could be detected anymore (Figure 2b).

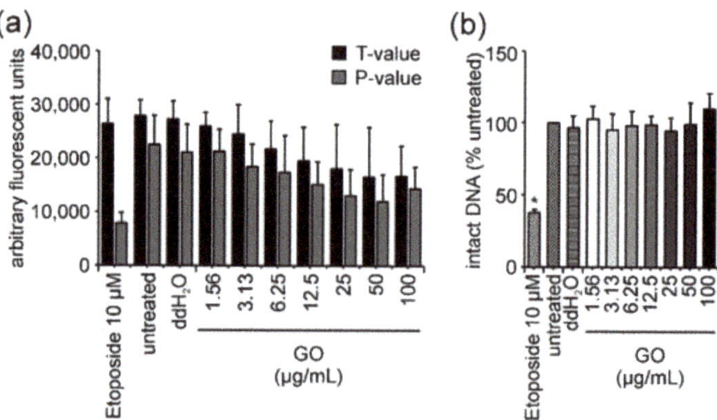

Figure 2. GO-induced interference and interference correction in the FADU assay after 3 h of incubation in Jurkat E6-I cells. Following 3 h exposure of Jurkat E6-I cells to GO, the FADU assay was performed and revealed a dose-dependent decrease in fluorescence of T- and *p*-values (**a**). After correction of the observed interference, no reduction in intact DNA was observed for any concentration of GO (**b**). Only incubation with 10 µM etoposide for 30 min, which served as the positive control, induced genotoxic effects. Data shown represent the mean of three independent experiments and the corresponding standard deviation. (* $p \leq 0.05$).

After 3 h exposure to Au-NP I, a slight dose-dependent decrease in cell viability was observed (Figure 3a). The highest concentration of 100 µg/mL led to a moderate and statistically significant reduction in cell viability to 74%. A stronger decrease in cell viability was observed for both MWNTs. Cytotoxic effects started at concentrations of 12.5 and 25 µg/mL and declined to 32 and 44% for 50 µg/mL MWNT A and MWNT C, respectively. Additionally, ZnO-NP and PS-NP treatment influenced cell viability negatively and both materials caused similar dose-dependent effects. The highest concentration of 100 µg/mL reduced cell viability to 70 and 66% for ZnO-NP and PS-NP, respectively. Subsequently, the DNA-damaging potential of these ENMs was addressed in the comet assay. In comparison to the untreated and vehicle controls (ddH$_2$O for Au-NP I, ZnO-NP and PS-NP; Pluronic F-127 for MWNT A and MWNT C) an increase in DNA damage expressed as tail intensity percentage was observed for Au and in particular for ZnO-NP (Figure 3b). Au treatment resulted in 20 and 16% tail intensity for 50 and 100 µg/mL, respectively. The highest concentration of ZnO-NP reached values of approximately 45%. A very slight and insignificant dose-dependent increase was observed for sublethal concentrations of ZnO-NP (1.56, 3.13 and 6.25 µg/mL) with tail intensities of 14, 16 and 18%, respectively. DNA damage induction was not detected for the two MWNTs or for PS-NP. Results of the FADU assay revealed no reduction in intact DNA for Au-NP I, PS-NP, MWNT A and C (Figure 3c). The strongest, yet still weak dose-dependent effect, was observed for ZnO-NP where the highest concentration resulted in 69% of intact DNA compared to the untreated control (Figure 3c).

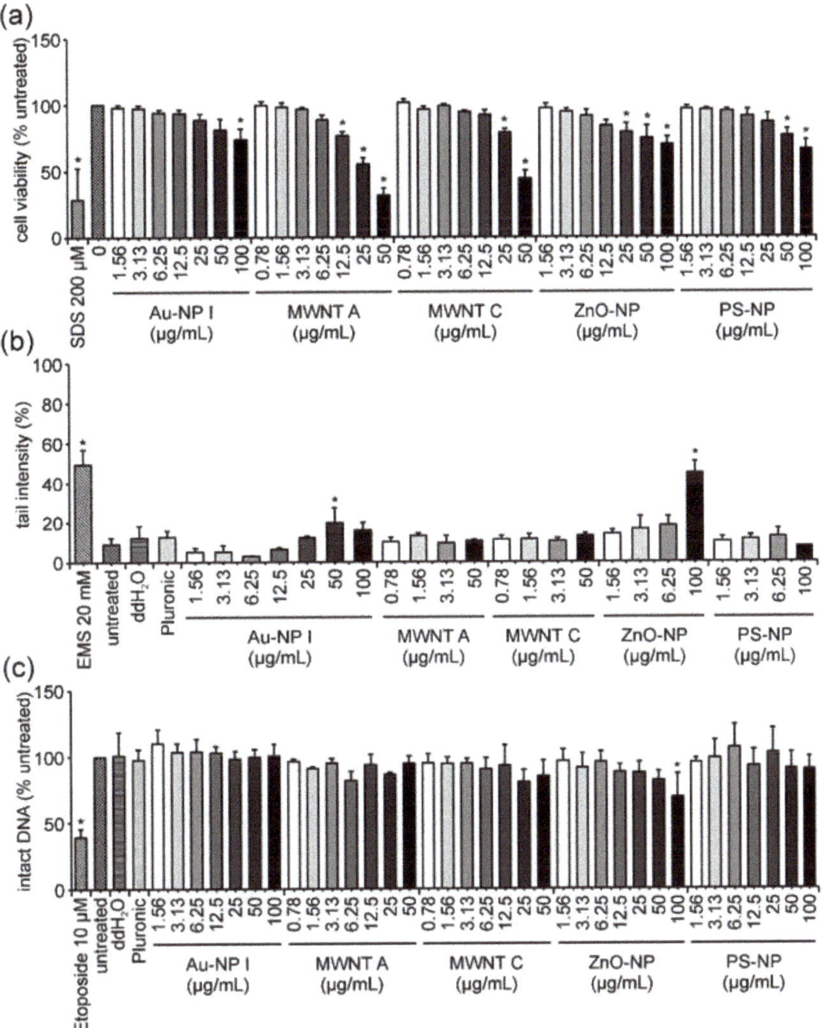

Figure 3. **Influence of Au-NP I, MWNT A, MWNT C, ZnO-NP and PS-NP on Jurkat E6-I cell viability and DNA damage induction after 3 h of incubation.** Following incubation of Jurkat E6-I cells with different concentrations of Au-NP I, MWNT A, MWNT C, ZnO-NP and PS-NP for 3 h, cell viability was determined by MTT assay (**a**). As a positive control, cells were incubated with 200 μM SDS (3 h). DNA damage expressed as tail intensity percentage was assessed by alkaline comet assay (**b**). EMS (30 min) served as the positive control. The FADU assay was performed as a second method for genotoxicity assessment (**c**). Treatment with etoposide (30 min) served as the positive control. Results represent the mean and corresponding standard deviations from at least three independent experiments. (* $p \leq 0.05$).

The same set of experiments was conducted after 24 h of incubation with all ENMs under investigation. Regarding cell viability, no significant reduction was observed for SiO_2-160, MS-SiO_2-140, MSHT-SiO_2-300, TiO_2 and GO after 24 h (Figure 4a). Using the comet assay, no formation of DNA damage could be detected for any of these ENMs after 24 h of incubation (Figure 4b). Results of the FADU assay confirmed this observation for

SiO$_2$-160, MSHT-SiO$_2$-300 as well as TiO$_2$-NPs (Figure 4c). A minor, yet significant effect could only be detected for treatment with 12.5, 25 and 50 µg/mL MS-SiO$_2$-160 resulting in 79.9% and 72% intact DNA, respectively. At the highest concentration of 100 µg/mL, the level of intact DNA increased again to 88%. Similarly, MSHT-SiO$_2$-300-treatment resulted in 79% intact DNA at concentrations of 25 and 50 µg/mL, respectively, increasing again to 85% at 100 µg/mL. Considering the rather significant variability (indicated as standard deviation in Figure 4c), results have to be interpreted with caution. Since no increase in tail intensity could be detected with the more sensitive comet assay, the massive DNA-damaging potential of the silica particles could be excluded. However, interference reactions in the FADU assay should be addressed in more detail in the future. Comparable to the 3 h time point, GO-treatment for 24 h again displayed the most pronounced effects in the FADU assay. However, as shown for the 3 h time point, the dose-dependent decrease was, once more, not only detectable for the actual samples (*p*-values) but also for the interference control samples (T-values) (Supplementary Figure S1a). Therefore, interference correction of GO results was performed, which consequently eliminated the initially observed reduction in intact DNA (Supplementary Figure S1b). For all other analyzed ENMs, T-values revealed no decrease in fluorescence, indicating the absence of quenching effects. Hence, interference correction was not essential for those samples.

Significant dose-dependent effects on cell viability were detected after 24 h of exposure to Au-NP I, MWNT A, MWNT C, ZnO-NP and PS-NP (Figure 5a). The reduction in cell viability was more pronounced for all analyzed ENMs after 24 h in comparison to the 3 h measurement. The weakest effect was observed for Au-NP I, while similar results were obtained for both MWNTs. The strongest effects were observed for ZnO-NP and PS-NP. The highest concentration of Au-NP I, MWNT A, MWNT C and ZnO-NP resulted in values of 54, 30, 25 and 2% cell viability, respectively. In the case of PS-NP, no detectable signal could be obtained for the concentration of 100 µg/mL anymore. The results gained from the comet assay revealed no increase in tail intensity for both MWCNTs (Figure 5b). Increased tail intensity values were observed for the highest concentration of Au-NP I, ZnO-NP and PS-NP. For Au-NP I, only a moderate increase of 35% tail intensity was observed, while ZnO-NP- and PS-NP-treatment reached values of 64 and 86%, respectively. As cell viability was barely or not at all detectable for ZnO-NP and PS-NP after 24 h of incubation, respectively, these high tail intensities in the comet assay were caused by the massive fragmentation of DNA during the process of cell death. This also became apparent upon examination of the microscopic appearance of the comets, which were highly damaged and only detectable with the Comet IV software by additional manual adjustments. Hence, the number of detectable comets used for the evaluation was below the number of comets selected for the analysis of all other samples (100 comets per sample). Results of the FADU assay uncovered a dose-dependent decrease in intact DNA after 24 h of incubation with ZnO-NP and PS-NP (Figure 5c).

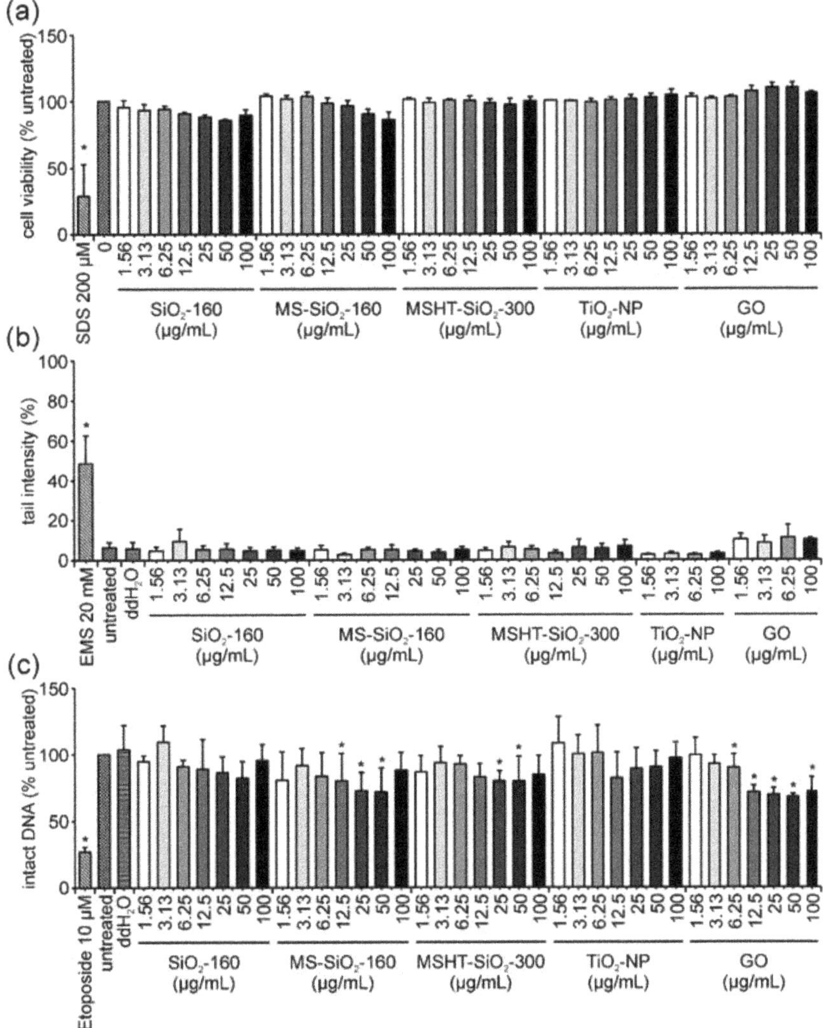

Figure 4. Influence of SiO$_2$-160, MS-SiO$_2$-160, MSHT-SiO$_2$-300, TiO$_2$-NP and GO on Jurkat E6-I cell viability and DNA damage induction after 24 h of incubation. Following incubation of Jurkat E6-I cells with different concentrations of SiO$_2$-160, MS-SiO$_2$-160, MSHT-SiO$_2$-300, TiO$_2$-NP and GO for 24 h, cell viability was determined by MTT assay (**a**). As a positive control, cells were incubated with 200 µM SDS (24 h). DNA damage expressed as tail intensity percentage was assessed by alkaline comet assay (**b**). EMS (30 min) served as the positive control. The FADU assay was performed as a second method for genotoxicity assessment (**c**). Treatment with etoposide (30 min) served as the positive control. Results represent the mean and corresponding standard deviations from at least three independent experiments. (* $p \leq 0.05$).

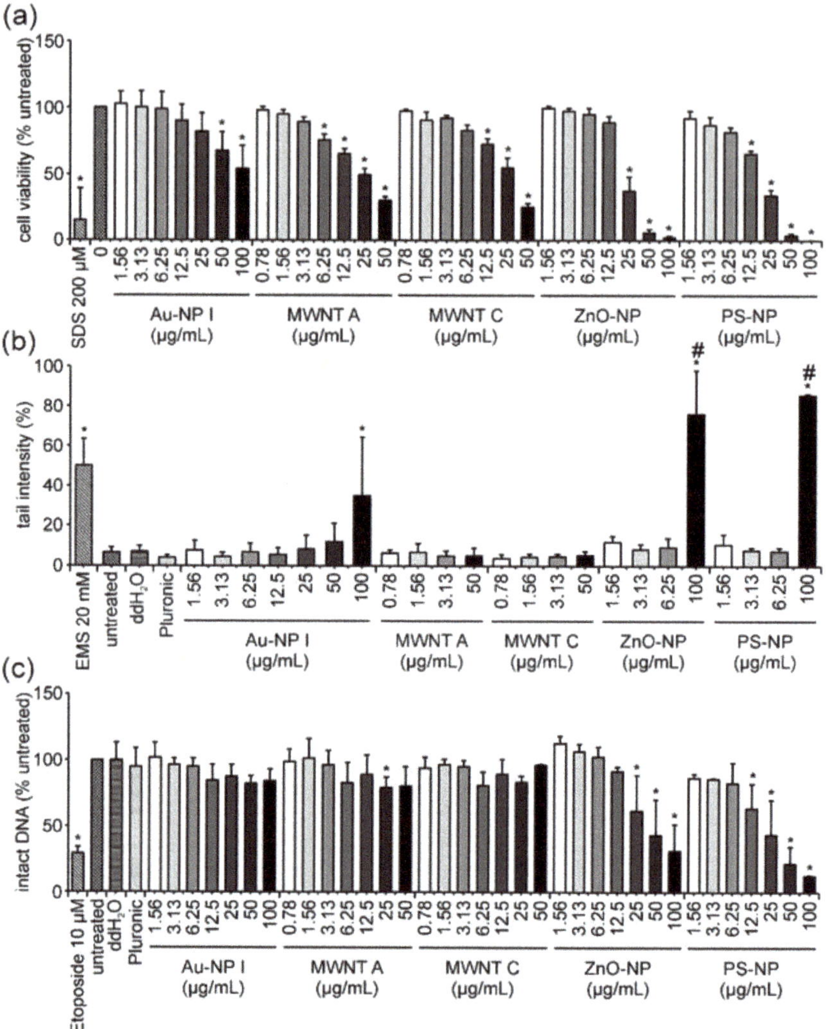

Figure 5. Influence of Au-NP I, MWNT A, MWNT C, ZnO-NP and PS-NP on Jurkat E6-I cell viability and DNA damage induction after 24 h of incubation. Following incubation of Jurkat E6-I cells with different concentrations of Au-NP I, MWNT A, MWNT C, ZnO-NP and PS-NP for 24 h, cell viability was determined by MTT assay (**a**). As a positive control, cells were incubated with 200 μM SDS (24 h). DNA damage expressed as tail intensity percentage was assessed by alkaline comet assay (**b**). EMS (30 min) served as the positive control. The FADU assay was performed as a second method for genotoxicity assessment (**c**). Treatment with etoposide (30 min) served as the positive control. Results represent the mean and corresponding standard deviations from three independent experiments. # Only a reduced number of comets (i.e., less than 100) could be counted per experiment in these samples. (* $p \leq 0.05$).

For both ENMs, the examination of T-values revealed a strong dose-dependent decrease, similar to what has been described for GO previously. However, in the case of ZnO-NP and PS-NP, this dose-dependent decrease in T-values was not due to interference but caused by cell death and associated DNA fragmentation (Figures 6c,d and 7c,d). This conclusion is supported by two facts: (i) After 3 h of ZnO-NP- and PS-NP-treatment T-values did not decline, even though the same amount of ENMs was present during the reaction and measurement, therefore, confirming the absence of ENM-induced interferences on fluorescence (Figures 6a,b and 7a,b); (ii) Only a few cells were present on the comet assay slides, indicating significant cell loss upon ZnO-NP- and PS-NP-treatment. For cells treated with Au-NP I and MWNT C no significant effects could be observed in the FADU assay (Figure 5c).

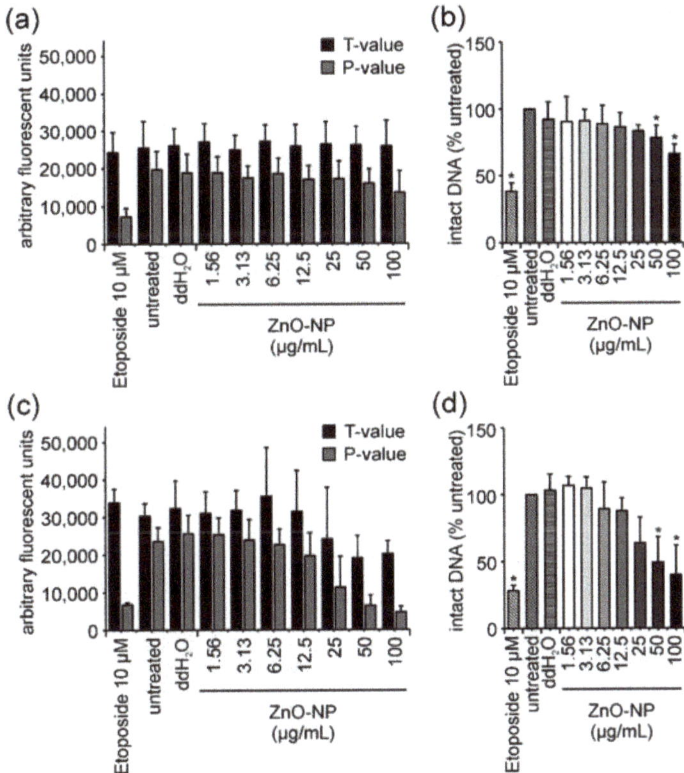

Figure 6. Influence of ZnO-NPs on T- and *p*-values in the FADU assay and corresponding interference correction. Following 3 h (**a,b**) and 24 h (**c,d**) of exposure of Jurkat E6-I cells to ZnO-NP, the FADU assay was performed. After 3 h of incubation, only *p*-values decreased dose-dependently (**a**), while after 24 h, T- and *p*-values decreased with increasing ZnO-NP concentrations (**c**). Results following interference correction for the 3-h (**b**) and 24-h (**d**) time point are shown. Data shown represent the mean of at least three independent experiments and the corresponding standard deviation. (* $p \leq 0.05$).

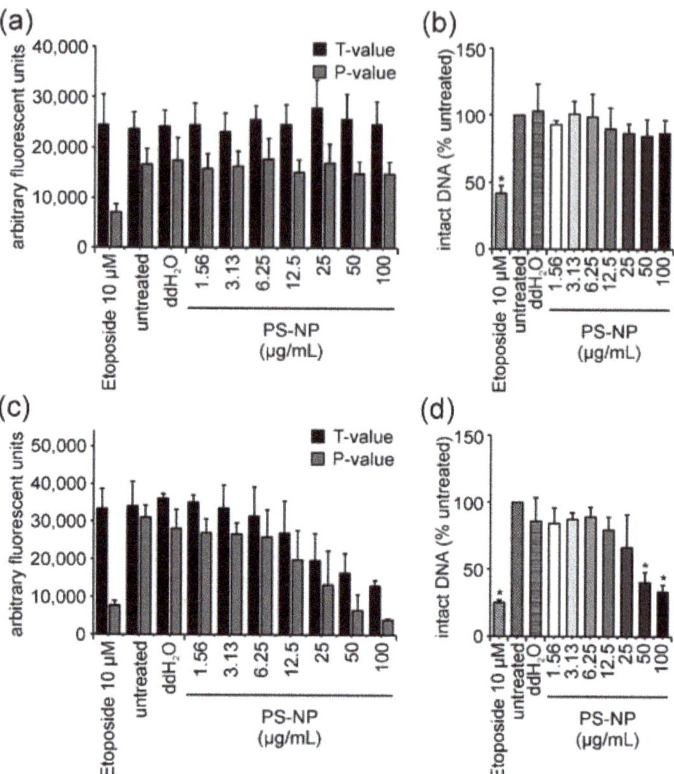

Figure 7. Influence of PS-NPs on T- and *p*-values in the FADU assay and corresponding interference correction. Following 3 h (**a,b**) and 24 h (**c,d**) of exposure of Jurkat E6-I cells to PS-NP, the FADU assay was performed. After 3 h of incubation, only *p*-values decreased dose-dependently (**a**), while after 24 h, T- and *p*-values decreased with increasing PS-NP concentrations (**c**). Results following interference correction for the 3-h (**b**) and 24-h (**d**) time point are shown. Data shown represent the mean of at least three independent experiments and the corresponding standard deviation. (* $p \leq 0.05$).

It has previously been shown that the toxicity of many metal-based ENMs, including ZnO, TiO_2, Ag and CeO_2, is at least partially caused by their specific properties related to their small size and high surface reactivity. However, in some cases, toxicity may be triggered or further enhanced by the release of free metal ions [40]. The identical batch of ZnO-NP has previously been thoroughly investigated in Jurkat A3 cells and it was shown that the observed cytotoxicity was fundamentally dependent on the fast extracellular dissolution of ZnO-NP and resulting high concentrations of Zn^{2+} ions, which were taken up by the cells [41]. To understand the role of free zinc with the observed ZnO-NP cytotoxicity and genotoxicity in Jurkat E6-I cells, the effect of equimolar concentrations of $ZnCl_2$ was analyzed with the same set of assays.

Exposure of Jurkat E6-I cells for 3 h to $ZnCl_2$ resulted in a likewise dose-dependent decrease in cell viability in comparison to equimolar concentrations of ZnO-NP-treated cells (Figure 3a and Supplementary Figure S2a). After 24 h of incubation, $ZnCl_2$ reduced cell viability at the corresponding concentration of 12.5 µg/mL ZnO-NP slightly more (compare with Figure 5a and Supplementary Figure S2b). However, overall, very similar results were obtained, which suggest that the observed ZnO-NP-induced cytotoxicity is, also in these cells, mainly caused by the release of Zn^{2+}. Equivalent results were also obtained in the comet assay where the highest concentration of $ZnCl_2$ induced tail intensity values of 56 and 74% after 3 and 24 h of treatment, respectively (Supplementary Figures S2c and S2d), which is comparable to the data acquired for 100 µg/mL ZnO-NP (Figure 3b). Likewise, in the FADU assay, $ZnCl_2$- and ZnO-NP-treatment for 24 h dose-dependently reduced the level of intact DNA to the same extent (Figure 3c and Supplementary Figure S2f). Furthermore, T-value controls for both treatment conditions and time points behaved in a comparable manner (Figure 6 and Supplementary Figure S3). The only difference between $ZnCl_2$- and ZnO-NP-treatment became apparent after 3 h of treatment in the FADU assay. While ZnO-NPs induced a true genotoxic response at 100 µg/mL (Figures 3c and 6b), no reduction in intact DNA could be observed for equivalent amounts of $ZnCl_2$ (Supplementary Figure S2e) indicating an as yet uncharacterized additional nano-effect at early time points in the FADU assay.

4. Discussion

At present, there is only limited comparability and consequently a great uncertainty regarding the genotoxic potential of various ENMs. Despite the growing literature on ENM genotoxicity in human cell lines, the published results are quite often controversial due to the different cell types, different ENM types and different readout systems used. In general, including regulatory purposes, the evaluation and interpretation of in vitro data should rely on standardized and solid experimental results. To avoid experimental, assay specific artifacts, it is mandatory to assess each endpoint by two independent, complementary methods that rely on different measurement principles. Such an approach minimizes the probability of systematic or assay-intrinsic errors, which is of particular importance concerning the detection and avoidance of ENM-induced interferences [37]. Not all modes of genotoxic action can be addressed with this study; therefore, we focused, in particular, on a reliable and reproducible methodology for the detection of DNA strand breaks. We investigated the DNA damaging potential of ten different ENMs with distinct properties, thus, aiming for a first preliminary structure–activity relationship analysis. The two independent methods used were the alkaline comet assay as the most frequently used method to assess DNA strand breaks and the novel FADU technology for efficient and semi-automated DNA damage detection [42]. Even though both assays measure DNA strand breaks as the endpoint, only the alkaline comet assay is able to detect additional ALS [35]. Therefore, a careful comparison of results obtained with the comet and the FADU technique is of great importance; in particular, to gain first insights into the mode of action of a certain genotoxicant. In the following paragraphs, the genotoxicity results obtained with the two methods are discussed for each type of ENM in relation to cell viability data.

4.1. TiO₂-NP: The "Easy One" Neither Induces Cyto- Nor Genotoxicity and Does Not Interfere in the FADU Assay

4.1. TiO_2-NP: The "Easy One" Neither Induces Cyto- Nor Genotoxicity and Does Not Interfere in the FADU Assay

Cell viability of Jurkat E6-I cells was not affected by treatment with TiO_2-NP after 3 and 24 h (Figures 1a and 4a). The data on ENM genotoxicity obtained by the alkaline comet assay and the FADU assay are in good agreement showing no genotoxic potential in both assays (Figures 1b,c and 4b,c). Furthermore, no interference with the fluorescent readout in the FADU assay could be observed. While these results are highly consistent in themselves, contradictory findings concerning TiO_2-NP (geno)toxicity have been published. For instance, Gosh and coworkers and Khan and coworkers [43,44] reported the genotoxic effects of two different types of TiO_2-NP in human lymphocytes analyzed by comet assay. In contrast, a study by Hackenberg et al., 2011, reported no genotoxicity for these cells after treatment with a distinct type of TiO_2-NP. Likewise, contradictory results were reported for various cell lines. In human nasal mucosa cells and TK6 cells, no TiO_2-NP induced genotoxicity could be determined by comet assay, whereas Jugan and coworkers [45] could demonstrate that several different TiO_2-NPs elicited genotoxic effects on A549 cells [45–47]. Since distinct types of TiO_2-NPs have been used in these studies, a direct comparison is not feasible and a generalized conclusion for all TiO_2-NP subtypes regarding their genotoxicity and cytotoxicity is still not possible today. Here, we could demonstrate the absence of genotoxicity by means of two independent methods. Additionally, no genotoxic effects following TiO_2-NP-treatment could be detected in A549 cells (data not shown). This is in agreement with results from Hackenberg and colleagues (2011), who applied the same type of TiO_2-NP. This allows for the conclusion that this particular type of TiO_2-NP induces neither cytotoxicity nor genotoxicity in different cell lines in vitro. Further SAR studies are still urgently needed to reduce the need for time-consuming case-by-case evaluations. The assay combination presented here could serve as an ideal platform suitable for diverse cellular models.

4.2. GO: The "Interfering One" Does Not Induce DNA Damage but Showcases Interference Reactions in the FADU Assay

Cell viability of Jurkat E6-I cells remained unaffected at all GO concentrations analyzed and up to 24 h of treatment (Figures 1a and 4a). Likewise, tail intensity values measured in the comet assay did not exceed those of the untreated control cells (Figures 1b and 4b). Surprisingly, a dose-dependent decrease in the percentage of intact DNA was observed for increasing concentrations ($\geq$12.5 µg/mL) of GO when analyzed in the FADU assay. This effect was observed for both time points in a very similar manner (Figure 1c and Supplementary Figure S1). GO has previously been shown to interfere with different fluorescence-based assays due to nanoscale-surface energy transfer effects from fluorophores to GO [48–52]. Therefore, analogous interference reactions with the SybrGreen® fluorophore used in the FADU assay could explain the discrepancy between FADU and comet assay results.

In the FADU assay, T-values represented the total amount of DNA in each sample. The pH in these control samples was kept constant to avoid alkaline unwinding even at sites of DNA breakage. Therefore, T-values were directly proportional to the number of cells per sample and were expected to remain constant upon treatment with true genotoxicants, which induce DNA damage but do not affect cell viability, i.e., the number of viable cells. It shall be stressed again that concomitant cell death analyses were indispensable to assure unchanged numbers of viable cells upon treatment.

GO-treatment reduced T-values in a dose-dependent manner (Figure 2a and Supplementary Figure S1a) and "in parallel" to p-value reduction. Together with unchanged cell viability (Figure 1a), this indicated that interference reactions indeed took place in the FADU assay and could be quantified using the T-value controls. Mathematical interference correction, as described in Section 2, eradicated the ostensible genotoxic effect on Jurkat E6-I cells (Figure 2b and Supplementary Figure S1b). Thus, we could establish a straightforward and easy-to-use experimental setup to quantify and mathematically correct for ENM-induced interferences in the FADU assay making it a suitable tool to screen for ENM-induced DNA damage. Moreover, we could clearly demonstrate that GO is neither cytotoxic nor induces DNA strand breaks in Jurkat E6-I cells under the experimental conditions chosen.

4.3. SiO₂-NP of Different Porosities: The "Unclear Ones" Neither Induce Cytotoxicity Nor DNA Damage but Lead to Unclear Results in the FADU Assay

Different types of SiO_2-NPs have previously been shown to induce negative and positive results in different genotoxicity assays, including the in vitro micronucleus, the comet and the mutation assay with various cell systems [53–59]. The panel of ENMs analyzed in this study comprised three different SiO_2-NPs of distinct porosity. As demonstrated by the results of the MTT assay, none of the SiO_2-NPs reduced cell viability after 3 and 24 h of incubation (Figures 1a and 4a). Likewise, no increase in tail intensity could be observed in the comet assay at either time point (Figures 1b and 4b). The absence of DNA damage was further confirmed after 3 h of treatment by results of the FADU assay (Figure 1c). However, the mesoporous silica sample, MS-SiO_2-160, led to a minor reduction in intact DNA after 24 h of incubation with 25 and 50 µg/mL. No dose-dependency could be observed, and the level of intact DNA increased again at 100 µg/mL, making a cellular response rather unlikely and hinting toward an interference phenomenon. Mesoporous ENMs are frequently used as carrier systems for different molecules, drugs or dyes. Therefore, it is possible that with an increasing concentration of ENMs an increasing proportion of SybrGreen® was trapped within the pores of the particles [60,61]. This effect might be covered at the highest applied concentration of mesoporous SiO_2-NPs due to their intrinsic scattering properties, which could prevail at this high density of ENMs [62]. However, a comparable reaction would also be expected after 3 h of incubation and for T-values—both of which was not the case. T-values did not change upon MS-SiO_2-160-treatment at both time points analyzed (Supplementary Figure S4), indicating no classical interference with the fluorescence signal. Nevertheless, scattering properties of the ENMs could change over time in a protein-containing cell culture medium due to the formation of a protein corona and particle agglomeration leading to an as yet unidentified interference reaction. Taking into account the rather high variability of the FADU assay results, the comparably small effect of MS-SiO_2-NP on intact DNA and the absence of DNA strand breaks in the (more sensitive) comet assay, the validity of the FADU results for this particular type of ENM is questionable. Furthermore, the absence of DNA strand breaks, as demonstrated in the comet assay, is in accordance with the vast majority of published genotoxicity studies on various SiO_2-NPs [59,63,64]. Thus, it is evident that more in-depth studies are required to elucidate the mode of action of different silica particles in the FADU assay to eventually make it a reliable screening tool for this particular type of ENM.

4.4. PS-NP: The "Purely Cytotoxic" One Induces High Levels of Cytotoxicity thereby Generating False-Positive DNA Damage Results and Could Serve as a Benchmark Material

Exposure of Jurkat E6-I cells to PS-NP for 3 h resulted in a slight dose-dependent decrease in cell viability at the two highest concentrations (Figure 3a). After 3 h, neither in the comet nor FADU assay, was an induction of genotoxicity observed at any concentrations analyzed (Figure 3b,c). This indicates, that in these cells, a reduction in cell viability down to 66%, as measured by the MTT assay, does not trigger considerable DNA fragmentation. DNA fragmentation due to cell death is known as a potential cause of false-positive genotoxicity results in different in vitro assays (e.g., [65,66]). We can, therefore, conclude that PS-NPs induce cytotoxicity but do not induce DNA strand breaks at early time points. Following 24 h of PS-NP incubation, cell viability was strongly reduced in a dose-dependent manner (Figure 5a). For the highest concentration, almost no viable cells could be detected anymore. This was also reflected in a reduced number of nuclei available for comet assay analysis. The remaining nuclei showed a strong increase in tail intensity which was not observed for sublethal concentrations (Figure 5b). Consistently, a significant dose-dependent reduction in the percentage of intact DNA was obtained in the FADU assay (Figure 5c). These observations demonstrate that at least high concentrations of PS-NPs induce pure cytotoxicity and that the observed DNA damage can be attributed to DNA fragmentation secondary to cell death. An important question is how to interpret the dose-dependent effects observed at lower concentrations in the FADU assay. Looking at T and p values after 24 h of PS-NP-treatment (Figure 7c) shows an equivalent dose-dependent reduction in T-value signals indicating either a nanomaterial induced interference (as described for GO) or a loss in cell number. Since T values did not change in the 3 h samples, classical nanomaterial interference can be excluded. Hence, the reduction in intact DNA is not based on genotoxic effects but merely on the destruction of DNA integrity due to cell death.

It is still unknown at which level of cytotoxicity false-positive effects appear in genotoxicity assays and how these are affected by the mode of cell death [67]. Different suggestions for adequate substance concentrations to be used in genotoxicity assessments exist in the literature and cell viability ranges from 70 to 90% [68]. Furthermore, optimal concentration ranges can be affected by the choice of assay used for determination of cell viability, as a reduction in metabolic activity (e.g., by MTT assay) might occur at lower concentrations in comparison to, for example, cytolysis.

Thus, PS-NPs could serve as a benchmark material to determine the level of cytotoxicity at which false-positive genotoxicity results can be expected in the cell type under investigation and the corresponding cytotoxicity assay used.

4.5. ZnO-NP: True Genotoxicity vs. Pure Cytotoxicity

A plethora of in vitro studies on ZnO-NP toxicity has previously shown significant effects on the viability of cancer cell lines of the immune system, lung, kidney, skin, and the gut, as well as in primary cells such as neural stem cells, T-lymphocytes or fibroblasts [41,69]. ROS formation and a severe oxidative stress response [70–77] as well as DNA damage in various cell types [73,78,79] have been reported. Furthermore, studies described the dissolution of ZnO-NP and associated Zn^{2+} ion toxicity and/or genotoxicity [41,80,81].

In this study, following exposure to ZnO-NP, the cell viability of Jurkat E6-I cells decreased in a dose-dependent manner after 3 h to approximately 70% for the highest concentration (Figure 3a). For the same exposure time, results of the comet assay revealed a slight trend in DNA damage induction at sublethal concentrations (3.13 and 6.25 µg/mL) and a significant induction of 45% tail intensity at the highest concentration of 100 µg/mL (Figure 3b). Consistently, the amount of intact DNA as measured in the FADU assay, was significantly reduced at 100 µg/mL. In comparison to PS-NP results, this increase in DNA damage upon ZnO-NP-treatment in both assays was rather unexpected. While PS-NP and ZnO-NP show very similar cytotoxicity profiles, no induction of DNA damage was observed for PS-NP after 3 h of treatment. This comparison indicates that even though cell viability (i.e., metabolic activity) is already affected, DNA damage is still induced simultaneously and independently of cell death and can, thus, be interpreted as a real genotoxic effect at early time points. This conclusion is further supported by the analysis of T- and *p*-values in the FADU assay (Figure 6). After 3 h of incubation, T-values were not affected by ZnO-NP treatment—not even at the highest concentrations analyzed. Therefore, the reduction in *p*-values can be considered as a real genotoxic event.

In contrast, after 24 h of ZnO-NP-treatment, a massive reduction in cell viability was observed reaching values of only 2% for the highest concentration (Figure 5a). This led to a strong increase in tail intensity in the comet assay (Figure 5b) and a significant reduction in the percentage of intact DNA in the FADU assay (Figure 5c). All dose–response curves are highly similar to those observed for 24-h PS-NP-treatment. Accordingly, the number of analyzable nuclei at 100 µg/mL of ZnO-NP was markedly reduced and the respective T-values in the FADU assay declined with increasing ZnO-NP concentrations. Nanomaterial-induced interferences could be excluded due to the constant 3-h T-values. Thus, we can conclude that after prolonged exposure to ZnO-NP the observed DNA damage is due to DNA fragmentation in the process of cell death.

As described earlier, in vitro cytotoxicity of ZnO-NPs can be caused by the high solubility and release of free Zn^{2+}, which can lead to the disruption of cellular Zn homeostasis associated with the loss of cell viability, oxidative stress and mitochondrial dysfunction [69,82,83]. The same batch of ZnO-NP induced a caspase-independent alternative apoptosis pathway independent of ROS formation in the Jurkat subclone A3 [41]. This was a consequence of the extracellular release of high amounts of Zn^{2+} followed by rapid cellular uptake. In this study, it could be shown that $ZnCl_2$-treatment induced the same dose–response curve in the MTT assay as ZnO-NPs, suggesting that Zn^{2+} ions are responsible for ZnO-NP-induced cell death. Similarly, ZnO-NP- and $ZnCl_2$-treatment induced comparable effects on DNA damage in the comet and FADU assay with one exception: the 3-h real genotoxic effect of ZnO as measured in the FADU assay. This was not observed upon $ZnCl_2$-treatment. The most likely explanation is that ZnO-NPs and Zn^{2+} ions induce distinct types of DNA lesions at early time points of treatment. As previously analyzed in great detail [35], ALS can only be detected using the alkaline comet assay. Therefore, our results indicate that ZnO-NPs induce DNA strand breaks and, in addition, ALS. In contrast, $ZnCl_2$-treatment results in ALS only. Since these specific lesions cannot be detected by the FADU assay, no reduction in intact DNA could be observed upon 3 h of $ZnCl_2$-treatment.

4.6. Au-NP: True Genotoxicity vs. Pure Cytotoxicity and a Potential Mechanism of Action

Gold is generally considered an inert and biocompatible material. Consequently, for a long time, Au-NPs were expected to behave similarly and to be non-toxic [84,85]. However, various publications with often contradictory findings on cytotoxicity and genotoxicity of Au-NPs appeared over the past years. For example, Au-NPs with a size of 1–2 nm were reported to induce a high cytotoxicity, while 15 nm-sized Au-NPs were non-toxic in different cell lines [86]. Similarly, 5 nm-sized Au-NPs induced genotoxic effects, whereas 50 nm-sized Au-NPs did not [87]. While such a size-dependency for cytotoxicity as well as for genotoxicity was observed in different studies [88,89], opposing reports have also been published [90,91], thereby explaining the need for a reliable genotoxicity assessment of Au-NPs.

Our results demonstrate that the cell viability of Jurkat E6-I cells was slightly reduced after 3 h of treatment with Au-NP I (Figure 3a). This dose-dependent effect was more pronounced after 24 h of incubation (Figure 5a). Interestingly, the comet assay revealed an increase in tail intensity after 3 h and 24 h, indicating that DNA damage induction by these ENMs (Figures 3b and 5b). However, results of the FADU assay showed no significant reduction in intact DNA at both time points (Figures 3c and 5c). This discrepancy can be explained by differences in the sensitivity of these two methods regarding certain types of DNA lesions. As published by Singh et al., the detection limit of the comet assay is 0.03 Grey, while the FADU assay reaches a detection limit of 0.1 Grey [92]. While X-ray-induced DNA damage is already measured with greater sensitivity in the comet assay, differences for other genotoxic stimuli (causing distinct kinds of lesions) might be even more pronounced. Both methods detect DNA strand breaks as general endpoints but they are based on different measurement principles and do not necessarily detect the same spectra of DNA lesions.

Our previous study [35] demonstrated the importance of pH to also detect ALS. While the same Au-NP I induced very high DNA damage levels in A549 cells in the alkaline comet assay (performed at pH 13.2–13.7), only a weak DNA damage induction was observed in the neutral comet and the FADU assay (performed at pH 12.5–12.9; [20]). These results led to the conclusion that mainly ALS, which can be detected as strand breaks under extremely high pH of 13 and above, are induced by these NPs. The same phenomenon could explain the results observed in Jurkat E6-I. However, another remaining question is whether the assumed DNA-damaging effect is due to genuine genotoxicity and in consistency with previously published data [35] or due to cell death, which was not observed in A549 cells. Comparing the Au-NP cytotoxicity results to the proposed benchmark material PS-NP, the reduction in cell viability after 3 h of treatment was still in a range where no DNA fragmentation would be expected. Since genotoxicity was still detectable in the comet assay, this can be attributed to genuine DNA damage. In addition, Au-NPs did not affect T-values in the FADU assay at any concentrations and time points analyzed. This further supports the conclusion that the observed increase in tail intensity is not primarily due to cell death but rather a true genotoxic event. However, after 24 h of exposure at least the highest concentration (100 µg/mL) of Au-NP I would be expected to lead to DNA fragmentation due to considerable cell death (Figure 5). The lack of a genotoxic effect measured by the FADU assay indicates that the DNA fragmentation did not yet reach a significant level. The increase in tail intensity as measured by the comet assay could, thus, be due to ALS, as previously shown [35] and discussed above.

4.7. MWNTs: Cytotoxicity without DNA Damage—Is That Possible?

Following 3 h of incubation with MWNT A and C, cell viability was reduced dose-dependently to 32 and 44% for the highest concentration, respectively (Figure 3a). This effect increased after 24 h of incubation down to 30 and 25%, respectively (Figure 5a). At such low levels of cell viability and in relation to the proposed benchmark material (PS-NP), DNA fragmentation was expected to influence the comet as well as FADU assay results. However, in neither was assay DNA damage was observed (Figures 3 and 5).

Furthermore, no influence on T-values could be detected (Supplementary Figure S5). On the one hand, this indicates that MWNTs do not interfere with the fluorescence readout of the FADU assay. On the other hand, a massive loss of viable cells can also be excluded. This would suggest that MWNTs influence metabolic activity to a greater extent without impacting on actual cell death as, for example, PS-NP-treatment. Furthermore, MWNTs are known to interfere with the MTT assay and even though interference controls have been run in this study (data not shown), an overestimation of the cytotoxic potential is still possible [93]. Further analysis utilizing additional cell viability assays and a more detailed analysis on genotoxicity are needed to elucidate the remaining ambiguities. Nevertheless, the data presented here allow for the conclusion that, at sublethal concentrations, none of the investigated MWNTs induced DNA strand breaks in Jurkat E6-I cells.

5. Conclusions

With the set of ENM provided and corresponding results in combination with two complementary, yet independent, genotoxicity assays and the implemented controls, we believe that genotoxicity assessment can be improved and brought to the next level of reliability. False-positive genotoxicity results can be avoided, and true genotoxicity can be detected with a high level of confidence.

Even though a classical SAR could not be deduced from the data set provided, we can still conclude that porosity in the case of silica particles neither influenced assay results (interference) nor cytotoxicity and DNA damage in Jurkat E6-I cells. Likewise, shape (2D vs. spheroidal vs. tubes) does not seem to be the decisive factor in terms of cytotoxicity and DNA damage under the experimental conditions chosen. We can further conclude that TiO_2-NPs, GO, all three types of SiO_2-NPs as well as MWNT A and C do not induce DNA strand breaks in Jurkat E6-I cells. Using PS-NPs as a purely cytotoxic benchmark material allowed us to identify the true genotoxic potential for ZnO-NP at a short (i.e., 3 h) exposure time as well as for Au-NP I at both exposure times analyzed. Further relating the ZnO-NP results to $ZnCl_2$ data and previously published findings on Au-NP I revealed a potential nano-specific genotoxic effect for ZnO-NP that was not caused by the release of Zn^{2+} ions. Based on GO results, we established an easy-to-use quantitative interference control for the FADU assay making it a promising screening tool for ENM genotoxicity. We believe that the approach described here will be applicable for any cell type of interest, given that a suitable cell-type-specific, purely cytotoxic benchmark material is available.

Supplementary Materials: The following are available online at https://www.mdpi.com/article/10.3390/nano12020220/s1, Figure S1: GO-induced interference and interference correction in the FADU assay after 24 h of incubation in Jurkat E6-I cells, Figure S2: Influence of $ZnCl_2$ on Jurkat E6-I cell viability and DNA damage induction after 3 h and 24 h of incubation, Figure S3: Influence of $ZnCl_2$ on T- and p-values in the FADU assay and corresponding interference correction, Figure S4: No influence of MS-SiO_2-160 on T-values in the FADU assay, Figure S5: No influence of MWNT A and MWNT C on T-values in the FADU assay.

Author Contributions: Conceptualization, C.H., A.B. and P.W.; methodology, S.M. and C.H.; formal analysis, S.M. and C.H.; investigation, S.M. and A.R.; resources, P.W.; data curation, S.M. and C.H.; writing—original draft preparation, S.M.; writing—review and editing, C.H. and P.W.; visualization, S.M. and C.H.; supervision, C.H., P.W. and A.B.; project administration, C.H. and P.W.; funding acquisition, P.W. All authors have read and agreed to the published version of the manuscript.

Funding: The authors acknowledge funding from the NanoScreen Materials Challenge co-funded by the Competence Centre for Materials Science and Technology (CCMX). Non-financial support in the form of equipment supply for the duration of the study was provided by 3T Analytik (3T GmbH & Co. KG, Germany).

Institutional Review Board Statement: Not applicable.

Informed Consent Statement: Not applicable.

Data Availability Statement: The data presented in this study are available in the Supplementary Materials: Supplementary_RawData_MTT.xlsx; Supplementary_RawData_FADU.xlsx; Supplementary_RawData_Comet.xlsx.

Acknowledgments: We thank BioRender, the graphical abstract was created using BioRender.com.

Conflicts of Interest: A. Bürkle declares no known competing interest. S. May, C. Hirsch, A. Rippl and P. Wick report non-financial support from 3T Analytik (3T GmbH & Co. KG, Germany) in the form of equipment supply for the duration of the study. The funders had no role in the design of the study; in the collection, analyses, or interpretation of data; in the writing of the manuscript, or in the decision to publish the results.

References

1. Zhu, S.; Li, L.; Gu, Z.; Chen, C.; Zhao, Y. 15 Years of Small: Research Trends in Nanosafety. *Small* **2020**, *16*, e2000980. [CrossRef]
2. Ayres, J.G.; Borm, P.; Cassee, F.R.; Castranova, V.; Donaldson, K.; Ghio, A.; Harrison, R.M.; Hider, R.; Kelly, F.; Kooter, I.M.; et al. Evaluating the toxicity of airborne particulate matter and nanoparticles by measuring oxidative stress potential—A workshop report and consensus statement. *Inhal. Toxicol.* **2008**, *20*, 75–99. [CrossRef]
3. Ju-Nam, Y.; Lead, J.R. Manufactured nanoparticles: An overview of their chemistry, interactions and potential environmental implications. *Sci. Total Environ.* **2008**, *400*, 396–414. [CrossRef]
4. Klaper, R.D. The Known and Unknown about the Environmental Safety of Nanomaterials in Commerce. *Small* **2020**, *16*, 2000690. [CrossRef] [PubMed]
5. Singh, N.; Manshian, B.; Jenkins, G.J.S.; Griffiths, S.M.; Williams, P.M.; Maffeis, T.G.G.; Wright, C.J.; Doak, S.H. NanoGenotoxicology: The DNA damaging potential of engineered nanomaterials. *Biomaterials* **2009**, *30*, 3891–3914. [CrossRef] [PubMed]
6. Xia, T.; Li, N.; Nel, A.E. Potential Health Impact of Nanoparticles. *Annu. Rev. Public Health* **2009**, *30*, 137–150. [CrossRef]
7. Royal Society. *Nanoscience and Nanotechnologies: Opportunities and Uncertainties*; Royal Society: London, UK, 2004.
8. Doak, S.H.; Griffiths, S.M.; Manshian, B.; Singh, N.; Williams, P.M.; Brown, A.P.; Jenkins, G.J.S. Confounding experimental considerations in nanogenotoxicology. *Mutagenesis* **2009**, *24*, 285–293. [CrossRef]
9. Gonzalez, L.; Lison, D.; Kirsch-Volders, M. Genotoxicity of engineered nanomaterials: A critical review. *Nanotoxicology* **2008**, *2*, 252–273. [CrossRef]
10. Landsiedel, R.; Kapp, M.D.; Schulz, M.; Wiench, K.; Oesch, F. Genotoxicity investigations on nanomaterials: Methods, preparation and characterization of test material, potential artifacts and limitations-Many questions, some answers. *Mutat. Res. Rev. Mutat.* **2009**, *681*, 241–258. [CrossRef]
11. Wani, M.R.; Shadab, G. Titanium dioxide nanoparticle genotoxicity: A review of recent in vivo and in vitro studies. *Toxicol. Ind. Health* **2020**, *36*, 514–530. [CrossRef]
12. Yazdimamaghani, M.; Moos, P.J.; Dobrovolskaia, M.A.; Ghandehari, H. Genotoxicity of amorphous silica nanoparticles: Status and prospects. *Nanomedicine* **2019**, *16*, 106–125. [CrossRef] [PubMed]
13. Guadagnini, R.; Kenzaoui, B.H.; Walker, L.; Pojana, G.; Magdolenova, Z.; Bilanicova, D.; Saunders, M.; Juillerat-Jeanneret, L.; Marcomini, A.; Huk, A.; et al. Toxicity screenings of nanomaterials: Challenges due to interference with assay processes and components of classic in vitro tests. *Nanotoxicology* **2015**, *9*, 13–24. [CrossRef]
14. Karlsson, H.L. The comet assay in nanotoxicology research. *Anal. Bioanal. Chem.* **2010**, *398*, 651–666. [CrossRef]
15. Stone, V.; Johnston, H.; Schins, R.P.F. Development of in vitro systems for nanotoxicology: Methodological considerations. *Crit. Rev. Toxicol.* **2009**, *39*, 613–626. [CrossRef]
16. Magdolenova, Z.; Bilanicova, D.; Pojana, G.; Fjellsbo, L.M.; Hudecova, A.; Hasplova, K.; Marcomini, A.; Dusinska, M. Impact of agglomeration and different dispersions of titanium dioxide nanoparticles on the human related in vitro cytotoxicity and genotoxicity (vol 14, pg 455, 2012). *J. Environ. Monitor.* **2012**, *14*, 3306. [CrossRef]
17. Moller, P. Assessment of reference values for DNA damage detected by the comet assay in human blood cell DNA. *Mutat. Res. Rev. Mutat.* **2006**, *612*, 84–104. [CrossRef] [PubMed]
18. Azqueta, A.; Dusinska, M. The use of the comet assay for the evaluation of the genotoxicity of nanomaterials. *Front. Genet.* **2015**, *6*, 239. [CrossRef] [PubMed]
19. Mack, M.; Schweinlin, K.; Mirsberger, N.; Zubel, T.; Burkle, A. Automated screening for oxidative or methylation-induced DNA damage in human cells. *Altex* **2021**, *38*, 63–72. [CrossRef]
20. Moreno-Villanueva, M.; Pfeiffer, R.; Sindlinger, T.; Leake, A.; Muller, M.; Kirkwood, T.B.; Burkle, A. A modified and automated version of the 'Fluorimetric Detection of Alkaline DNA Unwinding' method to quantify formation and repair of DNA strand breaks. *BMC Biotechnol.* **2009**, *9*, 39. [CrossRef]
21. Ostling, O.; Johanson, K.J. Microelectrophoretic Study of Radiation-Induced DNA Damages in Individual Mammalian-Cells. *Biochem. Biophys. Res. Commun.* **1984**, *123*, 291–298. [CrossRef]
22. Singh, N.P.; Mccoy, M.T.; Tice, R.R.; Schneider, E.L. A Simple Technique for Quantitation of Low-Levels of DNA Damage in Individual Cells. *Exp. Cell Res.* **1988**, *175*, 184–191. [CrossRef]
23. Organisation for Economic Co-operation and Development (OECD). *Test No. 489: In Vivo Mammalian Alkaline Comet Assay*; OECD Publishing: Paris, France, 2016.

24. Collins, A.R.; Oscoz, A.A.; Brunborg, G.; Gaivao, I.; Giovannelli, L.; Kruszewski, M.; Smith, C.C.; Stetina, R. The comet assay: Topical issues. *Mutagenesis* **2008**, *23*, 143–151. [CrossRef] [PubMed]
25. Magdolenova, Z.; Collins, A.; Kumar, A.; Dhawan, A.; Stone, V.; Dusinska, M. Mechanisms of genotoxicity. A review of in vitro and in vivo studies with engineered nanoparticles. *Nanotoxicology* **2014**, *8*, 233–278. [CrossRef]
26. Collins, A.R.; El Yamani, N.; Lorenzo, Y.; Shaposhnikov, S.; Brunborg, G.; Azqueta, A. Controlling variation in the comet assay. *Front. Genet.* **2014**, *5*, 359. [CrossRef] [PubMed]
27. Forchhammer, L.; Ersson, C.; Loft, S.; Moller, L.; Godschalk, R.W.; van Schooten, F.J.; Jones, G.D.; Higgins, J.A.; Cooke, M.; Mistry, V.; et al. Inter-laboratory variation in DNA damage using a standard comet assay protocol. *Mutagenesis* **2012**, *27*, 665–672. [CrossRef]
28. Forchhammer, L.; Johansson, C.; Loft, S.; Moller, L.; Godschalk, R.W.; Langie, S.A.; Jones, G.D.; Kwok, R.W.; Collins, A.R.; Azqueta, A.; et al. Variation in the measurement of DNA damage by comet assay measured by the ECVAG inter-laboratory validation trial. *Mutagenesis* **2010**, *25*, 113–123. [CrossRef]
29. Johansson, C.; Moller, P.; Forchhammer, L.; Loft, S.; Godschalk, R.W.; Langie, S.A.; Lumeij, S.; Jones, G.D.; Kwok, R.W.; Azqueta, A.; et al. An ECVAG trial on assessment of oxidative damage to DNA measured by the comet assay. *Mutagenesis* **2010**, *25*, 125–132. [CrossRef]
30. Birnboim, H.C.; Jevcak, J.J. Fluorometric method for rapid detection of DNA strand breaks in human white blood cells produced by low doses of radiation. *Cancer Res.* **1981**, *41*, 1889–1892. [PubMed]
31. Dusinska, M.; Slamenova, D. Application of alkaline unwinding assay for detection of mutagen-induced DNA strand breaks. *Cell Biol. Toxicol.* **1992**, *8*, 207–216. [CrossRef] [PubMed]
32. Baumstark-Khan, C.; Hentschel, U.; Nikandrova, Y.; Krug, J.; Horneck, G. Fluorometric analysis of DNA unwinding (FADU) as a method for detecting repair-induced DNA strand breaks in UV-irradiated mammalian cells. *Photochem. Photobiol.* **2000**, *72*, 477–484. [CrossRef]
33. Daniel, F.B.; Chang, L.W.; Schenck, K.M.; Deangelo, A.B.; Skelly, M.F. The Further Development of a Mammalian DNA Alkaline Unwinding Bioassay with Potential Application to Hazard Identification for Contaminants from Environmental-Samples. *Toxicol. Ind. Health* **1989**, *5*, 647–665. [CrossRef] [PubMed]
34. Moreno-Villanueva, M.; Eltze, T.; Dressler, D.; Bernhardt, J.; Hirsch, C.; Wick, P.; von Scheven, G.; Lex, K.; Burkle, A. The automated FADU-assay, a potential high-throughput in vitro method for early screening of DNA breakage. *ALTEX* **2011**, *28*, 295–303. [CrossRef] [PubMed]
35. May, S.; Hirsch, C.; Rippl, A.; Bohmer, N.; Kaiser, J.P.; Diener, L.; Wichser, A.; Burkle, A.; Wick, P. Transient DNA damage following exposure to gold nanoparticles. *Nanoscale* **2018**, *10*, 15723–15735. [CrossRef]
36. Himly, M.; Geppert, M.; Hofer, S.; Hofstatter, N.; Horejs-Hock, J.; Duschl, A. When Would Immunologists Consider a Nanomaterial to be Safe? Recommendations for Planning Studies on Nanosafety. *Small* **2020**, *16*, e1907483. [CrossRef]
37. Bohmer, N.; Rippl, A.; May, S.; Walter, A.; Heo, M.B.; Kwak, M.; Roesslein, M.; Song, N.W.; Wick, P.; Hirsch, C. Interference of engineered nanomaterials in flow cytometry: A case study. *Colloids Surf. B Biointerfaces* **2018**, *172*, 635–645. [CrossRef]
38. Zook, J.M.; MacCuspie, R.I.; Locascio, L.E.; Halter, M.D.; Elliott, J.T. Stable nanoparticle aggregates/agglomerates of different sizes and the effect of their size on hemolytic cytotoxicity. *Nanotoxicology* **2011**, *5*, 517–530. [CrossRef]
39. Cassano, J.C.; Roesslein, M.; Kaufmann, R.; Luethi, T.; Schicht, O.; Wick, P.; Hirsch, C. A novel approach to increase robustness, precision and high-throughput capacity of single cell gel electrophoresis. *ALTEX* **2020**, *1*, 95–109. [CrossRef]
40. Auffan, M.; Rose, J.; Wiesner, M.R.; Bottero, J.Y. Chemical stability of metallic nanoparticles: A parameter controlling their potential cellular toxicity in vitro. *Environ. Pollut.* **2009**, *157*, 1127–1133. [CrossRef]
41. Buerki-Thurnherr, T.; Xiao, L.; Diener, L.; Arslan, O.; Hirsch, C.; Maeder-Althaus, X.; Grieder, K.; Wampfler, B.; Mathur, S.; Wick, P.; et al. In vitro mechanistic study towards a better understanding of ZnO nanoparticle toxicity. *Nanotoxicology* **2013**, *7*, 402–416. [CrossRef] [PubMed]
42. Wick, P.; Franz, P.; Huber, S.M.; Hirsch, C. Innovative Techniques and Strategies for a Reliable High-Throughput Genotoxicity Assessment. *Chem. Res. Toxicol.* **2020**, *33*, 283–285. [CrossRef] [PubMed]
43. Ghosh, M.; Bandyopadhyay, M.; Mukherjee, A. Genotoxicity of titanium dioxide (TiO$_2$) nanoparticles at two trophic levels: Plant and human lymphocytes. *Chemosphere* **2010**, *81*, 1253–1262. [CrossRef]
44. Khan, M.; Naqvi, A.H.; Ahmad, M. Comparative study of the cytotoxic and genotoxic potentials of zinc oxide and titanium dioxide nanoparticles. *Toxicol. Rep.* **2015**, *2*, 765–774. [CrossRef] [PubMed]
45. Jugan, M.L.; Barillet, S.; Simon-Deckers, A.; Herlin-Boime, N.; Sauvaigo, S.; Douki, T.; Carriere, M. Titanium dioxide nanoparticles exhibit genotoxicity and impair DNA repair activity in A549 cells. *Nanotoxicology* **2012**, *6*, 501–513. [CrossRef]
46. Hackenberg, S.; Friehs, G.; Froelich, K.; Ginzkey, C.; Koehler, C.; Scherzed, A.; Burghartz, M.; Hagen, R.; Kleinsasser, N. Intracellular distribution, geno- and cytotoxic effects of nanosized titanium dioxide particles in the anatase crystal phase on human nasal mucosa cells. *Toxicol. Lett.* **2010**, *195*, 9–14. [CrossRef] [PubMed]
47. Hackenberg, S.; Friehs, G.; Kessler, M.; Froelich, K.; Ginzkey, C.; Koehler, C.; Scherzed, A.; Burghartz, M.; Kleinsasser, N. Nanosized titanium dioxide particles do not induce DNA damage in human peripheral blood lymphocytes. *Environ. Mol. Mutagenesis* **2011**, *52*, 264–268. [CrossRef]
48. Chen, Y.; Star, A.; Vidal, S. Sweet carbon nanostructures: Carbohydrate conjugates with carbon nanotubes and graphene and their applications. *Chem. Soc. Rev.* **2013**, *42*, 4532–4542. [CrossRef]

49. Chung, C.; Kim, Y.K.; Shin, D.; Ryoo, S.R.; Hong, B.H.; Min, D.H. Biomedical applications of graphene and graphene oxide. *Acc. Chem. Res.* **2013**, *46*, 2211–2224. [CrossRef] [PubMed]
50. Feng, L.Y.; Wu, L.; Qu, X.G. New Horizons for Diagnostics and Therapeutic Applications of Graphene and Graphene Oxide. *Adv. Mater.* **2013**, *25*, 168–186. [CrossRef]
51. Loh, K.P.; Bao, Q.L.; Eda, G.; Chhowalla, M. Graphene oxide as a chemically tunable platform for optical applications. *Nat. Chem.* **2010**, *2*, 1015–1024. [CrossRef]
52. Wick, P.; Louw-Gaume, A.E.; Kucki, M.; Krug, H.F.; Kostarelos, K.; Fadeel, B.; Dawson, K.A.; Salvati, A.; Vazquez, E.; Ballerini, L.; et al. Classification Framework for Graphene-Based Materials. *Angew. Chem. Int. Ed.* **2014**, *53*, 7714–7718. [CrossRef]
53. Downs, T.R.; Crosby, M.E.; Hu, T.; Kumar, S.; Sullivan, A.; Sarlo, K.; Reeder, B.; Lynch, M.; Wagner, M.; Mills, T.; et al. Silica nanoparticles administered at the maximum tolerated dose induce genotoxic effects through an inflammatory reaction while gold nanoparticles do not. *Mutat. Res. Genet. Toxicol. Environ.* **2012**, *745*, 38–50. [CrossRef]
54. Gonzalez, L.; Thomassen, L.C.J.; Plas, G.; Rabolli, V.; Napierska, D.; Decordier, I.; Roelants, M.; Hoet, P.H.; Kirschhock, C.E.A.; Martens, J.A.; et al. Exploring the aneugenic and clastogenic potential in the nanosize range: A549 human lung carcinoma cells and amorphous monodisperse silica nanoparticles as models. *Nanotoxicology* **2010**, *4*, 382–395. [CrossRef]
55. Guidi, P.; Nigro, M.; Bernardeschi, M.; Scarcelli, V.; Lucchesi, P.; Onida, B.; Mortera, R.; Frenzilli, G. Genotoxicity of amorphous silica particles with different structure and dimension in human and murine cell lines. *Mutagenesis* **2013**, *28*, 171–180. [CrossRef]
56. Lankoff, A.; Arabski, M.; Wegierek-Ciuk, A.; Kruszewski, M.; Lisowska, H.; Banasik-Nowak, A.; Rozga-Wijas, K.; Wojewodzka, M.; Slomkowski, S. Effect of surface modification of silica nanoparticles on toxicity and cellular uptake by human peripheral blood lymphocytes in vitro. *Nanotoxicology* **2013**, *7*, 235–250. [CrossRef]
57. Park, M.V.D.Z.; Verharen, H.W.; Zwart, E.; Hernandez, L.G.; van Benthem, J.; Elsaesser, A.; Barnes, C.; Mckerr, G.; Howard, C.V.; Salvati, A.; et al. Genotoxicity evaluation of amorphous silica nanoparticles of different sizes using the micronucleus and the plasmid lacZ gene mutation assay. *Nanotoxicology* **2011**, *5*, 168–181. [CrossRef]
58. Tarantini, A.; Huet, S.; Jarry, G.; Lanceleur, R.; Poul, M.; Tavares, A.; Vital, N.; Louro, H.; Silva, M.J.; Fessard, V. Genotoxicity of Synthetic Amorphous Silica Nanoparticles in Rats Following Short-Term Exposure. Part1: Oral Route. *Environ. Mol. Mutagenesis* **2015**, *56*, 218–227. [CrossRef] [PubMed]
59. Uboldi, C.; Giudetti, G.; Broggi, F.; Gilliland, D.; Ponti, J.; Rossi, F. Amorphous silica nanoparticles do not induce cytotoxicity, cell transformation or genotoxicity in Balb/3T3 mouse fibroblasts. *Mutat. Res. Genet. Toxicol. Environ.* **2012**, *745*, 11–20. [CrossRef]
60. Vallet-Regi, M.; Colilla, M.; Izquierdo-Barba, I.; Manzano, M. Mesoporous Silica Nanoparticles for Drug Delivery: Current Insights. *Molecules* **2018**, *23*, 47. [CrossRef] [PubMed]
61. Zhang, Y.Z.; Zhi, Z.Z.; Jiang, T.Y.; Zhang, J.H.; Wang, Z.Y.; Wang, S.L. Spherical mesoporous silica nanoparticles for loading and release of the poorly water-soluble drug telmisartan. *J. Control. Release* **2010**, *145*, 257–263. [CrossRef] [PubMed]
62. Fu, Y.H.; Kuznetsov, A.I.; Miroshnichenko, A.E.; Yu, Y.F.; Luk'yanchuk, B. Directional visible light scattering by silicon nanoparticles. *Nat. Commun.* **2013**, *4*, 1527. [CrossRef] [PubMed]
63. Barnes, C.A.; Elsaesser, A.; Arkusz, J.; Smok, A.; Palus, J.; Lesniak, A.; Salvati, A.; Hanrahan, J.P.; de Jong, W.H.; Dziubaltowska, E.; et al. Reproducible Comet assay of amorphous silica nanoparticles detects no genotoxicity. *Nano Lett.* **2008**, *8*, 3069–3074. [CrossRef] [PubMed]
64. Tavares, A.M.; Louro, H.; Antunes, S.; Quarre, S.; Simar, S.; De Temmerman, P.J.; Verleysen, E.; Mast, J.; Jensen, K.A.; Norppa, H.; et al. Genotoxicity evaluation of nanosized titanium dioxide, synthetic amorphous silica and multi-walled carbon nanotubes in human lymphocytes. *Toxicol. Vitro* **2014**, *28*, 60–69. [CrossRef]
65. Hartmann, A.; Agurell, E.; Beevers, C.; Brendler-Schwaab, S.; Burlinson, B.; Clay, P.; Collins, A.; Smith, A.; Speit, G.; Thybaud, V.; et al. Recommendations for conducting the in vivo alkaline Comet assay. 4th International Comet Assay Workshop. *Mutagenesis* **2003**, *18*, 45–51. [CrossRef]
66. Meintieres, S.; Biola, A.; Pallardy, M.; Marzin, D. Apoptosis can be a confusing factor in in vitro clastogenic assays. *Mutagenesis* **2001**, *16*, 243–250. [CrossRef] [PubMed]
67. Kirkland, D.; Pfuhler, S.; Tweats, D.; Aardema, M.; Corvi, R.; Darroudi, F.; Elhajouji, A.; Glatt, H.; Hastwell, P.; Hayashi, M.; et al. How to reduce false positive results when undertaking in vitro genotoxicity testing and thus avoid unnecessary follow-up animal tests: Report of an ECVAM Workshop. *Mutat. Res. Genet. Toxicol. Environ.* **2007**, *628*, 31–55. [CrossRef] [PubMed]
68. Huk, A.; Collins, A.R.; El Yamani, N.; Porredon, C.; Azqueta, A.; de Lapuente, J.; Dusinska, M. Critical factors to be considered when testing nanomaterials for genotoxicity with the comet assay. *Mutagenesis* **2015**, *30*, 85–88. [CrossRef]
69. Vandebriel, R.J.; De Jong, W.H. A review of mammalian toxicity of ZnO nanoparticles. *Nanotechnol. Sci. Appl.* **2012**, *5*, 61–71. [CrossRef] [PubMed]
70. Heng, B.C.; Zhao, X.X.; Xiong, S.J.; Ng, K.W.; Boey, F.Y.C.; Loo, J.S.C. Toxicity of zinc oxide (ZnO) nanoparticles on human bronchial epithelial cells (BEAS-2B) is accentuated by oxidative stress. *Food Chem. Toxicol.* **2010**, *48*, 1762–1766. [CrossRef]
71. Huang, C.C.; Aronstam, R.S.; Chen, D.R.; Huang, Y.W. Oxidative stress, calcium homeostasis, and altered gene expression in human lung epithelial cells exposed to ZnO nanoparticles. *Toxicol. Vitro* **2010**, *24*, 45–55. [CrossRef] [PubMed]
72. Kim, I.S.; Baek, M.; Choi, S.J. Comparative Cytotoxicity of Al_2O_3, CeO_2, TiO_2 and ZnO Nanoparticles to Human Lung Cells. *J. Nanosci. Nanotechnol.* **2010**, *10*, 3453–3458. [CrossRef]
73. Lin, W.S.; Xu, Y.; Huang, C.C.; Ma, Y.F.; Shannon, K.B.; Chen, D.R.; Huang, Y.W. Toxicity of nano- and micro-sized ZnO particles in human lung epithelial cells. *J. Nanopart. Res.* **2009**, *11*, 25–39. [CrossRef]

74. Moos, P.J.; Chung, K.; Woessner, D.; Honeggar, M.; Cutler, N.S.; Veranth, J.M. ZnO Particulate Matter Requires Cell Contact for Toxicity in Human Colon Cancer Cells. *Chem. Res. Toxicol.* **2010**, *23*, 733–739. [CrossRef]
75. Premanathan, M.; Karthikeyan, K.; Jeyasubramanian, K.; Manivannan, G. Selective toxicity of ZnO nanoparticles toward Gram-positive bacteria and cancer cells by apoptosis through lipid peroxidation. *Nanomed. Nanotechnol.* **2011**, *7*, 184–192. [CrossRef]
76. Pujalte, I.; Passagne, I.; Brouillaud, B.; Treguer, M.; Durand, E.; Ohayon-Courtes, C.; L'Azou, B. Cytotoxicity and oxidative stress induced by different metallic nanoparticles on human kidney cells. *Part Fibre Toxicol.* **2011**, *8*, 10. [CrossRef] [PubMed]
77. Xia, T.; Kovochich, M.; Liong, M.; Madler, L.; Gilbert, B.; Shi, H.B.; Yeh, J.I.; Zink, J.I.; Nel, A.E. Comparison of the Mechanism of Toxicity of Zinc Oxide and Cerium Oxide Nanoparticles Based on Dissolution and Oxidative Stress Properties. *ACS Nano* **2008**, *2*, 2121–2134. [CrossRef]
78. Osman, I.F.; Baumgartner, A.; Cemeli, E.; Fletcher, J.N.; Anderson, D. Genotoxicity and cytotoxicity of zinc oxide and titanium dioxide in HEp-2 cells. *Nanomedicine* **2010**, *5*, 1193–1203. [CrossRef] [PubMed]
79. Yin, H.; Casey, P.S.; McCall, M.J.; Fenech, M. Effects of Surface Chemistry on Cytotoxicity, Genotoxicity, and the Generation of Reactive Oxygen Species Induced by ZnO Nanoparticles. *Langmuir* **2010**, *26*, 15399–15408. [CrossRef] [PubMed]
80. Eixenberger, J.E.; Anders, C.B.; Hermann, R.J.; Brown, R.J.; Reddy, K.M.; Punnoose, A.; Wingett, D.G. Rapid Dissolution of ZnO Nanoparticles Induced by Biological Buffers Significantly Impacts Cytotoxicity. *Chem. Res. Toxicol.* **2017**, *30*, 1641–1651. [CrossRef]
81. Scherzad, A.; Meyer, T.; Kleinsasser, N.; Hackenberg, S. Molecular Mechanisms of Zinc Oxide Nanoparticle-Induced Genotoxicity Short Running Title: Genotoxicity of ZnO NPs. *Materials* **2017**, *10*, 1427. [CrossRef]
82. Kao, Y.Y.; Chen, Y.C.; Cheng, T.J.; Chiung, Y.M.; Liu, P.S. Zinc Oxide Nanoparticles Interfere With Zinc Ion Homeostasis to Cause Cytotoxicity. *Toxicol. Sci.* **2012**, *125*, 462–472. [CrossRef]
83. Osmond, M.J.; McCall, M.J. Zinc oxide nanoparticles in modern sunscreens: An analysis of potential exposure and hazard. *Nanotoxicology* **2010**, *4*, 15–41. [CrossRef]
84. Cobley, C.M.; Chen, J.Y.; Cho, E.C.; Wang, L.V.; Xia, Y.N. Gold nanostructures: A class of multifunctional materials for biomedical applications. *Chem. Soc. Rev.* **2011**, *40*, 44–56. [CrossRef] [PubMed]
85. Connor, E.E.; Mwamuka, J.; Gole, A.; Murphy, C.J.; Wyatt, M.D. Gold nanoparticles are taken up by human cells but do not cause acute cytotoxicity. *Small* **2005**, *1*, 325–327. [CrossRef]
86. Pan, Y.; Neuss, S.; Leifert, A.; Fischler, M.; Wen, F.; Simon, U.; Schmid, G.; Brandau, W.; Jahnen-Dechent, W. Size-dependent cytotoxicity of gold nanoparticles. *Small* **2007**, *3*, 1941–1949. [CrossRef] [PubMed]
87. Lebedova, J.; Hedberg, Y.S.; Odnevall Wallinder, I.; Karlsson, H.L. Size-dependent genotoxicity of silver, gold and platinum nanoparticles studied using the mini-gel comet assay and micronucleus scoring with flow cytometry. *Mutagenesis* **2018**, *33*, 77–85. [CrossRef]
88. Shang, L.; Nienhaus, K.; Nienhaus, G.U. Engineered nanoparticles interacting with cells: Size matters. *J. Nanobiotechnol.* **2014**, *12*, 5. [CrossRef] [PubMed]
89. Xia, Q.Y.; Li, H.X.; Liu, Y.; Zhang, S.Y.; Feng, Q.Y.; Xiao, K. The effect of particle size on the genotoxicity of gold nanoparticles. *J. Biomed. Mater. Res. A* **2017**, *105*, 710–719. [CrossRef]
90. Di Bucchianico, S.; Fabbrizi, M.R.; Cirillo, S.; Uboldi, C.; Gilliland, D.; Valsami-Jones, E.; Migliore, L. Aneuploidogenic effects and DNA oxidation induced in vitro by differently sized gold nanoparticles. *Int. J. Nanomed.* **2014**, *9*, 2191–2204. [CrossRef]
91. Li, J.J.; Zou, L.; Hartono, D.; Ong, C.N.; Bay, B.H.; Yung, L.Y.L. Gold nanoparticles induce oxidative damage in lung fibroblasts in vitro. *Adv. Mater.* **2008**, *20*, 138. [CrossRef]
92. Singh, N.P. Microgels for estimation of DNA strand breaks, DNA protein crosslinks and apoptosis. *Mutat. Res. Fund. Mol. Mech.* **2000**, *455*, 111–127. [CrossRef]
93. Belyanskaya, L.; Manser, P.; Spohn, P.; Bruinink, A.; Wick, P. The reliability and limits of the MTT reduction assay for carbon nanotubes-cell interaction. *Carbon* **2007**, *45*, 2643–2648. [CrossRef]

 *nanomaterials*

Article

Comparison of Metal-Based Nanoparticles and Nanowires: Solubility, Reactivity, Bioavailability and Cellular Toxicity

Johanna Wall [1,†], Didem Ag Seleci [2,†], Feranika Schworm [1], Ronja Neuberger [1], Martin Link [1], Matthias Hufnagel [1], Paul Schumacher [1], Florian Schulz [3], Uwe Heinrich [4], Wendel Wohlleben [2] and Andrea Hartwig [1,*]

[1] Department of Food Chemistry and Toxicology, Faculty of Chemistry and Biosciences, Institute of Applied Biosciences, Karlsruhe Institute of Technology (KIT), 76131 Karlsruhe, Germany; Johanna.Wall@kit.edu (J.W.); fschworm@gmail.com (F.S.); Ronja.Neuberger@kit.edu (R.N.); Martin.Link@kit.edu (M.L.); Matthias.Hufnagel@gmail.com (M.H.); Paul.Schumacher@kit.edu (P.S.)
[2] BASF SE, 67063 Ludwigshafen, Germany; Didem.Ag-Seleci@basf.com (D.A.S.); wendel.wohlleben@basf.com (W.W.)
[3] Fraunhofer ITEM, 30625 Hannover, Germany; Florian.Schulz@item.fraunhofer.de
[4] ToxConsultant, 30625 Hannover, Germany; uwe.heinrich1512@gmail.com
* Correspondence: Andrea.Hartwig@kit.edu
† These authors contributed equally to this work.

Citation: Wall, J.; Seleci, D.A.; Schworm, F.; Neuberger, R.; Link, M.; Hufnagel, M.; Schumacher, P.; Schulz, F.; Heinrich, U.; Wohlleben, W.; et al. Comparison of Metal-Based Nanoparticles and Nanowires: Solubility, Reactivity, Bioavailability and Cellular Toxicity. *Nanomaterials* **2022**, *12*, 147. https://doi.org/10.3390/nano12010147

Academic Editors: Bing Yan and Zhanjun Gu

Received: 22 November 2021
Accepted: 28 December 2021
Published: 31 December 2021

Publisher's Note: MDPI stays neutral with regard to jurisdictional claims in published maps and institutional affiliations.

Abstract: While the toxicity of metal-based nanoparticles (NP) has been investigated in an increasing number of studies, little is known about metal-based fibrous materials, so-called nanowires (NWs). Within the present study, the physico-chemical properties of particulate and fibrous nanomaterials based on Cu, CuO, Ni, and Ag as well as TiO_2 and CeO_2 NP were characterized and compared with respect to abiotic metal ion release in different physiologically relevant media as well as acellular reactivity. While none of the materials was soluble at neutral pH in artificial alveolar fluid (AAF), Cu, CuO, and Ni-based materials displayed distinct dissolution under the acidic conditions found in artificial lysosomal fluids (ALF and PSF). Subsequently, four different cell lines were applied to compare cytotoxicity as well as intracellular metal ion release in the cytoplasm and nucleus. Both cytotoxicity and bioavailability reflected the acellular dissolution rates in physiological lysosomal media (pH 4.5); only Ag-based materials showed no or very low acellular solubility, but pronounced intracellular bioavailability and cytotoxicity, leading to particularly high concentrations in the nucleus. In conclusion, in spite of some quantitative differences, the intracellular bioavailability as well as toxicity is mostly driven by the respective metal and is less modulated by the shape of the respective NP or NW.

Keywords: metal-based nanoparticles and nanowires; solubility; intracellular bioavailability; oxidative reactivity

1. Introduction

Engineered metal-based nanomaterials are used in many consumer products, such as textiles, electronics, or medicinal products [1–4]. Besides metal-based nanoparticles, fibrous materials (nanowires) are also gaining increasing attention, e.g., for transparent conductive layers on displays [5,6]. Thus, a wide range of nanomaterials is available, differing in physicochemical properties such as size, shape, and surface chemistry.

Inhalation is the most crucial route of exposure and uptake of nanomaterials, especially at the workplace [7]. Depending on the size and shape, nanomaterials can reach different areas in the lung, including the alveolar region [8]. However, a general toxicological evaluation seems difficult, due to the diversity of these nanomaterials, and considering their broad range of physicochemical properties. Therefore, a grouping approach based on several physicochemical properties appears to be promising to elucidate the toxicological potential of nanomaterials [9–11].

For metal-based nanomaterials, the generation of ROS and thus oxidative stress has been identified as an important mechanism leading to genotoxicity and cytotoxicity [12]. Hereby, two scenarios are conceivable. First, the intracellular release of metal ions appears to be of major importance [13–15]. The so-called 'Trojan horse type' mechanism describes an endocytotic uptake, followed by lysosomal dissolution of the metal-based materials, resulting in high metal ion concentrations within the cells. This observation has already been made for Cu-, Ag-, and Ni-based nanoparticles (NP) [16–19]. Second, the materials themselves are able to catalyze the formation of ROS, due to their specific surface reactivity. Recent studies already demonstrated a high oxidative reactivity of different metal-based nanoparticles [20].

While there are increasing numbers of studies conducted with particulate nanomaterials, little is known about the reactivity of metal-based nanowires (NW). Depending on the cell type, previous studies indicate that metal-based fibrous nanomaterials are taken up into cells [21–24]. Furthermore, for Ag NW an uptake by endosomal and lysosomal vesicles has already been postulated [25]. However, the questions whether or not their fibrous structure contributes to toxicity and whether or not metal ions are intracellularly released as well remain. Therefore, within the present study, we comprehensively characterized particulate and fibrous metal-based nanomaterials with respect to their physicochemical properties. Besides nanomaterials known to potentially release toxic metal ions (Cu, CuO, Ni, and Ag), rather non-reactive materials (CeO$_2$ and TiO$_2$) with particulate and fibrous morphologies were also included. We compared them with respect to abiotic metal ion release in different physiologically relevant media and to acellular reactivity. Subsequently, four different cell lines were applied to compare cytotoxicity as well as intracellular metal ion release in the cytoplasm and nucleus. Since inhalation is likely to be the most critical route of exposure, two human lung epithelial cell lines were chosen, representing different regions of the respiratory tract, namely A549 cells (alveolar region) and BEAS-2B (normal human bronchial epithelium obtained from non-cancerous individuals). Furthermore, differentiated THP-1 cells (human peripheral blood monocytes) were chosen as a model for human macrophages. Finally, since inhalation studies are mostly performed in rats, RLE-6TN, an alveolar epithelial cell line derived from the rat lung was included as well, to compare the toxicity and bioavailability derived for human cells to rat cells.

2. Materials and Methods

2.1. Materials

All materials were purchased (purity > 99%) from Sigma Aldrich (Darmstadt, Germany) or Carl Roth GmbH (Karlsruhe, Germany). The nanomaterials included are listed in Table 1.

Table 1. Sources of all nanomaterials included in this study.

Material	Form	Source	Name/Item Number
Ag	NP	RAS AG	Agpure W10 (NM300K)
	NW	RAS AG	ECOS HC
CeO$_2$	NP	JRC	NM212
Cu	NP	Io-li-tec	NM-0016-HP
	NW	PlasmaChem	PL-CuW50
CuO	NP	BASF	CUO_1_NP_PROD *
Ni	NP	Sigma-Aldrich	577995
	NW	PlasmaChem	PL-NiW200
TiO$_2$	NP	JRC	NM105
	NW	PlasmaChem	PL-TiOW50

* from Sustainable Nanotechnologies Project [26]. NP: nanoparticle; NW: nanowire.

2.2. Physicochemical Characterization

In a first step, the nanomaterials were dispersed freshly according to the NANOGENO-TOX protocol to a concentration of 2.56 mg/mL in 0.05% BSA. Sonication was performed using a Branson Analog Sonifier 450 (Brookfield, CT, USA) for 13:25 min at 10% amplitude (7179 J). After sonication, the stock solution was diluted in supplemented RPMI-1640 to achieve the respective incubation concentrations. To investigate the impact of a freeze-thawing protocol [27] on the particle properties of the NP, samples were treated according to Keller and colleagues prior to DLS analyzes. Briefly, immediately after sonication, samples were frozen in liquid nitrogen at -196 °C. For DLS analyzes, samples were sonicated at 60 °C for at least 1 min or until completely thawed. Subsequently, NP were diluted to 100 µg/mL in supplemented RPMI-1640, incubated at 37 °C for 1 h, and analyzed [27].

Measurements of hydrodynamic diameter, zeta potential, and polydispersity index (PDI) of the NM dispersions were performed at a concentration of 100 µg/mL in RPMI supplemented with 10% FBS using a Zetasizer Nano ZS (Malvern Instruments Ltd., Worcestershire, UK).

Electron microscopy to obtain primary particle diameters as well as fiber diameters and lengths was performed in cooperation with the Laboratory for Electron Microscopy at KIT. Here, either a transmission electron microscope (CM200 FEG/ST, Philips, Amsterdam, The Netherlands) for NP or a scanning electron microscope (1530 Gemini with Schottky field emitter, LEO, Zeiss, Oberkochen, Germany) for nanowire was used. Subsequently, at least 300 particles or fibers were analyzed regarding their diameter and length using the ImageJ software (Rasband, National Institutes of Health, Bethesda, MD, USA).

The effective densities of the NP in supplemented RPMI-1640 were obtained as previously described by DeLoid and colleagues [28]. Briefly, 1 mL of the respective sample were transferred into VCM tubes (87007, TPP Techno Plastic Products AG, Trasadingen, Switzerland) and centrifuged for 1 h at $3000 \times g$. Subsequently, the volume of the pelleted NP was determined using a VCM "easy read" measuring device (87010, TPP Techno Plastics Products AG, Trasadingen, Switzerland). Effective densities were finally calculated according to De Loid and colleagues [28].

Data on deposited NP was obtained in silico using the Distorted Grid (DG) model as published previously by DeLoid and colleagues [29,30]. The model was performed according to Keller and colleagues using an auxiliary MATLAB macro to integrate data input via Microsoft Excel and allowing batch processing of multiple materials. The applied parameters for modeling particle deposition are listed in Table S1, the respective effective densities are stated in Table 2. The purity of the nanomaterials was determined in cooperation with the Institute of Applied Geosciences at KIT using ICP-MS (X-Series2 with collision cell technology, Thermo Fisher, Langenselbold, Germany).

2.3. Cell Culture

A549 cells (human adenocarcinoma cells) were kindly provided by Dr. Roel Schins (Leibniz Research Institute for Environmental Medicine, Düsseldorf, Germany). A549 cells were cultured in RPMI-1640 with 10% FBS, 100 U/mL penicillin, and 100 µg/mL streptomycin.

THP-1 cells (human peripheral blood monocytes, ATCC TIN-202) were kindly provided by Dr. Richard Gminski (Albert-Ludwig-University Freiburg, Department of Environmental Health Sciences and Hygiene, Freiburg, Germany) and cultured in supplemented RPMI-1640 composed as described above. Prior to experiments, THP-1 cells were seeded in cell culture dishes and differentiated with 30 ng/mL phorbol 12-myristate 13-acetate (PMA, diluted in DMSO) for 4 days. After 4 days, the differentiated cells (dTHP-1 cells) were cultured for further 3–5 days in fresh cell culture media [31].

RLE-6TN cells (alveolar epithelial cells derived from the rat lung, ATCC CRL-2300) were cultured in Hams-F12 media supplemented with 10% FBS, 100 U/mL penicillin, 100 µg/mL streptomycin, 10 µg/mL bovine pituitary extract, 5 µg/mL bovine insulin, 2.5 ng/mL epidermal growth factor, 2.5 ng/mL insulin like growth factor I and 1.25 µg/mL transferrin. Cells were subcultured twice a week.

Beas-2B cells (human lung bronchial epithelial cells, ATCC CRL-9609) were kindly provided by Dr. Carsten Weiss (Karlsruhe Institute of Technology, Karlsruhe, Germany). They were grown in precoated cell culture dishes (10 µg/mL fibronectin, 30 µg/mL collagen and 10 µg/mL bovine serum albumin in PBS) and cultured in keratinocyte growth media.

All cells were cultured under a humidified atmosphere with 5% CO_2 in air (HeraSafe, Thermo Scientific, Langenselbold, Germany). Except of the A549 cells, accutase was used instead of trypsin to detach adherent cells. For in vitro experiments, the cell number for each cell line was adjusted to achieve a confluent monolayer at the time of incubation, with the exception of dTHP-1 cells which were not able to form a confluent monolayer and therefore were seeded at the same cell number as the A549 cells.

2.4. Dissolution

2.4.1. Static Dissolution

To investigate the solubility of the nanomaterials, the dissolution rate of the materials in different physiologically relevant media as well as in the experimental setup was determined as previously described [15]. Briefly, the nanomaterial stock solution was diluted to a concentration of 100 µg/mL either in artificial alveolar fluid (AAF; pH 7.4 (composed of magnesium chloride (0.0952 g/L), sodium chloride (6.0193 g/L), potassium chloride (0.2982 g/L), disodium hydrogen phosphate (0.1420 g/L), sodium sulfate (0.0710 g/L), calcium chloride dihydrate (0.3676 g/L), sodium acetate trihydrate (0.9526 g/L), sodium hydrogen carbonate (2.6043 g/L), trisodium citrate dehydrate (0.0970 g/L), lecithin (0.1000 g/L)), artificial lysosomal fluid (ALF; pH 4.5 (composed of sodium chloride (3.210 g/L), sodium hydroxide (6.000 g/L), citric acid (20.800 g/L), calcium chloride dihydrate (0.1285 g/L), disodium hydrogen phosphate (0.0710), sodium sulfate (0.0390 g/L), magnesium chloride (0.0476 g/L), glycine (0.0590 g/L), sodium citrate dehydrate (0.0770 g/L), sodium tartrate dihydrate (0.0900 g/L), sodium lactate (0.0850 g/L), sodium pyruvate (0.0860 g/L)), or cell culture media and incubated for 24 h or 168 h at 37 °C and 150 rpm in a centrifuge tube. After incubation, the suspension was centrifuged at $3000 \times g$ for 1 h, followed by repeated centrifugation of the supernatants at $16,000 \times g$ for 1 h. Subsequently 2 mL of the supernatant were collected and centrifuged again at $16,000 \times g$ for 1 h. The supernatant was checked to exclude residual particles via dynamic light scattering and subsequently metal analyzes was performed. Subsequently, 1 mL of the supernatant was heated stepwise to 95 °C to dry up. The remnants were further digested with 1:1 HNO_3 (69%)/H_2O_2 (31%) (v/v) by repeated stepwise heating to 95 °C. The residue was then solubilized in 1 mL HNO_3 (0.2%) and metal ion content was measured by either GF-AAS (PinAAcle 900 T, Perkin Elmer, Rodgau, Germany) or ICP-MS (iCAP RQ with collision cell technology, Thermo Fisher, Langenselbold, Germany).

2.4.2. Dynamic Dissolution and Transformation

The flow-through setup which implements a "continuous flow system" (ISO 19057:2017) was used to detect nanoparticles and nanowire dissolution [32]. Briefly, NP/NW mass of 1 mg was weighed onto a membrane (cellulose triacetate, Sartorius Stedim Biotech GmbH, Göttingen, Germany: 47 mm diameter, 5 kDa pore size), topped by another membrane, and enclosed in flow-through cells. The flow-through cells were kept upright within a thermostatically controlled water bath to ensure that emerging air bubbles were able to leave the system without accumulating within the cell. The phagolysosomal simulant fluid (PSF) pH 4.5, which is an acidic buffer simulating the phagolysosomal compartment of macrophages [33], was employed at 37 ± 0.5 °C. The programmable sampler drew 10 mL eluates once per day from the total 100 mL collected. The ion concentration in the eluates was determined by ICP optical emission spectrometry (ICP-OES, Agilent 5100, Agilent Technologies, Santa Clara, CA, USA). After the experiment, the cells were flushed with deionized water before opening them to rinse the remaining solids off the membrane. The resulting suspension was then pelleted onto a transmission electron microscopy (TEM) grid held at the bottom of a centrifuge vial within 30 min and dried subsequently. By this procedure, the morphology of the remaining solids could be inspected with a reduction of interference from drying artifacts of PSF salts, which are removed by this preparation. Particle morphology was analyzed by TEM with a Tecnai G2-F20ST or Tecnai Osiris Microscope (FEI Company, Hillsboro, OR, USA) at an acceleration voltage of 200 keV under bright-field conditions. X-ray photoelectron spectroscopy (XPS) was performed using a Phi Versa Probe 5000 spectrometer (Physical Electronics, Feldkirchen, Germany) applying monochromatic Al Kα radiation.

2.5. Abiotic Reactivity (FRAS Assay)

The SOP, which described the multi-dose protocol of the Ferric Reduction Ability of Serum (FRAS) assay, published in 2017 by BASF [34], was used for reactivity testing of samples under physiological conditions.

Table 2. Summarized physicochemical properties of the investigated nanoparticles. d_p: primary particle diameter determined by transmission electron microscopy, d_h: hydrodynamic diameter, PDI: polydispersity index, SSA: specific surface area.

	Cu NP	CuO NP	Ni NP	TiO$_2$ NP (NM105)	CeO$_2$ NP (NM212)	Ag NP (NM300K)
d_p (nm)	55.2 ± 1.5	17.1 ± 0.4	21.4 ± 0.1	23.7 ± 0.5	21.5 ± 0.3	15.5 ± 0.04
d_h (nm)	308.2 ± 40.3	160.3 ± 42.1	388.0 ± 33.2	165.8 ± 14.2	187.0 ± 7.3	72.4 ± 10.0
PDI	0.23 ± 0.07	0.48 ± 0.05	0.67 ± 0.02	0.14 ± 0.01	0.20 ± 0.02	0.31 ± 0.06
ζ-potential (mV)	-15.3 ± 0.02	-14.8 ± 0.2	-15.7 ± 0.2	-14.8 ± 0.2	-15 ± 0.6	-11.2 ± 2.1
SSA (m^2/g)	10.7 ± 0.6	47 [#]	6.4 ± 0.3	46.2 *	27 *	N/A **
effective density (g/cm^3)	1.78 ± 0.02	1.98 ± 0.03	2.54 ± 9.14	1.38 ± 0.06	1.97 ± 0.14	2.07 ± 0.14
fraction of deposited dose in 24 h (%)	64	56	90	22	53	27
purity (% wt)	98.6 ± 0.4	98.7 ± 0.81	98.7 ± 0.86	91.5 ± 061	98.5	99.3 ± 0.08

* data taken from respective JRC report [35]. ** not quantified as material is only available as dispersion. # data taken from project data (SUN-project) [26].

Briefly, samples were incubated with Human Blood Serum (HBS) for 3 h at 37 °C. Before incubation, bath sonication for 1 min was applied to prevent the formation of large agglomerates and access whole surface area. NMs were separated from HBS via ultracentrifugation (AUC-Beckman XL centrifuge, Beckman Coulter, Brea, CA, USA) at $14,000 \times g$ for 150 min). Subsequently, 100 µL of NM-free HBS supernatant were incubated in the FRAS reagent that contains the Fe^{3+} complex. The total antioxidant depletion as a measure of the oxidative potential of NMs was determined by using UV-vis spectrum of the iron complex solution. Trolox, a water-soluble analog of vitamin E, was used as an antioxidant to calibrate the FRAS results. Different Trolox concentrations (from 0.001 to 0.1 g/L) were tested by FRAS assay to obtain FRAS absorption signals that were linearly fitted. Finally, the oxidative damage induced by NP and NW was calculated in Trolox equivalent units (TEUs).

Additionally, fresh NM samples were prepared to evaluate the ion contribution. After an ultracentrifugation step, the ion concentration in NM-free HBS supernatants was determined by ICP-MS (Nexion 2000b, Perkin Elmer, Waltham, MA, USA). Using water-soluble metal salts (CuSO$_4 \cdot$5H$_2$O and NiCl$_2$), ion solutions with equivalent concentrations were prepared and the associated oxidative damage in HBS was measured by FRAS method. For each NM and ion dose, triplicate measurements were performed.

2.6. Cytotoxicity, Bioavailability, and Intracellular Distribution

For cytotoxicity testing, the ATP content was quantified with the CellTiter-Glo® Luminescent Cell Viability Assay Kit (Promega GmbH, Bremen, Germany). Briefly, cells were seeded in 96-well plates and incubated with the respective nanomaterials or 500 nM staurosporine (positive control). After 24 h incubation the medium was removed and 100 µL fresh medium was added to the wells. After 30 min of equilibration, 100 µL of CellTiter-Glo® was added and chemiluminescence was measured on the Infinite® 200 Pro microplate reader (Tecan Group Ltd., Männedorf, Switzerland). For analyzing the relative cell count (RCC) and bioavailability, cells were seeded as a confluent monolayer and incubated with three different doses of the nanomaterials (Table S2). After 24 h, incubation was stopped by removing the incubation media from the cells. The cells were washed with PBS, collected, and analyzed via a Casy cell counter obtaining cell number and cell volume. Cell count was used to determine the RCC. To quantify the amount of bioavailable metal ions within the whole cell, cells were lysed in RIPA buffer (0.01 M Tris pH 7.6, 0.15 M NaCl, 0.001 M EDTA, 1% (v/v) Triton-X 100, 1% DOC, 0.01% SDS, 1 × protease-inhibitor) for 30 min followed by 1 h centrifugation at 14,000× g to remove the cell membrane and undissolved material residues [17]. The supernatant was then used for graphite furnace atomic absorption spectrometry (GF-AAS) as described above. Residues of particles were excluded using dynamic light scattering (data not shown). To investigate the intracellular distribution of the metal ions within the cytoplasm and the nucleus, cells were fractionated as described previously [15,17], using the Nuclear Extract Kit (ActiveMotif, Carlsbad, CA, USA).

3. Results

3.1. Physicochemical Characterization

All materials were characterized as raw material or dispersion in detail. For all particles, hydrodynamic (d_h) and primary diameter (d_p, obtained via TEM or SEM) as well as ζ-potential and PDI were determined. Additionally, all particles were investigated with respect to their specific surface area (SSA), effective density, and deposition efficiency for subsequent in vitro studies using the Distorted Grid (DG) model (Table 2). The primary particle size of all materials ranged between 15.5–55.2 nm, with Ag NP (NM 300K) showing the smallest size followed by CuO NP, Ni NP, CeO$_2$ NP (NM 212), and TiO$_2$ NP (NM 105). Cu NP showed the largest primary diameter of 55.2 nm. The hydrodynamic diameter of all materials was in the same order of magnitude, except for the Ni NP which showed a high d_h of 388 nm. Additionally, a rather high PDI for the Ni NP was observed. The latter, in combination with the high d_h, indicates that the Ni NP dispersion was very poly-disperse and that this particle species tended to agglomerate. The deposited dose ranged between 22% in case of TiO$_2$ NP and 90% in case of Ni NP. To facilitate the dispersion preparation for in vitro studies, a comparison between freshly prepared and thawed material dispersions was made. Hereby, d_h, Z-average and the deposited dose were investigated (Table S3, Figures S1 and S2). Altogether, no major differences were observed, indicating the applicability of a freeze-thaw protocol. For the fibrous materials, primary length and diameter, as well as ζ-potential and SSA were investigated (Table 3). All nanowire showed a mean length above 5 µm and a width below 500 nm, indicating WHO fibre-like properties with an aspect ratio higher than 3:1. To calculate the deposited dose for the NP, the DG model was used [29]. Since this model uses spherical structures to simulate the deposited dose it is not applicable for fibrous materials such as nanowire.

Table 3. Summarized physicochemical properties of the investigated nanowire. Length and width have been determined by scanning electron microscopy, SSA: specific surface area.

	Length (µm)	Width (nm)	ζ-Potential (mV)	SSA (nm^2/g)	Purity (% wt)
Cu NW	6.3 ± 0.4	300 ± 6	−14.1	1.49	>99.5 [#]
Ni NW	9.97 ± 0.29	280 ± 6	−14.5	1.61	99.1
Ag NW	10.6 ± 0.28	110 ± 1.6	−4.1 ± 0.1	3.47	99.1 ± 0.65
TiO$_2$ NW	7.3 ± 0.2	0.87 ± 0.03	n.a.	n.a.	n.a.

[#] data from supplier. n.a.: not available; experiments were not performed due to batch-to-batch variation of TiO$_2$ NW resulting in suspensions of poor quality.

3.2. Dissolution and Transformation

The solubility in physiologically relevant fluids was determined using two approaches, applying a static and a dynamic method. While the static approach was applied for artificial alveolar (AAF, pH 7.4) and artificial lysosomal (ALF, pH 4.5) fluid, the dynamic approach was conducted with a phagolysosomal simulant fluid (PSF, pH 4.5). The results are summarized as a proportion of dissolved material after seven days of incubation (Table 4). Most materials were not soluble in AAF. Only Cu NP and Cu NW showed a low and comparable solubility in AAF of 12%. The solubility in artificial lysosomal fluid (ALF) depended strongly on the material examined. High solubilities were observed in the case of Cu NP (64%) and Cu NW (57%), as well as for CuO NP (57%) in the static method. These materials also showed high solubility in the dynamic process, whereby the CuO NP exerted the highest solubility of 97%. The nickel materials exhibited a moderate (Ni NW) to high solubility (Ni NP) in the static process, while Ni NW showed a high solubility of 94% in the dynamic process. In both, static and dynamic approaches, no or only a very low solubility for the Ag NP was detected. The fibrous silver material (Ag NW) was also found to be insoluble by the static approach, while a low solubility of 11% was found applying the dynamic system. For particulate and fibrous TiO_2 as well as for particulate CeO_2 no solubility was observed in any of the media tested.

Table 4. Summary of the solubility in biologically relevant model fluids after seven days. AAF: Artificial alveolar fluid (pH 7.4); ALF: Artificial lysosomal fluid (pH 4.5); PSF: Phagolysosomal simulant fluid (pH 4.5).

Material and Form		Static Dissolution (% Dissolved)		Dynamic Dissolution (% Dissolved)
		AAF	ALF	PSF
Ag	NP (NM 300K)	0.5 ± 0.1	0.2 ± 0.0	1.6
	NW	0.5 ± 0.2	0.2 ± 0.0	11
CeO_2	NP (NM 212)	0.001	0.021	0.3
Cu	NP	12.0 ± 4.7	63.8 ± 3.4	45.5
	NW	6.0 ± 2.7	57 ± 3.1	35
CuO	NP	3.9 ± 1.8	57.2 ± 5.1	97.3
Ni	NP	3.7 ± 0.4	56.1 ± 15.5	63.2
	NW	0.9 ± 0.6	35.5 ± 10.1	94.4
TiO_2	NP (NM 105)	0.002	0.022	0.3
	NW	0.001	0.021	0

Next, the transformation of nanomaterial shape and speciation of NW after dynamic dissolution was investigated (Figure 1). For this purpose, the flow-through cells were flushed with water, opened and the remaining solids were rinsed onto a centrifuge vial with a TEM grid at the bottom. By centrifugation, all solid material > 10 nm was spun onto the TEM grid and the supernatant containing the buffer salts was discarded. Compared to non-treated Ag NW, the occurrence of particulate morphologies was observed, which represents the thermodynamically stable form. A smaller number of Cu NW, decorated by newly formed substructures, was found on TEM grids due to high solubility in PSF. The tendency of increased polymorphism from long fibers towards a higher number of small particle structures coincided with sulfidation (EDXS, data not shown), which may have formed passivating layers. Since Ni NW almost completely dissolved, no NW were detected during TEM measurements. As expected, undissolved TiO_2 NW stayed aggregated after dissolution.

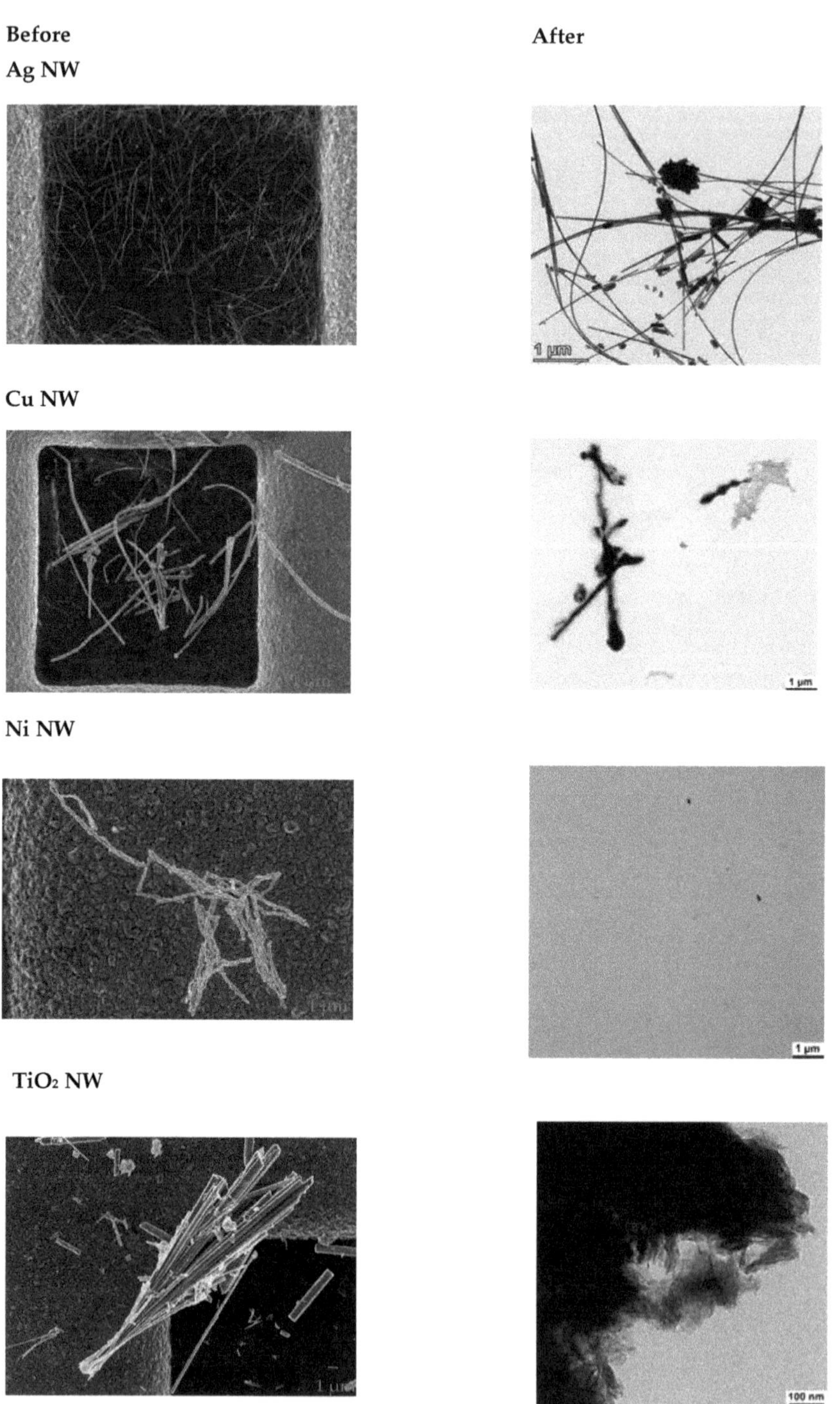

Figure 1. TEM images of NP and NW before and after treatment in the flow-through cells with phagolysosomal simulant fluid (PSF).

3.3. Abiotic Reactivity

To assess the abiotic reactivity, the so-called ferric reduction ability of serum (FRAS) assay was applied. This assay is based on the measurement of a mass-metric Biological Oxidative Damage (mBOD) of nanomaterials due to their oxidative potential by the reduction of human blood serum [34]. For each NP and NW, a dose-response was carried out and one concentration close to ~20% of maximum NM oxidative potential was selected for the evaluation of ion contribution (Figure 2).

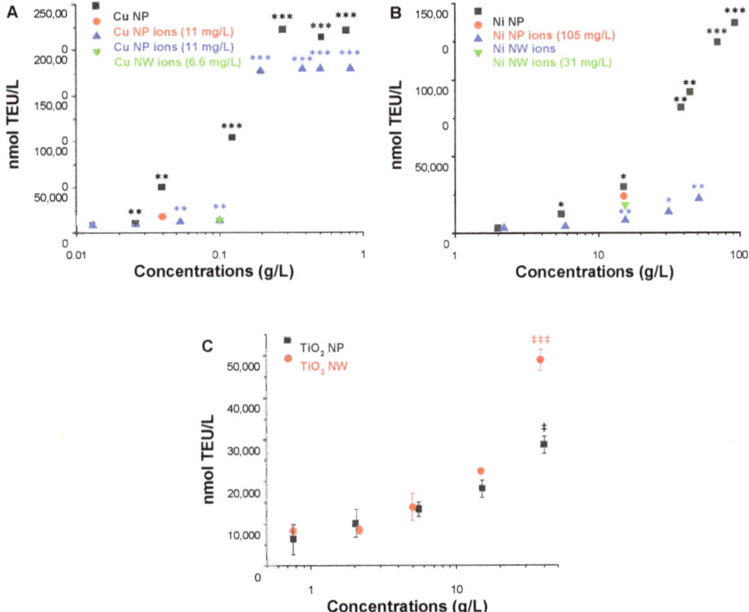

Figure 2. FRAS testing (**A**) Cu NP, Cu NP ions, Cu NW and Cu NW ions (**B**) Ni NP, Ni NP ions, Ni NW and Ni NW ions (**C**) TiO_2 NP and TiO_2 NW. Error bars indicate one standard deviation from triplicate testing, and are smaller than the size of the symbol in most cases. Statistics were performed using either ANOVA-Dunnett's T3 (* $p \leq 0.05$, ** $p \leq 0.01$, *** $p \leq 0.001$) or the 2-sided Dunnett's test (‡ $p \leq 0.05$, ‡‡‡ $p \leq 0.001$).

After incubation of the Cu NP in the FRAS assay media for the duration of the assay at a concentration of 0.04 g/L (~20% of the measured Cu NP oxidative potential), the actual Cu ion concentration was determined, being 11 mg/L. The ion oxidative potential was more than two times and thus significantly lower ($18,048 \pm 2863$ nmol TEU/L) than the response induced by the total Cu NP ($49,783 \pm 644$ nmol TEU/L); therefore the reactivity of Cu NP at 0.04 g/L was predominately assigned to the particle with a steep dose-response curve. Similarly, CuO NP presented high reactivity which was dominated by the particle itself (Figure S3A). The ion contribution of Cu NW was tested at a concentration of 0.1 g/L and the actual Cu ion concentration detected was 6.6 mg/L. The reactivity of Cu NW (0.1 g/L) originated completely from ions (Figure 2A). Moreover, FRAS mBOD values were calculated at concentrations of 0.22 g/L as 842 ± 10 and 922.0 ± 5.4 nmol TEU/mg for Cu NP and NW respectively, representing very high reactivity for both forms.

The ion contribution for Ni NP and NW was examined at a concentration of 15 g/L. Ni ion concentrations were determined to be 105 mg/L and 31 mg/L, respectively. Both Ni ions and NPs contributed significantly to the reactivity of Ni NP at 15 g/L (Figure 2B), with values of $23,897 \pm 1470$ nmol/L TEU (ions) and $30,063 \pm 441$ nmol/L TEU (particles). The mBOD values at concentration 15 g/L are for Ni NP 2.00 ± 0.03 nmol TEU/mg and for

NW 0.50 ± 0.01 nmol TEU/mg. Although the difference is metrologically significant, the values are very similar in comparison to the dynamic range of the assay as exemplified by the values of the Cu-based materials.

Since TiO_2 NP and NW were insoluble materials, ion contribution to the reactivity was not considered. Similar mBOD values at a concentration of 15 g/L were calculated for both NP: 1.20 ± 0.09 and NW: 1.5 ± 0.22 nmol TEU/mg (not all replicates were useful for the NW, and the error of the worst triplicate in the concentration series is given). Again, the difference is metrologically significant, but the values are very similar in comparison to the dynamic range of the assay.

Moreover, the reactivity of CeO_2 and Ag NP was evaluated (Figure S3B,C). As an insoluble material, CeO_2 presented very low reactivity (mBOD: 1.90 ± 0.09 nmol TEU/mg at conc. 5.6 g/L). Ag NP showed intermediate reactivity with a mBOD value of 22.3 ± 0.2 nmol TEU/mg (at conc. 3.0 g/L).

3.4. Cell Viability and Bioavailability

For all cellular experiments doses are stated as µg/mL to facilitate comparison between NP and NW, since the parameter of the deposited dose was only available for NP and not for NW as described in Section 3.1.

3.4.1. Cell Viability

Cell viability was investigated in four different cell lines, namely three lung epithelial cell lines (A549, Beas-2B and RLE-6TN (rat)) and one cell line with macrophage-like properties (dTHP-1). For each nanomaterial, five different doses were chosen and their impact on the ATP content was assessed. Furthermore, RCC after incubation with three different doses was analyzed. (Figure S4). As shown in the overview provided in Figure 3, all materials which were soluble in lysosomal fluid (Cu, CuO, Ni) showed a dose-dependent cytotoxic effect in all investigated cell lines. Additionally, a dose-dependent decrease of the viability of the cells was seen after incubation with the Ag-based materials, even though an ion release in the lysosomal fluid was not observed for this material. The dTHP-1 cells, as a cell culture model with macrophage-like properties, revealed the most sensitive reaction to all lysosomal-soluble nanomaterials. Comparing NP and NW of the same material at the same concentrations, with the exception of Cu-based materials, the application of NW resulted in a less pronounced cellular toxicity. For the insoluble materials, TiO_2 NP and CeO_2 NP no decrease in viability was observed, even for the highest applied dose of 100 µg/mL.

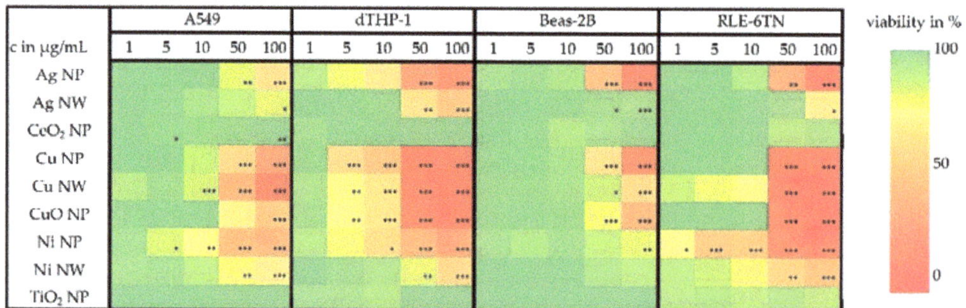

Figure 3. Impact of nanomaterials on the ATP content of A549, Beas-2B, RLE-6TN, and dTHP-1 cells after incubation with five different doses of the nanomaterials. The ATP content of incubated samples was normalized to an untreated control. Significantly different from negative controls: * $p \leq 0.05$, ** $p \leq 0.01$, *** $p \leq 0.001$ (ANOVA-Dunnet's *t*-test).

3.4.2. Bioavailability

Bioavailability of all nanomaterials was analyzed after 24 h incubation with three different concentrations (Figure 4) in four different cell lines (A549, Beas-2B, RLE-6TN, and dTHP-1). Doses were chosen in preliminary experiments and normalized to low, mid, and high cytotoxic effects. Cells were incubated with nanomaterials for 24 h in their respected cell culture media. Depicted are the means of three independently performed experiments ± standard deviation.

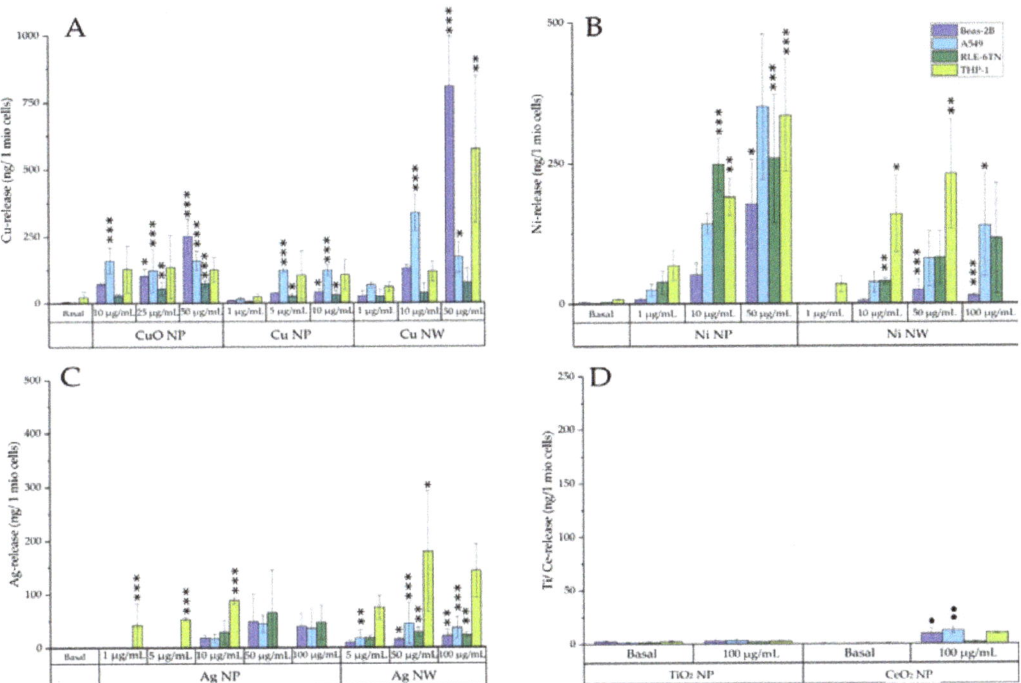

Figure 4. Bioavailability of Cu- (**A**), Ni- (**B**), Ag- (**C**), and Ce- and Ti-based materials (**D**) in A549 (light blue), Beas-2B (purple), RLE-6TN (dark green), and dTHP-1 (light green) cells. Bioavailability is displayed as ion release in ng/10^6 cells. Cells were incubated with nanomaterials for 24 h in their corresponding cell culture media. Depicted are the means of three independently performed experiments ± standard deviation. Statistics were performed using either ANOVA-Dunnett's test (* $p \leq 0.05$, ** $p \leq 0.01$, *** $p \leq 0.001$) or the unpaired t test (• $p \leq 0.05$, •• $p \leq 0.01$) to compare differences from basal concentration.

All materials which revealed a solubility in ALF (pH 4.5) also showed a dose-dependent increase of the intracellular ion release. The basal Cu content within the epithelial cell lines (A549, Beas-2B, RLE-6TN) was around 2–3 ng/10^6 cells (15 μM). Regarding the dTHP-1 cells, a basal Cu concentration of 20 ng/10^6 cells (30 μM) was observed. After incubation with the CuO NP, the intracellular content of released copper ions increased up to 250 ng/10^6 cells (1700 μM), being comparable for all cell lines except the RLE-6TN cells. Here, only a small increase of intracellular copper content after incubation with the CuO NP was observed (70 ng/10^6 cells (600 μM)). Comparing Cu NP and NW, the bioavailability of the Cu NW was considerably higher, especially in Beas-2B cells (up to 800 ng/10^6 cells (5500 μM)). This observation could be explained by the higher dissolution rate of the Cu NW in cell culture media used for Beas-2B cultivation (shown in Figure S5) and therefore a simultaneous uptake of Cu ions in Beas-2B cells. Besides Beas-2B cells, dTHP-1 cells also exerted a strong release of Cu ions after incubation of the Cu NW, with a maximum of 570 ng/10^6 cells (3157 μM).

For both Ni-based materials a strong dose-dependency in intracellular bioavailability was observed. Basal Ni concentrations ranged around 1–3 ng/10^6 cells (3–10 μM) for all epithelial cells and 7 ng/10^6 cells (26 μM) for the dTHP-1 cells. After incubation with 10 μg Ni NP/mL, intracellular Ni content increased up to 250 ng/10^6 cells (3000 μM). In comparison, the bioavailability of Ni NW was much lower. Intracellular Ni contents up to 44 ng/10^6 cells (485 μM) were seen for two epithelial cell lines (A549 and RLE-6TN) after an incubation dose of 10 μg Ni NW/mL After treatment with 50 μg Ni NW/mL and higher, Ni-ion content increased up to 140 ng/10^6 cells (950 μM). Regarding the Ni content in the dTHP-1 cells at an incubation dose of 10 μg Ni NW/mL, intracellularly dissolved Ni was found to be four times higher (160 ng/10^6 cells (600 μM)) compared to that of all epithelial cells. For Beas-2B cells only a low bioavailability was observed after incubation with Ni NW, leading to a maximum intracellular Ni content of 13 ng/10^6 cells (90 μM). A lower bioavailability of Ni in the Beas-2B cells was also observed after incubation with the Ni NP. Solubility of the Ni-based materials in cell culture media was very low, leading to a maximum dissolution rate of 4% in all cell culture media used in this study (Figure S5). Thus, the intracellular bioavailability of these materials can be correlated to the uptake of undissolved materials. Moreover, no differences in the solubility between the different cell culture media were seen. Therefore, differences in the intracellular bioavailability of the Ni-based materials appear to be dependent on the cell lines and their specific properties.

In acellular investigations, no dissolution of the Ag-based materials in ALF was seen. However, bioavailability was observed in the cellular studies. Incubation of the Ag NP resulted in a weak dose-dependent increase of the intracellular Ag content of up to 47 ng/10^6 cells (180 μM) at an incubation dose of 100 μg/mL for all epithelial cell lines. In contrast, Ag content in the dTHP-1 cells was much higher, reaching an intracellular Ag ion release of 94 ng/10^6 cells (175 μM) at 10 μg Ag NP/mL Incubation of the Ag NW led to a maximum Ag ion release of 34 ng/10^6 cells (102 μM) at the maximum dose of 100 μg/mL for the epithelial cells, whereas Ag ion content of the dTHP-1 cells was found to be around five times higher (150 ng/10^6 cells (330 μM)) after an incubation dose of 100 μg Ag NW/mL. For the insoluble materials TiO_2 NP and CeO_2 NP no bioavailability was seen even after treatment with the highest dose of 100 μg/mL.

3.4.3. Intracellular Distribution

Additionally, the intracellular distribution of all bioavailable materials was investigated by fractionating the cells into the soluble fractions of the cytoplasm and nucleus (Figure 5). Since treatment doses vary between the different materials, they are stated in Table S2. A dose-dependent increase of ion concentrations was seen for all of the materials in the cytoplasm as well as in the nucleus. For the Cu-based materials, a strong accumulation of intracellular dissolved Cu ions was found within the nucleus of all cell lines, reaching concentrations of 1 mM and higher. As already observed for the cellular bioavailability, the compartment-specific Cu concentration was much more pronounced after treatment with Cu NW when compared to the particulate Cu-based materials. Regarding the Ag-based materials, a nuclear accumulation was evident in all cell lines. Here, concentrations up to 4 mM were observed after incubation of Beas-2B cells with Ag NP and after applying Ag NW on dTHP-1 and RLE-6TN cells. For the Ni-based materials, a lower concentration of released Ni ions was found in the nucleus as compared to the cytoplasm of all cell lines.

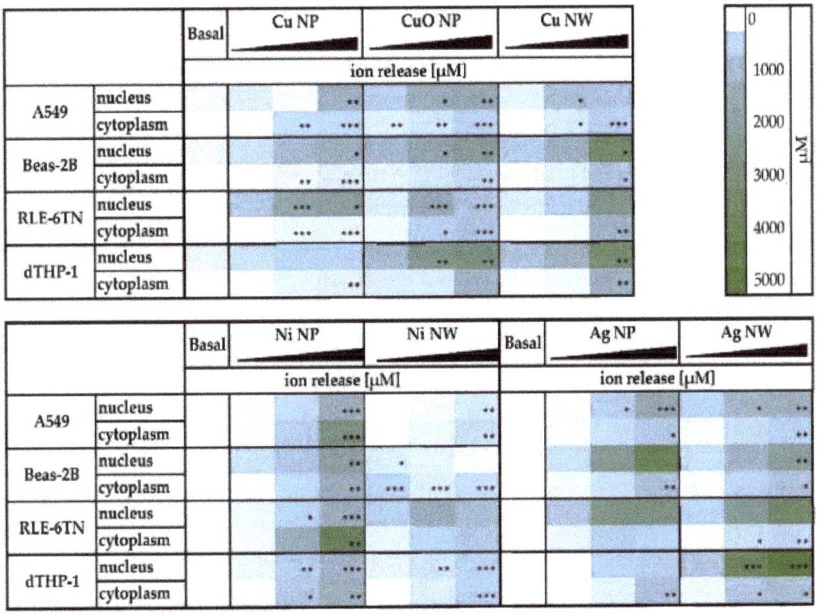

Figure 5 (heatmap, two panels):

Panel 1 — columns: Basal | Cu NP (increasing ion release [µM]) | CuO NP (increasing ion release [µM]) | Cu NW (increasing ion release [µM])

Cell line	Compartment	Basal	Cu NP	CuO NP	Cu NW
A549	nucleus		**	* **	*
A549	cytoplasm		** ***	** ** ***	* ***
Beas-2B	nucleus		*	* **	*
Beas-2B	cytoplasm		** ***	**	*
RLE-6TN	nucleus		*** *	*** ***	*
RLE-6TN	cytoplasm		*** ***	* ***	**
dTHP-1	nucleus			** **	**
dTHP-1	cytoplasm		**		**

Legend (µM): 0, 1000, 2000, 3000, 4000, 5000

Panel 2 — columns: Basal | Ni NP (increasing ion release [µM]) | Ni NW (increasing ion release [µM]) | Basal | Ag NP (increasing ion release [µM]) | Ag NW (increasing ion release [µM])

Cell line	Compartment	Basal	Ni NP	Ni NW	Basal	Ag NP	Ag NW
A549	nucleus		***	**		* ***	* **
A549	cytoplasm		***	**		*	**
Beas-2B	nucleus		**	*			**
Beas-2B	cytoplasm		**	*** *** ***		**	*
RLE-6TN	nucleus		* ***				
RLE-6TN	cytoplasm		**				* **
dTHP-1	nucleus		** ***	** ***			*** ***
dTHP-1	cytoplasm		* **	***		**	* *

Figure 5. Intracellular distribution in cytoplasm and nucleus of A549, Beas-2B, RLE-6TN and dTHP-1 cells after incubation with three doses metal-based nanomaterials. Significantly different from basal concentration: * $p \leq 0.05$, ** $p \leq 0.01$, *** $p \leq 0.001$ (ANOVA-Dunnet's *t*-test).

4. Discussion

In this study, nine different particulate or fibrous metal-based nanomaterials were investigated with respect to their physicochemical properties and solubility behavior in acellular fluids of different pH values. Furthermore, the cytotoxicity and intracellular bioavailability of the materials in four different cell lines relevant for inhalative exposure was determined. To the best of our knowledge, this is the first study systematically comparing the impact of different particulate and fibrous metal-based materials on all of these parameters in parallel, exerting some quantitative differences between nanomaterial shapes, but a more distinct impact of the respective metal species under investigation.

Concerning the acellular investigations, for all materials, solubility in AAF (pH 7.4) was not apparent or considered to be low. This suggests that the analyzed materials do not dissolve in the extracellular matrix of the respiratory tract which is in accordance with previous studies [24]. Therefore, a nanomaterial-cell interaction within the lung can be postulated. However, since nanomaterials are taken up via endocytosis and are subsequently transported to lysosomes, dissolution in this acidic environment appears to be relevant. This may result in higher solubility in this acidic cellular compartment, with potential intracellular metal ion release and thus potential metal-ion derived cellular toxicity. Therefore, two different approaches were chosen to determine solubility under acidic conditions and compared, namely, the static dissolution in artificial lysosomal fluid (ALF) and a dynamic dissolution approach in phagolysosomal simulant fluid (PSF), both pH 4.5. The dynamic approach was chosen additionally, since the lung is not a static system and dissolved ions are transported quickly to other compartments, rendering it as a more realistic approach [36,37]. Furthermore, the dissolution by the dynamic approach is not limited by saturation conditions and therefore an underestimated dissolution rate can be prevented [38]. Regarding the different materials under investigation, even after seven days no solubility in either experimental system was observed in case of TiO$_2$ NP or NW, CeO$_2$ NP nor in case of Ag NP. However, some solubility was observed for Ag NW in the dynamic system as opposed to no detectable dissolution under static conditions. Far higher solubility of around 50% and above was evident in case of Ni- and Cu-based materials. Here, both Ni NP and Ni NW, as well as CuO NP, exerted higher dissolution fractions in the dynamic system, while the opposite was observed in case of Cu NP and Cu NW. Nevertheless, in all cases except for the insoluble TiO_2, CeO_2, and Ag NP, solubility was highly accelerated under acidic conditions, evident by both experimental approaches. These differences in nanomaterial dissolution were also reflected in structural transformation as determined by TEM in the dynamic approach.

Additionally, the oxidative potentials of NP and NW and their free ions were evaluated by utilizing the FRAS assay, which measures biological oxidant damage in serum [15]. Here, NP and NW based on the same metals (Cu NP/NW, Ni NP/NW, TiO_2 NP/NW) demonstrated similar reactivity. However, the difference between the metals was more significant. While all Cu-based materials (NP/NW and CuO NP) exerted very high reactivity at low concentrations around and above 0.1 g/L, about 100-fold higher concentrations were required in case of TiO_2 NP/NW, Ni NP/NW, and CeO_2 NP to exert some but still low reactivity. With regard to the respective NP, the results confirm those obtained previously [20,39]. No such studies have been conducted for the NW analyzed within this study.

Since the respiratory tract is a complex system consisting of different cell types, the cytotoxicity and bioavailability of the nanomaterials in four different cell lines were investigated, all being relevant for the respiratory tract. Thus, three epithelial cell lines of human (A549, Beas-2B) or rat (RLE-6TN) origin, as well as a cell line with macrophage-like properties (differentiated THP-1) were applied. To assess the cytotoxicity, two parameters were chosen, namely RCC and ATP content, which were determined in all four cell lines. Based on the outcome, bioavailability studies were conducted at low, mid, and high cytotoxic doses of the respective materials, as stated in Table S2. To distinguish intracellular bioavailability from particles potentially stuck to the outer cell membrane, and to further discriminate between cytoplasmic and nuclear metal ion concentrations, two different fractionation protocols were applied as published previously [15,17]. Briefly, to assess bioavailability in whole cells, the cell membrane with potential material residues was separated by cell lysis followed by a centrifugation step. Besides the bioavailability in the whole cell, metal-ion concentrations in the cytoplasm and the nucleus were investigated. Here, cells were separated into the soluble fractions of cytoplasm and nucleus. In both approaches, metal-ion concentration was determined by atomic absorption spectrometry or ICP-MS afterwards.

In general, with the notable exception of Ag-based materials, both cytotoxicity as well as bioavailability reflected the acellular dissolution rates in physiological lysosomal media (pH 4.5), since materials that exhibited an acellular dissolution also showed a dose-dependent cytotoxicity and bioavailability within all tested cell lines. Here, highly elevated concentrations were seen in the cytoplasm and the nucleus; particularly high concentrations in the nucleus were found in the case of Cu- and Ag-based materials, reaching millimolar concentrations.

TiO_2 and CeO_2 NP, which were insoluble in acellular lysosomal fluid, also showed neither cytotoxicity nor intracellular metal ion release. This is in agreement with a previous study showing that resorbed TiO_2 NP remained within the phagosomes of the cells without measurable ion release in the cytoplasm and caused no cellular toxicity [40].

A good correlation between solubility in artificial lysosomal fluids, cytotoxicity, and intracellular bioavailability was also evident for Cu- and Ni-based materials, showing some cell line depending differences. For CuO NP, a pronounced and dose-dependent bioavailability was seen in the Beas-2B cells, followed by A549 and dTHP-1 cells. The correlation between the solubility at a low pH and intracellular bioavailability was already described for CuO NP in two previous studies, where the dissolution in ALF with the intracellular bioavailability in A549 and Beas-2B cells was compared [15,17]. Interestingly, the bioavailability of Cu NP in the Beas-2B cells, when compared to the A549 cells, was much lower. Comparing Cu NP and Cu NW at the same doses, a higher bioavailability was seen for Cu NW, however, the toxic effects of Cu NP and Cu NW were comparable. Additionally, high concentrations of released ions from Cu NW were found in the nuclei. The observation of higher Cu ion concentration within nuclei by Cu NW compared to Cu NP may result in a higher genotoxic potential of the fiber-shaped material due to the redox potential of Cu ions. However, this hypothesis needs to be further evaluated in subsequent studies.

After incubation with Ni NP, a dose-dependent intracellular nickel ion release was evident in all cell lines, with somehow less pronounced uptake in Beas-2B cells. This is in agreement with results presented previously by Capasso and colleagues, who demonstrated that the uptake of NiO NP in A549 cells is mainly endocytosis-related, while there was no evidence for endocytotic uptake of NiO NP in Beas-2B cells [41]. The same tendency was seen in the case of Ni NW, with lower levels of deliberated metal ions in all cell lines, possibly due to the branched structure of the fibers. Recent studies have already shown that Ni NW are taken up by different cell types, such as fibroblasts [42], colon cancer cells [43], and macrophages [44] causing different toxic effects. However, this study offers a quantitative comparison of the bioavailability of Ni NP and NW in different cell lines, which has, to that extent, not been published. Interestingly, Ni NW exhibited a higher bioavailability in dTHP-1 cells when compared to the epithelial cell lines. This indicates that macrophages rapidly start to take up nanowire via phagocytosis, which has already been observed in vivo [45]. Despite the fact that high concentrations of Ni ions were also found in the nucleus of all cells after incubation with the Ni materials, it can be stated that the intracellular released Ni ions mainly remain in the cytoplasm of all cell lines. This observation strengthens results reported by Schwerdtle and colleagues, who investigated the impact of NiO MP in A549 cells [46].

One very interesting example of differences in bioavailability observed in cells and suggested solubility from acellular studies is the case of Ag-based materials. While neither Ag NP nor Ag NW showed considerable ion release, even in acidic artificial lysosomal media, Ag NP, as well as Ag NW, revealed an intracellular bioavailability at all applied concentrations. The observed intracellular bioavailability is in agreement with recent studies [24,47]. Intracellular metal ion release was highest in dTHP-1 cells, with even higher metal ion concentrations in the nucleus when compared to the cytoplasm. The discrepancy between the acellular solubility and intracellular bioavailability appears to be unique for Ag-based materials and may be explained by the fact that silver forms insoluble secondary structures due to its affinity to S- and Cl-groups [48]. These secondary structures are not quantifiable by the static solubility approach used in this study, and may not be fully quantifiable with the dynamic approach either, even though some solubility was observed in the latter test system. Thus, it cannot be excluded that also in the acellular studies, there was a release of Ag ions which bound rapidly to buffer components resulting in the formation and precipitation of these insoluble secondary particles. Within the cell, however, Ag ions may be released, leading to a dynamic equilibrium between cellular reactants, and being quantifiable within the soluble fractions of the respective compartments.

5. Conclusions

While only minor differences were seen for acellular dissolution and abiotic oxidative reactivity detected by the FRAS assay when comparing NP and NW of the same metal, their reactivity and dissolution are mostly driven by the respective metal under investigation. High solubility in acidic fluids, as models for the lysosomal environment, and pronounced reactivity was seen for Cu-based particulate and fibrous materials. Similarly, high solubility but moderate reactivity was seen for Ni NP and NW. Interestingly, in the case of Ag, no dissolution in acellular fluids was observed, probably due to the formation of insoluble secondary structures; however, an intermediate oxidative reactivity was seen for the Ag NP. CeO_2- and TiO_2-based materials exhibited no acellular dissolution and no oxidative reactivity. The dissolution behavior of the metal-based nanomaterials was strongly reflected in cellular toxicity and intracellular bioavailability. Thus, CeO_2- and TiO_2-based materials showed neither cytotoxicity nor intracellular bioavailability in either cell line, while the bioavailability which was seen for the soluble materials also correlated with the cytotoxicity of these materials. Cytotoxic effects appear to be due to intracellular dissolved metal ions followed by a metal ion overload, and not due to nanomaterial-cell interactions. This is in line with the proposed Trojan-horse type mechanism [13,49]. An interesting exception was seen in the case of Ag-based materials; here the acellular dissolution was not predictive for its cellular toxicity and bioavailability. This may be due to the formation of secondary particles formed after the dissolution of the nanomaterials, which likely precipitate in acellular systems and thus remain undetectable in the soluble fraction, but which may add to the soluble and thus bioavailable fraction in the cellular system. Concerning the different cell lines applied, differences in toxicity and bioavailability were metal-dependent, with no common pattern across the metals. In the case of Ni NW and Ag NW, a comparatively high bioavailability was seen in THP-1 cells with macrophage-like properties, supporting their higher proficiency for phagocytotic uptake.

Supplementary Materials: The following supporting information can be downloaded at: https: //www.mdpi.com/article/10.3390/nano12010147/s1. Table S1: Parameters used in the DG-model simulations; Table S2: Dose selection for in vitro studies; Table S3: Comparison of freshly prepared and thawed particle dispersions with regard to Z-Average and deposited dose fraction calculated using the DG model; Figure S1: Size distribution of freshly prepared and thawed particle dispersions at concentrations of 100 μg/mL in supplemented RPMI-1640; Figure S2: Impact of freshly prepared and thawed particle dispersions at concentrations of 100 μg/mL in supplemented RPMI-1640 on the deposited dose fraction applying the DG model; Figure S3: FRAS-Testing; Figure S4: Impact of nanomaterials on the relative cell count of A549, Beas-2B, RLE-6TN and dTHP-1 cells after incubation with three different doses; Figure S5: Dissolution in cell culture media.

Author Contributions: Conceptualization, J.W., D.A.S., W.W., U.H. and A.H.; methodology, J.W., D.A.S., R.N., M.L., F.S. (Feranika Schworm), P.S., U.H., F.S. (Florian Schulz), M.H., W.W. and A.H.; validation, J.W., D.A.S., R.N., M.L., F.S. (Feranika Schworm), P.S., F.S. (Florian Schulz), M.H., W.W. and A.H.; formal analysis, J.W., D.A.S., R.N., M.L., F.S. (Feranika Schworm), P.S., F.S. (Florian Schulz), M.H., W.W. and A.H.; investigation, J.W., D.A.S., M.L., F.S. (Feranika Schworm), R.N. and M.H.; resources, A.H. and W.W.; data curation, J.W., D.A.S., W.W. and A.H.; writing—original draft preparation, J.W. and D.A.S.; writing—review and editing, J.W., D.A.S., R.N., M.L., F.S. (Feranika Schworm), P.S., U.H., F.S. (Florian Schulz), M.H., W.W. and A.H.; visualization, J.W., D.A.S., R.N., M.L., F.S. (Feranika Schworm), P.S., F.S. (Florian Schulz), M.H., W.W. and A.H.; supervision, A.H., W.W., M.H. and P.S.; project administration, A.H.; funding acquisition, A.H. All authors have read and agreed to the published version of the manuscript.

Funding: This research was funded by the German Federal Ministry of Education and Research (BMBF), grant number 03XP0211 (MetalSafety).

Informed Consent Statement: Not applicable.

Data Availability Statement: The data presented in this study are available on request from the first (J.W., D.A.S.) and corresponding author (A.H.) for researchers of academic institutes who meet the criteria for access to the confidential data.

Acknowledgments: We would like to thank Roel Schins (IUF, Düsseldorf, Germany) and Richard Gminski (University of Freiburg, Freiburg, Germany) for providing A549 cells and THP-1 cells. Furthermore, we would like to thank the Laboratories for Electron Microscopy at KIT and BASF, especially Heike Störmer and Volker Zibat, for taking the presented REM images, and Philipp Müller and Thorsten Wieczorek for TEM images, and the Institute of Applied Geosciences at KIT, particularly Elisabeth Eiche for performing the ICP-MS measurements. Lastly, we would like to thank Julian Oppler and Timur Okkali for their technical support.

Conflicts of Interest: The authors declare no conflict of interest.

Abbreviations

AAF	Artificial Alveolar Fluid
ALF	Artificial Lysosomal Fluid
BSA	Bovine serum albumin
DLS	Dynamic light scattering
dTHP-1	Differentiated monocytic THP-1 cells to macrophage-like cells
FRAS	Ferric Reduction Ability of Serum
GF-AAS	Graphite Furnace Atomic Absorption Spectrometry
ICP-MS	Inductively Coupled Plasma-Mass Spectrometry
NP	Nanoparticle
NW	Nanowire
PDI	Polydispersity index
PSF	Phagolysosomal Simulant Fluid
RCC	Relative Cell Count
ROS	Reactive oxygen species
REM	Raster electron microscopy
TEM	Transmission electron microscopy

References

1. Bapat, R.A.; Chaubal, T.V.; Joshi, C.P.; Bapat, P.R.; Choudhury, H.; Pandey, M.; Gorain, B.; Kesharwani, P. An overview of application of silver nanoparticles for biomaterials in dentistry. *Mater. Sci. Eng. C Mater. Biol. Appl.* **2018**, *91*, 881–898. [CrossRef] [PubMed]
2. Zhang, F.; Wu, X.; Chen, Y.; Lin, H. Application of silver nanoparticles to cotton fabric as an antibacterial textile finish. *Fibers Polym.* **2009**, *10*, 496–501. [CrossRef]
3. Ren, G.; Hu, D.; Cheng, E.W.C.; Vargas-Reus, M.A.; Reip, P.; Allaker, R.P. Characterisation of copper oxide nanoparticles for antimicrobial applications. *Int. J. Antimicrob. Agents* **2009**, *33*, 587–590. [CrossRef] [PubMed]
4. Beitollahi, H.; Nekooei, S. Application of a Modified CuO Nanoparticles Carbon Paste Electrode for Simultaneous Determination of Isoperenaline, Acetaminophen and N-acetyl-L-cysteine. *Electroanalysis* **2016**, *28*, 645–653. [CrossRef]
5. Xiong, X.; Zou, C.-L.; Ren, X.-F.; Liu, A.-P.; Ye, Y.-X.; Sun, F.-W.; Guo, G.-C. Silver nanowires for photonics applications. *Laser Photonics Rev.* **2013**, *7*, 901–919. [CrossRef]
6. Reich, D.H.; Tanase, M.; Hultgren, A.; Bauer, L.A.; Chen, C.S.; Meyer, G.J. Biological applications of multifunctional magnetic nanowires (invited). *J. Appl. Phys.* **2003**, *93*, 7275–7280. [CrossRef]
7. Kuhlbusch, T.A.; Wijnhoven, S.W.; Haase, A. Nanomaterial exposures for worker, consumer and the general public. *NanoImpact* **2018**, *10*, 11–25. [CrossRef]
8. Oberdörster, G.; Oberdörster, E.; Oberdörster, J. Nanotoxicology: An emerging discipline evolving from studies of ultrafine particles. *Environ. Health Perspect.* **2005**, *113*, 823–839. [CrossRef]
9. Arts, J.H.E.; Hadi, M.; Irfan, M.-A.; Keene, A.M.; Kreiling, R.; Lyon, D.; Maier, M.; Michel, K.; Petry, T.; Sauer, U.G.; et al. A decision-making framework for the grouping and testing of nanomaterials (DF4nanoGrouping). *Regul. Toxicol. Pharmacol. RTP* **2015**, *71*, S1–S27. [CrossRef]
10. Arts, J.H.E.; Hadi, M.; Keene, A.M.; Kreiling, R.; Lyon, D.; Maier, M.; Michel, K.; Petry, T.; Sauer, U.G.; Warheit, D.; et al. A critical appraisal of existing concepts for the grouping of nanomaterials. *Regul. Toxicol. Pharmacol. RTP* **2014**, *70*, 492–506. [CrossRef]
11. Oomen, A.G.; Bleeker, E.A.J.; Bos, P.M.J.; van Broekhuizen, F.; Gottardo, S.; Groenewold, M.; Hristozov, D.; Hund-Rinke, K.; Irfan, M.-A.; Marcomini, A.; et al. Grouping and Read-Across Approaches for Risk Assessment of Nanomaterials. *Int. J. Environ. Res. Public Health* **2015**, *12*, 13415–13434. [CrossRef]
12. Delaval, M.; Wohlleben, W.; Landsiedel, R.; Baeza-Squiban, A.; Boland, S. Assessment of the oxidative potential of nanoparticles by the cytochrome c assay: Assay improvement and development of a high-throughput method to predict the toxicity of nanoparticles. *Arch. Toxicol.* **2017**, *91*, 163–177. [CrossRef] [PubMed]
13. Cronholm, P.; Karlsson, H.L.; Hedberg, J.; Lowe, T.A.; Winnberg, L.; Elihn, K.; Wallinder, I.O.; Möller, L. Intracellular uptake and toxicity of Ag and CuO nanoparticles: A comparison between nanoparticles and their corresponding metal ions. *Small (Weinh. Bergstr. Ger.)* **2013**, *9*, 970–982. [CrossRef] [PubMed]
14. Strauch, B.M.; Hubele, W.; Hartwig, A. Impact of Endocytosis and Lysosomal Acidification on the Toxicity of Copper Oxide Nano- and Microsized Particles: Uptake and Gene Expression Related to Oxidative Stress and the DNA Damage Response. *Nanomaterials* **2020**, *10*, 679. [CrossRef]
15. Semisch, A.; Ohle, J.; Witt, B.; Hartwig, A. Cytotoxicity and genotoxicity of nano—And microparticulate copper oxide: Role of solubility and intracellular bioavailability. *Part Fibre Toxicol.* **2014**, *11*, 10. [CrossRef]
16. Hsiao, I.-L.; Hsieh, Y.-K.; Wang, C.-F.; Chen, I.-C.; Huang, Y.-J. Trojan-horse mechanism in the cellular uptake of silver nanoparticles verified by direct intra- and extracellular silver speciation analysis. *Environ. Sci. Technol.* **2015**, *49*, 3813–3821. [CrossRef]

17. Strauch, B.M.; Niemand, R.K.; Winkelbeiner, N.L.; Hartwig, A. Comparison between micro- and nanosized copper oxide and water soluble copper chloride: Interrelationship between intracellular copper concentrations, oxidative stress and DNA damage response in human lung cells. *Part Fibre Toxicol.* **2017**, *14*, 28. [CrossRef]

18. Latvala, S.; Hedberg, J.; Di Bucchianico, S.; Möller, L.; Odnevall Wallinder, I.; Elihn, K.; Karlsson, H.L. Nickel Release, ROS Generation and Toxicity of Ni and NiO Micro- and Nanoparticles. *PLoS ONE* **2016**, *11*, e0159684.

19. Gliga, A.R.; Skoglund, S.; Wallinder, I.O.; Fadeel, B.; Karlsson, H.L. Size-dependent cytotoxicity of silver nanoparticles in human lung cells: The role of cellular uptake, agglomeration and Ag release. *Part Fibre Toxicol.* **2014**, *11*, 11. [CrossRef] [PubMed]

20. Peijnenburg, W.J.G.M.; Ruggiero, E.; Boyles, M.; Murphy, F.; Stone, V.; Elam, D.A.; Werle, K.; Wohlleben, W. A Method to Assess the Relevance of Nanomaterial Dissolution During Reactivity Testing. *Materials* **2020**, *13*, 2235. [CrossRef]

21. Poland, C.A.; Byrne, F.; Cho, W.-S.; Prina-Mello, A.; Murphy, F.A.; Davies, G.L.; Coey, J.M.D.; Gounko, Y.; Duffin, R.; Volkov, Y.; et al. Length-dependent pathogenic effects of nickel nanowires in the lungs and the peritoneal cavity. *Nanotoxicology* **2012**, *6*, 899–911. [CrossRef]

22. Singh, M.; Movia, D.; Mahfoud, O.K.; Volkov, Y.; Prina-Mello, A. Silver nanowires as prospective carriers for drug delivery in cancer treatment: An in vitro biocompatibility study on lung adenocarcinoma cells and fibroblasts. *Eur. J. Nanomed.* **2013**, *5*, 195–204. [CrossRef]

23. Chung, K.F.; Seiffert, J.; Chen, S.; Theodorou, I.G.; Goode, A.E.; Leo, B.F.; McGilvery, C.M.; Hussain, F.; Wiegman, C.; Rossios, C.; et al. Inactivation, Clearance, and Functional Effects of Lung-Instilled Short and Long Silver Nanowires in Rats. *ACS Nano* **2017**, *11*, 2652–2664. [CrossRef]

24. Fizeşan, I.; Cambier, S.; Moschini, E.; Chary, A.; Nelissen, I.; Ziebel, J.; Audinot, J.-N.; Wirtz, T.; Kruszewski, M.; Pop, A.; et al. In vitro exposure of a 3D-tetraculture representative for the alveolar barrier at the air-liquid interface to silver particles and nanowires. *Part Fibre Toxicol.* **2019**, *16*, 14. [CrossRef]

25. Chen, S.; Goode, A.E.; Sweeney, S.; Theodorou, I.G.; Thorley, A.J.; Ruenraroengsak, P.; Chang, Y.; Gow, A.; Schwander, S.; Skepper, J.; et al. Sulfidation of silver nanowires inside human alveolar epithelial cells: A potential detoxification mechanism. *Nanoscale* **2013**, *5*, 9839–9847. [CrossRef] [PubMed]

26. Project Sustainable Nanotechnologies (SUN). *Deliverable D 1.4 Report on Characterization of Pristine Nanomaterials for (Eco)Toxicological Testing*; EU FP-7; SUN: Santa Clara, CA, USA, 2017; pp. 1–66.

27. Keller, J.G.; Quevedo, D.F.; Faccani, L.; Costa, A.L.; Landsiedel, R.; Werle, K.; Wohlleben, W. Dosimetry in vitro—Exploring the sensitivity of deposited dose predictions vs. affinity, polydispersity, freeze-thawing, and analytical methods. *Nanotoxicology* **2021**, *15*, 21–34. [CrossRef] [PubMed]

28. DeLoid, G.; Cohen, J.M.; Darrah, T.; Derk, R.; Rojanasakul, L.; Pyrgiotakis, G.; Wohlleben, W.; Demokritou, P. Estimating the effective density of engineered nanomaterials for in vitro dosimetry. *Nat. Commun.* **2014**, *5*, 3514. [CrossRef] [PubMed]

29. DeLoid, G.M.; Cohen, J.M.; Pyrgiotakis, G.; Pirela, S.V.; Pal, A.; Liu, J.; Srebric, J.; Demokritou, P. Advanced computational modeling for in vitro nanomaterial dosimetry. *Part Fibre Toxicol.* **2015**, *12*, 32. [CrossRef] [PubMed]

30. DeLoid, G.M.; Cohen, J.M.; Pyrgiotakis, G.; Demokritou, P. Preparation, characterization, and in vitro dosimetry of dispersed, engineered nanomaterials. *Nat. Protoc.* **2017**, *12*, 355–371. [CrossRef] [PubMed]

31. Braakhuis, H.M.; Giannakou, C.; Peijnenburg, W.J.G.M.; Vermeulen, J.; van Loveren, H.; Park, M.V.D.Z. Simple in vitro models can predict pulmonary toxicity of silver nanoparticles. *Nanotoxicology* **2016**, *10*, 770–779. [CrossRef]

32. International Organisation for Standardization. *ISO/TR19057: Nanotechnologies—Use and Application of Acellular In Vitro Tests and Methodologies to Assess Nanomaterial Biodurability*; International Organisation for Standardization: Geneva, Switzerland, 2007.

33. Wohlleben, W.; Waindok, H.; Daumann, B.; Werle, K.; Drum, M.; Egenolf, H. Composition, Respirable Fraction and Dissolution Rate of 24 Stone Wool MMVF with their Binder. *Part Fibre Toxicol.* **2017**, *14*, 29. [CrossRef]

34. Gandon, A.; Werle, K.; Neubauer, N.; Wohlleben, W. Surface reactivity measurements as required for grouping and read-across: An advanced FRAS protocol. *J. Phys. Conf. Ser.* **2017**, *838*, 12033. [CrossRef]

35. European Commission; Joint Research Centre; Institute for Health and Consumer Protection. *Titanium Dioxide, NM-100, NM-101, NM-102, NM-103, NM-104, NM-105: Characterisation and Physico Chemical Properties*; Publications Office of the European Union: Luxembourg, 2014; p. 77.

36. Keller, J.G.; Graham, U.M.; Koltermann-Jülly, J.; Gelein, R.; Ma-Hock, L.; Landsiedel, R.; Wiemann, M.; Oberdörster, G.; Elder, A.; Wohlleben, W. Predicting dissolution and transformation of inhaled nanoparticles in the lung using abiotic flow cells: The case of barium sulfate. *Sci. Rep.* **2020**, *10*, 458. [CrossRef]

37. Koltermann-Jülly, J.; Keller, J.G.; Vennemann, A.; Werle, K.; Müller, P.; Ma-Hock, L.; Landsiedel, R.; Wiemann, M.; Wohlleben, W. Abiotic dissolution rates of 24 (nano)forms of 6 substances compared to macrophage-assisted dissolution and in vivo pulmonary clearance: Grouping by biodissolution and transformation. *NanoImpact* **2018**, *12*, 29–41. [CrossRef]

38. Keller, J.G.; Peijnenburg, W.; Werle, K.; Landsiedel, R.; Wohlleben, W. Understanding Dissolution Rates via Continuous Flow Systems with Physiologically Relevant Metal Ion Saturation in Lysosome. *Nanomaterials* **2020**, *10*, 311. [CrossRef] [PubMed]

39. Bahl, A.; Hellack, B.; Wiemann, M.; Giusti, A.; Werle, K.; Haase, A.; Wohlleben, W. Nanomaterial categorization by surface reactivity: A case study comparing 35 materials with four different test methods. *NanoImpact* **2020**, *19*, 100234. [CrossRef]

40. Remzova, M.; Zouzelka, R.; Brzicova, T.; Vrbova, K.; Pinkas, D.; Rőssner, P.; Topinka, J.; Rathousky, J. Toxicity of TiO$_2$, ZnO, and SiO$_2$ Nanoparticles in Human Lung Cells: Safe-by-Design Development of Construction Materials. *Nanomaterials* **2019**, *9*, 968. [CrossRef] [PubMed]

41. Capasso, L.; Camatini, M.; Gualtieri, M. Nickel oxide nanoparticles induce inflammation and genotoxic effect in lung epithelial cells. *Toxicol. Lett.* **2014**, *226*, 28–34. [CrossRef]
42. Felix, L.P.; Perez, J.E.; Contreras, M.F.; Ravasi, T.; Kosel, J. Cytotoxic effects of nickel nanowires in human fibroblasts. *Toxicol. Rep.* **2016**, *3*, 373–380. [CrossRef]
43. Perez, J.E.; Contreras, M.F.; Vilanova, E.; Felix, L.P.; Margineanu, M.B.; Luongo, G.; Porter, A.E.; Dunlop, I.E.; Ravasi, T.; Kosel, J. Cytotoxicity and intracellular dissolution of nickel nanowires. *Nanotoxicology* **2016**, *10*, 871–880. [CrossRef]
44. Byrne, F.; Prina-Mello, A.; Whelan, A.; Mohamed, B.M.; Davies, A.; Gun'ko, Y.K.; Coey, J.; Volkov, Y. High content analysis of the biocompatibility of nickel nanowires. *J. Magn. Magn. Mater.* **2009**, *321*, 1341–1345. [CrossRef]
45. Schinwald, A.; Donaldson, K. Use of back-scatter electron signals to visualise cell/nanowires interactions in vitro and in vivo; frustrated phagocytosis of long fibres in macrophages and compartmentalisation in mesothelial cells in vivo. *Part Fibre Toxicol.* **2012**, *9*, 34. [CrossRef] [PubMed]
46. Schwerdtle, T.; Hartwig, A. Bioavailability and genotoxicity of soluble and particulate nickel compounds in cultured human lung cells. *Mat.-Wiss. U. Werkstofftech.* **2006**, *37*, 521–525. [CrossRef]
47. Theodorou, I.G.; Müller, K.H.; Chen, S.; Goode, A.E.; Yufit, V.; Ryan, M.P.; Porter, A.E. Silver Nanowire Particle Reactivity with Human Monocyte-Derived Macrophage Cells: Intracellular Availability of Silver Governs Their Cytotoxicity. *ACS Biomater. Sci. Eng.* **2017**, *3*, 2336–2347. [CrossRef] [PubMed]
48. Jiang, X.; Miclăuş, T.; Wang, L.; Foldbjerg, R.; Sutherland, D.S.; Autrup, H.; Chen, C.; Beer, C. Fast intracellular dissolution and persistent cellular uptake of silver nanoparticles in CHO-K1 cells: Implication for cytotoxicity. *Nanotoxicology* **2015**, *9*, 181–189. [CrossRef] [PubMed]
49. Limbach, L.K.; Wick, P.; Manser, P.; Grass, R.N.; Bruinink, A.; Stark, W.J. Exposure of engineered nanoparticles to human lung epithelial cells: Influence of chemical composition and catalytic activity on oxidative stress. *Environ. Sci. Technol.* **2007**, *41*, 4158–4163. [CrossRef] [PubMed]

 *nanomaterials*

Article

Agglomeration State of Titanium-Dioxide (TiO$_2$) Nanomaterials Influences the Dose Deposition and Cytotoxic Responses in Human Bronchial Epithelial Cells at the Air-Liquid Interface

Sivakumar Murugadoss [1,*,†], Sonja Mülhopt [2,*,†], Silvia Diabaté [3], Manosij Ghosh [1], Hanns-Rudolf Paur [2], Dieter Stapf [2], Carsten Weiss [3,†] and Peter H. Hoet [1,†]

[1] Laboratory of Toxicology, Unit of Environment and Health, Department of Public Health and Primary Care, KU Leuven, 3000 Leuven, Belgium; manosij.ghosh@kuleuven.be (M.G.); peter.hoet@kuleuven.be (P.H.H.)
[2] Institute for Technical Chemistry, Karlsruhe Institute of Technology, 76021 Karlsruhe, Germany; hanns@dr-paur.net (H.-R.P.); dieter.stapf@kit.edu (D.S.)
[3] Institute of Biological and Chemical Systems—Biological Information Processing, Karlsruhe Institute of Technology, 76021 Karlsruhe, Germany; sildia76344@gmx.de (S.D.); carsten.weiss@kit.edu (C.W.)
[*] Correspondence: sivakumar.murugadoss@kuleuven.be (S.M.); sonja.muelhopt@kit.edu (S.M.); Tel.: +49-721-608-23807 (S.M.)
[†] The authors equal to contribution.

Citation: Murugadoss, S.; Mülhopt, S.; Diabaté, S.; Ghosh, M.; Paur, H.-R.; Stapf, D.; Weiss, C.; Hoet, P.H. Agglomeration State of Titanium-Dioxide (TiO$_2$) Nanomaterials Influences the Dose Deposition and Cytotoxic Responses in Human Bronchial Epithelial Cells at the Air-Liquid Interface. *Nanomaterials* **2021**, *11*, 3226. https://doi.org/10.3390/nano11123226

Academic Editor: David M. Brown

Received: 28 October 2021
Accepted: 25 November 2021
Published: 27 November 2021

Publisher's Note: MDPI stays neutral with regard to jurisdictional claims in published maps and institutional affiliations.

Abstract: Extensive production and use of nanomaterials (NMs), such as titanium dioxide (TiO$_2$), raises concern regarding their potential adverse effects to humans. While considerable efforts have been made to assess the safety of TiO$_2$ NMs using in vitro and in vivo studies, results obtained to date are unreliable, possibly due to the dynamic agglomeration behavior of TiO$_2$ NMs. Moreover, agglomerates are of prime importance in occupational exposure scenarios, but their toxicological relevance remains poorly understood. Therefore, the aim of this study was to investigate the potential pulmonary effects induced by TiO$_2$ agglomerates of different sizes at the air–liquid interface (ALI), which is more realistic in terms of inhalation exposure, and compare it to results previously obtained under submerged conditions. A nano-TiO$_2$ (17 nm) and a non-nano TiO$_2$ (117 nm) was selected for this study. Stable stock dispersions of small agglomerates and their respective larger counterparts of each TiO$_2$ particles were prepared, and human bronchial epithelial (HBE) cells were exposed to different doses of aerosolized TiO$_2$ agglomerates at the ALI. At the end of 4h exposure, cytotoxicity, glutathione depletion, and DNA damage were evaluated. Our results indicate that dose deposition and the toxic potential in HBE cells are influenced by agglomeration and exposure via the ALI induces different cellular responses than in submerged systems. We conclude that the agglomeration state is crucial in the assessment of pulmonary effects of NMs.

Keywords: nanomaterials; titanium dioxide; agglomerates; air-liquid interface; pulmonary toxicity

1. Introduction

Nanotechnology is ubiquitous, brings novel advancements in all aspects of human life on a daily basis, and has a wide variety of applications, such as in consumer goods, electronics, communication, environmental treatments and remediations, agriculture, nanomedicine, water purification, textiles, aerospace industry, and efficient energy sources, among many others. The field of nanotechnology is one of the fastest expanding markets in the world and its global value is expected to exceed the USD 125 billion mark by 2024 [1].

Nanomaterials (NMs) are generally defined as a material with at least one dimension in the nanoscale (1–100 nm) range [2]. While NMs are abundant in nature and produced by various sources, such as forest fires and volcanic eruptions, they are also intentionally manufactured by nanotechnologies on a global scale for industrial and commercial purposes. EU recommended a definition for NM solely for regulatory purpose, which states that "natural, incidental or manufactured material containing particles, in an unbound state or as an aggregate or as an agglomerate and where, for 50% or more of the particles in the number size distribution, one or more external dimensions is in the size range 1 nm–100 nm" [3].

Among the manufactured NMs, titanium dioxide (TiO_2) is one of the widely used NMs in commercial applications and approximately four million tons of TiO_2 are produced annually worldwide [4–6]. Commercial TiO_2 NMs come in different crystalline forms such as anatase and rutile. As TiO_2 NMs reflect UV light, they are widely used in cosmetics and in paints as a UV filter [5] as well as in plastics [7]. TiO_2 NMs are also extensively used as food colourant (food additive E171) [8]. Due to their light dependent properties, TiO_2 NMs are being studied for potential medical and bio-medical applications such as antibacterial activity, biosensing, drug delivery, and implant applications [9,10]. The production of TiO_2 NMs is expected to expand continuously due to their potential in the energy sector and environmental based applications [11]. This clearly indicates that there is a potential for human exposure, particularly inhalation, as this is the major route of exposure to TiO_2 NMs in occupational settings and raises concerns about their safety and adverse pulmonary effects [12].

Toxicological evaluations of TiO_2 NMs are often performed using in vivo models such as mice and rats. Short and long term exposure to TiO_2 NMs via inhalation induced pulmonary inflammation, fibrosis and tumours [6,13–15]. A significant increase in cytotoxicity, inflammation, oxidative stress, and DNA damage was observed in mice exposed to high doses (10 mg/kg [16] and ~4 mg/kg [17]) of TiO_2 NMs. The studied endpoints are major key events identified to play essential roles in fibrosis and tumour development [14,18].

In vitro models are often employed as a first screening method, to unveil the mechanisms involved in the induction of adverse effects, and to prioritize NMs for further animal testing. Traditionally, submerged in vitro cell cultures are widely used to assess the adverse effects of NMs with a particular focus on the production of reactive oxygen species, which can be generated specially in case of TiO_2 NMs [19]. Submerged exposure to TiO_2 NMs induced cytotoxicity, oxidative stress, pro-inflammatory responses, and genotoxicity in lung derived immortalized cell lines [6]. In submerged exposure systems, the cells are covered with culture media to which NMs are added. The biomolecules present in the culture media can adsorb to the surface of the NMs to form a protein corona [20,21]. Such changes to the surface can potentially prevent the adverse effects of NMs [22], affect the physico-chemical properties relevant for toxicological assessment (size, surface area, surface composition, surface charge, and agglomeration, etc.) [23], and also effective density [24], an important parameter that determines the sedimentation of NMs. However, these modifications of NMs in cell culture medium often do not reflect the conditions upon inhalation in real life situations.

Recently, exposure at the air liquid interface (ALI) has been evolving as a potential alternative to conventional submerged in vitro exposure systems. At the ALI, cells grown on transwell plates are directly exposed to aerosolized particles and gases, which better reflects the exposure in vivo via inhalation [25–28]. Previously, we have developed, validated, and used a fully integrated ALI exposure system for the assessment of toxicological effects of various NMs and aerosols [29–33].

It is well known that the physicochemical properties of NMs influence their toxicity [34]. Among all the properties, the influence of agglomeration on the toxicity of NMs is less well studied and poorly understood. In our previous study, we assessed the influence of TiO_2 NM agglomeration on (cyto) toxicity and biological responses [35] using a human bronchial epithelial (HBE) cell line. However, the entire study was carried out in submerged exposure conditions. While there is only a limited number of toxicological investigations addressing adverse effects of TiO_2 NMs using ALI exposure systems [31,36,37], the impact of agglomeration has not been researched. Here, we prepared TiO_2 NM agglomerates of different sizes and performed toxicological studies employing ALI exposure. The aim of the present work was to investigate the cytotoxicity and biological responses in HBE cells after ALI exposure to different doses of TiO_2 agglomerates of different sizes and compare the results to those previously obtained under submerged conditions [35].

2. Materials and Methods

2.1. Preparation of Dispersions and Characterization of TiO₂ NMs

Two TiO_2 NMs (representative test materials) of different primary sizes were kindly provided by the European Commission's Joint Research Centre (JRC, Ispra, Italy). Mean primary size of TiO_2-JRCNM10202a was determined as 17 nm and TiO_2-JRCNM10200a as 117 nm. Therefore, the two NMs are indicated as 17 nm and 117 nm sized TiO_2 in the text. Both TiO_2 NMs are pristine and anatase in nature. Detailed physicochemical characterization of these NMs were provided in a previously published JRC report [38].

Detailed information on the development of the dispersion protocol to obtain two different agglomeration states (small and large agglomerates) of both TiO_2 NMs were published elsewhere [17,35]. Briefly, to obtain agglomerates of different sizes, particles were dispersed in different pH conditions (2 and 7), the dispersions were probe sonicated (7056 J) and stabilized with 1% bovine serum albumin (BSA). After stabilization, the suspensions at pH 2 were readjusted to pH 7–7.5 by slowly adding sodium hydroxide solution (NaOH). While the original dispersion protocol was developed to prepare 10 mL of stock dispersions and intended for submerged exposure, for ALI exposure, the quantity was scaled up to 120 mL to provide sufficient quantity of dispersions for the aerosolization of TiO_2 agglomerates during the 4 h exposure period. Each dispersion was freshly prepared before each exposure. Table 1 shows the nomenclature of different dispersions.

Table 1. Nomenclature of TiO_2 agglomerate dispersions.

	17 nm-SA	17 nm-LA	117 nm-SA	117 nm-LA
Particle type	JRCNM10202a	JRCNM10202a	JRCNM10200a	JRCNM10200a
Primary particle diameter	17 nm	17 nm	117 nm	117 nm
Agglomeration state	Small agglomerates	Large agglomerates	Small agglomerates	Large agglomerates

2.2. Cell Culture Maintenance

The human bronchial epithelial cell line (16HBE14o- or HBE) was kindly provided by Dr. Gruenert (University of California, San Francisco, CA, USA). HBE cells were cultured in DMEM/F12 supplemented with 5% fetal bovine serum (FBS), 1% penicillin-streptomycin (P-S) (100 U/mL), 1% L-glutamine (2 mM) and 1% fungizone (2.5 g/mL). All cell culture supplements were purchased from Invitrogen (Merelbeke, Belgium) unless otherwise stated. Cells were cultured in T75 flasks at 37 °C in a 100% humidified air containing 5% CO_2. Medium was changed every 2 or 3 days and cells were passaged every week (7 days). Cells from passage 4 to 8 were used for the experiments.

2.3. Air–Liquid Interface Exposure

For ALI exposure, 3.5×10^5 HBE cells/mL were seeded on the apical side of a 6-well transwell plate (Corning Costar Transwell insert membranes type 3450, culture area 4.67 cm^2, pore size 0.4 µm, cat no 10619141, Fischer scientific, Schwerte, Germany) with 1.5 mL of cell culture medium on the basolateral side and incubated overnight at 5% CO_2 and at 37 °C. Before ALI exposure, the apical and the basolateral media were removed. Then, cell culture medium was added into the basolateral compartment and the apical side was left uncovered (no medium). Uncovered cells were exposed to "clean air" (humidified synthetic air as negative control) or in parallel to airborne TiO_2 agglomerates at low dose without electrostatic deposition and at different levels of electrostatic deposition (400 V, 800 V, and 1200 V) for 4h to facilitate dose–response evaluation of different biological endpoints. After exposure, the medium at the basal side of the transwell inserts was collected for LDH analysis.

2.4. Aerosol Generation and Characterization

Figure 1 shows the layout of the ALI exposure system. For TiO_2 aerosol generation, a setup according to the VDI guideline 3491 (Technical Division Environmental Measurement Technologies, 2016) was used. The TiO_2 dispersions, continuously stirred during the experiment, were sprayed in a drying reactor with a silica gel fill along the walls using a two-substance nozzle (model 970, Düsen-Schlick GmbH, Untersiemau/Coburg, Germany). The dry TiO_2 aerosol was regularly characterized for the number size distribution in the drying reactor using a Scanning Mobility Particle Sizer SMPS + C (Grimm Aerosol GmbH, Ainring, Germany) and directed to the ALI exposure system, as described by Mülhopt et al. [30]. In the conditioning reactor of the ALI exposure system, the TiO_2 aerosol is tempered to 37 °C and humidified to 85% r. h., and then sampling streams are directed to the single exposure chambers containing the cell cultures using an exposure flow rate of 100 mL/min. For describing the aerosol state as exposed to the cell cultures, the number size distribution was also measured by a Scanning Mobility Particle Sizer U-SMPS (Palas GmbH, Karlsruhe, Germany) sampled in the aerosol conditioning reactor. Every 5 min a scan was performed; means and standard deviation were calculated from all scans of an experiment in each channel and corrected regarding sampling losses according to Asbach et al. [39].

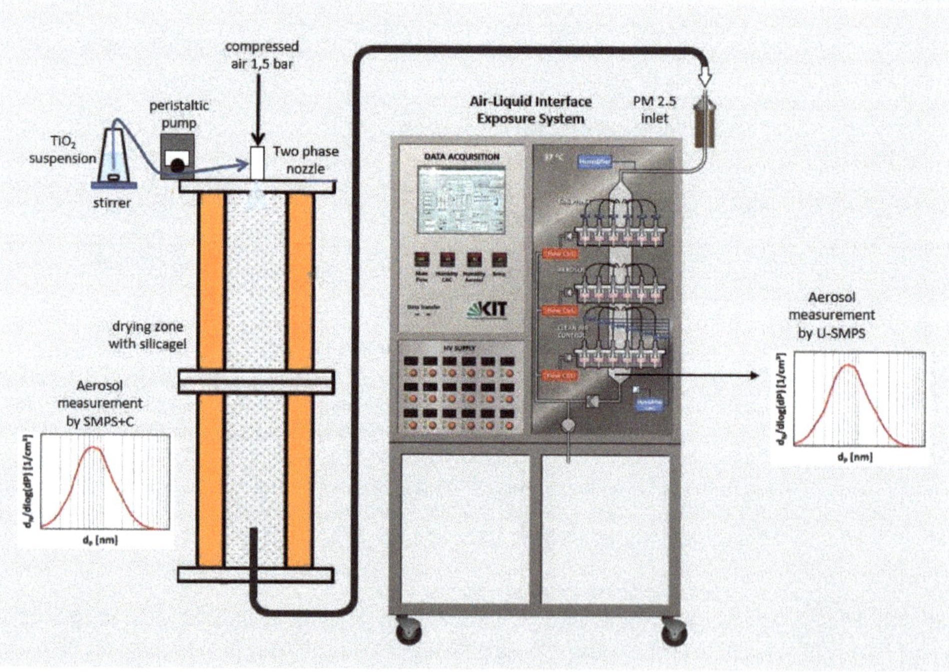

Figure 1. Experimental setup: generation of airborne TiO_2 agglomerates in dry air according to VDI guideline 3491 and exposure of human lung cells in the Air–Liquid-Interface Exposure System with accompanying measurement of particle size distribution using Scanning Mobility Particle Sizer (SMPS) at the dry stage in the reactor as well as for the aerosol inside the exposure system.

2.5. Determination of the Deposited Dose

The deposited cell culture surface dose is reflected by the deposited fraction of the TiO_2 agglomerates exposed as aerosol towards the cells and not easy to determine. For this reason, three different methods were applied: the online monitoring of mass dose using the quartz crystal microbalance QCM (Vitrocell Systems GmbH, Waldkirch, Germany) [29], the image analysis of exposed TEM grids as presented in [40] and the calculation from the SMPS measured number size distribution as shown in [30]. The effective density of all TiO_2 agglomerates were used as determined and reported in [35], and did not differ much between the dispersions ((17nm-SA: 1.55 g/cm^3), (17nm-LA: 1.48 g/cm^3), (117nm-SA: 1.78 g/cm^3), and (17nm-LA: 1.78 g/cm^3)).

2.6. Metabolic Activity

Metabolic activity was evaluated as a measure of cell viability using the WST-1 assay (Merck, Overijse, Belgium). At the end of ALI exposure, cells were washed with HBSS and incubated with 500 µL of WST1 reagent (diluted in HBSS at the ratio of 1:10) for 45 min. At the end of incubation, 100 µL was transferred to a 96 well plate and optical density was recorded at 450 nm. Sample OD values were subtracted from blank OD values and results were expressed as percentage of negative control cells. Cells exposed to clean air were treated as negative control and Triton X-100 (0.1%) lysed cells were treated as positive control (data not shown).

2.7. Membrane Integrity

Lactate dehydrogenase (LDH) activity in the cell culture supernatant was measured as an indicator of membrane damage. Briefly, 100 µL of cell culture medium collected at the basal side at the end of ALI exposure were transferred to a 96 well plate and incubated with LDH mixture (prepared as indicated in the manufacturer's protocol, Sigma-Aldrich, Taufkirchen, Germany, cat no 11644793001) and the optical density (OD) was recorded at 490 nm. Sample OD values were subtracted from blank OD values and results were expressed as percentage of Triton X-100 (0.1%) lysed cells. Cells exposed to clean air were treated as negative control.

2.8. Total Glutathione Measurements

Reduced glutathione (GSH) depletion was measured as an indicator of oxidative stress induction. Briefly, exposed cells were scraped, transferred into Eppendorf tubes and centrifuged at $150\times g$ for 5 min. Then, the supernatants were discarded and cells were resuspended in 1 mL of phosphate-buffered saline (PBS). After centrifugation, PBS was removed and 450 µL of 10 mM hydrochloric acid (HCL) was added to each tube. Cell lysis was performed by the freeze thawing procedure (15 min freezing, 15 min thawing for two times) and immediately protein content analysis (by BCA assay) was performed using 10 µL of the cell lysate. Then, the lysate was resuspended in 6.5% 5-sulfosalicylic acid (SSA), incubated on ice for 10 min and centrifuged at $20{,}800\times g$ (14,000 rpm) for 10 min at 4 °C to precipitate the proteins. The supernatants were stored at -80 °C for later GSH determination. GSH was measured using a glutathione detection kit (Enzo life sciences, Brussels, Belgium). Cells exposed to clean air were treated as negative control.

2.9. DNA Damage

Briefly, at the end of the ALI exposure, the cells were detached with trypsin, centrifuged at $250\times g$ for 5 min, suspended in the storage buffer, composed of sucrose 85.5 g/L, dimethyl sulfoxide (DMSO) 50 mL/L prepared in citrate buffer (11.8 g/L), pH 7.6, and immediately frozen at -80 °C. DNA strand breaks were measured using the alkaline comet assay kit (Trevigen, C.No.4250–050-K, Gaithersburg, MD, USA) according to the manufacturer's protocol. Fifty cells per gel were measured. Cells exposed to clean air were treated as negative control and cells treated with methyl methane sulfonate (Merck, Overijse, Belgium; 100 µM for 1h) served as positive control (data not shown). Results were expressed as percentage of DNA in the tail. NMs can interfere with the comet assay [41]. To evaluate this, we mixed TiO_2 NMs with clean air exposed cells (negative control) and cells treated with MMS (positive control), performed the comet assay, and compared the results with negative and positive controls prepared without TiO_2 NMs. The results indicated that the TiO_2 NMs at high concentrations (100 µg/mL) did not interfere with the assay.

2.10. Statistical Analysis

Two or three independent experiments were performed with six replicates each and data was presented as mean ± standard deviation (SD). Using GraphPad prism 7.04 for windows, GraphPad Software (7.04, La Jolla, CA, USA), www.graphpad.com (accssed on 24 November 2021), the results were analysed with one-way ANOVA followed by a Dunnett's multiple comparison test to determine the significance of differences compared with control.

3. Results

3.1. Size Characterization in Stock Suspensions

We obtained four agglomerate dispersions from two TiO_2 NMs of different sizes. Detailed information on the physicochemical characterization and methods used to characterize the agglomerates in stock dispersions were published elsewhere [17,35]. Using a standardized TEM technique in our previous study [42], we measured the size of several thousand agglomerates in each dispersion. The TEM based determination of the diameter (median feret min) indicated that the size of 17 nm sized TiO_2 NMs in their least agglomerated form (indicated as 17 nm-SA) was 33 nm, while it was 120 nm for the strongly agglomerated condition (17 nm-LA). The sizes of small (117 nm-SA) and large agglomerates (117 nm-LA) of 117 nm sized TiO_2 NMs were 148 and 309 nm, respectively (see Figure S1 and Table S1). In summary, at low pH agglomeration of TiO_2 NMs was modest whereas at neutral pH strong agglomeration of the small (17 nm) and less pronounced for the larger (117 nm) TiO_2 NM is observed [17,35].

3.2. Aerosol Characterization and Determination of Deposited Dose

The particle number size distributions (Figure 2) showed nearly the same characteristics for 17 nm-SA and 117 nm-LA with modal values x_M of 72 and 71 nm, respectively. The other titania 17 nm-LA and 117 nm-SA were also nearly identical with a size of x_M = 144 and 139 nm, respectively. All particle number size distributions have a typical geometric standard deviation σ_{geo} in the range of 2. These results show a similar trend as our previously reported TEM sizes (Supplementary Material Figure S1, [35]) for the different agglomerates in the stock solutions except for 117 nm-LA. Comparing SMPS and primary TEM data from stock solutions, the aerosol processing may cause differences for the size determination of 117 nm-LA agglomerates as the aerosol is characterized with the SMPS under nearly dry conditions, whereas for submerged exposure and subsequent TEM analysis, the water content and media components might increase the size of agglomerates. In addition, the SMPS measurements only cover the range of 10 to 800 nm and neglect possible larger agglomerates.

In Table 2, the summary of all measurements regarding TiO_2 aerosol characteristics is listed. The mass concentration is calculated from the number size distribution of the SMPS measurements. The QCM was operated without electrostatic deposition delivering the diffusional doses as listed. Image evaluation of exposed TEM grids provides deposited surface doses for the 17 nm-LA and 117 nm-SA, both corresponding very well with the QCM data in the case of diffusional deposition (Supplementary Material, Figure S2). The enhanced doses for electrostatic deposited agglomerates were evaluated for the 17 nm-LA and 117 nm-SA from the exposed TEM grids. For these two types of agglomerates, the dose enhancement factors were determined. In case of 117 nm-SA, deposition enhancement factors of 5 for 400 V, 12 for 800 V, and 13 for 1200 V were calculated. Similarly, in case of 17 nm-LA deposition enhancement factors of 4 for 400 V, 9 for 800 V, and 9 for 1200 V are derived. For both types of agglomerates, the relative increase in deposition becomes less with the increase of the electrostatic field strength. This saturation behavior has been also shown earlier [40] and occurs when all charged agglomerates are deposited. In the

TEM images of 17 nm-SA and 117 nm-LA individual particles could not be unambiguously identified due to a strong background signal, (Supplementary Material Figure S2), the doses listed in the summary table were calculated on the basis of the QCM data (measured at 0 V) multiplied by the enhancement factors derived at the different voltages for the corresponding particle types 17 nm-LA and 117 nm-SA.

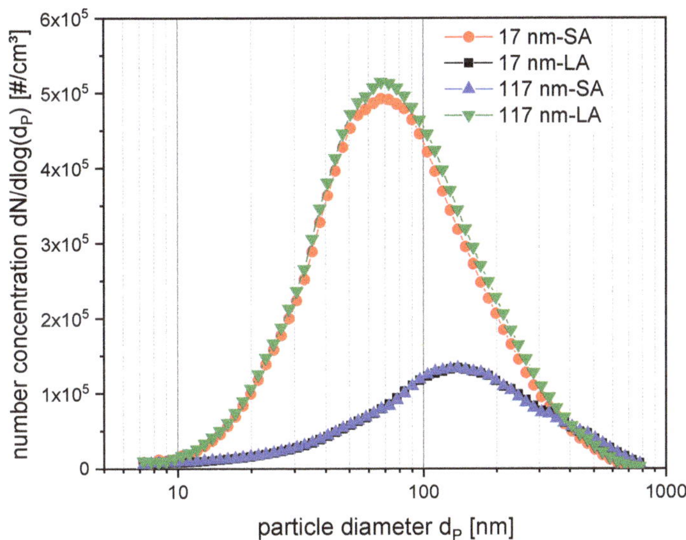

Figure 2. Particle size distributions of TiO_2 agglomerates measured by Scanning Mobility Particle Sizer U-SMPS in the range of 8 to 800 nm. Each curve shows the means of number size distributions in dependence of particle type and agglomeration state. Red circles: 17 nm-SA; black squares: 17 nm-LA; blue triangles: 117 nm-SA; and green inverted triangles: 117 nm-LA.

Table 2. Characteristics of TiO_2 aerosols and measured or calculated deposited surface doses on cell cultures.

Material	17 nm-SA	17 nm-LA	117 nm-SA	117 nm-LA
Modal value x_M [nm]	72 ± 7	144 ± 12	139 ± 10	71 ± 2
Geometric standard deviation σ_{geo}	2.0 ± 0.04	2.0 ± 0.12	2.0 ± 0.1	2.1 ± 0.04
Total number concentration [#/cm³]	$4.0 \times 10^5 \pm 7.3 \times 10^4$	$1.1 \times 10^5 \pm 3.1 \times 10^4$	$1.1 \times 10^5 \pm 1.5 \times 10^4$	$4.25 \times 10^5 \pm 2.5 \times 10^4$
Mass concentration [a] c_m [mg/m³]	1.6 ± 0.58	1.7 ± 0.45	1.9 ± 0.48	2.0 ± 0.47
Diffusional dose (0 V EF) QCM-signal [µg/cm²]	3.40 ± 0.89	1.62 ± 0.40	1.88 ± 0.62	2.89 ± 0.49
Diffusional dose (0 V EF) TEM analysis [µg/cm²]	n.a. [b]	1.78 ± 0.73	1.84 ± 0.58	n.a. [b]
Increased dose (400 V EF) TEM analysis [µg/cm²]	12.2 [c]	6.6 ± 0.82	9.15 ± 1.44	14.4 [d]

[a] calculated from SMPS data, [b] n.a. = not analyzed as particles could not be clearly identified by TEM analysis, [c] calculated on the basis of QCM data (0 V EF) multiplied with the corresponding factor for enhanced deposition at the different voltages as determined for the 17 nm-LA, [d] calculated on the basis of QCM data (µg/cm², 0 V EF) multiplied with the corresponding factor for enhanced deposition at the different voltages as determined for the 117 nm-SA.

3.3. Cytotoxicity: Effect on Metabolic Activity and LDH Release

After 4 h exposure to aerosolized TiO_2 agglomerates at the ALI, we measured the effect on cell metabolic activity using the WST-1 assay (Figure 3). Significant loss of metabolic activity (~20%) was observed in cell cultures exposed to smaller agglomerates of 17 nm sized TiO_2 (17 nm-SA) at the highest dose deposited ~30 µg/cm² (800 and 1200 V) while their larger counterparts (17 nm-LA) induced significant increase in metabolic activity (~25%) at the lowest (~1.6 µg/cm²) and at the highest doses (~16 µg/cm²) deposited. Although an increasing trend in the metabolic activity can be seen (Figure 3C,D) compared to their clean air controls, small (117 nm-SA) and large agglomerates (117 nm-LA) of 117 nm sized TiO_2 did not affect the metabolic activity significantly even at the highest doses deposited (~24 and 38 $\mu g/cm^2$, respectively). Subsequently, we measured the LDH activity in the supernatant (basal media) of the cells exposed at the ALI (Figure 4). Compared to clean air exposed controls, a trend of dose dependent increase in LDH activity was noticed for all TiO_2 agglomerates.

3.4. Oxidative Stress and DNA Damage

We measured GSH depletion as an indicator of oxidative stress induction (Figure 5). We detected significant and similar decrease of GSH for 17 nm-SA at doses ~12 and 30 $\mu g/cm^2$ while a slight but statistically significant increase in glutathione was noticed for 17 nm-LA only at the highest dose (~16 $\mu g/cm^2$) deposited. There was a non-significant increase of glutathione for both agglomerates of 117 nm sized TiO_2 (Figure 5C,D). We assessed the DNA strand breaks as a measure of DNA damage using the alkaline comet assay (Figure 6). An increasing trend in DNA damage was noticed for 17 nm-SA and 17 nm-LA which, however, was not significant compared to unexposed controls (Figure 6A,B). Both SA and LA of 17 nm TiO_2 NMs induced significant increase in DNA damage only at mass doses ~24 and 38 $\mu g/cm^2$, respectively (Figure 6C,D).

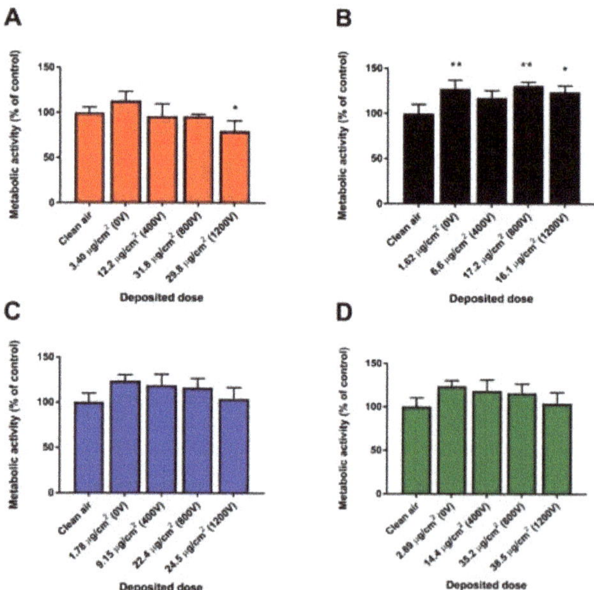

Figure 3. Effect on metabolic activity of HBE cells after 4 h exposure to TiO_2 agglomerates at the ALI. 17 nm-SA (**A**), 17 nm-LA (**B**), 117 nm-SA (**C**), and 117 nm-LA (**D**). Data are expressed as means $\pm$ SD from three independent experiments with six replicates each. $p < 0.05$ (*) and $p < 0.01$ (**) represent significant difference compared to control (One-way ANOVA followed by Dunnett's multiple comparison test).

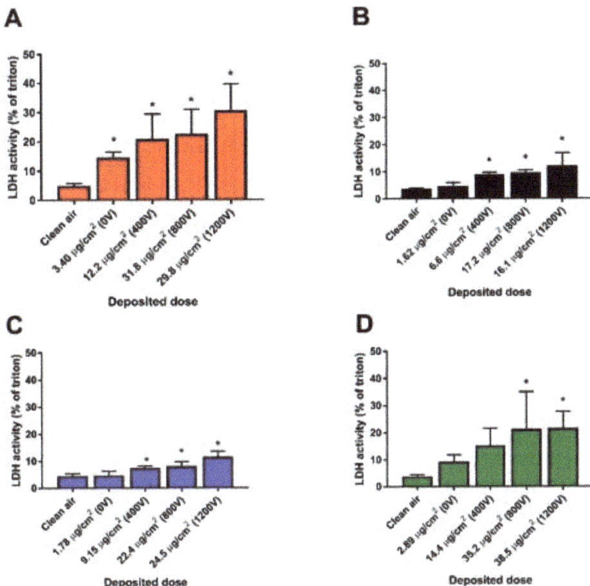

Figure 4. LDH activity measured in HBE cell supernatants after 4 h exposure to TiO_2 agglomerates at the ALI. 17 nm-SA (**A**), 17 nm-LA (**B**), 117 nm-SA (**C**), and 117 nm-LA (**D**). Data are expressed as means ± SD from three independent experiments with six replicates each. $p < 0.05$ (*) represent significant difference compared to control (One-way ANOVA followed by Dunnett's multiple comparison test).

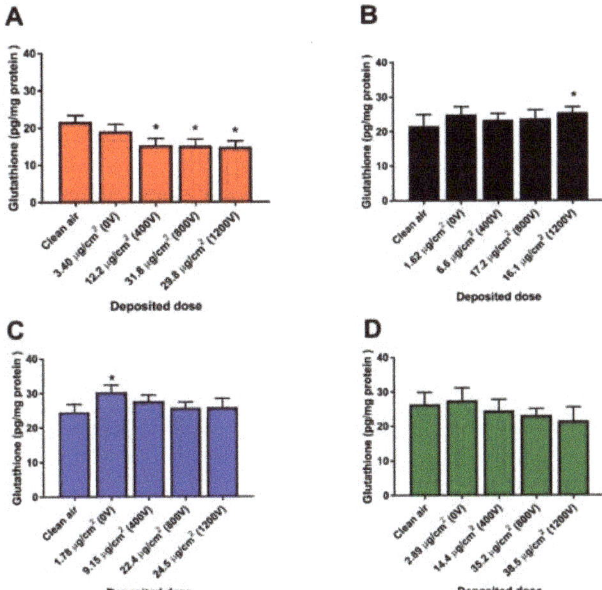

Figure 5. Glutathione levels measured in HBE cells after 4 h exposure to TiO_2 agglomerates at the ALI. 17 nm-SA (**A**), 17 nm-LA (**B**), 117 nm-SA (**C**), and 117 nm-LA (**D**). Data are expressed as means ± SD from two independent experiments with six replicates each. $p < 0.05$ (*) represent significant difference compared to control (One-way ANOVA followed by Dunnett's multiple comparison test).

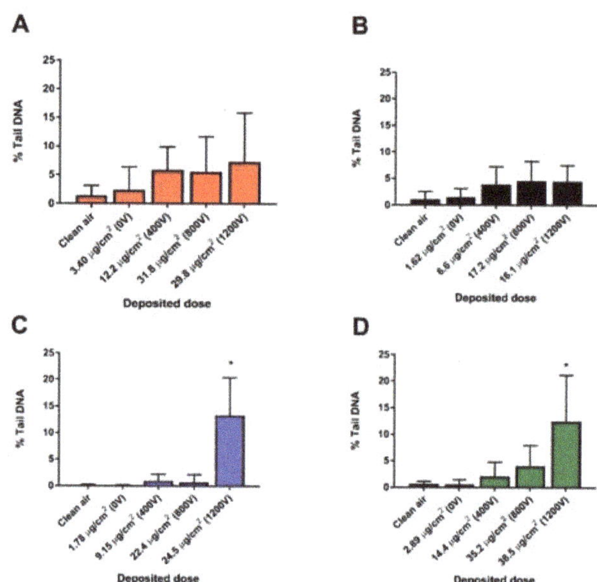

Figure 6. DNA damage measured in HBE cells after 4 h exposure to TiO_2 agglomerates at the ALI. 17 nm-SA (**A**), 17 nm-LA (**B**), 117 nm-SA (**C**), and 117 nm-LA (**D**). Data are expressed as means ± SD from two independent experiments with six replicates each. $p < 0.05$ (*) represent significant difference compared to control (One-way ANOVA followed by Dunnett's multiple comparison test).

3.5. Summary of Biological Responses

Although there is an overlap of the different deposited doses of the various TiO_2 agglomerates, for a convenient comparison, we rather considered the significant lowest observed adverse effect concentration for each endpoint to determine the toxic potency of agglomerates (Table 3). Increase in LDH activity and decrease in glutathione was noticed for 17 nm-SA even at low doses (3.4 and 12.2 $\mu g/cm^2$, respectively), which could possibly lead to a decrease in metabolic activity at the higher dose (30 $\mu g/cm^2$), but no such effects were noted for other agglomerates. These results indicate that the smaller agglomerates of nano-sized TiO_2 are more potent in terms of cytotoxicity and oxidative stress induction at the ALI. However, when considering DNA damage at the different deposited doses, agglomerates of non-nano sized TiO_2, small agglomerates in particular, appear to be more potent compared to agglomerates of nano-sized TiO_2.

Table 3. Significant lowest observed adverse effect concentration of different TiO_2 agglomerates observed for different biological endpoints at the ALI. "-" indicates no significant effect could be detected.

Dispersions	Highest Dose Deposited ($\mu g/cm^2$)	Decrease in Metabolic Activity ($\mu g/cm^2$)	Increase in LDH Activity ($\mu g/cm^2$)	Decrease in Glutathione ($\mu g/cm^2$)	Increase in DNA Damage ($\mu g/cm^2$)
17nm-SA	30	30	3.4	12.2	-
17nm-LA	16	-	6.5	-	-
117nm-SA	24.5	-	9	-	24.5
117nm-LA	38.5	-	35	-	38.5

Table 4 shows significant lowest observed adverse effect concentrations determined from our previously published study [35] for different endpoints in HBE cells exposed in submerged conditions. None of the agglomerates did induce significant cytotoxic effects at the tested doses but significant decrease in glutathione was noticed for all the agglomerates only at the dose of 155 $\mu g/cm^2$. Large agglomerates of 117 nm-SA induced DNA damage at the dose of 13 $\mu g/cm^2$ while other agglomerates induced such effects at the dose $\geq$ 26 $\mu g/cm^2$, indicating that the large agglomerates of non-nano sized TiO_2 are more potent in terms of DNA damage.

Table 4. Significant lowest observed adverse effect concentration of different TiO_2 agglomerates observed for different biological endpoints at the submerged exposure system (from our previously published study). "-" indicates no significant effect could be detected.

Dispersions	Highest Dose Tested ($\mu g/cm^2$)	Decrease in Metabolic Activity ($\mu g/cm^2$)	Increase in LDH Activity ($\mu g/cm^2$)	Decrease in Glutathione ($\mu g/cm^2$)	Increase in DNA Damage ($\mu g/cm^2$)
17nm-SA	155	-	-	155	52
17nm-LA	155	-	-	155	26
117nm-SA	155	-	-	155	26
117nm-LA	155	-	-	155	13

4. Discussion

Poor correlation of conventional in vitro and in vivo nanotoxicological exposure studies has been urging the development and validation of models that more closely represent the physiological responses of inhalation exposure. Compared to conventional submerged in vitro systems, air–liquid interface (ALI) exposures are shown to better mimic the inhalation exposure as cell cultures grown at the ALI are exposed to aerosolized particles. However, a deeper understanding of the behavior of NMs in relation to their physicochemical characteristics within the ALI system is essential. In this study, we investigated the influence of TiO_2 NM agglomeration on their deposition and cytotoxic potency in the ALI system. Our results indicated that dose deposition and their cytotoxic potential are influenced by TiO_2 agglomeration, particularly for nano-sized TiO_2.

In the current study, we could deposit in the absence of an EF mass doses of 1.6–3.4 $\mu g/cm^2$, which could be further enhanced in the presence of an EV to 29.8–38.5 $\mu g/cm^2$. Hence, in contrast to submerged exposure where the deposited dose of nano- and non-nano-sized NMs varies drastically, at the ALI similar doses independent of particle size could be deposited as agglomerates. This allows a more direct comparison of dose–response relationships without the need of additional modelling as required under submerged conditions. In a previous study, using the same ALI system, 0.17 $\mu g/cm^2$ and nearly 1.14 $\mu g/cm^2$ was deposited at 0 and 1000 V, respectively, for the same exposure duration using another non-agglomerated nano TiO_2 (NM-105) [31]. This indicates that the type of TiO_2 NMs and their agglomeration state influences the dose which is deposited.

We noticed that the smaller agglomerates of nano-sized TiO_2 NMs induced significant cytotoxicity and oxidative stress at the ALI at low doses (dose < 13 $\mu g/cm^2$) while agglomerates of 17 or 117 nm sized TiO_2 NMs induced oxidative stress, but no cytotoxicity, under submerged exposure conditions only at the highest dose tested (~155 $\mu g/cm^2$). In the case of DNA damage, small agglomerates of non-nano sized TiO_2 NMs appear to be more potent at the ALI while large agglomerates of non-nano sized TiO_2 NMs were found to be the most potent in submerged exposure conditions. These results indicate that the degree of agglomeration influences the potency of TiO_2 NMs to damage DNA in HBE cells differentially at the ALI and in submerged conditions.

Here, we found that the small agglomerates of nano-sized TiO_2 NMs (agglomerate size < 100 nm) are more potent in terms of cytotoxicity and oxidative stress induction at the ALI compared to submerged exposure conditions. Noël et al. exposed rats to 7 mg/m^3 of small (31 nm) and large agglomerates (194 nm) of TiO_2 NMs for 6h and noticed a significant increase in LDH activity and 8-isoprostane concentration in BALF, which are markers for cytotoxic and oxidative stress effects, respectively [43]. In another study of the same group, rats were exposed to 20 mg/m^3 of small (29, 28 and 35 nm) and large (156, 128 and 135 nm, respectively) agglomerates obtained from differently sized TiO$_2$ NMs (primary size of 5, 20, and 50 nm, respectively) for 6 h [44]. The results indicated that, only the small agglomerates (size < 100 nm) of 5 nm sized TiO$_2$ NMs caused a significant increase in cytotoxic effects while the small agglomerates (size < 100 nm) of all TiO$_2$ NMs, irrespective of primary particle size, induced a significant increase in oxidative damage compared to larger agglomerates (size > 100 nm), which showed no significant effects for these endpoints. These in vivo results are in agreement with our recent but also previous findings [31], indicating that ALI exposure systems are more suitable than submerged exposure assays to recapitulate adverse effects upon inhalation of NMs. Furthermore, it is of utmost importance to deposit doses in the range of ng to maximally a few ug of nanomaterials per cm^2 cellular surface area to recapitulate exposure of humans upon inhalation as outlined previously [25].

Recently, the European Food Safety Authority (EFSA) concluded that the use of TiO$_2$ as a food additive is no longer considered safe, which highlights the importance to investigate adverse effects of nano-TiO$_2$, genotoxic effects in particular [45]. In our previous study, we noticed that the small and large agglomerates of both TiO$_2$ NMs used in this study induced DNA damage in HBE cell cultures exposed at submerged conditions at a dose range of <50 µg/cm^2 (see Table 3) without inducing a significant increase in cytotoxicity and oxidative stress [35]. In this study, both agglomerates of 117 nm TiO$_2$ NMs induced DNA damage at the ALI within this dose range also without inducing a significant increase in cytotoxicity and oxidative stress. Our previous study and others have shown that the TiO$_2$ NMs were internalized by bronchial epithelial cells in submerged culture [35,46] and such internalized NMs can induce primary DNA damage by directly interacting with DNA, without the induction of cytotoxicity or oxidative stress. In the case of ALI, post exposure incubation for longer periods (such as 24 h) are needed to verify whether the induced DNA damage causes a difference in cell viability or oxidative stress. In contrast, small agglomerates of 17 nm sized TiO$_2$ NMs provoked cytotoxicity and oxidative stress but no DNA damage at the same doses. This indicates that genotoxic effects of TiO$_2$ NMs are impacted by their agglomeration state, as non-agglomerated TiO$_2$ NMs of a modal diameter of 47 nm induced DNA damage already at 1.12 µg/cm^2 [31]. Moreover, our results further suggest that the genotoxicity of submicron sized TiO$_2$ particles or their agglomerates should be also considered in the future.

5. Conclusions

In this study, we investigated the influence of agglomeration on the deposition and cytotoxic potency of TiO$_2$ NMs at the ALI. Our results indicate that dose deposition and the cytotoxic potential are influenced by agglomeration, particularly for nano-sized TiO$_2$ particles. This suggests that the agglomeration state of NMs is crucial in the assessment of pulmonary effects of NMs. Our findings also show that exposure via the ALI induces different cellular responses compared to exposure in submerged systems. More attention should be paid to the methods used to prepare the dispersions of TiO$_2$ NMs, specifically concerning agglomeration, in order to assess the (nano) effects at the air-liquid interface and to better predict the hazardous potential of NMs upon inhalation.

Supplementary Materials: The following are available online at https://www.mdpi.com/article/10.3390/nano11123226/s1. Figure S1: Representative TEM micrographs of freshly prepared TiO$_2$ stock dispersions, Figure S2: TEM micrographs of aerosolized TiO$_2$ agglomerates. Table S1: Characterization of freshly prepared TiO$_2$ stock dispersions.

Author Contributions: Conceptualization, S.M. (Sivakumar Murugadoss), S.M. (Sonja Mülhopt), S.D., M.G., H.-R.P., C.W. and P.H.H.; Data curation, S.M. (Sivakumar Murugadoss); Formal analysis, S.M. (Sivakumar Murugadoss); Funding acquisition, P.H.H.; Investigation, S.M. (Sivakumar Murugadoss); Methodology, S.M. (Sivakumar Murugadoss) and S.M. (Sonja Mülhopt); Project administration,

P.H.H.; Software, S.M. (Sivakumar Murugadoss); Supervision, D.S. and P.H.H.; Visualization, S.M. (Sivakumar Murugadoss), S.M. (Sonja Mülhopt), and P.H.H.; Writing—original draft, S.M. (Sivakumar Murugadoss), S.M. (Sonja Mülhopt), C.W. and P.H.H.; and Writing—review and editing, S.M. (Sivakumar Murugadoss), S.M. (Sonja Mülhopt), S.D., M.G., H.-R.P., C.W. and P.H.H. All authors have read and agreed to the published version of the manuscript.

Funding: This work was funded by Post-Doctoral Mandates KU Leuven internal funding (PDM/20/162), the Belgian Science Policy (BELSPO) program "Belgian Research Action through Interdisciplinary Network (BRAIN-be)" for the project "Towards a toxicologically relevant definition of nanomaterials (To2DeNano)" and EU H2020 project (H2020-NMBP-13-2018 RIA): RiskGONE (Science-based Risk Governance of NanoTechnology) under grant agreement no 814425.

Data Availability Statement: The primary data processed and presented in this study are available on request from the corresponding author.

Acknowledgments: We thank Sonja Oberacker for her assistance within this study. The authors thank JRC Nanomaterials Repository, Italy for providing the TiO_2 NMs. We acknowledge support by the KIT-Publication Fund of the Karlsruhe Institute of Technology.

Conflicts of Interest: The authors declare no conflict of interest.

References

1. Grewal, D.S. Funding Nanotechnology—A Comparative Study of Global and National Funding. *J. Nanom. Nanos. Tech.* **2019**, *105*, 1–10.
2. Oberdörster, G.; Oberdörster, E.; Oberdörster, J. Nanotoxicology: An Emerging Discipline Evolving from Studies of Ultrafine Particles. *Environ. Health Perspect.* **2005**, *113*, 823–839. [CrossRef] [PubMed]
3. Potocnik, J. Air quality: Meeting the challenge of protecting our health and environment. In *International Symposium on Ultrafine Particles*; Bruxelles, B., European Federation of Clean Air and Environmental Protection Associations, Eds.; EFCA: Brussels, Belgium, 2013.
4. Vance, M.; Kuiken, T.; Vejerano, E.P.; McGinnis, S.P.; Hochella, M.F., Jr.; Rejeski, D.; Hull, M.S. Nanotechnology in the real world: Redeveloping the nanomaterial consumer products inventory. *Beilstein J. Nanotechnol.* **2015**, *6*, 1769–1780. [CrossRef] [PubMed]
5. Musial, J.; Krakowiak, R.; Mlynarczyk, D.T.; Goslinski, T.; Stanisz, B.J. Titanium Dioxide Nanoparticles in Food and Personal Care Products—What Do We Know about Their Safety? *Nanomaterials* **2020**, *10*, 1110. [CrossRef]
6. Shi, H.; Magaye, R.; Castranova, V.; Zhao, J. Titanium dioxide nanoparticles: A review of current toxicological data. *Part. Fibre Toxicol.* **2013**, *10*, 15–33. [CrossRef] [PubMed]
7. Hufnagel, M.; May, N.; Wall, J.; Wingert, N.; Garcia-Käufer, M.; Arif, A.; Hübner, C.; Berger, M.; Mülhopt, S.; Baumann, W.; et al. Impact of Nanocomposite Combustion Aerosols on A549 Cells and a 3D Airway Model. *Nanomaterials* **2021**, *11*, 1685. [CrossRef] [PubMed]
8. EFSA, ANS. Re-evaluation of titanium dioxide (E 171) as a food additive. *EFSA J.* **2016**, *14*, e04545. [CrossRef]
9. Jafari, S.; Mahyad, B.; Hashemzadeh, H.; Janfaza, S.; Gholikhani, T.; Tayebi, L. Biomedical Applications of TiO2 Nanostructures: Recent Advances. *Int. J. Nanomed.* **2020**, *15*, 3447–3470. [CrossRef] [PubMed]
10. Ziental, D.; Czarczynska-Goslinska, B.; Mlynarczyk, D.T.; Glowacka-Sobotta, A.; Stanisz, B.; Goslinski, T.; Sobotta, L. Titanium Dioxide Nanoparticles: Prospects and Applications in Medicine. *Nanomaterials* **2020**, *10*, 387. [CrossRef]
11. Ali, I.; Suhail, M.; Alothman, Z.A.; Alwarthan, A. Recent advances in syntheses, properties and applications of TiO2nanostructures. *RSC Adv.* **2018**, *8*, 30125–30147. [CrossRef]
12. Oberdörster, G.; Ferin, J.; Lehnert, B.E. Correlation between Particle Size, In Vivo Particle Persistence, and Lung Injury. *EHP* **1994**, *102*, 173–179.
13. Heinrich, U.; Fuhst, R.; Rittinghausen, S.; Creutzenberg, O.; Bellmann, B.; Koch, W.; Levsen, K. Chronic Inhalation Exposure of Wistar Rats and two Different Strains of Mice to Diesel Engine Exhaust, Carbon Black, and Titanium Dioxide. *Inhal. Toxicol.* **1995**, *7*, 533–556. [CrossRef]
14. Braakhuis, H.M.; Gosens, I.; Heringa, M.B.; Oomen, A.G.; Vandebriel, R.J.; Groenewold, M.; Cassee, F.R. Mechanism of Action of TiO2: Recommendations to Reduce Uncertainties Related to Carcinogenic Potential. *Annu. Rev. Pharmacol. Toxicol.* **2021**, *61*, 203–223. [CrossRef] [PubMed]
15. Cosnier, F.; Seidel, C.; Valentino, S.; Schmid, O.; Bau, S.; Vogel, U.; Devoy, J.; Gaté, L. Retained particle surface area dose drives inflammation in rat lungs following acute, subacute, and subchronic inhalation of nanomaterials. *Part. Fibre Toxicol.* **2021**, *18*, 1–21. [CrossRef]
16. Relier, C.; Dubreuil, M.; Garcia, O.L.; Cordelli, E.; Mejia, J.; Eleuteri, P.; Robidel, F.; Loret, T.; Pacchierotti, F.; Lucas, S.; et al. Study of TiO 2 P25 nanoparticles genotoxicity on lung, blood and liver cells in lung overload and non-overload conditions after repeated respiratory exposure in rats. *Toxicol. Sci.* **2017**, *156*, 527–537. [CrossRef]
17. Murugadoss, S.; Godderis, L.; Ghosh, M.; Hoet, P. Assessing the Toxicological Relevance of Nanomaterial Agglomerates and Aggregates Using Realistic Exposure In Vitro. *Nanomaterials* **2021**, *11*, 1793. [CrossRef] [PubMed]

18. Halappanavar, S.; Brule, S.V.D.; Nymark, P.; Gaté, L.; Seidel, C.; Valentino, S.; Zhernovkov, V.; Danielsen, P.H.; De Vizcaya-Ruiz, A.; Wolff, H.; et al. Adverse outcome pathways as a tool for the design of testing strategies to support the safety assessment of emerging advanced materials at the nanoscale. *Part. Fibre Toxicol.* **2020**, *17*, 1–24. [CrossRef]

19. Mülhopt, S.; Diabaté, S.; Dilger, M.; Adelhelm, C.; Anderlohr, C.; Bergfeldt, T.; De La Torre, J.G.; Jiang, Y.; Valsami-Jones, E.; Langevin, D.; et al. Characterization of Nanoparticle Batch-To-Batch Variability. *Nanomaterials* **2018**, *8*, 311. [CrossRef] [PubMed]

20. Monopoli, M.P.; Pitek, A.S.; Lynch, I.; Dawson, K.A. Formation and Characterization of the Nanoparticle–Protein Corona. In *Springer Protocols Handbooks*; Springer: Singapore, 2013; Volume 1025, pp. 137–155.

21. Ruh, H.; Kühl, B.; Brenner-Weiss, G.; Hopf, C.; Diabaté, S.; Weiss, C. Identification of serum proteins bound to industrial nanomaterials. *Toxicol. Lett.* **2012**, *208*, 41–50. [CrossRef] [PubMed]

22. Panas, A.; Marquardt, C.; Nalcaci, O.; Bockhorn, H.; Baumann, W.; Paur, H.-R.; Mülhopt, S.; Diabate, S.; Weiss, C. Screening of different metal oxide nanoparticles reveals selective toxicity and inflammatory potential of silica nanoparticles in lung epithelial cells and macrophages. *Nanotoxicology* **2012**, *7*, 259–273. [CrossRef]

23. Magdolenova, Z.; Bilaničová, D.; Pojana, G.; Fjellsbø, L.M.; Hudecova, A.; Hasplova, K.; Marcomini, A.; Dusinska, M. Impact of agglomeration and different dispersions of titanium dioxide nanoparticles on the human related in vitro cytotoxicity and genotoxicity. *J. Environ. Monit.* **2012**, *14*, 455–464. [CrossRef] [PubMed]

24. DeLoid, G.; Cohen, J.M.; Darrah, T.; Derk, R.; Rojanasakul, L.; Pyrgiotakis, G.; Wohlleben, W.; Demokritou, P. Estimating the effective density of engineered nanomaterials for in vitro dosimetry. *Nat. Commun.* **2014**, *5*, 1–10. [CrossRef] [PubMed]

25. Paur, H.-R.; Cassee, F.R.; Teeguarden, J.; Fissan, H.; Diabate, S.; Aufderheide, M.; Kreyling, W.G.; Hänninen, O.; Kasper, G.; Riediker, M.; et al. In-vitro cell exposure studies for the assessment of nanoparticle toxicity in the lung—A dialog between aerosol science and biology. *J. Aerosol Sci.* **2011**, *42*, 668–692. [CrossRef]

26. Lacroix, G.; Koch, W.; Ritter, D.; Gutleb, A.; Larsen, S.T.; Loret, T.; Zanetti, F.; Constant, S.; Chortarea, S.; Rothen-Rutishauser, B.; et al. Air–Liquid InterfaceIn VitroModels for Respiratory Toxicology Research: Consensus Workshop and Recommendations. *Appl. Vitr. Toxicol.* **2018**, *4*, 91–106. [CrossRef]

27. OECD. Guidance Document on Good In Vitro Method Practices (GIVIMP); Series on Testing and Assessment No. 286. Available online: http://www.oecd.org/officialdocuments/publicdisplaydocumentpdf/?cote=ENV/JM/MONO(2018)19&doclanguage= en (accessed on 23 August 2018).

28. Sandra, V.; An, J.; Jo, V.L.; Masha, V.D.; Diane, B.; Witters, H.; Sylvie, R.; Lieve, G.; Lize, D.; Evelien, F. Alternative air–liquid interface method for inhalation toxicity testing of a petroleum-derived substance. *MethodsX* **2020**, *7*, 101088. [CrossRef] [PubMed]

29. Mülhopt, S.; Diabaté, S.; Krebs, T.; Weiss, C.; Paur, H.-R. Lung toxicity determination byin vitroexposure at the air liquid interface with an integrated online dose measurement. *J. Phys. Conf. Ser.* **2009**, *170*, 012008. [CrossRef]

30. Mülhopt, S.; Dilger, M.; Diabaté, S.; Schlager, C.; Krebs, T.; Zimmermann, R.; Buters, J.; Oeder, S.; Wäscher, T.; Weiss, C.; et al. Toxicity testing of combustion aerosols at the air–liquid interface with a self-contained and easy-to-use exposure system. *J. Aerosol Sci.* **2016**, *96*, 38–55. [CrossRef]

31. Diabaté, S.; Armand, L.; Murugadoss, S.; Dilger, M.; Fritsch-Decker, S.; Schlager, C.; Béal, D.; Arnal, M.-E.; Biola-Clier, M.; Ambrose, S.; et al. Air–Liquid Interface Exposure of Lung Epithelial Cells to Low Doses of Nanoparticles to Assess Pulmonary Adverse Effects. *Nanomaterials* **2020**, *11*, 65. [CrossRef]

32. Sapcariu, S.C.; Kanashova, T.; Dilger, M.; Diabaté, S.; Oeder, S.; Passig, J.; Radischat, C.; Buters, J.; Sippula, O.; Streibel, T.; et al. Metabolic Profiling as Well as Stable Isotope Assisted Metabolic and Proteomic Analysis of RAW 264.7 Macrophages Exposed to Ship Engine Aerosol Emissions: Different Effects of Heavy Fuel Oil and Refined Diesel Fuel. *PLoS ONE* **2016**, *11*, e0157964. [CrossRef]

33. Kanashova, T.; Sippula, O.; Oeder, S.; Streibel, T.; Passig, J.; Czech, H.; Kaoma, T.; Sapcariu, S.C.; Dilger, M.; Paur, H.-R.; et al. Emissions from a Modern Log wood Masonry Heater and Wood Pellet Boiler: Composition and Biological Impact on Air-Liquid Interface Exposed Human Lung Cancer Cells. *JMCM* **2018**, *1*, 23–35.

34. Murugadoss, S.; Lison, D.; Godderis, L.; Van Den Brule, S.; Mast, J.; Brassinne, F.; Sebaihi, N.; Hoet, P.H. Toxicology of silica nanoparticles: An update. *Arch. Toxicol.* **2017**, *91*, 2967–3010. [CrossRef] [PubMed]

35. Murugadoss, S.; Brassinne, F.; Sebaihi, N.; Petry, J.; Cokic, S.M.; Van Landuyt, K.L.; Godderis, L.; Mast, J.; Lison, D.; Hoet, P.H.; et al. Agglomeration of titanium dioxide nanoparticles increases toxicological responses in vitro and in vivo. *Part. Fibre Toxicol.* **2020**, *17*, 1–14. [CrossRef] [PubMed]

36. Medina-Reyes, E.I.; Delgado-Buenrostro, N.L.; Leseman, D.L.; Déciga-Alcaraz, A.; He, R.; Gremmer, E.R.; Fokkens, P.H.; Flores-Flores, J.O.; Cassee, F.R.; Chirino, Y.I. Differences in cytotoxicity of lung epithelial cells exposed to titanium dioxide nanofibers and nanoparticles: Comparison of air-liquid interface and submerged cell cultures. *Toxicol. Vitr.* **2020**, *65*, 104798. [CrossRef] [PubMed]

37. Leroux, M.M.; Doumandji, Z.; Chézeau, L.; Gaté, L.; Nahle, S.; Hocquel, R.; Zhernovkov, V.; Migot, S.; Ghanbaja, J.; Bonnet, C.; et al. Toxicity of TiO2 Nanoparticles: Validation of Alternative Models. *Int. J. Mol. Sci.* **2020**, *21*, 4855. [CrossRef] [PubMed]

38. Rasmussen, K.; Mast, J.; Temmerman, P.-J.D.; Verleysen, E.; Waegeneers, N.; Van Steen, F.; Van Doren, E.; Jensen, K.A.; Birkedal, R.; Levin, M.; et al. *Titanium Dioxide, NM-100, NM-101, NM-102, NM-103, NM-104, NM-105: Characterisation and Physico-Chemical Properties*; Publications Office of the European Union: Luxembourg, 2014.

39. Asbach, C.; Kaminski, H.; Fissan, H.; Monz, C.; Dahmann, D.; Mülhopt, S.; Paur, H.-R.; Kiesling, H.J.; Herrmann, F.; Voetz, M.; et al. Comparison of four mobility particle sizers with different time resolution for stationary exposure measurements. *J. Nanoparticle Res.* **2009**, *11*, 1593–1609. [CrossRef]
40. Mülhopt, S.; Schlager, C.; Berger, M.; Murugadoss, S.; Hoet, P.; Krebs, T.; Paur, H.-R.; Stapf, D. A novel TEM grid sampler for airborne particles to measure the cell culture surface dose. *Sci. Rep.* **2020**, *10*, 8401. [CrossRef]
41. Ferraro, D.; Anselmi-Tamburini, U.; Tredici, I.G.; Ricci, V.; Sommi, P. Overestimation of nanoparticles-induced DNA damage determined by the comet assay. *Nanotoxicology* **2016**, *10*, 861–870. [CrossRef]
42. De Temmerman, P.-J.; Verleysen, E.; Lammertyn, J.; Mast, J. Semi-automatic size measurement of primary particles in aggregated nanomaterials by transmission electron microscopy. *Powder Technol.* **2014**, *261*, 191–200. [CrossRef]
43. Noël, A.; Maghni, K.; Cloutier, Y.; Dion, C.; Wilkinson, K.; Hallé, S.; Tardif, R.; Truchon, G. Effects of inhaled nano-TiO2 aerosols showing two distinct agglomeration states on rat lungs. *Toxicol. Lett.* **2012**, *214*, 109–119. [CrossRef]
44. Noël, A.; Charbonneau, M.; Cloutier, Y.; Tardif, R.; Truchon, G. Rat pulmonary responses to inhaled nano-TiO2: Effect of primary particle size and agglomeration state. *Part. Fibre Toxicol.* **2013**, *10*, 48. [CrossRef]
45. (Faf), E.P.O.F.A.A.F.; Younes, M.; Aquilina, G.; Castle, L.; Engel, K.; Fowler, P.; Fernandez, M.J.F.; Fürst, P.; Gundert-Remy, U.; Gürtler, R.; et al. Safety assessment of titanium dioxide (E171) as a food additive. *EFSA J.* **2021**, *19*, e06585. [CrossRef]
46. Ghosh, M.; Öner, D.; Duca, R.-C.; Cokic, S.M.; Seys, S.; Kerkhofs, S.; Van Landuyt, K.; Hoet, P.; Godderis, L. Cyto-genotoxic and DNA methylation changes induced by different crystal phases of TiO_2-np in bronchial epithelial (16-HBE) cells. *Mutat. Res. Mol. Mech. Mutagen.* **2017**, *796*, 1–12. [CrossRef] [PubMed]

 *nanomaterials*

Article

Assessing the Toxicological Relevance of Nanomaterial Agglomerates and Aggregates Using Realistic Exposure In Vitro

Sivakumar Murugadoss [1], Lode Godderis [2,3], Manosij Ghosh [1] and Peter H. Hoet [1,*]

[1] Laboratory of Toxicology, Unit of Environment and Health, Department of Public Health and Primary Care, KU Leuven, 3000 Leuven, Belgium; sivakumar.murugadoss@kuleuven.be (S.M.); manosij.ghosh@kuleuven.be (M.G.)
[2] Laboratory for Occupational and Environmental Hygiene, Unit of Environment and Health, Department of Public Health and Primary Care, KU Leuven, 3000 Leuven, Belgium; lode.godderis@kuleuven.be
[3] IDEWE, External Service for Prevention and Protection at Work, Interleuvenlaan 58, 3001 Heverlee, Belgium
* Correspondence: peter.hoet@kuleuven.be; Tel.: +32-1633-0197

Citation: Murugadoss, S.; Godderis, L.; Ghosh, M.; Hoet, P.H. Assessing the Toxicological Relevance of Nanomaterial Agglomerates and Aggregates Using Realistic Exposure In Vitro. *Nanomaterials* **2021**, *11*, 1793. https://doi.org/10.3390/nano11071793

Academic Editors: Andrea Hartwig and Christoph Van Thriel

Received: 18 June 2021
Accepted: 8 July 2021
Published: 9 July 2021

Publisher's Note: MDPI stays neutral with regard to jurisdictional claims in published maps and institutional affiliations.

Abstract: Low dose repeated exposures are considered more relevant/realistic in assessing the health risks of nanomaterials (NM), as human exposure such as in workplace occurs in low doses and in a repeated manner. Thus, in a three-week study, we assessed the biological effects (cell viability, cell proliferation, oxidative stress, pro-inflammatory response, and DNA damage) of titanium-di-oxide nanoparticle (TiO_2 NP) agglomerates and synthetic amorphous silica (SAS) aggregates of different sizes in human bronchial epithelial (HBE), colon epithelial (Caco2), and human monocytic (THP-1) cell lines repeatedly exposed to a non-cytotoxic dose (0.76 μg/cm^2). We noticed that neither of the two TiO_2 NPs nor their agglomeration states induced any effects (compared to control) in any of the cell lines tested while SAS aggregates induced some significant effects only in HBE cell cultures. In a second set of experiments, HBE cell cultures were exposed repeatedly to different SAS suspensions for two weeks (first and second exposure cycle) and allowed to recover (without SAS exposure, recovery period) for a week. We observed that SAS aggregates of larger sizes (size ~2.5 μm) significantly affected the cell proliferation, IL-6, IL-8, and total glutathione at the end of both exposure cycle while their nanosized counterparts (size less than 100 nm) induced more pronounced effects only at the end of the first exposure cycle. As noticed in our previous short-term (24 h) exposure study, large aggregates of SAS did appear to be similarly potent as nano sized aggregates. This study also suggests that aggregates of SAS of size greater than 100 nm are toxicologically relevant and should be considered in risk assessment.

Keywords: nanotoxicology; titanium dioxide; synthetic amorphous silica; agglomerates and aggregates; realistic exposure in vitro

1. Introduction

Manufactured nanomaterials (NMs) are, due to their unique physico-chemical properties, used in a large variety of applications. Nowadays, at least 1800 products containing NMs, ranging from personal care products to sporting goods, are in circulation in the global market [1]. Concerns regarding the human health effects of NMs are gradually increasing due to their increased production and use [2–5].

In the real world, such as in occupational exposure settings, NMs exist as primary particles, agglomerates, aggregates, or as a mixture thereof [6–8]. In agglomerates, the particles are loosely bound by weak forces such as Van der Waals in a reversible manner, while in aggregates, particles are irreversibly fused together by chemical bonding such as covalent or ionic bonding [9]. The term agglomerates and aggregates (AA) is included in the definition of NMs recommended by the European Union [10]. It states that "manufactured material containing particles, in an unbound state or as an aggregate or as an agglomerate and where, for 50% or more of the particles in the number size distribution, one or more external dimensions is in the size range 1 nm–100 nm". However, the definition was recommended solely for regulatory applications without any regard for hazard. Moreover, the relevance of AA in terms of toxicological perspectives is still largely unknown.

Titanium-di-oxide (TiO_2) and synthetic amorphous silica (SAS) are among the most widely used NMs. Due to their unique properties, they have found applications in food, cosmetics, paints, etc. [2,11,12]. TiO_2 NMs are well known for their tendency to agglomerate [13], while SAS NMs are known to aggregate easily during their production for industrial/commercial applications [14]. Thus, to determine the influence of agglomeration and aggregation on NM toxicity, we investigated and compared in our previous studies the acute (24 h) toxicological effects of TiO_2 NMs in different agglomeration states [11] or SAS in different aggregation states [12] in three different cell lines. The results suggested that in most cases, large agglomerates or aggregates were not less potent compared to their smaller counterparts. This indicated that the toxicity of tested NMs was not mitigated by their agglomeration/aggregation state, and therefore AA of NMs of larger size (size greater than 100 nm) appear to be toxicologically relevant.

To date, most studies have evaluated the toxic potential of NMs after short-term exposure [15–17]. Recently, long-term and repeated low dose exposure studies for the hazard assessment have been set up for NMs, better mimicking the real life exposure (e.g., workers in production) that occurs (often) at low doses. Biological effects induced by NMs have also shown to be different between short-term versus (relatively) long-term exposure [18–20]. Xi et al., performed a 21 d (3w) exposure study using vanadium dioxide (VO_2) nanoparticles (NPs) [19]. In his study, A549 cells were repeatedly exposed to a low dose (0.2 µg/mL) of VO_2 NPs and the authors observed a 50% decreased proliferation during sub-culturing at the end of every week. Similarly, Chen et al. (2016) also performed a 21 d exposure study and noticed that the proliferation of Caco2 cells were reduced up to 50% when repeatedly exposed to 0.5 µg/mL of silver (Ag) NPs [20]. In both studies, an increase in cytokines and reactive oxygen species (ROS) generation were associated with decreased proliferation.

In this study, we aimed to determine how different AA suspensions influence the biological responses in cell cultures repeatedly exposed to a low dose (three week study). There is no consensus to estimate the dose for long-term exposure. We estimated 0.76 µg/cm^2 as an appropriate dose based on OELs for TiO_2 and SAS [21,22], which corresponds to a concentration of 2 µg/mL. This dose was also determined as non-cytotoxic in short-term experiments (data not shown).

2. Materials and Methods

2.1. Preparation of Dispersions and Size Characterization

Two TiO_2 NPs of different sizes (17 nm and 117 nm) in different agglomeration states (small and large agglomerates) were freshly prepared during each exposure as described in [11] (p. 9) and details of methods used for size characterization in stock are provided in (p. 10). Two different suspensions of SAS in different aggregation states (indicated as DE-AGGR and AGGR) were prepared. In addition, we also studied the two identified subfractions in the AGGR suspension (SuperN and PREC) as described in [12]. All suspensions were freshly prepared as described in [12] (pp. 8–9) and details of methods used for size characterization in stock are provided in (p. 9).

2.2. Cell Culture

The human bronchial epithelial cell line (16HBE14o- or HBE) and the human monocytic cell line (THP-1) were kindly provided by Dr. Gruenert (University of California, San Francisco, CA, USA), and the Caucasian colon adenocarcinoma cell line (Caco2) (P.Nr: 86010202) was purchased from Sigma-Aldrich (Overijse, Belgium). HBE cells were cultured in DMEM/F12 supplemented with 5% FBS, 1% penicillin-streptomycin (P-S)

(100 U/mL), 1% L-glutamine (2 mM) and 1% fungizone (2.5 g/mL) while RPMI 1640 supplemented with 10% FBS, 1% P-S (100 U/mL), 1% L-glutamine (2 mM) and 1% fungizone (2.5 g/mL) was used for THP-1. DMEM/HG supplemented with 10% FBS, 1% P-S (100 U/mL), 1% L-glutamine (2 mM), 1% fungizone (2.5 g/mL) and 1% non-essential amino acids (NEAA) was used for Caco2 cells. All cell culture supplements were purchased from Invitrogen (Merelbeke, Belgium) unless otherwise stated. Cells were cultured in T75 flasks (FALCON, Corning, NY, USA) at 37 °C in 100% humidified air containing 5% CO_2. Fresh medium was changed every 2 or 3 d and cells were passaged every week (7 d). Cells from passage 3–6 were used for experiments.

2.3. In Vitro Exposure Conditions

The experimental design used in this study was adapted from [19,20]. For the first exposure cycle (seven days), HBE cells, Caco2 cells, and THP-1 cells were seeded at a density of 10,000 cells/cm², 5000 cells/cm², and 10,000 cells/mL, respectively in six well plates (day 0). Based on cell doubling time, the cell numbers for each cell line were adjusted to attain optimal confluency at the end of the first exposure cycle. After overnight incubation (day 1), the cells were exposed to cell culture media containing 2 µg/mL or 0.76 µg/cm² of different suspensions of TiO_2 and SAS for 48 h (day 2 and 3). On day 3 and 5, the supernatant was removed; cell cultures were rinsed with warm HBSS twice and exposed to fresh cell culture media containing 2 µg/mL or 0.76 µg/cm² of NMs for 48 h. On day seven, the supernatants were collected and the cell cultures were washed and trypsinized (subculturing). The cell number and viability were determined immediately and the same number of cells (10,000 cells/cm², 5000 cells/cm² and 10,000 cells/mL for HBE, Caco2 and THP-1, respectively) were seeded for the second exposure cycle. The remaining cells were processed/stored for further analysis such as glutathione measurements and DNA damage. The steps were repeated for second (7–14 d) and third exposure cycle (14–21 d).

2.4. Cell Viability and Number Determination

During each subculture step, about 10 µL of cell suspension-trypan blue mix (1:1 ratio) was loaded into the counting chamber slides and cell viability and number was determined by the countess™ automated cell counter (Invitrogen, Merelbeke, Belgium). The results are expressed relative to control.

2.5. Total Glutathione Measurements

Reduced glutathione (GSH) was measured using a glutathione detection kit (Enzo life sciences, Brussels, Belgium) according to the manufacturer's protocol and the protein content was estimated using bicinchoninic acid (BCA) protein assay kit (Thermo Scientific Pierce, Merelbeke, Belgium). GSH was normalized to the total protein content and the results were expressed relative to control (untreated cells).

2.6. Cytokine Quantification

Interleukin (IL)-8 and IL-6 were quantified using ELISA kits (Sigma Aldrich, Overijse, Belgium). The cytokines were measured in the supernatants (collected during glutathione measurement experiments and stored at −20 °C) according to the manufacturer's protocol and the results were expressed relative to control (untreated cells).

2.7. Comet Assay

An alkaline comet assay kit [(Trevigen (C.No. 4250-050-K), Gaithersburg, MD, USA)] was used to quantify DNA strand breaks as a measure of DNA damage according to manufacturer's protocol. Cells treated with methyl methane sulfonate (MMS) (Sigma-Aldrich, Overijse, Belgium) 100 µM for 1–2 h served as positive control.

2.8. Statistical Analysis

Two independent experiments were performed in triplicate or duplicate, and data were presented as mean ± standard deviation (SD). Using GraphPad prism 7.04 for windows, GraphPad Software, La Jolla, CA, USA, www.graphpad.com, the results were analyzed with one-way ANOVA followed by a Dunnett's multiple comparison test to determine the significance of differences compared with control.

3. Results

3.1. Dispersion and Size Characterization

3.1.1. TiO$_2$ Suspensions

The results of size characterization and zeta potential of TiO$_2$ suspensions were already published in [11], and are therefore provided in the Supplementary Materials. Supplementary Figure S1 shows electron microscopy (TEM) micrographs of small (SA) and large agglomerates (LA) of 17 and 117 nm sized TiO$_2$ NPs and Table S1 shows the sizes of different TiO$_2$ suspensions characterized by different techniques. We used a standardized TEM technique in our previous study [23], which enabled us to measure the size of several thousand agglomerates in each suspension. The TEM characterization (median feret min) indicated that the size of 17 nm sized TiO$_2$ in their least agglomerated condition (indicated as 17 nm-SA) was 33 nm while it was 120 nm in their strongly agglomerated condition (17 nm-LA). The sizes of small (117 nm-SA) and large agglomerates (117 nm-LA) of 117 nm sized TiO$_2$ were 148 and 309 nm, respectively, indicating that there were also clear differences in sizes between SA and LA of both TiO$_2$ NPs. Although differences between SA and LA were observed in the sizes measured by dynamic light scattering (DLS) and particle tracking analysis (PTA), technical issues involved in observing larger sizes were discussed in [11] (p. 8). To verify the stability of agglomerates, TiO$_2$ stock suspensions were diluted to 100 µg/mL in complete culture medium (CCM) and sizes were measured using DLS at 0 h and 24 h (Supplementary Table S2). The sizes of all agglomerates remained similar at 0 and 24 h, indicating their good stability over time.

3.1.2. SAS Suspensions

The results of size characterization and zeta potential of SAS suspensions were already published in [12], and are therefore provided in the Supplementary Materials. Figure S2 shows the bright field (BF) microscopic image of different SAS suspensions and Table S3 shows the sizes of different SAS suspensions characterized by different techniques. SAS is a material with aggregates of broad size range (few hundred nm to few tenths µm). Thus we used different techniques (such as sonication and vortexing) to obtain suspensions with different sizes. The TEM characterization of sonicated suspension (de-aggregated, indicated as DE-AGGR) was quite straightforward and their mean feret min size was determined as 28 nm. However, using TEM and DLS, we were not able to determine the difference in sizes of other suspensions such as a vortexed suspension (aggregated, AGGR) or a suspension fractionated from AGGR [non-precipitating fraction (SuperN) and precipitating fraction (PREC)]. Thus, we used bright field microscopy and sizes of SuperN and PREC aggregates were roughly determined as 2.5 and 25 µm, respectively. By combining different techniques, we were able identify the differences in sizes between these SAS suspensions. To verify the stability of aggregates, SAS stock suspensions were diluted to 100 µg/mL in CCM and sizes were measured using DLS at 0 and 24 h. Despite knowing that AGGR and PREC sizes were not reflecting the realistic size distribution due to their quick sedimentation while performing DLS measurements, we provided the results in Supplementary Table S4. Thus, we only consider the sizes of DE-AGGR and SuperN aggregates. The sizes of DE-AGGR remained similar at 0 and 24 h, while the size of SuperN aggregates slightly reduced after 24 h.

3.2. Comparison of Biological Responses

3.2.1. TiO$_2$ Suspensions

The proliferation profiles and viability of cell cultures determined at the end of every week in three different cell lines is shown in Figure 1. None of the TiO$_2$ suspensions did affect the cell proliferation and viability at the end of any exposure cycles. Compared to control, no significant effects for any of these suspensions were noticed for glutathione depletion, IL-8 and IL-6 increase, or DNA damage (Figure 2), which were evaluated after the third exposure cycle only.

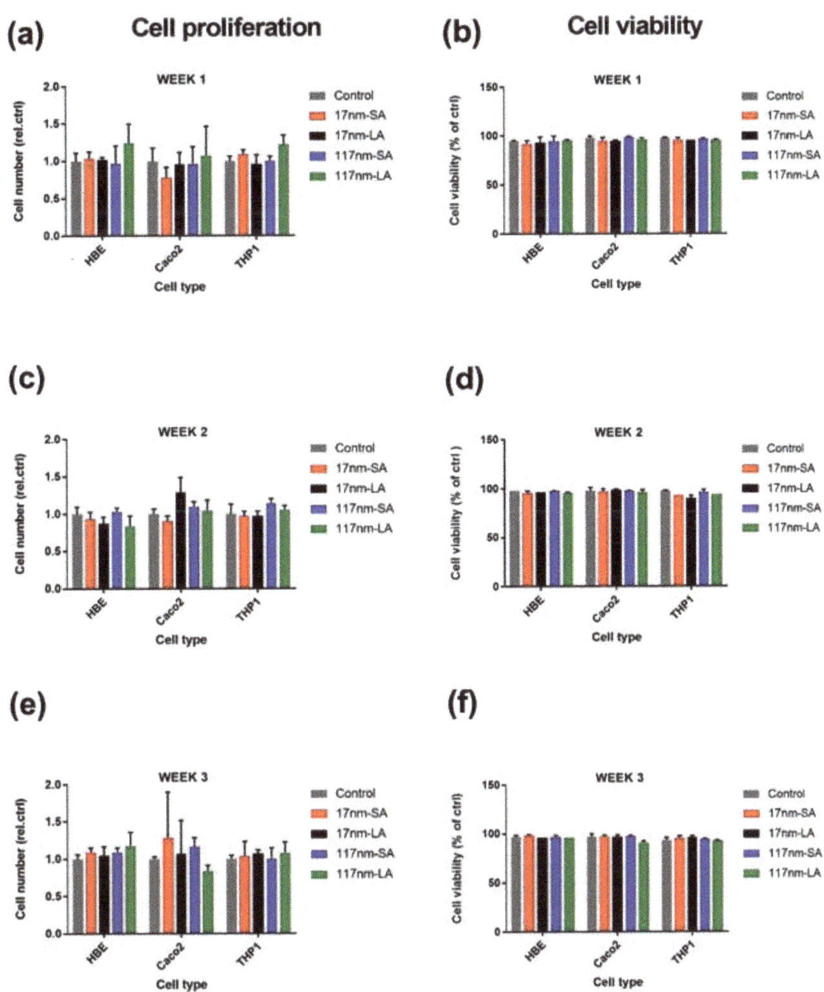

Figure 1. Effect of repeated low dose exposure to TiO$_2$ suspensions on cell proliferation and viability. Cell proliferation profiles (**a,c,e**) and cell viability (**b,d,f**) was measured in different cell cultures after first (**a,b**), second (**c,d**), and third exposure cycle (**e,f**). Data are expressed as means ±SD from two independent experiments performed in duplicates. SA—small agglomerates; LA—large agglomerates.

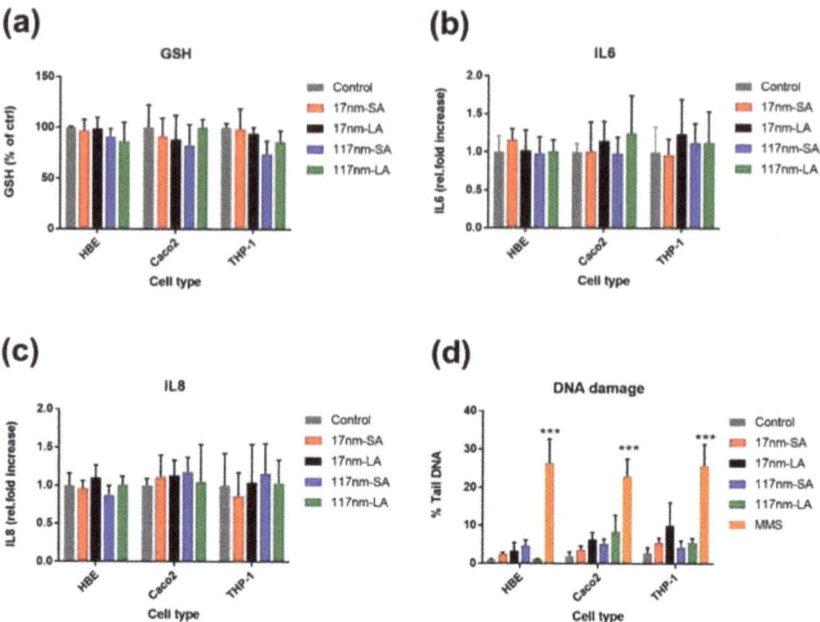

Figure 2. Effect of repeated exposure to TiO_2 suspensions (0.76 µ/cm^2) on biological responses. Total glutathione (GSH) (**a**), IL-6 (**b**), IL-8 (**c**), and DNA damage (**d**) was measured in different cell cultures after third exposure cycle. Data are expressed as means ±SD from two independent experiments performed in duplicate. $p < 0.001$ (***) represents significant differences compared to control (One-way ANOVA followed by Dunnett's multiple comparison test). SA—small agglomerates; LA—large agglomerates.

3.2.2. SAS Suspensions

Figure 3 shows the summary of biological responses evaluated in cell cultures exposed to SAS after the third exposure cycle. DE-AGGR reduced HBE cell number significantly compared to control but AGGR did not. DE-AGGR and AGGR induced a significant increase in IL-8 and IL-6 only in HBE cell cultures. As observed for TiO_2, SAS did not induce significant DNA damage at the tested dose. Importantly, no significant effects were noticed in the Caco2 or THP-1 cell lines in any of the biological endpoints measured. These preliminary results suggest that SAS induces biological responses at the tested dose, and it would be interesting to study and compare all fractions of the AGGR suspensions of SAS. In a set of follow-up experiments, we used only HBE cells to investigate other SAS suspensions for their effect on cell number, viability, GSH, IL-6, and IL-8. We planned two exposure cycles (two weeks) with a view to the potential recovery after discontinuing exposure, the third observation week was a recovery period without SAS exposure.

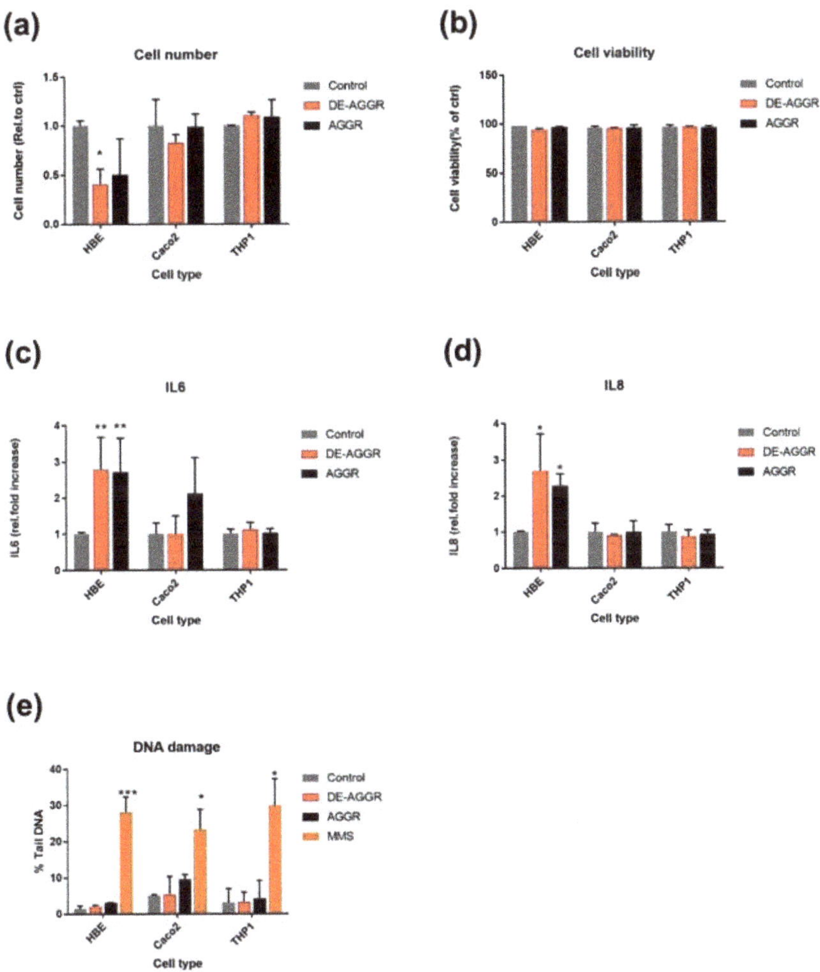

Figure 3. Effect of repeated exposure to SAS suspensions (0.76 μ/cm^2) on biological responses. Cell proliferation (**a**), viability (**b**), IL-6 (**c**), IL-8 (**d**), and DNA damage (**e**) was measured in different cell cultures after third exposure cycle. Data are expressed as means ± SD from two independent experiments performed in duplicate. $p < 0.05$ (*), $p < 0.01$ (**) and $p < 0.001$ (***) represent significant differences compared to control (One-way ANOVA followed by Dunnett's multiple comparison test). DE-AGGR—de-aggregated suspension; AGGR—aggregated suspension.

Effect on proliferation and viability: To determine the effect on cell proliferation, we measured cell number and cell viability at the end of each exposure cycle and recovery period (Figure 4). DE-AGGR and SuperN fractions strongly affected the cell growth at the end of the first exposure cycle. Compared to untreated cells, the DE-AGGR and SuperN fractions decreased the cell growth to about 65 and 50%, respectively (Figure 4a). AGGR, on the other hand, inhibited cell growth by about 20%. Surprisingly, DE-AGGR and AGGR exposed cell cultures recovered and remained similar compared to controls at the end of the second exposure cycle, but SuperN exposed cell cultures still exhibited decreased cell growth (about 35%). Despite a mild and non-significant decreasing trend observed at the end of the second exposure cycle, PREC fractions did not affect the cell growth significantly after both exposure cycles. After a week of recovery, all cell cultures exhibited similar growth to control. Compared to untreated controls, none of these suspensions affected the cell viability significantly after exposure cycles and recovery cycle (Figure 4b).

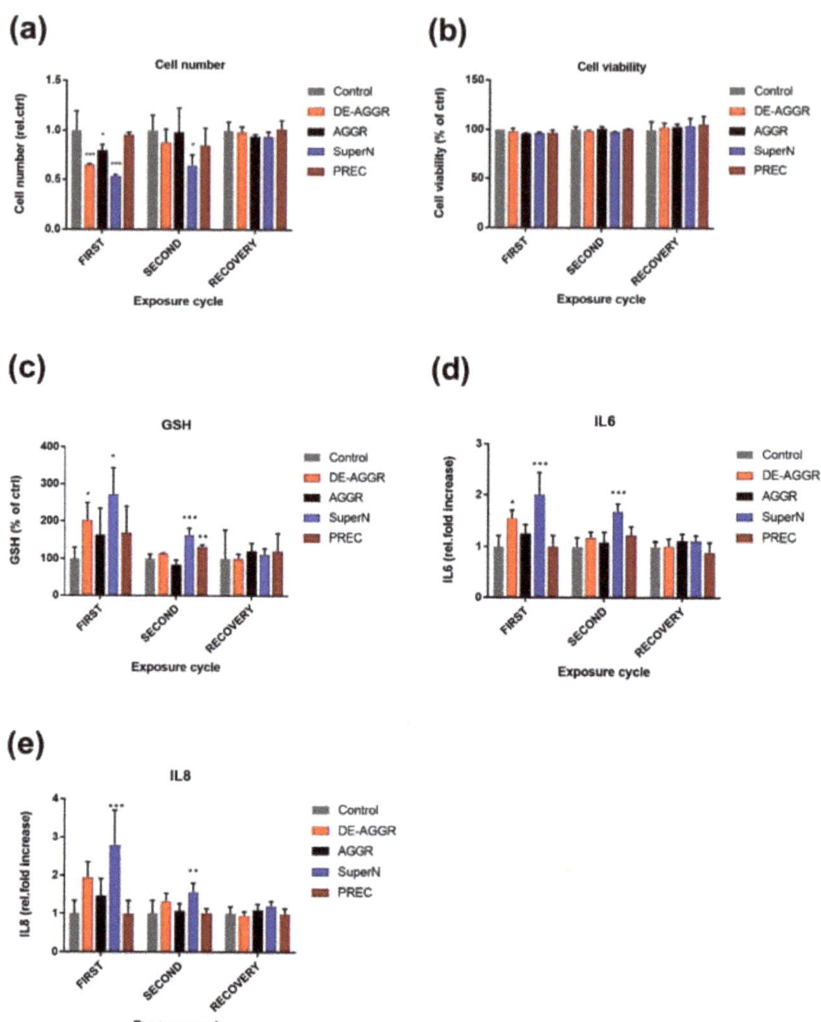

Figure 4. Effect of repeated exposure to different SAS suspensions (0.76 μ/cm^2) on biological responses. Cell proliferation (**a**), cell viability (**b**), total glutathione levels (GSH) (**c**), IL6 (**d**), and IL8 (**e**) were measured in HBE cell cultures after different exposure cycles. Recovery denotes a week of exposure to cell culture medium without SAS. Data are expressed as means ±SD from two independent experiments performed in duplicate. $p < 0.05$ (*), $p < 0.01$ (**) and $p < 0.001$ (***) represent significant differences compared to control (One-way ANOVA followed by Dunnett's multiple comparison test). DE-AGGR—de-aggregated suspension; AGGR—aggregated suspension; SuperN—non-precipitating suspension; PREC—precipitating suspension.

Effect on total glutathione: At the end of the first exposure cycle, we observed that the GSH levels had increased to about 200 (±47) and 270 (±71) % in DE-AGGR and SuperN exposed cells, respectively, compared to untreated cells (Figure 4c). Additionally, an upward trend was noticed for AGGR and PREC fractions but was not significant.

The GSH levels in DE-AGGR exposed cell cultures returned to normal after the second exposure cycle while the GSH levels were still high in SuperN exposed cells (about $160 \pm 18\%$). Interestingly, cell cultures exposed to PREC also showed mild but significantly increased GSH levels (about $130 \pm 5\%$). The GSH levels in all the exposed cell cultures returned to normal after seven days of recovery period.

Effect on cytokine secretion: After each cycle, cytokines such as IL-6 and IL-8 were quantified in the supernatant of cell cultures (Figure 4d,e, respectively). SuperN fractions resulted in a nearly 2-fold increase in IL6 and IL8 after one week exposure and remained significantly increased at the end of second week. Like at other endpoints, DE-AGGR fractions induced a significant increase only at the end of the first week of exposure. Compared to controls, AGGR and PREC did not affect the levels of IL-6 and IL-8. After a week of recovery, no differences between suspensions were found.

4. Discussion

In this study, we aimed to determine how different AA suspensions influence the biological responses in cell cultures repeatedly exposed (3w study) to a dose of $0.76 \, \mu g/cm^2$. Neither of the TiO_2 dispersions induced significant effects, while SAS suspensions generated by sonication (DE-AGGR) induced some effects compared to control and vortexed suspensions (AGGR), mainly in HBE cells. In an additional study comparing two weeks' exposure of HBE cells with four different SAS suspensions (AGGR, DE-AGGR, SuperN or PREC), it appears that SuperN did not appear to be less potent compared to De-AGGR, which is in line with the acute effects (24 h) described in our previous study [12].

In our recent study [11], we showed that TiO_2 agglomeration influences the toxicity/biological responses in high dose short-term exposure (24 h), while in this repeated low dose study, neither TiO_2 exposure nor their agglomeration influences the biological responses. In a three week exposure experiment, no cytotoxic effects were observed in human mesenchymal stem cells although nano-TiO_2 was detected in the cytoplasm [24]. Kocbek et al. (2010) did not notice any significant effects in keratinocytes repeatedly exposed to $10 \, \mu g/mL$ of TiO_2 NPs for three months, while at the same concentration ZnO NPs induced a decrease in mitochondrial activity, abnormal cell morphology, and disturbances in cell-cycle [25]. Vales et al. (2014) suggested that BEAS2B cells repeatedly exposed to $20 \, \mu g/mL$ for four weeks showed potential for carcinogenicity (soft agar assay) [26]. These results suggest that the TiO_2 dose used in our experiments ($2 \, \mu g/mL$) might not be sufficient to induce adverse effects. We based the choice of $2 \, \mu g/mL$ on our earlier 'acute' exposure experiments without cyto/genotoxicity, which now appears to be a relatively safe dose after three weeks of exposure.

In this study, DE-AGGR, the least aggregated and nano-sized SAS, induced a more pronounced effect than AGGR at the same mass concentrations. Our characterization revealed that 75% of total mass of AGGR was composed of PREC aggregates, which is about $25 \, \mu m$ in size [12]. PREC aggregates, when studied separately, did not induce any effects. Given their larger size, such aggregates are less likely to be taken up by the cells, and therefore induced no effects. This indicates that overall biological activity of SAS NMs in their manufactured form was reduced due to aggregation.

Similar to acute studies, SuperN fractions of AGGR suspension exhibited noticeable biological activity in a low dose repeated exposure study. The most quoted nanotoxicity paradigm is "the smaller the size of the NPs the greater the toxicity/biological responses". Likewise, several short-term cytotoxicity studies showed that nano-sized particles are more biologically active than micron-sized studies [27–29]. In a recent study, bronchial cells repeatedly exposed to a low dose of VO_2 NPs for three weeks showed greater adverse response for nano-sized particles than micron-sized particles [19]. In contrast to these observations, we observed that SuperN aggregates of size about $2.5 \, \mu m$ showed similar biological activity to nano-sized fractions. This suggests that larger aggregates of NP may not necessarily be considered biologically less active and highlights the need for a case-by-case analysis.

The size of SuperN aggregates (2.5 µm) is far greater than DE-AGGR aggregates (100 nm) yet falls under the category of respirable particles [30]. Therefore, exposure and hazard assessment of such fractions is valuable since commercially available SAS can be composed of small and large aggregates. We also observed that PREC was the least biologically active. Considering the size of the aggregates play a key role in determining its toxicological relevance, these findings could also contribute to the "safe-by-design" of SA, by considering aggregation as a critical factor.

Studies have indicated that the effects induced by NMs were different for short-term and long-term exposure [18–20]. In this study, we noticed an increase in glutathione levels after the first exposure cycle (one week) for both DE-AGGR and SuperN. Therefore, in addition to a three week exposure study, we also investigated the in vitro effects after short-term high dose exposure to SAS exposure under the same experimental conditions (Supplementary Figure S3). In short-term exposure (24 h), mild cytotoxicity (Figure S3b) and total glutathione depletion (Figure S3c) was observed at high concentrations of DE-AGGR and SuperN. Glutathione depletes when excessive ROS is produced. Several short-term studies have shown that SAS reduced glutathione levels [31–34], which is in agreement with our findings. This indicates that glutathione depletion is an earlier effect of short-term cytotoxicity while increased glutathione production is possibly a sign of a protective effect to prevent further damage. Further, decreased cell proliferation in a three week study is also consistent with an increase in IL-8 and IL-6, while only IL-8 was consistent with short-term cytotoxicity (Figure S3e). These results indicate that cell cultures may respond to NM differently depending on the modes of exposure (short-term high dose or low dose repeated exposure).

To have a view on the potential role of survival cells from first cycle exposure, the cells from the first cycle exposure were passaged and repeatedly exposed in the second cycle. At the end of second exposure cycle, we noticed that the increase in glutathione, IL-6, and IL-8 was somewhat less compared to the first exposure cycle in cell cultures exposed to DE-AGGR and SuperN suspensions. It appears that the cells stressed during first exposure cycle, undergoing recovery probably due to protective effects induced during the first exposure cycle. Moreover, cell viability at the end of both exposure cycles remained similar to control. Further research is needed to verify whether the decreased cell growth was the result of cell cycle arrest and/or cell death (apoptosis). Nevertheless, the cells recovered similarly to the control one week after exposure was discontinued, indicating that the response observed was due to continuous exposure to SAS. This finding is particularly important as this indicates that continuous human exposure to SAS results in elevated levels of biological responses, which could lead to adverse effects.

Numerous studies have reported that short-term in vivo exposure to SAS elevated the levels of LDH, IL-6, IL-8, and GSH depletion in the lung [15]. However, long-term and repeated exposure in vivo studies for SAS are scarce. In a study [34], rats were exposed to 50 mg/m^3 of SAS for 6 h/day, 5 days/week for 13 weeks and effects were characterized after 6.5 weeks and 13 weeks of exposure, and after three and eight months of recovery. An increase in cytotoxicity biomarkers (LDH) and inflammatory cells was noticed after 6.5 and 13 weeks, but the effects were significantly mitigated after both recovery periods. Genotoxicity was not observed at any of these time points. In another study [35], rats were exposed to 50 mg/m^3 of SAS for 6 h/day for five days and adverse effects were characterized after last exposure or one or three months later. SAS induced elevated levels of cytotoxicity biomarkers and lung damage after last exposure, but the effects were reversed three months post exposure. In our study, we observed that the effects induced by DE-AGGR and PREC were reversed after a one week recovery period. This suggests that our long-term exposure design may be appropriate to predict the in vivo adverse outcome of repeated exposure to NMs.

5. Conclusions

In this study, we demonstrated the toxicological relevance of AA in a repeated low dose in vitro exposure study. Neither TiO_2 exposure nor their agglomeration state affected the measured biological endpoints, possibly due to insufficient applied dose. On the other hand, we noticed that a fraction of SAS aggregates in their manufactured form (2.5 μm) did not appear biologically less active compared to nano-sized SAS produced by sonication. Apparently, in vitro studies with more biological endpoints and animal studies are required to verify these results. Moreover, further characterization is needed to reveal properties other than size that make SuperN fractions biologically more active. Since SAS used in this study is a representative of SAS approved as a food additive (E551), more attention needs to be paid in the future to the possible adverse effects of SuperN fractions, particularly their long-term effects. The results of this study also might spur toxicologists to perform more long-term studies in the future to reveal the toxicological relevance of other NMs that are agglomerated/aggregated in their manufactured form.

Supplementary Materials: The following are available online at https://www.mdpi.com/article/10.3390/nano11071793/s1: Figure S1: Representative TEM micrographs of freshly prepared TiO_2 stock suspensions, Figure S2: Representative bright field microscopic images of freshly prepared SAS stock suspensions, Figure S3: Influence of SAS aggregation on cytotoxicity and biological responses, Table S1: Size characterization of freshly prepared TiO_2 stock suspensions, Table S2: Size characterization of freshly prepared TiO_2 stock suspensions, Table S3: Characterization of freshly prepared SAS stock suspensions, Table S4: Z-average sizes (measured by DLS) of SAS suspensions in different cell culture medium (100 μg/mL).

Author Contributions: Conceptualization, S.M., L.G., M.G. and P.H.H.; methodology, S.M., M.G. and P.H.H.; investigation, S.M.; writing—original draft preparation, S.M.; review and editing, S.M., L.G., M.G. and P.H.H.; project administration and funding acquisition, P.H.H. All authors have read and agreed to the published version of the manuscript.

Funding: This work was funded by the Belgian Science Policy (BELSPO) program "Belgian Research Action through Interdisciplinary Network (BRAIN-be)" for the project "Towards a toxicologically relevant definition of nanomaterials (To2DeNano)" and EU H2020 project (H2020-NMBP-13-2018 RIA): RiskGONE (Science-based Risk Governance of NanoTechnology) under grant agreement No. 814425.

Institutional Review Board Statement: Not applicable.

Informed Consent Statement: Not applicable.

Data Availability Statement: The data presented in this study are available on request from the corresponding authors.

Acknowledgments: The authors thank JRC Nanomaterials Repository, Italy for providing the TiO_2 NPs and SAS.

Conflicts of Interest: The authors declare no conflict of interest.

References

1. Vance, M.E.; Kuiken, T.; Vejerano, E.P.; McGinnis, S.P.; Hochella, M.F.; Hull, D.R. Nanotechnology in the real world: Redeveloping the nanomaterial consumer products inventory. *Beilstein J. Nanotechnol.* **2015**, *6*, 1769–1780. [CrossRef]
2. Musial, J.; Krakowiak, R.; Mlynarczyk, D.T.; Goslinski, T.; Stanisz, B.J. Titanium dioxide nanoparticles in food and personal care products—What do we know about their safety? *Nanomaterials* **2020**, *10*, 1110. [CrossRef] [PubMed]
3. Napierska, D.; Thomassen, L.C.; Lison, D.; Martens, J.A.; Hoet, P.H. The nanosilica hazard: Another variable entity. *Part Fibre Toxicol.* **2010**, *7*, 39. [CrossRef]
4. Fruijtier-Pölloth, C. The safety of nanostructured synthetic amorphous silica (SAS) as a food additive (E 551). *Arch. Toxicol.* **2016**, *90*, 2885–2916. [CrossRef]
5. Efsa. Titanium Dioxide: E171 No Longer Considered Safe When Used as a Food Additive. European Food Safety Authority. 2021. Available online: https://www.efsa.europa.eu/en/news/titanium-dioxide-e171-no-longer-considered-safe-when-used-food-additive (accessed on 3 June 2021).

6. Mihalache, R.; Verbeek, J.; Graczyk, H.; Murashov, V.; Van Broekhuizen, P. Occupational exposure limits for manufactured nanomaterials, a systematic review. *Nanotoxicology* **2017**, *11*, 7–19. [CrossRef] [PubMed]

7. Kuhlbusch, T.A.; Asbach, C.; Fissan, H.; Göhler, D.; Stintz, M. Nanoparticle exposure at nanotechnology workplaces: A review. *Part. Fibre Toxicol.* **2011**, *8*, 22. [CrossRef] [PubMed]

8. Debia, M.; Bakhiyi, B.; Ostiguy, C.; Verbeek, J.H.; Brouwer, D.H.; Murashov, V. A Systematic Review of Reported Exposure to Engineered Nanomaterials. *Ann. Occup. Hyg.* **2016**, *60*, 916–935. [CrossRef] [PubMed]

9. Walter, D. Primary Particles–Agglomerates–Aggregates. *Nanomaterials* **2013**, 9–24. [CrossRef]

10. Potočnik, J. Commission recommendation of 18 October 2011 on the definition of nanomaterial (2011/696/EU). *Off. J. Eur. Union.* **2011**, *L275*, 38–40.

11. Murugadoss, S.; Brassinne, F.; Sebaihi, N.; Petry, J.; Cokic, S.M.; Van Landuyt, K.L.; Godderis, L.; Mast, J.; Lison, D.; Hoet, P.H.; et al. Agglomeration of titanium dioxide nanoparticles increases toxicological responses in vitro and in vivo. *Part. Fibre Toxicol.* **2020**, *17*, 1–14. [CrossRef] [PubMed]

12. Murugadoss, S.; Brule, S.V.D.; Brassinne, F.; Sebaihi, N.; Mejia, J.; Lucas, S.; Petry, J.; Godderis, L.; Mast, J.; Lison, D.; et al. Is aggregated synthetic amorphous silica toxicologically relevant? *Part. Fibre Toxicol.* **2020**, *17*, 1–12. [CrossRef] [PubMed]

13. Shi, H.; Magaye, R.; Castranova, V.; Zhao, J. Titanium dioxide nanoparticles: A review of current toxicological data. *Part. Fibre Toxicol.* **2013**, *10*, 15. [CrossRef] [PubMed]

14. Fruijtier-Pölloth, C. The toxicological mode of action and the safety of synthetic amorphous silica—A nanostructured material. *Toxicology* **2012**, *294*, 61–79. [CrossRef] [PubMed]

15. Murugadoss, S.; Lison, D.; Godderis, L.; Brule, S.V.D.; Mast, J.; Brassinne, F.; Sebaihi, N.; Hoet, P.H. Toxicology of silica nanoparticles: An update. *Arch. Toxicol.* **2017**, *91*, 2967–3010. [CrossRef] [PubMed]

16. Akter, M.; Sikder, M.T.; Rahman, M.M.; Ullah, A.A.; Hossain, K.F.; Banik, S.; Hosokawa, T.; Saito, T.; Kurasaki, M. A systematic review on silver nanoparticles-induced cytotoxicity: Physicochemical properties and perspectives. *J. Adv. Res.* **2018**, *9*, 1–16. [CrossRef]

17. Vandebriel, R.J.; De Jong, W.H. A review of mammalian toxicity of ZnO nanoparticles. *Nanotechnol. Sci. Appl.* **2012**, *5*, 61–71. [CrossRef]

18. Annangi, B.; Rubio, L.; Alaraby, M.; Bach, J.; Marcos, R.; Hernández, A. Acute and long-term in vitro effects of zinc oxide nanoparticles. *Arch. Toxicol.* **2015**, *90*, 2201–2213. [CrossRef] [PubMed]

19. Xi, W.-S.; Song, Z.-M.; Chen, Z.; Chen, N.; Yan, G.-H.; Gao, Y.; Cao, A.; Liu, Y.; Wang, H. Short-term and long-term toxicological effects of vanadium dioxide nanoparticles on A549 cells. *Environ. Sci. Nano* **2019**, *6*, 565–579. [CrossRef]

20. Chen, N.; Song, Z.-M.; Tang, H.; Xi, W.-S.; Cao, A.; Liu, Y.; Wang, H. Toxicological Effects of Caco-2 Cells Following Short-Term and Long-Term Exposure to Ag Nanoparticles. *Int. J. Mol. Sci.* **2016**, *17*, 974. [CrossRef]

21. CDC-NIOSH. Pocket Guide to Chemical Hazards—Titanium Dioxide. 2019. Available online: https://www.cdc.gov/niosh/npg/npgd0617.html (accessed on 11 June 2021).

22. CDC-NIOSH. Pocket Guide to Chemical Hazards—Silica, Amorphous. 2019. Available online: https://www.cdc.gov/niosh/npg/npgd0552.html (accessed on 11 June 2021).

23. De Temmerman, P.-J.; Verleysen, E.; Lammertyn, J.; Mast, J. Semi-automatic size measurement of primary particles in aggregated nanomaterials by transmission electron microscopy. *Powder Technol.* **2014**, *261*, 191–200. [CrossRef]

24. Hackenberg, S.; Scherzed, A.; Technau, A.; Froelich, K.; Hagen, R.; Kleinsasser, N. Functional responses of human adipose tissue-derived mesenchymal stem cells to metal oxide nanoparticles in vitro. *J. Biomed. Nanotechnol.* **2013**, *9*, 86–95. [CrossRef] [PubMed]

25. Kocbek, P.; Teskač, K.; Kreft, M.E.; Kristl, J. Toxicological aspects of long-term treatment of keratinocytes with ZnO and TiO$_2$ nanoparticles. *Small* **2010**, *6*, 1908–1917. [CrossRef]

26. Vales, G.; Rubio, L.; Marcos, R. Long-term exposures to low doses of titaniumx dioxide nanoparticles induce cell transformation, but not genotoxic damage in BEAS-2B cells. *Nanotoxicology* **2015**, *9*, 568–578. [CrossRef]

27. Gliga, A.R.; Skoglund, S.; Wallinder, I.O.; Fadeel, B.; Karlsson, H.L. Size-dependent cytotoxicity of silver nanoparticles in human lung cells: The role of cellular uptake, agglomeration and Ag release. *Part. Fibre Toxicol.* **2014**, *11*, 11. [CrossRef]

28. Rahman, Q.; Lohani, M.; Dopp, E.; Pemsel, H.; Jonas, L.; Weiss, D.G.; Schiffmann, D. Evidence that ultrafine titanium dioxide induces micronuclei and apoptosis in Syrian hamster embryo fibroblasts. *Environ. Health Perspect.* **2002**, *110*, 797–800. [CrossRef]

29. Napierska, D.; Thomassen, L.C.J.; Rabolli, V.; Lison, D.; Gonzalez, L.; Kirsch-Volders, M.; Martens, J.; Hoet, P.H. Size-dependent cytotoxicity of monodisperse silica nanoparticles in human endothelial cells. *Small* **2009**, *5*, 846–853. [CrossRef]

30. Brown, J.S.; Gordon, T.; Price, O.; Asgharian, B. Thoracic and respirable particle definitions for human health risk assessment. *Part. Fibre Toxicol.* **2013**, *10*, 12. [CrossRef] [PubMed]

31. Mendoza, A.; Torres-Hernandez, J.A.; Ault, J.G.; Pedersen-Lane, J.H.; Gao, D.; Lawrence, D.A. Silica nanoparticles induce oxidative stress and inflammation of human peripheral blood mononuclear cells. *Cell Stress Chaperon* **2014**, *19*, 777–790. [CrossRef] [PubMed]

32. Liang, H.; Jin, C.; Tang, Y.; Wang, F.; Ma, C.; Yang, Y. Cytotoxicity of silica nanoparticles on HaCaT cells. *J. Appl. Toxicol.* **2013**, *34*, 367–372. [CrossRef]

33. Guo, C.; Xia, Y.; Niu, P.; Jiang, L.; Duan, J.; Yu, Y.; Zhou, X.; Li, Y.; Sun, Z. Silica nanoparticles induce oxidative stress, inflammation, and endothelial dysfunction in vitro via activation of the MAPK/Nrf2 pathway and nuclear factor-κB signaling. *Int. J. Nanomed.* **2015**, *10*, 1463–1477. [CrossRef] [PubMed]
34. Johnston, C.J.; Driscoll, K.E.; Finkelstein, J.N.; Baggs, R.; O'Reilly, M.A.; Carter, J.; Gelein, R.; Oberdörster, G. Pulmonary chemokine and mutagenic responses in rats after subchronic inhalation of amorphous and crystalline silica. *Toxicol. Sci.* **2000**, *56*, 405–413. [CrossRef] [PubMed]
35. Arts, J.H.; Muijser, H.; Duistermaat, E.; Junker, K.; Kuper, C.F. Five-day inhalation toxicity study of three types of synthetic amorphous silicas in Wistar rats and post-exposure evaluations for up to 3 months. *Food Chem. Toxicol.* **2007**, *45*, 1856–1867. [CrossRef] [PubMed]

 *nanomaterials*

Review

The State of the Art and Challenges of In Vitro Methods for Human Hazard Assessment of Nanomaterials in the Context of Safe-by-Design

Nienke Ruijter [1], Lya G. Soeteman-Hernández [1], Marie Carrière [2], Matthew Boyles [3], Polly McLean [3], Julia Catalán [4,5], Alberto Katsumiti [6], Joan Cabellos [7], Camilla Delpivo [7], Araceli Sánchez Jiménez [8], Ana Candalija [7], Isabel Rodríguez-Llopis [6], Socorro Vázquez-Campos [7], Flemming R. Cassee [1,9] and Hedwig Braakhuis [1,*]

[1] National Institute for Public Health & the Environment (RIVM), 3721 MA Bilthoven, The Netherlands
[2] Univ. Grenoble-Alpes, CEA, CNRS, SyMMES-CIBEST, 17 rue des Martyrs, 38000 Grenoble, France
[3] Institute of Occupational Medicine (IOM), Edinburgh EH14 4AP, UK
[4] Finnish Institute of Occupational Health, 00250 Helsinki, Finland
[5] Department of Anatomy, Embryology and Genetics, University of Zaragoza, 50013 Zaragoza, Spain
[6] GAIKER Technology Centre, Basque Research and Technology Alliance (BRTA), 48170 Zamudio, Spain
[7] LEITAT Technological Center, 08225 Barcelona, Spain
[8] Instituto Nacional de Seguridad y Salud en el Trabajo (INSST), 48903 Barakaldo, Spain
[9] Institute for Risk Assessment Sciences (IRAS), Utrecht University, 3584 CS Utrecht, The Netherlands
* Correspondence: hedwig.braakhuis@rivm.nl

Citation: Ruijter, N.; Soeteman-Hernández, L.G.; Carrière, M.; Boyles, M.; McLean, P.; Catalán, J.; Katsumiti, A.; Cabellos, J.; Delpivo, C.; Sánchez Jiménez, A.; et al. The State of the Art and Challenges of In Vitro Methods for Human Hazard Assessment of Nanomaterials in the Context of Safe-by-Design. *Nanomaterials* **2023**, *13*, 472. https://doi.org/10.3390/nano13030472

Academic Editors: Andrea Hartwig and Christoph Van Thriel

Received: 23 December 2022
Revised: 16 January 2023
Accepted: 18 January 2023
Published: 24 January 2023

Abstract: The Safe-by-Design (SbD) concept aims to facilitate the development of safer materials/products, safer production, and safer use and end-of-life by performing timely SbD interventions to reduce hazard, exposure, or both. Early hazard screening is a crucial first step in this process. In this review, for the first time, commonly used in vitro assays are evaluated for their suitability for SbD hazard testing of nanomaterials (NMs). The goal of SbD hazard testing is identifying hazard warnings in the early stages of innovation. For this purpose, assays should be simple, cost-effective, predictive, robust, and compatible. For several toxicological endpoints, there are indications that commonly used in vitro assays are able to predict hazard warnings. In addition to the evaluation of assays, this review provides insights into the effects of the choice of cell type, exposure and dispersion protocol, and the (in)accurate determination of dose delivered to cells on predictivity. Furthermore, compatibility of assays with challenging advanced materials and NMs released from nano-enabled products (NEPs) during the lifecycle is assessed, as these aspects are crucial for SbD hazard testing. To conclude, hazard screening of NMs is complex and joint efforts between innovators, scientists, and regulators are needed to further improve SbD hazard testing.

Keywords: nanomaterials; safe-by-design; hazard testing; in vitro methods; SAbyNA; advanced materials

1. Introduction

The rapid expansion of the field of nanotechnology and its ever-growing number of applications has created a challenge for toxicologists and risk assessors. The continuous uncertainties surrounding nanomaterial (NM) safety, as well as the pace at which new NMs are developed, call for a more prevention-oriented strategy. The Safe-by-Design (SbD) concept is increasingly applied within the field of nanotechnology, as can be seen by the high number of EU funded nano-projects addressing SbD over the past years [1], and by its adoption in the EU Chemical Strategy for Sustainability as a strategy to meet the EU Green Deal ambitions [2,3].

SbD aims to reduce the human and environmental risk of a substance throughout its entire life cycle by minimizing or eliminating the hazard and/or by reducing exposure [4]. The concept of SbD consists of three pillars: safer materials and products, safer production, and safer use and end-of-life. For NMs, these were first described in the NanoReg2 project [5], and later in an internationally accepted working description of the OECD Safe Innovation Approach Report [6]. In practice, SbD is a two-step process: the first step is an early hazard and/or risk screening during the design phase of the innovation process of a new substance, NM, or product [7,8]. The second step is to take actions (SbD interventions) to reduce or minimize hazard, exposure, or both.

For NMs and nano-enabled products (NEPs), SbD interventions can be achieved in different ways. One option is to modify a NM in order to improve its safety profile. For example, Xia et al. (2011) showed that doping ZnO nanoparticles with iron reduces the shedding of harmful ions and reduces the toxicity of the particles upon pulmonary exposure [9]. Another example of a SbD intervention is applying a surface treatment to minimise NM biological reactivity, as has been successfully achieved for nano-SiO_2 by adding silanol groups to the silica surface [10]. Reducing exposure is also a fundamental part of SbD and can be achieved by implementing procedural changes such as working in closed systems or using wet synthesis methods [5]. Reduced release and therefore minimized exposure can also be achieved by altering the design of the NEP, for example by improving the immobilization between the NM and the matrix, as was conducted for silver NMs onto cotton fabrics [11].

The above-mentioned examples can only be achieved after first assessing hazard and risk early in the innovation process, and then using this knowledge to integrate safety into the design of the NM, NEP, or production process. For many NMs, and especially for novel ones, hazards are largely unknown [12], and cannot be predicted only based on physicochemical (PC) characterisation. Therefore, carrying out suitable hazard testing at the early stages of product development is of utmost importance for SbD applicability. SbD hazard testing aims to identify hazard warnings in the early stages of the innovation process using simple in vitro methods. Once a product is designed and produced, the manufacturer should comply to the regulations and perform hazard and risk assessment accordingly.

Many strategies and frameworks for hazard assessment of NMs in the context of SbD have been proposed in recent years [13–18], some proposing specific in vitro assays, and some based only on a selection of toxicological endpoints to consider. However, no comprehensive investigation of the suitability of currently available in vitro assays for such strategies has been conducted thus far.

From previous studies on the mechanisms of action of NMs, it is known that transformation (e.g., dissolution), reactivity, inflammation, cytotoxicity, and genotoxicity are among the most important parameters and endpoints to evaluate when assessing the hazard of a NM, and therefore these are suggested to be measured in many available strategies and frameworks [13,19–21]. Selecting in vitro tests suitable for SbD hazard testing is not trivial. Only a few OECD test guidelines and ISO standards are available specifically for testing NMs. Due to the interfering behaviour of NMs with their surrounding environment and with the assay readout, routinely used toxicity assays (i.e., those used to test soluble chemicals) may prove unsuitable or may require optimizations and inclusion of extra controls [22]. In contrast with hazard assessment for soluble chemicals, NM testing requires additional steps, such as dispersion protocols and determining the dose delivered to the cells in submerged cell culture experiments [23]. Specifically for the purpose of SbD hazard testing, since it is performed early in the development of a NM/NEP, assays will not only have to be compatible with the NM to be tested, but should preferably also be fast, cost-effective, and able to correctly indicate hazard warnings.

This work provides a practical and critical evaluation of the suitability of most frequently used in vitro toxicity assays and the challenges for their use in NM SbD hazard testing. For this purpose, criteria for the suitability of methods for application in a SbD hazard testing strategy are established, leading to an evaluation of the methods currently in use for the parameters and endpoints identified as important for the mechanisms of action of NMs. This work is conducted under the umbrella of the Horizon2020 project SAbyNA which aims to develop a user-friendly platform for industry with optimal workflows to support the development of SbD NMs and NEPs. For this purpose, existing resources, such as in vitro assays are identified, distilled, and streamlined. This state of the art and evaluation of in vitro assays for SbD applicability can be used as an outlook for innovators, regulators, industry, and scientists of how early hazard testing of NMs and NEPs can be put into practice to eventually contribute to the design of SbD NEPs.

2. Criteria

A set of performance criteria is proposed to evaluate the suitability of in vitro methods for SbD hazard testing. The criteria were adapted from the widely used Good In Vitro Method Practices (GIVIMP) [24] and tailored to suit SbD hazard testing for NMs specifically. Figure 1 shows several key considerations for assay selection for SbD hazard testing.

Figure 1. Considerations for assay selection for SbD hazard testing.

Predictive: The first criterium is that an in vitro assay should be sufficiently predictive of the in vivo situation. This comparison is preferably made with human data, or alternatively using animal data. SbD hazard testing is carried out in an early stage of product development and is considered a first screening. The aim of SbD hazard testing is to detect **early hazard warnings** and not to derive a point of departure for risk assessment. Therefore, assays are sufficiently predictive for SbD hazard testing when they are able to indicate hazard warnings. Assays that are able to **accurately rank** NMs/NEPs based on their hazard potency are of extra value for SbD hazard testing, as this will allow comparison of candidate NMs and comparison with benchmark NMs.

Predictivity can be assessed by looking into the prediction accuracy of the assay. An assay's **accuracy** to predict in vivo effects is a combination of its **sensitivity** and **specificity**. Sensitivity is the ability of the assay to detect true positives and specificity is the ability of the assay to detect true negatives.

Simple and cost effective: Simplicity and cost effectiveness are key for SbD hazard testing since these assays are to be performed in an early stage of NM/NEP product development. Ideally, an assay should be easy to perform, time-efficient and cost-effective.

Robust: An assay should give consistent and repeatable results between experimental repetitions and between different labs.

Compatible: An assay should preferably be compatible with a wide range of NMs, or at least its compatibility domain should be identified. Assays with optimized protocols specifically for NMs are preferred.

Readiness: Methods that are considered 'ready to use' and already standardized or (pre-)validated for NMs are prioritized.

3. Challenges of Testing NMs In Vitro for SbD Applicability

NMs are particulate matter, making NM in vitro testing by default more challenging than testing soluble chemicals. Several additional aspects need considering when testing NMs in vitro, including determining the behaviour of the NM in exposure medium, selecting a dispersion protocol to create stable suspensions which preferably mimic human exposure as much as possible, and assessing the potential interference of the NM with the assay components or optical readouts. Furthermore, elaborate characterization of the NM is required [25], but this will not be discussed further in this review. The fact that SbD hazard testing needs to be as simple as possible creates an important predicament that needs addressing. An overview of key aspects that should be taken into account when performing SbD hazard testing of NMs is shown in Figure 2. The most important aspects are discussed below.

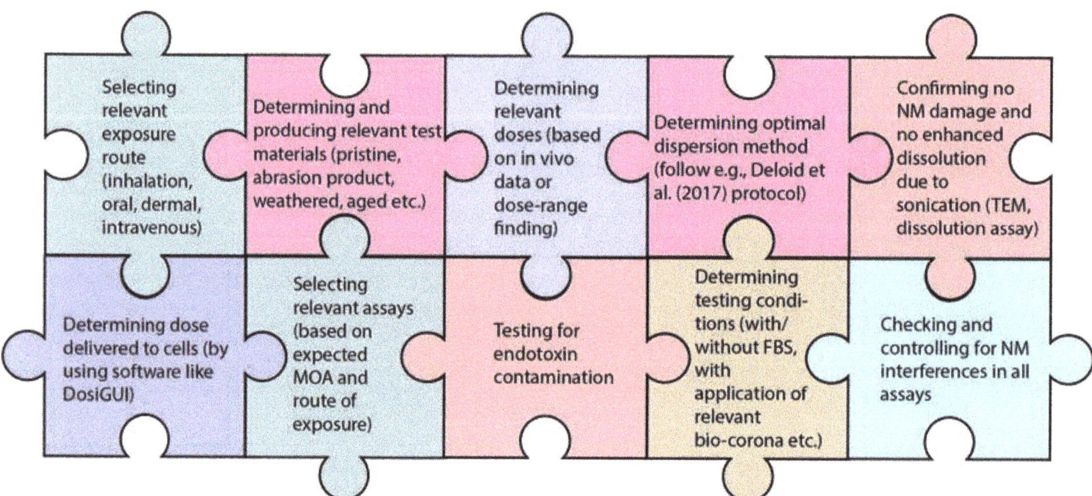

Figure 2. Overview of aspects that might have to be considered when performing SbD hazard testing, showing that simple testing can be challenging to achieve.

3.1. Choice of Dispersion Protocol

Classically, in vitro toxicity evaluation is performed in cultured cells maintained in submerged conditions. To ensure reproducible and controlled exposure from one replicate to another, stable suspensions of well-dispersed NMs are prepared, sometimes requiring energy input to disrupt particle agglomerates. For SbD hazard testing, a prerequisite for the suitability of a dispersion procedure is that the SbD properties (e.g., coatings or surface treatments) of the dispersed NM/NEP are preserved, and that the resulting dispersed NM/NEP is relevant for human exposure in terms of size and other physicochemical properties.

The most commonly used dispersion procedure for NMs is via sonication, using either an ultrasonic bath, a probe, or a cup-type sonicator [26,27]. For SbD hazard testing, sonication would not be the preferred option in some specific cases. For instance, sonication can break multi-walled carbon nanotubes (MWCNT), causing a reduction of their length [27], and therefore leading to different toxicity profiles than the MWCNT that humans would be exposed to. Sonication has been used to produce MWCNTs with different lengths from a same initial batch of MWCNTs [28,29], and Hadrup et al. (2021) concluded that the length of the MWCNT is a major determinant of its toxicity [29]. Therefore, when assessing hazard properties of CNTs in vitro, sonication should be limited as much as possible, and in case sonication is used, NM physicochemical properties should be verified to ensure they still maintain exposure-relevant characteristics.

Another example where sonication would have to be carefully considered is when testing specific synthetic amorphous silicas, which are in some cases intentionally produced as agglomerates. Sonication can disrupt most of the agglomerates, reducing the overall hydrodynamic diameter, which constitutes a substantial modification of the initial material [30]. In an inhalation exposure scenario, a person would be exposed to these agglomerates, and therefore sonication would not be the preferred option. However, there are indications that these agglomerates disintegrate in the intestine [31]. In the case of ingestion, the gut epithelium is exposed to nano-sized silica, and sonication would result in an exposure-relevant material.

Dissolution is a major determinant of the toxicity of some NMs (e.g., release of toxic metal ions such as silver or copper ions) and decreasing the NM dissolution potential can be considered a SbD intervention. Sonication has been shown to enhance the dissolution of some metallic NMs, such as Cu, Mn, and Co [32], and the dissolution can be further increased when proteins are present in the solution during sonication [33]. Thus, dispersion procedures that involve sonication of NMs, especially in a medium that contains proteins, such as the procedure optimized within the Nanogenotox project [34], should be carefully considered in view of exposure scenarios when testing NMs that can potentially dissolve.

Some NMs are designed as core-shell structures (e.g., quantum dots (QDs)) where the shell reduces dissolution and leaching of potentially toxic elements from the core. The design of more robust shells, used as a SbD intervention, reduces the QDs dissolution rate and thereby their toxicity [35–37]. However, the core-shell boundary is a region of fragility and sonication could promote shell fragmentation and the release of core contents. Thus, sonication could result in a reduced effect of the SbD intervention, potentially resulting in an overestimation of toxicity in SbD hazard testing. Therefore, for core-shell NMs, sonication should not be recommended, unless humans are exposed to fragmented QDs.

Coating NMs with surface ligands or grafting them on an inert matrix such as cellulose has also been tested as a SbD intervention to produce safe(r) photocatalytic paints containing TiO_2 NMs. Coating TiO_2 NMs with polyethylene glycol (PEG), poly(acrylic acid) (PAA), or 3,4-dihydroxy-L-phenylalanine (DOPA) increased the stability of the doped paint and its resistance to weathering and abrasion, while their grafting on cellulose fibres enhanced their photocatalytic properties, thereby allowing for the reduction of the amount of NMs necessary to reach efficient photocatalysis [38]. Again, sonicating these surface-coated TiO_2 NMs or TiO_2-containing composites might lead to the reduction of the effect of SbD interventions. In addition to that, extensive sonication has been shown to alter the zeta potential of TiO_2 and CeO_2 NMs [39,40] and to cause re-agglomeration of Cu or Mn NMs [33,41]. It should in each case be investigated what the exposure-relevant form of the NM or NEP is.

Moreover, samples could be contaminated by the release of Al and Ti from the sonication probe upon extensive sonication, potentially leading to toxicity [30,42]. Finally, extensive sonication of NMs in a growth medium containing proteins or in water with bovine serum albumin (BSA) (as in the procedure optimized within the Nanogenotox project [34]) could promote the degradation of proteins, leading to the formation of large aggregates of degraded proteins [43].

To conclude, for SbD hazard testing sonication should be carefully considered, and exposure relevancy should always be kept in mind. If exposure-relevant and stable dispersions in the exposure medium are obtained using simple methods such as vortexing, dispersion via sonication might not be needed. In the case of NMs that quickly agglomerate and form large clumps, more controlled sonication methods might be appropriate. For example, a protocol using minimum material-specific energy to reach a stable dispersion as described by DeLoid et al. (2017) could be used [23]. NMs should subsequently be characterized to ensure that no PC changes were produced that deviate from the exposure-relevant material. The PC properties of the NM tested should reflect the exposure conditions, whether it be the pristine NM with SbD interventions or the agglomerated NM. However, it has to be noted that unstable suspensions could lead to difficulties with reproducibility and/or interferences with the assay readout.

Finally, it might be recommended to also perform in vitro assays after extensive sonication, as this might be required for regulatory risk assessment. By doing this, the first steps towards compiling a dossier for regulatory compliance are made, and this might already indicate if any issues can be foreseen for market entry. Additionally, extensive sonication may provide a worst-case scenario in in vitro assays, which could fit in a precautionary approach. OECD guidance on sample preparation [44] is currently being revised to include considerations for the choice of a specific dispersion protocol rather than applying extensive sonication by default.

Box 1

> For SbD hazard testing sonication should be carefully considered, and exposure relevancy should always be kept in mind. If exposure-relevant and stable dispersions in the exposure medium are obtained using simple methods such as vortexing, dispersion via sonication might not be needed. In the case of NMs that quickly agglomerate and form large clumps, more controlled sonication methods might be appropriate. NMs should subsequently be characterized to ensure that no PC changes were produced that deviate from the exposure-relevant material.

3.2. Influence of Medium Components

Supplementation of cell-culture medium with serum (i.e., foetal calf serum (FCS) and foetal bovine serum (FBS)) is common practice in cell culture procedures as it is required for cell growth and maintenance. When exposing the NMs in a test medium, constituents of the medium including proteins, amino acids, lipids, and sugars adsorb on the surface of NMs, leading to the formation of the so-called biomolecular corona [45]. This corona is highly dynamic and may change upon changing the composition of the test medium [46,47]. This dense layer of biological molecules can modify NM toxicity in several ways. Firstly, it could do so by masking the surface reactive sites of the NM [48]. Secondly, serum may stabilize the NM dispersion, leading to a lower dose delivered to the cells in in vitro assays, as has been shown for TiO_2 NMs for example [49]. Lastly, a biomolecular corona may reduce NM surface energy, and thereby its cellular uptake via adhesion-induced endocytosis, as has been shown for SiO_2 NMs [50,51].

These effects are clear when comparing the toxicity of NMs tested with and without serum. For instance, the cytotoxic potency of polystyrene NMs was found to decrease 2-fold when the exposure medium contained serum [52]. Similarly, the cytotoxicity of SiO_2 NMs decreased up to 92%, and pro-inflammatory response decreased up to 87% when cells were exposed in medium with serum [53]. In addition, the species of origin of the serum could lead to different responses [54]. Addition of bovine serum albumin (BSA), a protein often used to help stabilize dispersions, has also been reported to reduce cytotoxicity [55,56].

In short, the addition of serum and BSA to the exposure medium may lead to lower toxicity in in vitro assays. Which approach is most suitable for SbD hazard testing should be explored further. Since SbD hazard testing is mostly focused on detecting hazard warnings, it could be argued that testing without serum is more appropriate, as it ensures a higher sensitivity. Additionally, when testing serum-free, a worst-case scenario could be mimicked without the protective effect of serum on NM-cell interaction. On the other hand, testing with serum is the more realistic approach as humans are rarely exposed to NMs without a biomolecular corona. Eventually, the route of (potential) human exposure should be taken into account when selecting an exposure protocol, as systemically injected NMs will immediately be covered by serum proteins, whereas inhaled NMs will come in contact with epithelial lining fluids, which contains a different set of biomolecules. For SbD hazard testing, exposure relevancy is important, and a biocorona could be applied which corresponds to the route of exposure, such as lung-lining fluid for pulmonary exposure. In the context of exposure relevance, and in the context of the 3Rs (replacement, reduction, and refining), the use of human serum or serum-free alternatives may be favoured over FBS.

3.3. Determining Dose Delivered to Cells

A particular challenge when testing NMs in vitro in adherent, submerged cell cultures is determining the delivered dose, i.e., the amount of material that reaches the cells. Settling of NMs depends on their density, size, and the properties of the cell-culture medium, as well as on their agglomeration state [57]. The latter is again influenced by the dispersion method used [58].

Determining the delivered dose in an in vitro experiment is an absolute requirement, even when performing a simple hazard screening for the purpose of SbD, and when performing high throughput screening (HTS) experiments. This is because the administered dose can differ substantially from the delivered dose that reaches the cells. For example, for particles that settle rapidly, the difference between administering 100 µL per well or 200 µL per well of the same concentration will mean a doubling of the amount of material per well, and thus a potential doubling of the delivered dose. Moreover, since sonication influences the agglomeration state, and the agglomeration state influences the settling rate and thus the dose delivered to cells, determining the delivered dose may help the comparison of data among independent experiments using different dispersion protocols. A visual representation of two example NMs with different settling rates is shown in Figure 3. For SbD hazard testing specifically, determination of delivered dose aids in a comparison to benchmark materials with known toxicity, as settling may differ greatly between NMs. The importance of determining the delivered dose was shown in a study by Pal et al. (2015), where a correction for the delivered dose led to a considerable change in the hazard ranking of a panel of NMs, after which the in vitro outcomes matched better with the in vivo results [59].

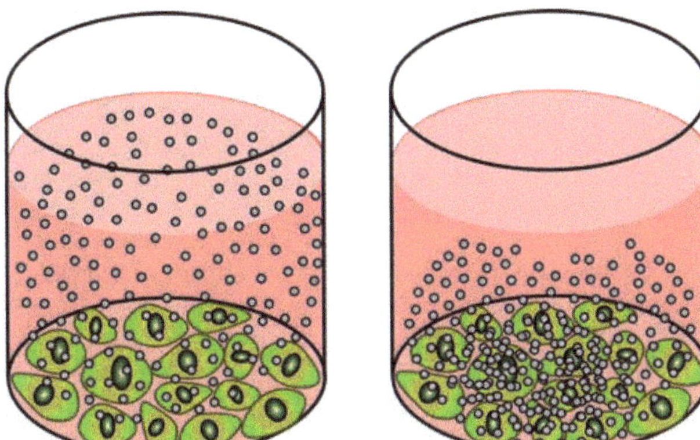

Figure 3. Visual representation of two NMs with different properties, resulting in different doses delivered to the cells, when administered doses are equal.

The delivered dose can be modelled using the DG [60] or ISD3 model [61], which are currently available in the DosiGUI software generated in the PATROLS project [62]. The DosiGUI is user friendly, however, these models also introduce uncertainty as they do not take into account some critical factors such as particle convection [63], or dispersion-stabilizing surface functionalization. Moreover, cell stickiness needs to be chosen from an arbitrary scale, which is often an unknown parameter that has a big effect on the modelled delivered dose [64].

These models require the effective density of the NM as input, which is the density of the NM in a dispersion. In the case of agglomerates, this includes the density of the medium trapped inside the agglomerate. The effective density can be measured using analytical ultracentrifugation (AUC) or using the volumetric centrifugation method (VCM). In order to adhere to the criteria of SbD hazard testing, the VCM is preferred over AUC, as it is easier, less costly, and does not require specialized equipment [65]. It is important to measure model input parameters precisely, as small differences in input as a result of instrument variation may lead to large differences in the modelled deposited dose [64]. Stable dispersions are a requirement for modelling the dose rate and final dose delivered to the cells, as the calculations are based on a one-size distribution. The accuracy of the model outcome—and thus the estimated deposited dose (rate)—is less accurate if the size distribution of the dispersion changes over time due to agglomeration or aggregation.

The need to determine the delivered dose adds an extra step to SbD hazard testing, leading to a reduction of the achievable simplicity. Determining the delivered dose is however a requirement, even for SbD hazard testing, as more precise dosimetry will allow for more informative hazard testing. This, however, only applies to submerged testing, and not to experiments in which, for instance, an air-liquid interface (ALI) exposure protocol is followed. For ALI exposures, a quartz crystal microbalance (QCM) may provide sensitive and accurate deposited-dose measurements [66].

3.4. SbD Hazard Testing of NEPs and NMs Released during the Life Cycle

One of the most important aspects of SbD is assessing the safety of a product along its entire life cycle (LC) [5]. Usually, only pristine NMs are included in toxicity assays. This could be insufficient, as humans are also exposed to NEPs, aged NM and/or NMs released during the product LC including production, use, and end-of-life. The physicochemical characteristics of the NMs released to the environment along the different stages of the LC can be very different in terms of shape, chemical composition, agglomeration state, and surface modification [67–71]. Moreover, NEPs and NMs released during the LC might pose a different hazard than pristine NMs [72–75]. Thus, gathering information on the characteristics and hazard of NEPs and NMs released during the LC is important for designing relevant SbD interventions.

Processes leading to the release of NMs from a NEP during the entire LC can be simulated under laboratory conditions, after which NMs can be collected (e.g., by using filters) and redispersed in liquid for toxicity testing [76] or collected in liquid suspensions directly [77]. Realistic, released NMs, relevant for consumer exposure during the use of NEPs, can be obtained by using standardized methods (e.g., abrasion and weathering) that are normally used to test the durability and performance of NEPs. In the case of abrasion processes, there are different instruments that can be used to simulate mild or hard abrasion. Experimental parameters (e.g., cycles of abrasion, abrasion materials, normal load at the top of the abrader, etc.) can be tuned to reflect closely the NEP use conditions.

Aging experiments simulate conditions to which a product could be exposed during its use phase and are usually performed in a weathering chamber under accelerated conditions of UV exposure and rain. The weathering conditions (e.g., duration of cycles of light and rain, duration of the experiment, etc.) can be selected to follow international standards or be customized. In order to obtain higher quantities of released material (worst-case scenario), NEPs can be fragmented and sieved [78].

Released aerosols can be size-separated by using e.g., a cascade impactor to ensure inhalable or respirable fractions of NMs; After which they are collected on filters [79]. Efforts should be made to ensure high extraction efficiency and minimal compositional alterations when extracting material from filters [80]. Another option which is less easy but mimics better a real-life exposure is the direct exposure of cells to the released material, as performed by Zarcone and colleagues [81] for diesel exhaust. This approach is however somewhat more labour-intensive for SbD hazard testing but might be useful for gaining a more fundamental insight into the toxicity of released materials (exposure-relevant material) without losing a fraction to filter extraction.

After obtaining and extracting the material from filters, the same actions should be taken as for pristine NM testing, such as an accurate dose determination, controlling for interference, endotoxin contamination testing, choosing an appropriate dispersion protocol, etc. For NEPs and NMs released during the LC, endotoxin contamination might pose an extra challenge as these materials are generally not produced in sterile environments. Finally, compatibility in submerged settings might pose additional challenges as a NEP matrix is often plastic-based and might float on culture medium.

Feasibility and Relevance

Obtaining sufficient amounts of NM that are released at a given stage of the LC can be challenging due to low emission rates, contamination with other substances (e.g., sanding material), as well as laboursome and time-consuming procedures. The question is whether performing SbD hazard testing on pristine NMs is sufficiently relevant when assessing the safety of a NM along its entire life cycle. There are some examples in literature in which both pristine and released NMs have been tested to assess and compare their hazards. In most cases, materials released during the LC induced less or equal toxicity as compared to the pristine NM, as has been shown in in vivo studies [75,82,83] as well as in vitro [84]. This means that testing pristine materials, albeit often far from representing the reality, can still represent a worst-case scenario. In this case, risk screening of NMs released from a NEP can be mainly based on emission rates combined with the hazard information of the pristine NM. However, it should be noted that there is very little known about the toxicity of released NMs as compared to pristine NMs, and the exception may prove the rule.

Similarly, testing pristine NMs may represent a worst-case scenario for aged NMs. For example, freshly ground quartz particles have been shown to induce higher levels of pulmonary inflammation and cytotoxicity as compared to aged quartz [85].

For now, it should be considered on a case-by-case basis whether testing forms other than the pristine NM is required. More research is needed to determine whether the use of the pristine NM in SbD hazard testing is sufficient due to its 'worst-case' nature, or whether testing aged NMs, and NMs released during LC is crucial for designing SbD interventions.

Box 2

> It should be considered on a case-by-case basis whether testing forms other than the pristine NM is required. More research is needed to determine whether the use of the pristine NM in SbD hazard testing is sufficient due to its 'worst-case' nature, or whether testing aged NMs, and NMs released during LC is crucial for designing SbD interventions.

3.5. Challenging NMs and Advanced Materials

Hydrophobic NMs, NMs with low material density, multi-component NMs, and other advanced materials yet to be invented may show poor compatibility with commonly used in vitro assays. Applying a single standardized exposure method to all types of NMs will inherently give biased outcomes. For SbD hazard testing it is therefore important to consider the compatibility of NMs with challenging physicochemical properties and to be prepared for future novel advanced materials.

3.5.1. Hydrophobic Particles

Since cells are always cultured in aqueous culture medium, hydrophobic particles can be of extra difficulty to test. Carbon nanotubes (CNTs) and graphene-based particles, for example, are notorious for being difficult to disperse in culture medium. Ethanol pre-wetting, using different dispersion media [86], and adjusting sonication time and frequency have been shown to improve dispersibility of NMs. However, for some NMs a stable dispersion can never be achieved in cell-culture medium. For example, some CNTs are specifically designed to agglomerate in order to reduce their dustiness and thereby improve their safety. For cases such as these, dry exposures at the air-liquid interface (ALI) should be considered when focussing on potentially respirable NMs. This requires a cell type that can be cultured on membranes in medium on the basal side, while being exposed to the air on the apical side. For the generation of a dry (dust) aerosol for ALI exposure, several methods are available [87]. However, it should be kept in mind that if a dust cannot be generated in a laboratory setting, inhalation of the NM is very unlikely. Thus, in these cases the relevance of an inhalation study should be reconsidered.

3.5.2. Buoyant NMs

NMs with a density lower than cell-culture medium (e.g., certain types of plastic particles or agglomerates with a low effective density) will float and do not settle over time, resulting in no contact with adhering cells in a classical, submerged in vitro setup. This will likely lead to an underestimation of the potency of NMs such as nano-plastics and liposomes [88]. A solution to solve the problem of buoyant particles is to perform an inverted 'overhead' cell culture, where the cells are not cultured on the bottom of a culture dish, but upside down on top of the exposure medium. With this approach, it was possible to produce a dose response for several floating particles, whereas the traditional approach did not show any results [88,89]. For buoyant NMs, where inhalation is the relevant exposure route, ALI exposures can also be an option.

3.5.3. Multicomponent NMs and Other Advanced Materials

In the past years, the more complex multicomponent nanomaterials (MCNM) have gained popularity. These next generation NMs consist of two or more materials or substances, giving rise to properties (e.g., reactivity) that are not equal to the sum of the properties of each component [90]. There are still many knowledge gaps when it comes to the toxicity of these and other novel advanced materials, which is why the concept of SbD is a suitable prevention-oriented approach. Whether these materials are compatible with the available toxicity assays is unknown, and this might pose challenges for future SbD hazard testing. Theoretically, MCNMs could exhibit multiple types of assay interference, attributed to the individual components of the MCNM. It is important to be aware of these challenges and to always assess interference.

4. Evaluation of In Vitro Methods for SbD Hazard Testing

4.1. Cytotoxicity

Measuring cell viability or cytotoxicity is a fundamental part of most hazard assessment strategies and integrated approaches to testing and assessment (IATA's) for several reasons. Firstly, cytotoxic potency (for example LC_{50}) gives an indication of the relative hazard of a NM. Secondly, cytotoxicity assays allow for the selection of appropriate sublethal doses for further mechanistic testing (e.g., genotoxicity and inflammation). Lastly, for several mechanistic assays such as genotoxicity assays, cytotoxicity measurements are a requirement for the correct interpretation of the results. Cell viability can be determined by the measurement of various cellular parameters, such as mitochondrial activity, lysosomal integrity, and membrane integrity. Different endpoints should be included to assess cytotoxicity [91–93], as results from different assays do not always correspond [94].

4.1.1. Most Frequently Used Assays, Strengths and Limitations

The most-used approaches for measuring cytotoxicity or cell viability in vitro include measuring mitochondrial activity (examples are MTT, MTS, XTT, and WST-1 assays), release of cytoplasm components (examples are LDH and AK), lysosomal integrity (Neutral red uptake), apoptosis markers (caspase 3/7), and stains that can specifically enter apoptotic and/or necrotic cells (Trypan blue, Propidium iodide, and Annexin V). Propidium iodide and Annexin V can be combined to determine plasma membrane restructuring which can be representative of either necrosis or apoptosis specifically. Most cytotoxicity assays are relatively simple, can be carried out in a 96-well microplate format, could be used for HTS, and have commercial kits available. An exception are the assays that require microscopic evaluation of a certain staining, as this is more labour-intensive.

For many cytotoxicity and viability assays, NMs can interfere with assay reagents and/or the optical readout [95–98]. Therefore, potential interference of the NMs with assays should always be assessed [92,99]. The elimination of NMs via high-speed centrifugation may reduce optical interference [92]. Alternatively, since mitochondrial activity is measured intracellularly, NMs can be washed away from cells prior to incubation with the reagent to avoid interaction of the NMs with the reagent [100]. For products measured in the supernatant (e.g., LDH and AK), washing is not feasible, but centrifugation can help remove larger NMs and thereby reduce optical interference. However, some NMs are known to inactivate or adsorb LDH directly [101]. If interference still occurs after taking precautions, it is advised to perform another type of cytotoxicity test, as the subtraction of the average background signal of the NMs will reduce the accuracy of the outcome [97,102,103]. For specific NMs, some assays might prove not to be compatible.

Much effort has been put into the optimization and standardization of in vitro cytotoxicity assays specifically for NMs in the past years. An ISO standard for the MTS assay was published in 2018 [104]. In 2021, an ISO standard was published for impedance measurements for NMs specifically [105]. This assay involves growing cells on an electrode during exposure to the NM. The detachment of cells, indicating cytotoxicity, is measured as a decrease in electrochemical impedance, as demonstrated in the assessment of polylactic acid NM-induced toxicity in A549 epithelial cells [106]. This assay is less prone to interferences as no optical readout and no assay reagents are required. However, it does require specialized equipment not available in many labs. Internationally standardized and harmonized standard operating procedures (SOPs) for other cytotoxicity assays have not been published to date.

In the NanoReg project, an interlaboratory study for the MTS assay was carried out, and acceptable robustness levels were found depending on the cell type. The human alveolar cell-line A549 showed a good agreement in cytotoxicity between labs, whereas the differentiated human monocyte cell-line THP-1 (dTHP-1) showed varying results and a poor robustness [107]. In a large interlaboratory study by Piret et al. (2017), a good robustness was found for the MTS assay and ATP content measurements. These comparisons were carried out using both A549 as well as dTHP-1 cell lines, and two different NMs. The authors stressed the importance of avoiding interference of the NM with the assay in order to obtain more reliable results, and a lower inter-laboratory variability. They also found that the caspase-3/7 assay showed a high inter-laboratory variability [100].

A large interlaboratory study of eight labs studied how to improve the robustness of the LDH and MTS assay. After a first round of experiments, adaptations to the protocols were made and robustness increased significantly within and between laboratories. Changes made to the protocols included the optimization of the differentiation of THP-1 cells and centrifugation after incubation with MTS reagent to remove NMs [108]. These findings on the MTS assay were confirmed in another interlaboratory study using the A549 cell line. Additional sources of variability were identified in this study. A549 cells from two different suppliers showed a large difference in cytotoxicity in response to polystyrene NMs. Also, the inclusion of serum effectuated large differences in cytotoxicity as compared with serum-free experiments. Moreover, differences in pipetting techniques (e.g., harsh aspiration vs. gentle pipetting and completely removing medium vs. partially removing medium before MTS incubation) and dispersion protocols were identified as causing differences in results between laboratories [52]. The importance of more elaborate and detailed SOPs was again stressed in a recent inter-laboratory study, where the inclusion of several acceptance criteria was found to improve the robustness of the MTS assay, such as maximum acceptable variations between replicates, minimum cell survival, and maximum interference levels [102].

4.1.2. Predictivity and Relevance

Whether in vitro cytotoxicity assays are predictive of in vivo acute toxicity has been studied for years for soluble chemicals. For NMs, however, there are only a few studies that correlate in vitro cytotoxicity with in vivo toxicity. Therefore, in this section we have included not only studies that correlate in vitro cytotoxicity with in vivo markers of cell death (apoptosis, necrosis etc.), but also with any type of in vivo toxicity.

In general, the predictivity of cytotoxicity assays depends on the mechanism of action of the NM, as well as on the cell type used for the in vitro study [109–111]. NMs which exert their effect through the shedding of toxic ions are usually also cytotoxic in vitro [100]. In a comprehensive comparison study, in vitro cytotoxicity was compared to in vivo lung inflammation for several different particles, using comparable doses. LDH release and trypan blue exclusion assays were able to predict the inflammation-inducing effects of ion-shedding NMs, but not of poorly soluble NMs [112]. However, in another study, in vitro LDH release in response to poorly soluble TiO_2 NMs correlated well to the in vivo number of polymorphonuclear cells (PMN) in BALF. This correlation was only present when the dose was expressed as surface area, and not when using mass as dose metric [113].

The toxic effects of CNTs in vivo upon inhalation are not easily predicted using in vitro cytotoxicity assays, unless the toxicity is caused by metal impurities [109]. Also for CNTs, the in vitro effects differ between cell types [114]. Other carbon-based materials, such as diesel exhaust also do not show an accurate correlation of cytotoxic response with in vivo effects. LDH release from A549 cells and LDH measured in BALF from rats upon instillation with diesel exhaust did not correspond, and even showed an opposite ranking in toxicity [115]. However, the suspensions used were not purely the particle fraction and contained other substances such as lube oil (which floats on culture medium), possibly causing the contrasting rankings.

The choice of cell type is crucial for performing a predictive in vitro cytotoxicity assay. For example, the WST-1 and NRU assays were able to establish an accurate ranking in toxicity of Ag, Au, SiO_2, and MWCNTs, but IC50 values differed between the cell types used [114]. In a study by Sayes and colleagues, in vitro LDH release did not correlate with rat pulmonary LDH release and inflammation (% PMNs) for rat primary pulmonary macrophages and rat pulmonary epithelial L2 cells grown in mono-cultures. However, when grown in co-culture, in vivo LDH release and inflammation were accurately predicted via the in vitro LDH release for crystalline silica and ZnO (but not for amorphous silica) [94]. A similar study also showed a good correlation for this co-culture model for ZnO NM, but only at the highest (particle overload) dose [116].

When choosing a cell line, immune cells are found to give a higher prediction accuracy than fibroblasts [114]. When macrophages are thought to be involved in the toxicity of NM, it is especially important to select an immune cell type for testing cytotoxicity. THP-1 cells which were differentiated to macrophages (dTHP-1) showed a higher sensitivity for cytotoxic effects as compared to A549 cells (alveolar cell line) for a panel of 24 NMs [117]. Cho and colleagues found that differentiated peripheral blood mononuclear cells and isolated lung macrophages performed better compared to cell lines such as dTHP-1, A549, and 16-HBE [112]. The fact that primary cells are more sensitive than cell lines is generally accepted. Despite this, primary cells are more difficult to work with and more expensive, and will therefore most likely be disfavoured for SbD hazard testing.

4.1.3. Overview of Needs and Knowledge Gaps

Table 1 shows a summary of how the different cytotoxicity assays perform in terms of the criteria for SbD hazard testing. As cytotoxicity measurements are a requirement for several mechanistic assays, they are crucial for SbD hazard testing. Simple cytotoxicity assays—although optimizations for NMs are needed—serve as a good starting point for detecting hazard warnings in SbD hazard testing. It is recommended to include at least two different cytotoxicity assays as different assays measure different mechanisms [92,118]. A combination of a mitochondrial activity assay and a membrane-integrity assay is recommended.

Table 1. Evaluation of suitability of cytotoxicity assays for SbD hazard testing.

Performance Criteria	Mitochondrial Activity (MTT, MTS, XTT, WST-1, Alamar Blue)	Cell Membrane Integrity (LDH)	Cell Membrane Integrity Staining (Trypan Blue, Propidium Iodide, Annexin V)	Lysosomal Integrity (Neutral Red Uptake)	Caspase 3/7 Assay
Simplicity and cost	Easy and cost-effective, commercial kits available.	Easy and cost-effective, commercial kits available.	Microscopic evaluation is time-consuming. Using flow cytometry increases time efficiency.	Easy and cost-effective, commercial kits available.	Easy and cost-effective, commercial kits available.
Predictivity (Sensitivity and Specificity)	Depends on the mechanism of toxicity of the particle, and the cell type used [110]. Macrophages seem more sensitive [114,117]. Assay better equipped to detect cytotoxicity of ion-shedding NMs [100]. Possibly suitable for making accurate rankings in toxicity [114].	Depends on the mechanism of toxicity of the particle, and the cell type used. LDH results have been shown to correlate with in vivo results for ion shedding NMs [112] as well as poorly soluble NMs [113].	Not assessed for NMs specifically.	Not assessed for NMs specifically	Not assessed for NMs specifically.
Robustness	For MTS assay, decent robustness but depending on cell type used [107], and only when interferences are correctly avoided [100,108]. More elaborate SOPs and harmonization between labs enhance assay robustness [52,102].	Similar robustness as MTS assay [108].	Not assessed for NMs specifically.	Not assessed for NMs specifically.	One study showed high inter-laboratory variability [100].
Compatibility	Many NMs interfere with the substrate, the product, or the optical readout. Can be overcome by washing cells before incubation with reagent, and centrifugation to get rid of NMs [100,108].	Many NMs interfere with the enzyme, the reagent, or the optical readout. Can be overcome via centrifugation. Washing not possible as LDH is measured in supernatant.	NMs may interfere with the dye.	NMs may interfere with the dye.	NMs may interfere with the dye or the readout.
Readiness	ISO protocol for MTS assay.	No NM-specific standardized protocol available.	No NM-specific standardized protocol available.	No NM-specific standardized protocol available.	No NM-specific standardized protocol available.

Prediction accuracy of cytotoxicity assays should be investigated further. It was shown that predictivity depends on the cell type used and the mode of action (MOA) of the NM. Assay applicability domains should be mapped in more detail to understand which toxic effects can be predicted with in vitro cytotoxicity assays and which cannot. We also found that protocol optimization improves assay robustness. Moreover, interferences are quite common for cytotoxicity assays, and they can be avoided by taking the right precautions and including the right controls, which is crucial even when performing SbD hazard testing. Together, this indicates the need for optimised and standardized protocols for NMs specifically. This will in turn also aid the determination of the prediction accuracy of assays.

Box 3

As cytotoxicity measurements are a requirement for several mechanistic assays, they are crucial for SbD hazard testing. Simple cytotoxicity assays—although optimizations for NMs are needed—serve as a good starting point for detecting hazard warnings in SbD hazard testing. It is recommended to include at least two different cytotoxicity assays as different assays measure different mechanisms A combination of a mitochondrial activity assay and a membrane integrity assay is recommended.

4.2. Dissolution

Although in vitro testing of dissolution is a measure of a PC property, and not directly a measure of toxicity, the results obtained can be used to infer potential toxicity, or even potential pathogenicity. This is through consideration of a material's biodurability or its transformation to ions or molecules. In this context, biodurability may be accompanied with biopersistence, which historically has been linked to the fibre pathogenicity paradigm such as that relating to asbestos, CNTs, and other respirable fibres [119], but also in relation to poorly soluble particles such as TiO_2. Long-term inhalation exposure to poorly soluble particles can induce impaired clearance and chronic inflammation that might even progress to cancer, as has been observed for TiO_2 in rats [120]. The human relevance of these results is a topic currently receiving a resurgence in interest within the scientific community [121]. Conversely, rapid dissolution of a substance can indicate exposure to potentially harmful soluble components, such as metal ions, which can be released in body compartments that are otherwise inaccessible.

Information on dissolution in relevant conditions is greatly beneficial for the hazard assessment of NMs and in defining SbD interventions. Information on dissolution is already a requirement of REACH and EFSA [21] and dissolution rates are a valuable criterion within all of the current risk assessment tools available for NM hazard assessment and can also be used for grouping/read-across [14,16,18].

4.2.1. Most Frequently Used Assays, Strengths and Limitations

There are various methods used in in vitro testing of dissolution, acellular and cellular, which have not changed significantly for some time. There are a number of guidance documents including ones from ISO (ISO 19057:2017) and OECD (OECD GD No. 318, specifically for environmental studies) which provide the start of standardization of these techniques. The output of various EU projects (e.g., GRACIOUS, Gov4Nano and BIORIMA) will also greatly impact the development of this methodology.

For acellular testing, it is possible to test within static systems [122] or flow-through (dynamic) systems [123,124]. The application of these methods is extremely diverse, as the formulation of different simulant fluids may facilitate the simulation of any biological compartment, including extracellular and intracellular compartments, and any exposure route of interest including oral, dermal, or inhalation [125], with recommendations made within an ISO technical report (ISO 19057:2017). There are a number of differences, both subtle and substantial, in the simulant fluids used that determine the accuracy of in vivo prediction. For example, components such as citric acid have a significant effect on the dissolution of certain metals, and inclusion of proteins/serums will also have an effect on dissolution [126]. These considerations have been recently reviewed [127]. Two recently completed projects (nanoGRAVUR and GRACIOUS) identify the abiotic flow-through system ISO/TR 19057:2017 as the most relevant system [124,128], with a technology readiness level (TRL) identified as high/medium for metals using Inductively Coupled Plasma Mass Spectroscopy (ICP-MS) analysis and medium/low for materials such as CNTs that require techniques such as Transmission Electron Microscopy (TEM) or X-Ray Photoelectron Spectroscopy (XPS).

Although there are currently no accepted methods for in vitro cellular dissolution testing, various studies have been conducted and these may be more reflective of the in vivo response following inhalation of particles, although there have been concerns raised with the cellular methods.

Acellular Methods

Acellular dissolution can certainly be considered simple, considering the practical requirements. However, each methodology has different demands and associated limitations. For example, the solutes released during dissolution within a static system, especially those of a basic nature, may cause enhanced nucleation, precipitation, changes in localised pH, and/or saturation effects preventing further dissolution [129–132]; this behaviour is unlikely to reflect the in vivo behaviour, demonstrating clear limitations of static systems. Although dynamic systems, by design, circumvent these issues, they are not without limitations, and there are a number of factors which may affect their reproducibility [132]. NMs may pass through filters used in flow-through systems, leading to misinterpretation of results and potential false-positive results [133], or filters may become blocked and ruptured by components of the more complex fluids such as proteins or lipids [134]. Practically, dynamic systems are cumbersome due to the high volume of liquids required in long-running tests [135]. Although a comparison of acellular methods is not often made, when done so the findings have been confounding. For example, the dissolution rate of gold nanoparticles has been found similar in static and dynamic systems [136], while the solubility of $BaSO_4$ has been shown to differ in static and dynamic systems [137].

It is often reported that distinction between different material forms is possible, with a level of sensitivity allowing for a distinction between dissolution rates leading to grouping, as has been suggested for fibrous NMs within the GRACIOUS project [138]. This approach has been aligned with previously established methodology for man-made vitreous fibres (MMVF) and asbestos, whereby NMs biodurability can be defined by the respective dissolution in alveolar fluid and lysosomal fluid. Good comparability has already been found for in vitro dissolution of MMVF materials and in vivo biopersistence [139], allowing confidence in this approach. Heavy influencers of sensitivity include the analytical method used to detect released ions, as well as high background measurements caused by the complexity of fluids used, although this can be alleviated through the removal (or reduction) of specific metal components within the fluid, in line with the solutes expected to be released from the test material [140]. The current use of dissolution within RA tools may not be so greatly impacted by sensitivity, as the thresholds used are very broad and/or rather elementary, using a ranking based on dissolution time (ANSES, Swiss Precautionary Matrix) or soluble concentration (GUIDEnano). Advances have been made recently to include threshold decisions based on dissolution rate [21,123]. Nevertheless, unless very significant changes are made to the particle to result in very different dissolution behaviour, it is unlikely that the sensitivity of the thresholds used will be dynamic enough to provide meaningful SbD decisions on dissolution behaviour.

In terms of compatibility, the potential of acellular tests is broad, and other than particular hydrophobic materials, it is difficult to list examples that could not be tested. In fact, these acellular methods have been used for some time to resolve the time-kinetic release of metals within complex materials, such as man-made fibres, or from occupational dusts such as welding fumes [141]. The biological predictivity of acellular tests, although not always established within the literature, has been demonstrated to a relatively high level, with various promising outcomes. For example, the solubility of $BaSO_4$ in the dynamic system was considered to reproduce what is known for the solubility of $BaSO_4$ in vivo, while the results of the static system underestimated this [137]. With the use of a lysosomal simulant fluid, the dynamic dissolution system was also shown to replicate cellular dissolution of $BaSO_4$ (in conjunction with $SrCO_3$ and ZnO) in rat macrophage models [123], and similarly acellular dissolution of MoO_3 in the same lysosomal simulant fluid was found comparable to dissolution within mouse macrophage models [142]. These studies, and others, have demonstrated that by using various simulated biological fluids of intracellular compartments and/or lung lining fluid, a number of correlations with either cellular assays or in vivo exposures can be attained [123,143–146]; however, it should be acknowledged that there is an equal number of studies that have shown no correlation, raising the concern for appropriate fluid selection [127].

Cellular Methods

The basic principle of the cellular method is simple and can be performed cheaply as typical assessments investigate dissolution within cells up to 24 h. The difficulty with this methodology is the success of analysing the ions released. There are various options for separating cells and supernatant, such as centrifugal ultrafiltration and cloud point extraction [147], however it is not always as straightforward as the acellular assessments, as released ions may form complexes with biomolecules and therefore separation may be hampered [148]. Additional concerns may arise from the complexing of ions to biomolecules. Therefore, studies have often opted to determine both the ion concentration and the NM concentration to aim to avoid false positives or false negatives [149].

Koltermann-Jülly et al. (2018) found that macrophage-assisted dissolution in vitro was only applicable for an exposure period of 1–2 days, which they believe is too short and may be responsible for the low amounts of ions detected for the NMs tested [123].

The authors state that using cellular systems gives no additional benefit to the abiotic flow-through system with regards to predicting the in vivo response. This conclusion, however, is based only on the three materials tested. Moreover, when a cellular method uses uptake as an inference of dissolution, overestimating particle concentrations may occur, when measurements are not only of internalised particles but also of those adhered to the cell membrane. However, this is likely to be resolved by following well-described methodology which includes steps to limit this interference such as ensuring thorough washing and etching of the cells prior to analysis to remove adhered particles [150]. Further promising methodologies for this include the isolation of NMs and ions from cells using Triton X-114-based cloud point extraction as has previously been conducted for intracellular Ag NMs and Ag+ isolation [147,149].

4.2.2. Overview of Needs and Knowledge Gaps

Table 2 shows a summary of how the different dissolution assays perform in terms of the criteria for SbD hazard testing. There is a wealth of studies available for interpretation of in vitro dissolution methods, and although there are promising findings, there are still too many uncertainties to be sure of which model is most appropriate or reliable for specific materials. For SbD hazard testing, the use of a static system would be preferred, due to its simplicity. Although correlations with in vivo outcomes have been shown for static and flow-through systems, this is not always the case, and therefore requires further attention. Going forward, assessing predictivity will be important, especially when assessing novel materials; however, auspiciously, as shown above, for some substances a strong relationship between acellular, cellular, and in vivo findings has already been observed. It has been previously suggested that specific in vitro methods (e.g., specific fluid choices) should be selected based on feasible degradation pathways [142] which could be dependent upon specific degradation routes e.g., complexation, protonation, or to establish robust and fully accurate biological simulations [127]. Moreover, if these methods are to be applied in grouping and read-across approaches, the development of reliable and robust methods for determining particle dissolution rates has been considered paramount [123].

Table 2. Evaluation of suitability of dissolution assays for SbD hazard testing.

Performance Criteria	Acellular Assays		Cellular Assays
	Static Dissolution (e.g., OECD Series on Testing and Assessment No. 29)	Flow-through/Dynamic Dissolution	Cellular In Vitro Dissolution
Simplicity and cost	This system is the simplest and could be conducted by commercial laboratories without extensive investment in equipment.	Requires much greater effort with regards to setup, and also requires a large volume of fluid.	The basic principle of this method is simple and can be performed cheaply. Would be considered high throughput but as cellular will typically incur higher costs than acellular.
Predictivity (Sensitivity and Specificity)	Static dissolution studies have been found to correlate with in vivo results in some instances [143,146,151], but in others poor correlation is observed [137,144,146,152,153]. Losses in sensitivity may arise due to any sample handling (e.g., acidifying the sample, filtration) or saturation of ions. For highly soluble materials, dissolution may continue during centrifugation steps, resulting in greater values of dissolution.	Good correlation observed between flow-through system using a specific simulant fluid (modified Gamble's) and intratracheal instillation in vivo [154], and for some particles dynamic dissolution in phagolysosomal simulate fluid (PSF) was a good predictor for short term inhalation study in rats [123] and intratracheal instillation in rats [137]. Losses in sensitivity may arise due to any sample handling (e.g., acidifying the sample). Additional concerns about losses in the system due to filtration.	Results do appear to correlate well with in vivo in some instances (e.g., fast dissolution of Ag NMs in vivo [155] and in vitro [149]). Study by Koltermann-Jülly et al. (2018) found very low levels of dissolution in macrophages compared with the abiotic flow-through system and clearance in vivo [123]. Sensitivity relies on the capability of analysing released material. Additional concerns may arise from complexing of ions to biomolecules.
Compatibility	The basic setup is compatible with many materials. Issues may arise with hydrophobic materials and with any material whereby sensitivity cannot be achieved for further analysis due to interference with components in the biofluid mixture (e.g., Ag NMs).	The basic setup is compatible with many materials. Issues may arise with any material whereby sensitivity cannot be achieved for further analysis due to interference with components in the biofluid mixture or membranes used (e.g., Ag NMs).	Most common analytical technique used is ICP-MS, therefore this methodology is the most compatible with metals. Carbon-based NMs such as CNT have used analytical techniques such as UV-Vis, Raman spectroscopy, and EM, however the sensitivity of these techniques is likely to be far less.

Table 2. *Cont.*

| Performance Criteria | Acellular Assays | | Cellular Assays |
	Static Dissolution (e.g., OECD Series on Testing and Assessment No. 29)	Flow-through/Dynamic Dissolution	Cellular In Vitro Dissolution
Robustness	Large variability between different biofluids.	Can result in false positives and false negatives due to issues with the filtering system (i.e., due to NMs passing through pores or causing blockages in filters).	No evidence of inter-laboratory comparisons. Issues may arise due to inclusion of particles on the surface of the cell rather than internalised particles only.
Readiness	OECD protocol but specifically for environmental studies. Various fluid compositions available.	ISO protocol outlining basic methodology. TRL identified as high/medium for metals and medium/low for organic materials (e.g., CNT) [124].	Validated assays available but no standardized method.

Box 4

There is a wealth of studies available for interpretation of in vitro dissolution methods, and although there are promising findings, there are still too many uncertainties to be sure of which model is most appropriate or reliable for specific materials. For SbD hazard testing, the use of a static system would be preferred, due to its simplicity.

4.3. Oxidative Potential and Oxidative Stress

The oxidative potential (OP) of a NM is a chemical property that defines the ability of a NM to form potentially toxic species such as hydroxyl ($^{\bullet}$OH) and superoxide (O_2^{-}) radicals and hydrogen peroxide (H_2O_2) (collectively called reactive oxygen species (ROS)), or reactive nitrogen species (RNS) through redox reactions. This parameter is part of many NM hazard assessment strategies and grouping approaches due to its potential as a predictor of toxicity [13–15,18]. The pros and cons of OP assays in NM research have extensively been reviewed previously [156,157].

Oxidative stress (OS) is a cellular state in which the amount of ROS, caused by NM OP or the release of reactive ions, overwhelms the cells' antioxidant capacity, potentially leading to the oxidation of biomolecules, inflammation, and oxidative DNA damage [158,159]. OS is seen as an important key event in the mode of action of many NMs and is therefore important to quantify as an early warning indictor [156,160,161].

4.3.1. Most Frequently Used Assays, Strengths and Limitations
Acellular Methods

In an acellular assay, OP is usually measured as a rate of depletion of a reductor. OP assays do not measure reductor depletion by OP only, since the release of reactive ions by dissolution will also lead to a depletion. Multiple acellular assays have been proposed to evaluate NM OP. The acellular dichloro-dihydro-fluorescein (DCFH) assay, electron paramagnetic resonance (EPR), electron spin resonance (ESR) spectroscopy, and the Ferric Reducing Ability of Serum (FRAS) assay are frequently used and have been evaluated extensively in literature [162–167]. An additional assay which is especially relevant for measuring NM OP is the haemolysis assay. For this assay, red blood cells are isolated from whole blood, and the ability of NMs to disturb their membranes is measured through absorbance. This assay requires whole blood, but no cell culturing, making it an easy and cost-effective method. No interferences have been reported in this assay, but it has not been studied extensively for NMs. There is no information available on robustness, and there is no publicly available standardized protocol.

Out of the other acellular assays, standardized protocols are only available for EPR/ESR (i.e., ISO 18827:2017). These assays require relatively expensive equipment, not always available in standard laboratories. FRAS and DCFH are considered simple and low-cost assays, requiring equipment that is present in any standard biological lab [165,166,168]. The FRAS method was originally developed to measure the ferric reduction in blood plasma (FRAP) [169]. It has been adapted to be used with serum, optimized for smaller volumes [165] and for multi-dose measurements, while showing good sensitivity and reproducibility for several metal-bearing NMs [163]. Recently, the FRAS protocol was successfully adapted to measure the reactivity of graphene-based materials by adding a filtration step for NMs of very low density [168]. Interlaboratory studies for the FRAS assay have not yet been performed.

The robustness of the acellular DCFH assay using different NMs was evaluated in a recent inter-laboratory study. A good robustness was found for the positive control NMs when normalizing fluorescence values between labs. However, for the other NMs, interlaboratory reproducibility differed per particle type [170]. Several papers reported NM interference with, for example, the fluorescent readout of the DCFH assay [162,166,171]. Zhao and Riediker (2014) identified several other factors that could reduce the reliability of the acellular DCFH assay, such as the use of different dispersing agents, as well as using too-high concentrations of NM [172]. Interference by way of NM flocculation or optical interference has been noted regarding the FRAS assay when testing various NM pigments [173], while no NM interference has been reported for ESR/EPR.

The EPR, DCFH, and FRAS assays have recently been evaluated for NM-grouping purposes. Results showed that the sensitivity of the methods greatly depends on the type of particle studied. For example, CuO, $BaSO_4$, and Mn_2O_3 were consistent in their reactivity level across the three methods, but ZnO and CeO_2 only showed a response in the FRAS assay, and not in EPR and DCFH measurements [173]. This might suggest that reactive species produced by certain NMs are captured better by some assays than by others, or that the FRAS assay is more sensitive in general. The latter has been confirmed in several studies that showed that the FRAS and ESR/EPR perform better than the DCFH assay in terms of sensitivity [163,171,174,175].

The choice of assay should depend on the goal of testing. For example, if one would like to know which types of radicals are formed in order to know what to change in the NM design as a SbD intervention, ESR/EPR measurements with different spin traps will provide the most informative results [176]. The FRAS assay can provide a more general image of ROS generating potential, as a result of the cocktail of antioxidants that is present in serum. The DCFH assay is especially sensitive for one-electron oxidizing species (such as hydroxyl radicals) [177]. It should be taken into account that these assays measure the OP of the NM in the specific environment required by the assay. OP is greatly influenced by the exposure environment, and therefore the OP measured in the assay may not fully reflect the OP in a real-life exposure scenario. A relevant protein corona could be applied to ensure exposure relevancy, as is described in Section 3.2.

Cellular Methods

Cell-based assays can directly measure the intracellular ROS, irrespective of their origin (i.e., as a result of the surface chemistry of the NM, as a cell-generated signalling molecule, or as a defence mechanism of the cell within the phagolysosome), for example through use of the cellular DCFH-DA assay. Other options include the assessment of the effect of these radicals on biomolecules such as lipids (e.g., lipid peroxidation) and proteins (e.g., protein carbonylation), cellular antioxidant status (e.g., glutathione (GSH:GSSG ratio)), and antioxidant gene regulation (HO-1 expression and Nrf-2 reporter cell lines), of which the latter two are extensively described in Boyles et al. (2016) [178].

It has been suggested that OS measured in a cell-system has advantages over measuring the OP in acellular systems. By measuring in a cellular environment, the cells' ability to defend itself against the induced OS is taken into account, the ROS' generated genotoxicity can be assessed, and other mechanisms other than the OP that lead to OS are captured as well [156]. Other mechanisms leading to OS, such as through mitochondrial perturbation, have been shown for chemicals extracted from diesel exhaust particles [179,180], yet there is no convincing evidence that NMs are capable of inducing OS through mechanisms other than OP or ion release.

4.3.2. Predictivity and Relevance

For SbD hazard testing, it would be desirable to be able to predict human health effects or at least effects that are observed in studies in experimental animals with simple, fast, and cheap acellular OP assays. The ability of acellular assays to predict cellular oxidative stress and in vivo oxidative stress markers is quite good, as shown in a comprehensive review by Moller et al. (2010), but not for all particles and all test systems [160]. It has been shown that NMs can induce different types of ROS [181] and therefore it depends on the type of NM and their MOA whether assays can predict cellular and in vivo effects. For example, data derived with the haemolysis assay correlate very well with in vivo pulmonary inflammation for a panel of 13 metal oxide NMs (92% prediction accuracy), whereas EPR (69% prediction accuracy) and DCFH (77% prediction accuracy) results showed lower correlations [164]. The haemolysis assay was able to predict in vivo pro-inflammatory responses of both NMs that act through soluble ions as well as NMs that act through surface reactivity with a prediction accuracy of 62.5% for a panel of eight NMs [112]. The FRAS assay has even been shown to be able to correctly distinguish OPs between several types of CNTs [175]. In a study comparing pulmonary inflammation (PMN influxes) upon inhalation of a range of NMs with acellular ESR and DCFH results, the correlation was reasonable; however, here it was concluded that ESR measurements in macrophages give a higher prediction accuracy than the acellular assays [182]. For SiO_2 NMs, EPR results correlated very well with in vitro cellular cytotoxicity [183]. ESR also correlated well with in vitro protein carbonylation for a large panel of NMs [184]. However, ESR as well as the FRAS assay were able to accurately predict only 50% of the in vivo outcomes for a panel of 35 NMs [162].

False positives in acellular OP assays (when compared to in vivo outcomes) can be explained by the fact that cells and organisms can resolve ROS to a certain extent. Therefore, effort should go towards establishing thresholds for these assays. False negatives in acellular OP assays can be explained by the fact that other mechanisms other than OP can lead to pulmonary inflammation as well, which cannot be detected by these assays. The large variation between prediction accuracies between studies could be explained by the differences between the NM panels tested. Each assay has a specific applicability domain and prediction accuracy will therefore depend on the NM types and the resulting types of ROS.

In general, cellular assays show a higher prediction accuracy than acellular assays, and a combination of both might perform even better [157,162,185]. However, for SbD hazard testing, acellular assays may already give a good indication of toxicity and could serve as a valuable initial screening in the very early stages of NM development.

4.3.3. Overview of Needs and Knowledge Gaps

Table 3 shows a summary of how the different OP assays perform in terms of the criteria for SbD hazard testing. There are clear indications that only measuring acellular reactivity would be sufficient for SbD hazard testing, when cellular testing is already performed for cytotoxicity, genotoxicity, and pro-inflammatory effects. OP assays can be used to categorize the materials and to explore in more depth if the OP can or should be reduced in a SbD intervention. Mapping the prediction accuracy for each assay, as well as an applicability domain will help understand which assays can be used to predict which specific effects.

Table 3. Evaluation of suitability of oxidative potential assays for SbD hazard testing.

Performance Criteria	FRAS	ESR/EPR	DCFH Acellular	Haemolysis Assay
Simplicity and cost	Very simple but needs large amounts of NM.	Very simple, yet might be difficult to find lab with specialized ESR/EPR equipment.	Very simple and only requires a fluorescence reader.	Very simple and only requires absorbance reader and whole blood.
Predictivity (Sensitivity and Specificity)	The assay is able to detect NMs' reactivity at low concentrations and in a dose-dependent manner with higher sensitivity compared to DCFH assay [163]. Could distinguish between CNT types [175]. Prediction accuracy reported: 50% [162].	Depending on spin trap used. Aids to identify specific ROS types, which could be useful for SbD interventions [176]. Prediction accuracies reported: 69% [164] and 50% [162]. Correlated well with in vitro cytotoxicity and protein carbonylation [183,184].	Lacks sensitivity as compared to FRAS and ESR/EPR [163,171,175]. However, protocol adaptations [172] show ameliorated sensitivity. Prediction accuracies reported: 77% [164].	Is thought to be able to detect OP of both surface reactive as well as ion-shedding NMs [112]. Showed very high prediction accuracy (92%) in one study [164].
Compatibility	Good compatibility with a wide range of NMs. Optical interferences are largely avoided using a centrifugation step but have been reported [173]. Adapted method suggested for graphene-based materials [168].	Good compatibility with a wide range of NMs [162]. No interferences reported.	High background signals resulted from dye auto-oxidation [171]. NM interferences reported [162,166,184]. Adapted DCFH protocol reduces interferences [172].	No interferences reported, yet might be expected due to absorbance readout.
Robustness	No interlaboratory study performed. Found to be reproducible and reliable within the same lab [163,165]	Not assessed for NMs specifically.	Previously lacked robustness [175]. Interlaboratory round robin tests in GRACIOUS project showed satisfactory reproducibility for positive control NMs using optimized SOP [170].	Not assessed for NMs specifically.
Readiness	No NM-specific standardized protocol available. Gandon et al. (2017) protocol available [163].	ISO protocol available (ISO 18827:2017)	No NM-specific standardized protocol available. Boyles et al. (2022) protocol available [170].	No NM-specific standardized protocol available.

Box 5

> There are clear indications that only measuring acellular reactivity would be sufficient for SbD hazard testing, when cellular testing is already performed for cytotoxicity, genotoxicity, and pro-inflammatory effects.

4.4. Inflammation

Many IATAs and testing strategies include the measurement of inflammatory potential using NMs since this is generally accepted as one of the key mechanisms of NM toxicity [15,19,20]. Pulmonary inflammation in response to NM exposure has been shown to lead to several adverse health effects, such as fibrosis as well as lung cancer in animal studies [120,186,187]. For oral exposure, inflammation is a key parameter in NM toxicity as well [188]. However, this section will focus on assays targeting the pulmonary route of exposure only.

4.4.1. Most Frequently Used Assays, Strengths and Limitations

It is impossible to capture the complexity of an in vivo inflammatory response in an in vitro model, where recruitment of inflammatory cells other than those already present cannot occur. It is however possible to detect the cytokines responsible for this recruitment in an in vitro experiment. The most widely used approach to assess inflammatory responses in in vitro assays is measuring cytokine production or secretion, using, for instance, an enzyme-linked immunosorbent assay (ELISA), RT-qPCR, or multiplex-based immunoassays [189] after exposing cultured cells to NMs. Measuring the levels of pro-inflammatory cytokines and other inflammatory mediators may give insight into the mechanisms of the immunomodulatory effects of NMs in vitro, such as inflammasome activation or dendritic cell maturation. Cytokines of specific interest for NM pulmonary toxicity are, amongst others, IL-8 as markers for neutrophil recruitment [190], IL-1β as a marker for NLRP3 inflammasome activation [191], and TNF-α as a marker for macrophage activation [192].

Cytokine release can be measured in e.g., epithelial cells, macrophages, and dendritic cells, cultured in mono- and co-cultures. Cells can be exposed in a submerged setup or at the air-liquid interface (ALI), where cells are cultured in contact with the air and exposed to aerosols on the apical side whilst kept in medium on the basal side, better resembling the physiological environment of cells in the respiratory tract (Figure 4).

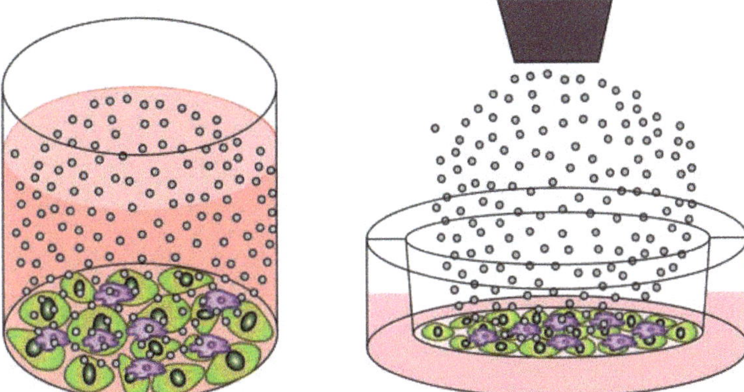

Figure 4. Submerged (**left**) and ALI exposures (**right**) to NMs. Submerged exposures are considered easier, whereas ALI exposures are considered more physiologically relevant for inhalation (and dermal and intestinal) exposures.

A critical factor when assessing pro-inflammatory responses in cell models is that some NMs can interfere with common in vitro assays. SWCNT and MWCNT can non-specifically adsorb TNF-α and IL-8 to their surface, and TiO_2 NMs have been described to be able to adsorb IL-8, thus causing a false-negative result in ELISA assays [99,118]. This effect has also been observed for Ag NM in combination with TNF-α and IL-8 [100]. NMs are also known to interfere with the components of the ELISA. This problem can be overcome by centrifugation to remove the NMs from the supernatant before performing the ELISA. It is also essential to test the NMs for endotoxin contamination, as endotoxins can induce inflammation at very low concentrations, leading to false positive results [193,194], especially since NMs are generally not produced in a sterile environment.

Despite the relevance of pro-inflammatory effects of NMs in human health, there is currently no validated test method available to investigate inflammatory responses in vitro. Submerged assays have been used far longer compared to the relatively new ALI models, and thus more advances in standardization and optimization have been accomplished. Only a few studies have been performed to show the robustness of one of these protocols. At the ALI, Calu-3 cells with and without macrophages (either differentiated THP-1 cells (dTHP-1) or primary cells) showed high reproducibility in seven participating labs based on measurements of membrane integrity and mitochondrial activity. Cytokine release however showed higher variability, although similar trends between the seven labs were observed [195]. The reproducibility between labs after exposure of A549 cells at the ALI to lipopolysaccharide (LPS) as a positive control was found to be quite low. However, after protocol optimizations, special training of personnel for cell handling, and homogenization of disposables and reagents, the reproducibility increased [196]. The reproducibility of results between labs when using dTHP-1 cells is a frequent topic of debate. Not only do they show varying responses to NMs, but also to a positive control such as LPS, as shown in a large inter-laboratory study [100]. In another large inter-laboratory study by

Xia et al. (2013), it was shown that good results can be obtained when using very detailed protocols and using the same batch of serum and cells. They also showed that cell-culture conditions and the duration of differentiation greatly affect the variability of dTHP-1 cells between labs [108].

A frequently used alternative for dTHP-1 macrophages are monocyte derived macrophages (MDMs), derived from donor blood. Even though they are considered more predictive of the in vivo situation, they are also known for their donor-to-donor variation. The same holds true for the use of commercially available primary epithelial cells, which are considered more relevant, but also show considerable variation [197].

In ALI exposure systems, there are many other factors that may contribute to an increased variability, such as the accuracy of the microbalance in the exposure system, the quality of the nebulizer used, the method of sample preparation, etc. A comprehensive overview of factors that can influence reliability and robustness can be found in Petersen et al. (2021) [198].

In terms of predictivity, commercially available primary cells generally give a good indication of in vivo effects for known inflammation-inducing particles. Studies have shown that the pro-inflammatory effects of quartz [197], Ag NMs [199], SiO_2 NMs [200], and Pd and Cu NMs [201] are accurately predicted using primary cell models. Co-cultures of cell lines are also able to predict in vivo responses in many cases, as has been shown for quartz [94,202], CuO [112], and ZnO [94]. In this latter study, it was shown that co-cultures perform better than the two cell types separately, showing the importance of interplay between epithelial cells and immune cells. The addition of macrophages seems to be crucial in order to capture a much wider domain of immunological responses as compared to epithelial cells only, as has been shown in multiple studies [108,203,204]. Epithelial cell lines in mono-cultures were not able to predict the toxic effects of quartz [205], but did accurately detect Ag NMs' pro-inflammatory effects [112].

In short, primary cells are the most sensitive, followed by co-culture systems with macrophages, and then mono-cultures. There are however some studies that prove otherwise. Cho et al. (2013) showed that cell lines performed similar to primary alveolar macrophages and differentiated PBMCs in terms of accuracy [112]. The A549 epithelial cell line in tri-culture with inflammatory cells did not pick up the pro-inflammatory effect of Ag NMs [206]. Mono-cultures of the epithelial cell line 16-HBE better predicted in vivo effects of Ag NMs than when in co-culture with dTHP-1 cells [207]. Furthermore, CeO_2, Co_3O_4, and NiO NMs induced an increase in granulocytes in BALF, whereas no pro-inflammatory effects were seen in submerged mono- and co-cultures [112]. Finally, BEAS-2B and dTHP-1 cells were able to predict a ranking in pro-inflammatory effects of several types of CNTs which corresponded to in vivo markers of lung fibrosis in two separate studies [208,209]. This could mean that cell lines could be suitable for SbD hazard testing. Likely, different modes of action of toxicity require different levels of complexity in a cell model. In order to be sure about the predictive capacity of the different cell types, more types of NMs should be tested.

The exposure method chosen will also impact the predictivity of the method. Exposing ALI-cultured cells is generally considered a more sensitive approach, since it is more physiologically relevant, as has been shown in multiple studies [210–212]. However, for SbD hazard testing it is desirable to work with a model that is as simple and cost-effective as possible. This disfavours the use of primary cells and favours simple submerged exposure systems as opposed to the more complex ALI cultures. There are strong indications that simple submerged models could be predictive enough for SbD hazard testing. For example, in a study by Loret et al. (2016) they concluded that, indeed, co-cultures were more sensitive than monocultures, and that ALI exposures were more sensitive than submerged cultures. However, the general ranking of the NMs in terms of their toxicity was similar across the various exposure methods [213]. A study by Di Ianni et al. (2021) showed a strong correlation between in vitro submerged co-cultures and in vivo results when testing CNTs [190]. In a study by Herzog et al. (2014), the pro-inflammatory effects of Ag

NMs were not detected in ALI culture conditions, but were detected under submerged conditions, suggesting a better performance of the submerged model [214]. Submerged and ALI exposures performed equally well for cytotoxicity in response to TiO_2 [215]. In a study by Panas et al. (2014), submerged conditions were more sensitive in detecting the pro-inflammatory effects of SiO_2 NMs as compared to ALI conditions [216]. Altogether, the potential for submerged experiments to predict in vivo responses has been shown in multiple studies, and is worth exploring further, especially for SbD hazard testing.

The differences in sensitivity between the ALI and submerged exposures can be explained by many (potentially confounding) factors. Firstly, certain cell types such as A549 are suggested to produce surfactant at the ALI, but not under submerged conditions, making them more vulnerable to toxic effects in a submerged experiment [217]. Secondly, the effective dose in submerged experiments is not always (correctly) calculated, and this may lead to a skewed comparison to ALI and in vivo results. Thirdly, the medium used in the submerged experiments may have an impact on NM behaviour in terms of the protein corona and dissolution rate, which does not occur, or occurs differently, in ALI experiments. Lastly, studies finding a good correlation between an in vitro model and in vivo results are more likely to be published, leading to publication bias. In a comprehensive overview of different cell types and exposure methods by McLean et al. (in preparation), it was shown that for quartz hazard prediction, the strength of in vitro prediction of in vivo responses was highly inconsistent, and largely dependent upon the data and study quality, which highlighted a need for robust SOPs which take into account numerous requirements for in vitro/in vivo extrapolation (McLean et al., in preparation).

There are several advantages of using ALI over submerged exposures. Particle alterations due to interaction with medium (such as the formation of a protein corona, dissolution, agglomeration) are no longer an issue, and calculating the deposited dose is much easier as compared to submerged experiments [212,218]. ALI exposures are compatible with a wider range of NMs, including hydrophobic particles, as exposures can be performed using a powder. Without having to take into consideration their behaviour in medium, abrasion products of NEPs can also directly be applied, making this type of model especially interesting for assessing life-cycle considerations, as is crucial for SbD hazard testing. Using ALI exposures is however more time-consuming and less high-throughput. Additionally, robustness of deposited dose after nebulization in an ALI setup may be low, depending on the NM used [219].

4.4.2. Overview of Needs and Knowledge Gaps

Table 4 shows a summary of how the different inflammation assays perform in terms of the criteria for SbD hazard testing. For SbD hazard testing, it is crucial to include tests that are predictive yet simple and cost-effective. Therefore, based on the current literature, the use of a submerged co-culture model including at least a type of macrophage might be the most suitable. However, more research is needed to confirm that simple methods are predictive enough for early hazard screening by testing data-rich NMs. Moreover, novel and advanced NMs should be tested in the available cell models in order to determine the compatibility of the cell models and readouts with different types of NMs. For a better predictivity, avoiding issues with dosimetry and medium interactions, and for hydrophobic particles, a simple ALI experiment can be set-up for SbD hazard testing.

A short-lived inflammatory response is beneficial to help clear NMs from the lung, and macrophage recruitment may not necessarily be a hazard warning. There is still a poor understanding of which amount of inflammation could be considered an adverse outcome, especially when measured in vitro. Establishing meaningful thresholds for these assays is important. Moreover, since pulmonary inflammation is mostly a chronic adverse effect, more work should be focusing on predicting chronic effects with in vitro assays, with which a promising start has been made in the PATROLS project [220].

Table 4. Evaluation of suitability of Inflammation assays for SbD hazard testing.

Performance Criteria	Submerged Cell Models		ALI Cell Models	
	Submerged Cytokine Release Mono-Culture	Submerged Cytokine Release Co-Culture	ALI Cytokine Release Mono-Cultures	ALI Cytokine Release Co-Cultures
Simplicity and cost	Simple and cost effective.	Simple and cost effective, however creating a co-culture requires more effort and experience than a mono-culture.	Requires specialized exposure equipment and a certain level of expertise.	Requires specialized exposure equipment and a certain level of expertise; creating a co-culture requires more effort and experience than a mono-culture.
Predictivity (Sensitivity and Specificity)	Good correlation with in vivo found [190]. Found to be more sensitive than ALI in several studies [214,216]. Accurate ranking found [208,209].	Combination of immune cell and epithelial cell more predictive than epithelial cell alone [94]. Accurate ranking found [213].	Generally good for primary cells. Lower predictivity of epithelial cell lines. ALI exposures found more sensitive than submerged in several studies [210–212].	Co-cultures perform better than two cell types separately [94]. BMDL of this model comes closer to the in vivo BMDL compared to submerged [211].
Robustness	Large inter-laboratory variability for THP-1 cells [100,108], no inter-laboratory data on other cell types.	Not assessed for NMs specifically.	Low reproducibility but similar trends between labs [195]. Low reproducibility improved after protocol optimizations [196].	Low reproducibility but similar trends between labs [195].
Compatibility	NMs may interfere with ELISA [99,118].	NMs may interfere with ELISA [99,118].	Compatible with a wide range of materials, including hydrophobic and low-density NMs. However, NMs may interfere with ELISA [99,118].	Compatible with a wide range of materials, as exposures do not necessarily require a dispersion. However, NMs may interfere with ELISA. Might be more suitable for NMs released form NEPs.
Readiness	No NM-specific standardized protocol available.	No NM-specific standardized protocol available.	No NM-specific standardized protocol available.	No NM-specific standardized protocol available.

Box 6

> For SbD hazard testing, it is crucial to include tests that are predictive yet simple and cost-effective. Therefore, based on current literature, the use of a submerged co-culture model including at least a type of macrophage might be the most suitable. However, more research is needed to confirm that simple methods are predictive enough for early hazard screening.

4.5. Genotoxicity

One of the main safety concerns related to NMs is their possible genotoxicity [221,222]. Genotoxicity describes the capacity of a chemical or physical agent to produce genetic damage that, if left unrepaired, may lead to cancer [223]. Therefore, every mutagen is potentially carcinogenic [222].

Due to the important consequences to human health, mutagenicity is a hazard endpoint required in all product regulations (chemicals, biocides, pharmaceuticals, medical devices, food additives, cosmetics, etc.) [224]. The assessment of genotoxicity is based on validated in vitro assays, which can be followed up by validated in vivo assays, depending on the in vitro outcome and the regulation involved [225]. Therefore, genotoxicity assessment at an early stage of innovation is highly advised. In fact, genotoxicity is a key endpoint in most of the testing strategies developed for NMs [13,19,221,226,227].

4.5.1. Most Frequently Used Assays, Strengths and Limitations

The mutagenicity of chemicals is usually evaluated on the basis of a battery of standard genotoxicity assays, able to detect gene mutations, chromosomal damage, and aneuploidy, as all these different mechanisms need to be considered in the assessment [221,224,227]. A core in vitro battery comprising the Ames test (detecting bacterial gene mutations) plus the in vitro micronucleus test (detecting chromosomal damage and aneuploidy) was already proposed 10 years ago for soluble chemicals [228]. Nearly 100% (958 out of 962) of rodent carcinogens or in vivo genotoxins were correctly detected with these two tests, which makes this battery a particularly sensitive combination [224]. However, the specificity of both assays together was unacceptably low (12.0%), giving rise to a high rate of false-positive results [229]. Hence, most of the EU regulations require a follow-up in vivo study when in vitro positive results are obtained [224]. In the case of NMs, the Ames test does not appear to be a suitable method as some NMs may not be able to penetrate through the bacterial wall, whereas others may kill the bacteria due to their bactericidal effects [230–234]. Based on this evidence, results obtained with this method should be followed up with other gene mutation assays using mammalian cells [21,235], or better yet, the Ames test should be avoided for NMs.

A roadmap for the genotoxicity testing of NMs was suggested some years ago [236], followed by guidance and common considerations [237]. There are two OECD TGs for assessing in vitro mammalian gene mutations: the In vitro Mammalian Cell Gene Mutation Tests using the Hprt and xprt genes (OECD TG 476), and the In vitro Mammalian Cell Gene Mutation Tests Using the Thymidine Kinase Gene (OECD TG 490). The latter, also called the mouse lymphoma assay (MLA), can detect a broader spectrum of genetic damage than the former, including chromosome rearrangements, deletions, and mitotic recombination [236]. Both assays are time-consuming, requiring long culture times (e.g., 10–14 days before counting colony formation), which has probably precluded an extensive use of these assays. For soluble chemicals, the sensitivity and specificity of the MLA assay was reported to be 73.1 and 39.0%, respectively, resulting in a prediction accuracy of 62.9 % [224]. In the case of NMs, given the low number of studies performed with these assays, and the wide variety of NMs included in these few studies, it has not been possible to draw any conclusions concerning the relative sensitivity of the various reporter genes to the potential mutagenicity of NMs [236]. Nevertheless, there are ongoing efforts to adapt the HPRT assay for use with NMs, e.g., within the EU H2020 RiskGone project, where round robin activities are ongoing.

Among the assays detecting chromosome damage, the In vitro Mammalian Cell Micronucleus (MN) Test (OECD TG 487) has been the most extensively used in nanotoxicology [221,230]. The assay detects chromosome mutations induced by either clastogenic or aneugenic agents. In addition, it also detects most mutagenic events as most mechanisms leading to gene mutation also induce chromosome mutations [228]. For soluble chemicals, the sensitivity and specificity of the MN assay was reported to be 78.7 and 30.8%, respectively [229], resulting in a concordance of 67.8%. In the case of NMs, there are few papers that evaluated a similar material, and there is a substantial variation in the methodology applied, which precludes raising conclusions on the reproducibility and predictability of this assay [236].

The most classically used version of the MN assay is the cytochalasin-blocked MN assay, which includes the use of cytochalasin B, a cytokinesis blocking agent that enables the identification of dividing cells [238]. However, as cytochalasin B may impair NM intracellular internalization, leading to false-negative results in the MN assay [239], it is recommended to successively treat the cells with NMs, and then with this agent [240,241]. Since cells should undergo mitosis for binucleated cells to form, the use of serum in exposure medium is recommended. The need of both proliferating cells and cells accumulating NMs in the MN assay has been illustrated recently while optimizing the MN assay on 3D cell models. The 3D EpiDerm™ skin model accumulates less NMs than 2D skin cells, resulting in less MNs upon exposure to genotoxic ZnO NMs [242]. Moreover, HepG2 spheroids still hold the capacity to proliferate while HepaRG spheroids do not, and genotoxic ZnO NMs do not show a positive outcome in the MN assay in 3D HepaRG while they do on the 3D HepG2 model [242,243]. An adaptation of this TG for NMs, within the OECD project 4.95 ('Guidance Document on the Adaptation of In vitro Mammalian Cell Based Genotoxicity TGs for Testing of Manufactured Nanomaterials'), is currently ongoing based on previous recommendations [236]. One of the first round robin studies on the in vitro

MN assay was performed within the NanoGenotox project [244], involving 12 laboratories, and comparing the genotoxicity of three reference materials. Relatively reproducible results were obtained in some cases, but they were material- and cell line-specific. A similar conclusion was reached by Louro et al. (2016) on four benchmark MWCNTs in two lung epithelial cell lines [245]. One reason could be the low fold increase over control values, as was also pointed out in the genotoxicity assessment performed by Elespuru et al. (2018) [236]. Currently, round robin tests are planned within the OECD project 4.95 and the EU H2020 RiskGone project.

Lately, the MN assay has been applied in co-culture systems involving inflammatory cells (e.g., THP-1 cells) and target cells (e.g., lung epithelial cells) allowing the evaluation of the mechanisms of action (primary vs. secondary) underlying NMs' genotoxicity [246,247]. Genotoxins operating by a secondary mechanism of action, mediated by inflammation, are assumed to have a threshold response [248].

Classically, the MN assay has involved a labour-intensive manual scoring under the microscope. However, the speed of the analyses can nowadays be increased by using automated microscope scoring platforms [249], or flow cytometry [250–252]. The latter has recently been adapted to NMs [253].

The other validated method for assessing chromosome damage is the In vitro Mammalian Chromosome Aberration (CA) Test (OECD TG 473). This assay has been used much less because it is more time-consuming and requires a significant level of expertise to score the aberrations [236]. Furthermore, the CA assay does not detect aneugens, while the MN assay does (OECD TG 473, 2016). Hence, the CA assay would not be recommended for SbD hazard testing.

Furthermore, HTS approaches that could be applied to the testing of NMs are currently in development. These methods are non-OECD-guideline methods but proved efficient to detect potential NM genotoxicity. The comet assay is by far the most employed among these assays, and it could complement the recommended in vitro mutagenicity assays [236]. During the past decade, some effort has been dedicated to increase its throughput, with the highest achieved in the 96-minigel version using gel bond films [254]. It has been optimized and successfully applied to assess the genotoxicity of NMs within the FP7 NanoREG project [255].

Lastly, one commonly used genotoxicity assay for NMs is the immunolabelling of DNA repair protein foci, such as gamma-H2AX, which form during DNA double-strand break repair. The background of DNA double-strand breaks in cells is generally very low (although some exceptions exist, such as in some cancer cell lines), which makes these assays very sensitive. High-throughput versions of the assays exist. Foci can be counted using automated microscopy platforms or flow cytometry, with the advantage of their rapidity and possibility of analysing other cell parameters such as cell viability or apoptosis simultaneously [256]. Such high-throughput methods have rarely been applied on advanced 3D cell models for assessing NM genotoxicity because they necessitate additional steps. For example, a high-throughput comet assay can be performed on 3D cells after enzymatic and mechanical dissociation of the spheroid [257], reducing simplicity. Still, such advanced models could increase the predictivity of the assay. For example, Ag NMs cause significant DNA damage in a 2D liver-cell system, while the outcome of the comet assay is insignificant in 3D HepG2 spheroids, which is similar to most of the in vivo studies published up to now reporting Ag NM genotoxicity via comet assay [258].

4.5.2. Overview of Needs and Knowledge Gaps

Table 5 shows a summary of how the different genotoxicity assays perform in terms of the criteria for SbD hazard testing. One of the main problems for determining the sensitivity and predictability of the genotoxicity assays when assessing NMs is the absence of nano-sized particulate controls. NM-specific controls have rarely been demonstrated [236,259], making comparisons among labs difficult. Furthermore, it is not possible to establish historical positive control ranges that would confirm the sensitivity of the tests [259]. Based on the currently available information, we follow the recommendations by Elespuru et al. (2018) [236] to use the MN assay in combination with a gene-mutation assay (HPRT or MLA). In the meantime, further optimizations of genotoxicity assays for testing NMs are ongoing.

Table 5. Evaluation of suitability of genotoxicity assays for SbD hazard testing.

Performance Criteria	Simple Cell Models		More Complex Cell Models	
	Gene Mutations in Cell Lines (OECD TGs 476 and 490)	Chromosome Damage in Cell Lines (OECD TGs 487 MN Assay)	Gene Mutations in Advanced Models	Chromosome Damage in Advanced Models (OECD TG 487)
Simplicity and cost	(+/−) Time consuming, requiring long culture times (e.g., 10–14 days before counting colony formation). Relatively cheap.	(+) Simple and relatively cheap. Analyses can be sped up using automatic image analysis systems and flow-cytometry.	(○) Not used up to now, would necessitate 3D model dissociation before cell plating, i.e., simplicity reduced as compared to simple models.	(+/−) Relatively simple and cheap for advanced models. Would be more time consuming and expensive than 2D models [260].
Predictivity (Sensitivity and Specificity)	(○) Conventional chemicals: adequate (62.9%) [224,229]. NMs: no conclusions can be reached [236,259].	(○) Conventional chemicals: adequate (67.8%) [224,229] NMs: no conclusions can be reached [236,259].	(○) Not used up to now, no conclusion can be reached.	(○) Co-culture systems may allow the evaluation of the involved genotoxicity mechanisms of action [246,247]. They may be more predictive of an in vivo-like response [260]. 3D models do not seem to be appropriate for applying this assay due to the lack of cell proliferation [242,243]. When the 3D model involves proliferating cells, it is more sensitive than 2D models, due to higher metabolic activity [260].
Robustness	(○) No inter-laboratory comparisons available for NMs. Ongoing comparisons within the EU H2020 RiskGone project.	(+/−) Relatively reproducible results in some cases, but material- and cell line-specific [244]. Future inter-laboratory comparisons under the OECD project 4.95.	(○) Not used up to now, no conclusion can be reached.	(○) Not enough studies available yet to allow reaching conclusions.
Compatibility	(○) Too low number of studies to reach conclusions [236].	(+) Suitable for different NMs (no interferences reported). No information about adequacy for complex materials.	(○) No conclusion can be reached. Still, for NMs could prove unsuitable since only the cells at the periphery of the spheroid/organoid would be exposed to NMs.	(+) Suitable for different NMs (no interferences reported). No information about adequacy for complex materials.
Readiness	(−) No NM-specific standardized protocol available.	(−) No NM-specific standardized protocol available.	(−) No NM-specific standardized protocol available.	(−) No NM-specific standardized protocol available.

For advanced models, such as 3D models, the 3D skin comet and micronucleus assays are sufficiently validated for conventional chemicals and individual OECD TGs could start being developed [260]. However, the 3D airway and liver models are still lacking assays that could measure micronuclei and gene mutations, respectively [260]. Working with NMs raises additional technical hurdles that need to be overcome. Nevertheless, advanced models will offer advantages over current assays, especially by mimicking better the human body response and being able to evaluate modes of actions, e.g., secondary genotoxicity [261].

Box 7

Based on the currently available information, we follow the recommendations by Elespuru et al. (2018) to use the MN assay in combination with a gene mutation assay (HPRT or MLA). In the meantime, further optimizations of genotoxicity assays for testing NMs are ongoing.

5. Discussion and Outlook

The development, manufacturing, and use of NMs with novel properties and potentially undesired health risks are growing at a rapid rate. One way to reduce potential adverse effects caused by NMs is to incorporate SbD in NM development processes. Within a SbD approach, the potential hazards of a NM throughout the life cycle are assessed at an early stage of product innovation. Although advancements have been made in terms of nano-specific risk assessment strategies [13,19,138,227], several challenges remain when applying these strategies to SbD:

1. Current hazard and risk-assessment strategies are not easily applied in an early hazard assessment for SbD applicability, as the proposed and required assays are too time-consuming and costly to be performed early in the development process of a NM.
2. The suitability for SbD hazard testing of currently available in vitro assays in terms of predictivity, cost-effectiveness, sensitivity, specificity, robustness, and compatibility are largely unclear.

In this review, in vitro toxicity assays have been critically assessed in terms of their suitability for SbD hazard testing. The main purpose of SbD hazard testing is the identification of early hazard warnings and obtaining a general idea of the potential hazards of a novel NM, NEP, or components released thereof during the LC. It therefore serves as a first screening during the early stages of the development of a new NM or NEP. For SbD hazard testing, a balance needs to be sought between simplicity and comprehensive testing that addresses all concerns (Figure 5). The more elaborate the assessment, the more uncertainty is minimized, and the more the testing becomes too complex for the purpose of SbD.

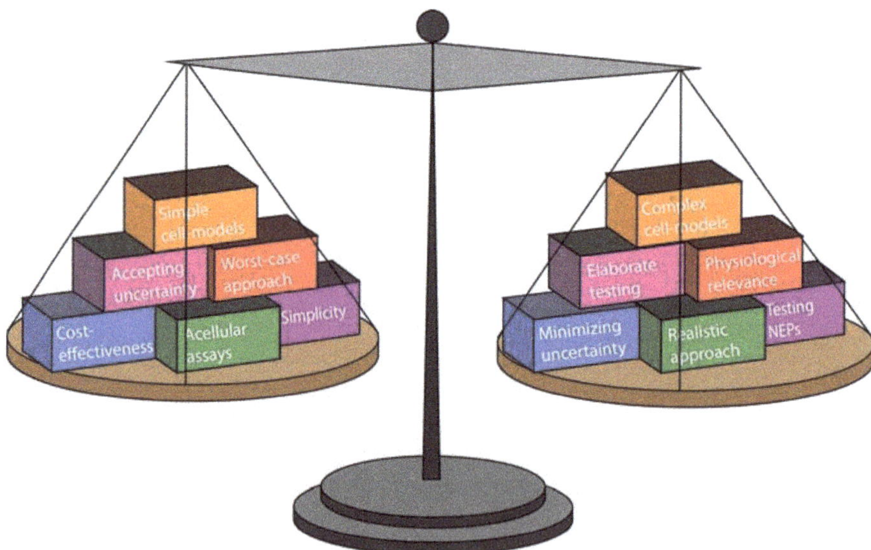

Figure 5. The balance of SbD hazard testing. SbD aims to address safety at an early stage in the product development process. On the one hand, SbD tries to be comprehensive to address all concerns, while on the other hand the approach should be simple.

5.1. Assay Predictivity

5.1.1. Early Hazard Warnings

Correlating in vitro effects to in vivo potency has not yet been possible [109] and is not a requirement for SbD hazard testing. The identification of hazard warnings and detecting the most potent NMs is more relevant, and in this review it was shown that many assays are capable of doing this. The prediction accuracy of the evaluated assays in many cases depends on the type of particle, sample preparation, as well as the type of cell system used. Effects of NMs which assert their effect through ion shedding, such as Ag and ZnO NMs, were most accurately predicted across all toxicity endpoints. Current in vitro assays were less capable of predicting the effects of NMs acting through surface reactivity, and fibre-like NMs. Another important finding across several toxicity outcomes was the better predictivity of primary cells as compared to cell lines, as well as better predictivity of macrophage-like cells as compared to other cell types. However, there are clear indications that simple submerged assays might be suitable for prediction of adverse effects in vivo, as was also shown in a recent review by Di Ianni et al. (2022) [262].

5.1.2. Hazard Ranking

For an assay to be able to establish an accurate ranking in toxicity is valuable for SbD hazard testing, as this allows the use of the assay for comparison between candidate NMs, and for comparison to benchmark materials with known toxicity. There are indications that in vitro cytotoxicity assays are able to predict an adequate ranking in toxicity which corresponds to in vivo pulmonary inflammation [114]. Also, the detection of cytokines at the ALI could potentially detect a ranking that corresponds to in vivo pro-inflammatory mediators [211]. However, both of these studies could not draw any definitive conclusions on comparable rankings. Two studies showed that simple submerged cell lines are able to produce accurate rankings in pro-inflammatory effects that corresponded to in vivo markers of fibrosis [208,209]. It is important to note that the accuracy of toxicity rankings is hugely dependent on the calculation of the dose delivered to the cells, and this should therefore always be carried out [59].

5.1.3. Applicability Domains

A low prediction accuracy of an assay could be improved by exploring the exact applicability of the assay. Ensuring that the specific MOA that caused the in vivo toxicity can be detected using the assay will reduce the rate of false negative outcomes. The applicability domain of each assay should be well-understood (which assay can predict what kind of in vivo (human) toxicity) to be able to use assays that are fit-for-purpose. When looking at the transition to animal free testing in general, this is one of the issues that needs addressing for soluble chemicals as well [263]. In that respect, in vitro toxicity testing of NMs should be mechanism-based by looking for specific effects or MOAs [264]. If the applicability domain of assays and cell models is established with more certainty, it is possible to combine assays into a strategy to holistically assess NM toxicity in vitro. In such a strategy, assay and cell type selection would be facilitated by a combination of applicability domains and the most relevant exposure route.

5.1.4. Prediction Accuracy

Prediction accuracy in terms of sensitivity and specificity should be determined for more assays, to increase knowledge about assay reliability. Determination of accuracy is a crucial step during the validation process of in vitro assays in general [265]. For example, for regulatory genotoxicity testing, it is known that in vitro assays have a high sensitivity but low specificity. This means that there is a high chance of false positives, and positives should always be confirmed in an in vivo study. A better understanding of the prediction accuracies of in vitro assays used for SbD hazard testing would help enormously with the interpretation of results.

5.1.5. Challenges in Assessing Predictivity

Although in vitro assays have been used for some time to test NM toxicity, not all the criteria could be evaluated properly due to the lack of or limited availability of high-quality data. Especially for predictivity, there is a data gap that needs to be filled in order to correctly interpret results. Several factors complicate the assessment of prediction accuracy of in vitro assays. Since there is a lack of human data on NM toxicity, predictivity of assays is at present evaluated compared to in vivo data derived from studies in experimental animals. This means that an in vitro model comprised of human cells is being compared to animal data, to predict a human response. The relevance of this approach is questionable, due to the differences between humans and experimental animals [266,267]. The lack of deposited dose calculations, interference controls, proper characterization, and varying sample preparation protocols (e.g., the use of serum, different media, different dispersion techniques) across the literature add another layer of complexity to assessing the predictivity of assays. Since these factors can have such an impact on assay outcome, assay standardization will aid in determining assay predictivity for adverse human health effects. Additionally, the lack of clear positive and negative controls for NMs hampers the assessment of prediction accuracy.

Finally, it must be noted that in the papers reviewed here, an optimistic perspective is given about the predictivity of in vitro assays for toxicity and adverse health outcomes in vivo. We should however be cautious, as negative results or results with a low correlation with known in vivo or human health responses may not reach publication: a phenomenon known as publication bias.

5.2. Outlook for Innovators, Regulators, and Industry Based on Current Knowledge

An overview of the most important knowns and unknowns with regards to NM SbD hazard testing is summarized in Table 6. Figure 6 shows what we think is the road forward towards successfully putting SbD hazard testing into practice. The successful implementation of SbD hazard testing requires efforts from innovators, regulators, as well as from industry.

Table 6. Overview of the most important findings in this review, including knowns and needs for SbD hazard testing.

		What We Know for SbD Hazard Testing	What We Need for SbD Hazard Testing
NM treatment	Dispersion protocols	-Sonication can destroy intrinsic NM properties that might be part of its safer design. -Sonication can induce underestimation of toxicity by reducing the length of CNTs. -Sonication can enhance dissolution and release of (toxic) ions. -Sonication leads to a lower state of agglomeration.	-Consensus around dispersion protocols. -Dispersion guidance which covers all relevant exposure conditions and takes into account SbD interventions.
	Experimental design	-Testing NMs with serum results in lower in vitro toxicity. -Calculation of the dose delivered to the cells can have impact on toxicity ranking and is therefore also required for SbD hazard testing.	-Consensus and guidance for experimental design in the context of SbD.
	Compatibility and LC	-Humans are not only exposed to pristine NMs, but also to NEPs, aged NMs, and NMs released during the LC. -Testing NMs released from NEPs may pose challenges in terms of feasibility and compatibility -Compatibility of novel NMs with currently available in vitro assays unknown.	-Guidance on how to approach testing NMs with unknown compatibility. -More research towards determining whether testing pristine NMs is sufficient for SbD hazard testing.
Assay protocols	Cytotoxicity	-More elaborate SOPs enhance robustness. -Many NMs interfere with cytotoxicity assays, which should not be overlooked. -In vivo effect of ion shedding NMs is sufficiently accurately predicted. -Measuring cytotoxicity is useful for identifying hazard warnings.	-Further standardization and validation of cytotoxicity assays -Thresholds for cytotoxicity in the context of SbD -More focus on assays that do not pose interference issues. -To confirm predictivity of cytotoxicity assays
	Dissolution	-Dissolution rate may infer bio-persistency, which is important information for SbD hazard testing. -Predictivity largely depends on readout method as well as biological fluid choice. -Static acellular dissolution seems to be the most appropriate method for SbD hazard testing, especially as they are rather simple, however some studies indicate otherwise.	-To confirm that measuring static acellular dissolution is indeed sufficiently predictive for SbD hazard testing. -Meaningful thresholds for dissolution rates that allow for detection of differences that will lead to meaningful SbD decisions and interventions.

Table 6. *Cont.*

		What We Know for SbD Hazard Testing	What We Need for SbD Hazard Testing
Assay protocols	Oxidative Potential	-Acellular assays might be predictive enough for SbD hazard testing. -In vivo effects of ion-shedding NMs is sufficiently accurately predicted. -There are indications that the haemolysis assay can accurately predict effects of surface-reactive NMs. -FRAS and ESR assays are more sensitive than DCFH. -FRAS assay can provide accurate ranking.	-To confirm that measuring acellular OP is predictive enough for SbD testing. -Meaningful thresholds for OP in the context of SbD.
	Inflammation	-The use of a type of immune cell is crucial (using only epithelial cells is not sufficient). -Primary cell models have better predictivity, but (immune) cell lines may suffice for SbD hazard testing. -Co-cultures seem to perform better than mono-cultures. -More elaborate SOPs enhance robustness.	-More work needed to develop in vitro models that can predict chronic inflammation. -Thresholds for in vitro inflammation in the context of SbD. -To confirm that submerged mono-cultures of macrophage cell lines are predictive enough.
	Genotoxicity	-Prediction accuracies very well established for soluble chemicals, but not for NMs. -It is important that the cell model of choice is capable of NM uptake. -The absence of NM positive controls makes determination of prediction accuracy challenging.	-To determine prediction accuracies of assays for NM specifically. -Round robin initiatives to test robustness of assays.

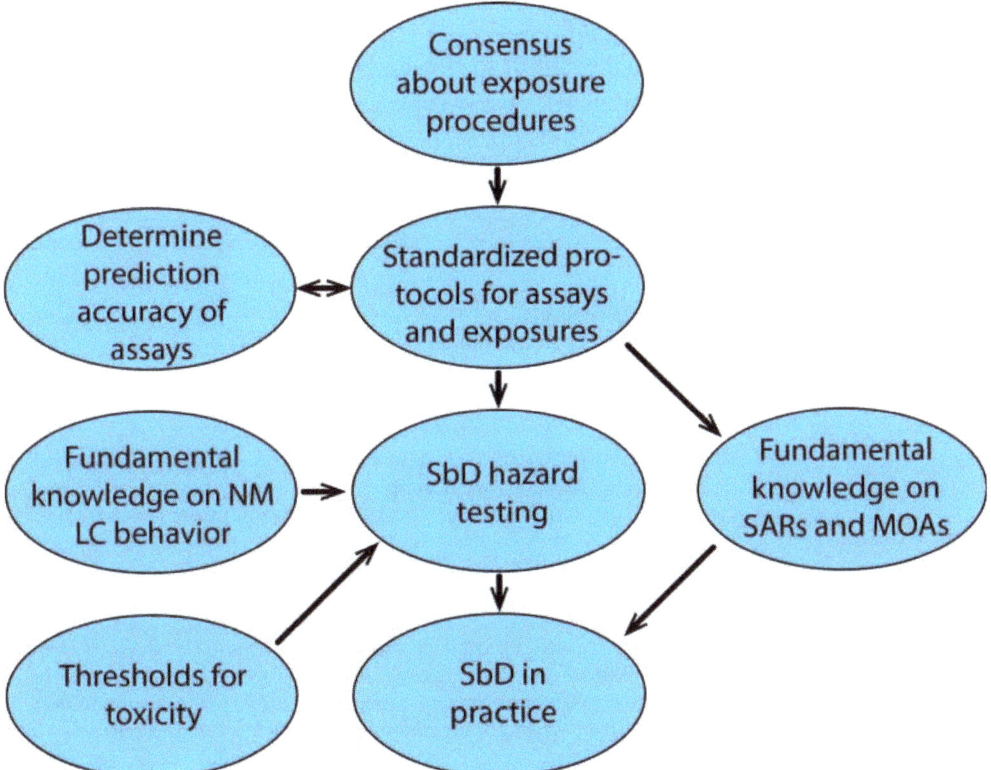

Figure 6. Factors that became evident throughout this review that are crucial for putting SbD hazard testing into practice. Protocol standardization is key for SbD hazard testing, as well as for better understanding structure–activity relationships and prediction accuracies of assays.

5.2.1. A Change in Mindset towards Purpose-Driven Innovations

The current European policy landscape (the European Green Deal, the European Chemical Strategy for Sustainability and the Zero Pollution Action Plan [2,3,268]) demands a new mindset for innovating. SbD provides an approach aiming at developing safer NMs and NEPs by integrating safety into the innovation process and material development in a LC thinking approach, from design to end-of-life. Any innovation that does not have a green or sustainable purpose will not survive.

5.2.2. Starting In Silico: Databases and SARs

SbD hazard assessment should first and foremost be based on material knowledge and material–activity relationships. Before even starting in vitro experiments, an elaborate evaluation of available physicochemical data should be performed [220]. Here, certain hazard warnings could already be noticed. For example, the structure–activity relationship (SAR) of high aspect ratio NMs (HARNs) and mesothelioma risk is widely accepted [138]. It would be unnecessary to perform hazard testing on HARNs, as it would already be clear beforehand that this material raises a hazard warning. Another potential hazard warning would be respirable crystalline silica particles, due to their structure–activity relationship with silicosis and lung cancer [269,270]. Knowledge on structure activity relationships is especially important for the identification of potential hazards and application of SbD interventions.

For novel advanced materials however, limited information on these tox-driving properties is available, and the SbD decisions are mostly based on SbD hazard-testing outcomes.

5.2.3. Importance of Experimental Design

The physical aspects of NMs add another dimension to the complexity of toxicity testing. It should always be considered that the way the experiment is carried out (dispersion protocol, medium type, addition of serum) affects the outcomes and that the behaviour of the particle in the culture dish (settling, agglomerating, floating, dissolution, formation of protein corona) should always be analysed [23,49,53,58,59]. Checking and accounting for assay interference is crucial, also for SbD hazard testing and high throughput screening, where it is often overlooked [97]. This makes SbD hazard testing for NMs more challenging than that of soluble chemicals.

5.2.4. Combinations of Assays

An integrated approach to testing and assessment (IATA) that can combine information from multiple sources (available data, in silico tools, in vitro assays) is the way forward towards an effective early hazard identification of NMs and NEPs and for the development of SbD interventions. This review discusses simple assays since it focusses on the initial stages of innovation. However, at more advanced stages, SbD hazard testing may also include approaches that are not as simple and cost-effective [264]. With regards to the transition to animal-free alternatives, the focus on simplicity as is required for SbD hazard testing should not create a barrier for the development of more realistic and innovative cell models with potentially better predictivity, such as induced pluripotent stem cells (iPSCs) and organoids.

For inflammatory potential, chronic inflammation (leading to tissue damage and remodelling as well as loss of functions) is the adverse outcome of concern, which is presently not captured by one or more in vitro tests. An acute pro-inflammatory effect in an in vitro assay as measured by cytokine secretion, in combination with slow dissolution, indicating high bio-persistency [14], might together indicate that the NM induces chronic inflammation. Combining assay outcomes in SbD hazard testing should be further explored.

5.2.5. Thresholds for Toxicity

In order to raise hazard warnings and to interpret results from combinations of assays, thresholds are needed. This is especially challenging for inflammatory potential assays. Macrophages are the major defence mechanism against foreign materials, and their activation is crucial for the clearance of NMs [192]. It is unclear when a beneficial immune response turns into persistent pulmonary inflammation in vivo, and how to predict this in vitro.

Previously established frameworks have made a step towards generating thresholds for toxicity. The Nanoreg2 framework and the Swiss Precautionary Matrix score NMs as low, medium, or high hazard according to their fold change increase as compared to a negative control [271]. The Nanoreg2 framework adds a scoring system that allows for combining outcomes of different assays, and subsequent comparison of different NMs. With both approaches, a significantly positive response in an assay might still lead to a classification as low hazard.

Since SbD hazard testing is performed as an early screening, and its main goal is determining early hazard warnings, a zero-tolerance principle might be more suitable in this case (as is common practice in the pharmaceutical industry). For primary genotoxicity, a zero-tolerance principle is already in place in regulatory risk assessment, as genotoxic carcinogens are regarded as having no threshold and thus an acceptable exposure level cannot be derived [224]. For SbD hazard testing, it could be argued that a worst-case approach would be suitable for the other endpoints as well, meaning that any indication of inflammation, reactivity, or cytotoxicity at relevant doses would raise a hazard warning. Here it is important to consider the possibility of false negatives produced in the assays.

For SbD hazard testing, the inclusion of benchmark NMs with known in vivo toxicity is recommended to compare the new NM to existing information. Thresholds could be set according to the response of the benchmark NM in a specific assay. Alternatively, an appropriate ranking in potency of NMs could be useful for making SbD decisions when comparing several candidate NMs.

5.2.6. Assay Standardization

In Figure 6, assay standardization is represented connecting many important aspects. As mentioned throughout this review, assay standardization is a key need for the further development of SbD hazard testing, as well as for putting SbD into practice. Firstly, we showed that in vitro-in vivo comparisons are hampered by the lack of standardized protocols. Moreover, fundamental research into structure–activity relationships will benefit from standardized protocols as well. Assay standardization will result in more high-quality fundamental data on the MOAs of toxicity of NMs, which will in turn aid the refining of SbD hazard testing. Ultimately, standardization will increase the chances of industrial use and acceptance of these assays into existing legal frameworks, which will make incorporating SbD approaches more appealing for manufacturers [272].

On the contrary, the complexity of NM toxicity testing hampers the standardization of assays. It is for example impossible to create one exposure method suitable for all NMs, especially considering NMs of the future which will possess yet unknown properties. A case-by-case or targeted approach will be needed for specific NMs with incompatible PC properties. In some cases, standardization may not be feasible, but guidance will be of great help.

Assay standardization should be followed by assay validation in order to improve our understanding of the robustness, predictivity, and compatibility of the assays. The ongoing work in the OECD's Working Party on Manufactured Nanomaterials [273], the Malta Initiative [274], and work in ongoing European projects such as NanoHarmony [275], Nanomet [276] and Gov4Nano [277] are currently supporting the standardization efforts.

5.2.7. Compatibility (NEPs and Novel Materials)

Safety along the LC as well as keeping pace with the rapid emergence of advanced materials are important hallmarks of SbD. Consumers are most likely exposed to NEPs and not pristine NMs. Therefore, assay optimization is needed to be able to test NEPs in an accurate way. Assay compatibility with NEPs and novel advanced materials needs to be studied further. More data on how NMs can change over the LC and the possible risks they may pose during this process is very much needed. This will help determine whether testing only pristine NMs may be sufficient for SbD hazard testing.

5.2.8. Gathering Experimental Data following FAIR Principles

Since SbD hazard testing will involve the generation of large datasets, it is important to ensure that the data gathered from the different in vitro assays are adequately collected using templates that support FAIR principles, and that the data is findable, accessible, interoperable and reusable. Guidance for finding these templates can be found in the GoFair initiative and guidance on experimental workflows design and implementation can be found within the NanoCommons initiative.

5.2.9. The Chemical Strategy for Sustainability

Although this review covers cytotoxicity, dissolution, oxidative potential, inflammatory potential, and genotoxicity, the Chemical Strategy for Sustainability has put forth extra endpoints to ensure the ambition towards a toxic-free environment and protection against the most harmful chemicals is fulfilled [3]. One of these endpoints is endocrine disruption. Under REACH, endocrine disruptors are identified as substances of very high concern alongside chemicals known to cause cancer, mutations, and toxicity to reproduction. Work is ongoing by ECHA to develop classification and labelling criteria for endocrine disruption [278]. From a NM-perspective, there is increasing evidence showing endocrine disruption and reproductive impairments caused by NMs such as nano plastics [279,280], and this warrants further attention.

Although this review is only focused on SbD, sustainability impacts should also be considered early in the innovation process. Safe-and-sustainable-by-design is a central element of the European Chemical Strategy for Sustainability and it demands the optimization of safety and sustainability interventions in the design of NMs, NEPs, and all processes in a life-cycle approach.

6. Conclusions

This review provides the first building blocks towards an early hazard testing strategy for SbD applicability and is the first detailed state of the art analysis of in vitro assays against performance criteria (simplicity and cost effectiveness, predictivity, robustness, compatibility, and readiness) for SbD hazard testing. The most important conclusions are:

- Based on current knowledge, primary cell models and more physiologically relevant exposure methods provide better predictions of in vivo results. However, the aim of SbD hazard testing is to detect early hazard warnings using simple methods. There are strong indications that simpler assays, such as acellular OP assays, static dissolution assays, and simple submerged cell-based assays for cytotoxicity, genotoxicity, and inflammation give sufficiently accurate information for identifying early hazard warnings or even hazard rankings, when carried out correctly.
- The suitability of these simple assays for SbD hazard testing has to be further confirmed in future studies. More model comparisons between simple, complex, and in vivo models are needed to investigate whether simple in vitro models are indeed sufficiently predictive and suitable for SbD hazard testing, preferably using standardized methods. Additionally, the applicability domain of in vitro assays to detect NM toxicity should be mapped more precisely to correctly interpret results.
- Assay standardization proved to be critical for the progression of SbD hazard testing as it will improve in vitro-in vivo comparisons, improve fundamental knowledge on NM toxicity, support industrial use, and is a first step towards regulatory acceptance.
- Simplicity is not always feasible when testing NMs, even though it has been put forward as one of the criteria for SbD hazard testing. Dispersion protocols, dose delivered to cells, compatibility issues, interferences, testing NEPs and NMs released along the LC, etc., all complicate SbD hazard testing of NMs and reduce achievable simplicity. Innovators, industry, regulators, and policymakers should realize that

Nanomaterials **2023**, *13*, 472

the hazard assessment of NMs and advanced materials is complex and that in vitro tests need to be further developed, tested, and evaluated to assess their suitability in identifying potential hazards.

Author Contributions: All authors contributed to the conceptualization of the manuscript. The preparation of the different sections of the original draft was done by the following authors: Introduction by N.R. and L.G.S.-H.; Criteria by N.R.; Choice of dispersion protocol by M.C. and J.C. (Joan Cabellos); the use of serum and stabilizers by N.R.; determining dose delivered to cells by N.R.; SbD hazard testing of NEPs and NMs released during the life cycle by M.C., C.D., A.S.J. and N.R.; challenging NMs and advanced materials by N.R.; cytotoxicity by H.B. and N.R.; Dissolution by M.B. and P.M.; oxidative potential and oxidative stress by A.K., J.C (Joan Cabellos) and N.R.; inflammation by A.K., A.C. and N.R.; genotoxicity by M.C. and J.C. (Julia Catalán); discussion and outlook by N.R., H.B. and L.G.S.-H.; conclusions by N.R., L.G.S.-H., F.R.C. and H.B. Afterwards, in several iterations all authors read, commented, and proposed revisions to the entire manuscript, which were compiled by N.R. All authors have read and agreed to the published version of the manuscript.

Funding: This research was funded by the SAbyNA project, European Union's Horizon 2020 research and innovation program under grant agreement No 862419, and by the Dutch Ministry of Infrastructure and Water Management.

Institutional Review Board Statement: Not applicable.

Informed Consent Statement: Not applicable.

Acknowledgments: The authors would like to thank Rob Vandebriel and Agnes Oomen for their valuable suggestions and critical review of the manuscript.

Conflicts of Interest: The authors declare no conflict of interest. The funders had no role in the design of the study; in the collection, analyses, or interpretation of data; in the writing of the manuscript; or in the decision to publish the results.

References

1. Falk, A.; Cassee, F.R.; Valsami-Jones, E. *Safe-by-Design and EU Funded NanoSafety Projects*; European Commission: Brussels, Belgium, 2021. [CrossRef]
2. European Commission. *Communication from the Commission to the European Parliament, the Council, the European Economic and Social Committee and the Committee of the Regions: The European Green Deal*; European Commission: Brussels, Belgium, 2019.
3. European Commission. *Chemicals Strategy for Sustainability—Towards a Toxic-Free Environment*; European Commission: Brussels, Belgium, 2020.
4. Gottardo, S.; Alessandrelli, M.; Amenta, V.; Atluri, R.; Barberio, G.; Bekker, C.; Bergonzo, P.; Bleeker, E.; Booth, A.M.; Borges, T.; et al. *NANoREG Framework for the Safety Assessment of Nanomaterials*; European Commission Joint Research Centre: Brussels, Belgium, 2017.
5. Sánchez Jiménez, A.; Puelles, R.; Pérez-Fernández, M.; Gómez-Fernández, P.; Barruetabeña, L.; Jacobsen, N.R.; Suarez-Merino, B.; Micheletti, C.; Manier, N.; Trouiller, B.; et al. Safe(r) by design implementation in the nanotechnology industry. *NanoImpact* **2020**, *20*, 100267. [CrossRef]
6. OECD. *Moving Towards a Safe(r) Innovation Approach (SIA) for More Sustainable Nanomaterials and Nano-Enabled Products*; Series on the Safety of Manufactured Nanomat; OECD: Paris, France, 2020.
7. Schmutz, M.; Borges, O.; Jesus, S.; Borchard, G.; Perale, G.; Zinn, M.; Sips, Ä.A.J.A.M.; Soeteman-Hernandez, L.G.; Wick, P.; Som, C. A Methodological Safe-by-Design Approach for the Development of Nanomedicines. *Front. Bioeng. Biotechnol.* **2020**, *8*, 258. [CrossRef]
8. Soeteman-Hernandez, L.G.; Apostolova, M.D.; Bekker, C.; Dekkers, S.; Grafström, R.C.; Groenewold, M.; Handzhiyski, Y.; Herbeck-Engel, P.; Hoehener, K.; Karagkiozaki, V.; et al. Safe innovation approach: Towards an agile system for dealing with innovations. *Mater. Today Commun.* **2019**, *20*, 100548. [CrossRef]
9. Xia, T.; Zhao, Y.; Sager, T.; George, S.; Pokhrel, S.; Li, N.; Schoenfeld, D.; Meng, H.; Lin, S.; Wang, X.; et al. Decreased Dissolution of ZnO by Iron Doping Yields Nanoparticles with Reduced Toxicity in the Rodent Lung and Zebrafish Embryos. *ACS Nano* **2011**, *5*, 1223–1235. [CrossRef] [PubMed]
10. Rubio, L.; Pyrgiotakis, G.; Beltran-Huarac, J.; Zhang, Y.; Gaurav, J.; DeLoid, G.; Spyrogianni, A.; Sarosiek, K.A.; Bello, D.; Demokritou, P. Safer-by-design flame-sprayed silicon dioxide nanoparticles: The role of silanol content on ROS generation, surface activity and cytotoxicity. *Part. Fibre Toxicol.* **2019**, *16*, 40. [CrossRef]
11. Zhou, J.; Cai, D.; Xu, Q.; Zhang, Y.; Fu, F.; Diao, H.; Liu, X. Excellent binding effect of l-methionine for immobilizing silver nanoparticles onto cotton fabrics to improve the antibacterial durability against washing. *RSC Adv.* **2018**, *8*, 24458–24463. [CrossRef] [PubMed]

12. Hjorth, R.; van Hove, L.; Wickson, F. What can nanosafety learn from drug development? The feasibility of "safety by design". *Nanotoxicology* **2017**, *11*, 305–312. [CrossRef]

13. Dekkers, S.; Wijnhoven, S.W.P.; Braakhuis, H.M.; Soeteman-Hernandez, L.G.; Sips, A.J.A.M.; Tavernaro, I.; Kraegeloh, A.; Noorlander, C.W. Safe-by-Design part I: Proposal for nanospecific human health safety aspects needed along the innovation process. *NanoImpact* **2020**, *18*, 100227. [CrossRef]

14. Di Cristo, L.; Oomen, A.G.; Dekkers, S.; Moore, C.; Rocchia, W.; Murphy, F.; Johnston, H.J.; Janer, G.; Haase, A.; Stone, V.; et al. Grouping Hypotheses and an Integrated Approach to Testing and Assessment of Nanomaterials Following Oral Ingestion. *Nanomaterials* **2021**, *11*, 2623. [CrossRef]

15. Arts, J.H.E.; Hadi, M.; Irfan, M.-A.; Keene, A.M.; Kreiling, R.; Lyon, D.; Maier, M.; Michel, K.; Petry, T.; Sauer, U.G.; et al. A decision-making framework for the grouping and testing of nanomaterials (DF4nanoGrouping). *Regul. Toxicol. Pharmacol.* **2015**, *71*, S1–S27. [CrossRef]

16. Stone, V.; Gottardo, S.; Bleeker, E.A.J.; Braakhuis, H.; Dekkers, S.; Fernandes, T.; Haase, A.; Hunt, N.; Hristozov, D.; Jantunen, P.; et al. A framework for grouping and read-across of nanomaterials- supporting innovation and risk assessment. *Nano Today* **2020**, *35*, 100941. [CrossRef]

17. Rauscher, H.; Rasmussen, K.; Riego Sintes, J.; Sala, S. *Safe and Sustainable by Design Chemicals and Materials*; Publications Office of the EU: Luxembourg, 2022.

18. Braakhuis, H.M.; Murphy, F.; Ma-Hock, L.; Dekkers, S.; Keller, J.; Oomen, A.G.; Stone, V. An Integrated Approach to Testing and Assessment to Support Grouping and Read-Across of Nanomaterials After Inhalation Exposure. *Appl. In Vitro Toxicol.* **2021**, *7*, 112–128. [CrossRef] [PubMed]

19. Dekkers, S.; Oomen, A.G.; Bleeker, E.A.; Vandebriel, R.J.; Micheletti, C.; Cabellos, J.; Janer, G.; Fuentes, N.; Vázquez-Campos, S.; Borges, T.; et al. Towards a nanospecific approach for risk assessment. *Regul. Toxicol. Pharmacol.* **2016**, *80*, 46–59. [CrossRef] [PubMed]

20. Halappanavar, S.; Nymark, P.; Krug, H.F.; Clift, M.J.D.; Rothen-Rutishauser, B.; Vogel, U. Non-Animal Strategies for Toxicity Assessment of Nanoscale Materials: Role of Adverse Outcome Pathways in the Selection of Endpoints. *Small* **2021**, *17*, 2007628. [CrossRef]

21. EFSA Scientific Committee; More, S.; Bampidis, V.; Benford, D.; Bragard, C.; Halldorsson, T.; Hernández-Jerez, A.; Hougaard Bennekou, S.; Koutsoumanis, K.; Lambré, C. Guidance on risk assessment of nanomaterials to be applied in the food and feed chain: Human and animal health. *EFSA J.* **2021**, *19*, e06768.

22. Gulumian, M.; Cassee, F.R. Safe by design (SbD) and nanotechnology: A much-discussed topic with a prudence? *Part. Fibre Toxicol.* **2021**, *18*, 32. [CrossRef]

23. DeLoid, G.M.; Cohen, J.M.; Pyrgiotakis, G.; Demokritou, P. Preparation, characterization, and in vitro dosimetry of dispersed, engineered nanomaterials. *Nat. Protoc.* **2017**, *12*, 355–371. [CrossRef]

24. Pedersen, E.; Fant, K. *Guidance Document on Good In Vitro Method Practices (GIVIMP): Series on Testing and Assessment No. 286*; OECD: Paris, France, 2018.

25. Faria, M.; Björnmalm, M.; Thurecht, K.J.; Kent, S.J.; Parton, R.G.; Kavallaris, M.; Johnston, A.P.R.; Gooding, J.J.; Corrie, S.R.; Boyd, B.J.; et al. Minimum information reporting in bio–nano experimental literature. *Nat. Nanotechnol.* **2018**, *13*, 777–785. [CrossRef]

26. Taurozzi, J.S.; Hackley, V.A.; Wiesner, M.R. Ultrasonic dispersion of nanoparticles for environmental, health and safety assessment—Issues and recommendations. *Nanotoxicology* **2011**, *5*, 711–729. [CrossRef]

27. Kaur, I.; Ellis, L.-J.; Römer, I.; Tantra, R.; Carriere, M.; Allard, S.; Mayne-L'Hermite, M.; Minelli, C.; Unger, W.; Potthoff, A.; et al. Dispersion of Nanomaterials in Aqueous Media: Towards Protocol Optimization. *J. Vis. Exp.* **2017**, *130*, e56074. [CrossRef]

28. Simon-Deckers, A.; Gouget, B.; Mayne-L'Hermite, M.; Herlin-Boime, N.; Reynaud, C.; Carrière, M. In vitro investigation of oxide nanoparticle and carbon nanotube toxicity and intracellular accumulation in A549 human pneumocytes. *Toxicology* **2008**, *253*, 137–146. [CrossRef]

29. Hadrup, N.; Knudsen, K.B.; Carriere, M.; Mayne-L'Hermite, M.; Bobyk, L.; Allard, S.; Miserque, F.; Pibaleau, B.; Pinault, M.; Wallin, H.; et al. Safe-by-design strategies for lowering the genotoxicity and pulmonary inflammation of multiwalled carbon nanotubes: Reduction of length and the introduction of COOH groups. *Environ. Toxicol. Pharmacol.* **2021**, *87*, 103702. [CrossRef]

30. Retamal Marín, R.R.; Babick, F.; Lindner, G.-G.; Wiemann, M.; Stintz, M. Effects of Sample Preparation on Particle Size Distributions of Different Types of Silica in Suspensions. *Nanomaterials* **2018**, *8*, 454. [CrossRef]

31. Peters, R.; Kramer, E.; Oomen, A.G.; Herrera Rivera, Z.E.; Oegema, G.; Tromp, P.C.; Fokkink, R.; Rietveld, A.; Marvin, H.J.P.; Weigel, S.; et al. Presence of Nano-Sized Silica during In Vitro Digestion of Foods Containing Silica as a Food Additive. *ACS Nano* **2012**, *6*, 2441–2451. [CrossRef]

32. Cappellini, F.; Hedberg, Y.; McCarrick, S.; Hedberg, J.; Derr, R.; Hendriks, G.; Wallinder, I.O.; Karlsson, H.L. Mechanistic insight into reactivity and (geno)toxicity of well-characterized nanoparticles of cobalt metal and oxides. *Nanotoxicology* **2018**, *12*, 602–620. [CrossRef]

33. Pradhan, S.; Hedberg, J.; Blomberg, E.; Wold, S.; Wallinder, I.O. Effect of sonication on particle dispersion, administered dose and metal release of non-functionalized, non-inert metal nanoparticles. *J. Nanopart. Res.* **2016**, *18*, 285. [CrossRef]

34. Alstrup Jensen, K.; Kembouche, Y.; Christiansen, E.; Jacobsen, N.; Wallin, H.; Guiot, C.; Spalla, O.; Witschger, O. *Final Protocol for Producing Suitable Manufactured Nanomaterial Exposure Media*; NANOGENOTOX deliverable report 3; EAHC: Luxembourg, 2011.

35. Wegner, K.D.; Dussert, F.; Truffier-Boutry, D.; Benayad, A.; Beal, D.; Mattera, L.; Ling, W.L.; Carrière, M.; Reiss, P. Influence of the Core/Shell Structure of Indium Phosphide Based Quantum Dots on Their Photostability and Cytotoxicity. *Front. Chem.* **2019**, *7*, 466. [CrossRef]

36. Dussert, F.; Wegner, K.D.; Moriscot, C.; Gallet, B.; Jouneau, P.-H.; Reiss, P.; Carriere, M. Evaluation of the Dermal Toxicity of InZnP Quantum Dots Before and After Accelerated Weathering: Toward a Safer-By-Design Strategy. *Front. Toxicol.* **2021**, *3*, 636976. [CrossRef]

37. Tarantini, A.; Wegner, K.D.; Dussert, F.; Sarret, G.; Beal, D.; Mattera, L.; Lincheneau, C.; Proux, O.; Truffier-Boutry, D.; Moriscot, C.; et al. Physicochemical alterations and toxicity of InP alloyed quantum dots aged in environmental conditions: A safer by design evaluation. *NanoImpact* **2019**, *14*, 100168. [CrossRef]

38. Rosset, A.; Bartolomei, V.; Laisney, J.; Shandilya, N.; Voisin, H.; Morin, J.; Michaud-Soret, I.; Capron, I.; Wortham, H.; Brochard, G.; et al. Towards the development of safer by design TiO$_2$-based photocatalytic paint: Impacts and performances. *Environ. Sci. Nano* **2021**, *8*, 758–772. [CrossRef]

39. Roebben, G.; Ramirez-Garcia, S.; Hackley, V.A.; Roesslein, M.; Klaessig, F.; Kestens, V.; Lynch, I.; Garner, C.M.; Rawle, A.; Elder, A.; et al. Interlaboratory comparison of size and surface charge measurements on nanoparticles prior to biological impact assessment. *J. Nanopart. Res.* **2011**, *13*, 2675–2687. [CrossRef]

40. Bihari, P.; Vippola, M.; Schultes, S.; Praetner, M.; Khandoga, A.G.; Reichel, C.A.; Coester, C.; Tuomi, T.; Rehberg, M.; Krombach, F. Optimized dispersion of nanoparticles for biological in vitro and in vivo studies. *Part. Fibre Toxicol.* **2008**, *5*, 14. [CrossRef]

41. Sharma, G.; Kodali, V.; Gaffrey, M.; Wang, W.; Minard, K.R.; Karin, N.J.; Teeguarden, J.G.; Thrall, B.D. Iron oxide nanoparticle agglomeration influences dose rates and modulates oxidative stress-mediated dose–response profiles in vitro. *Nanotoxicology* **2014**, *8*, 663–675. [CrossRef]

42. Betts, J.N.; Johnson, M.G.; Rygiewicz, P.T.; King, G.A.; Andersen, C.P. Potential for metal contamination by direct sonication of nanoparticle suspensions. *Environ. Toxicol. Chem.* **2013**, *32*, 889–893. [CrossRef]

43. Gülseren, I.; Güzey, D.; Bruce, B.D.; Weiss, J. Structural and functional changes in ultrasonicated bovine serum albumin solutions. *Ultrason. Sonochem.* **2007**, *14*, 173–183. [CrossRef] [PubMed]

44. OECD. Guidance on sample preparation and dosimetry for the safety testing of manufactured nanomaterials. In *Proceedings of Tour de Table at the 7th Meeting of the Working Party on Manufactured Nanomaterials No. 36*; Series on the Safety of Manufactured Nanomaterials; OECD: Paris, France, 2012.

45. Shannahan, J.H. Nanoparticle–Biocorona. In *Encyclopedia of Nanotechnology*; Bhushan, B., Ed.; Springer: Dordrecht, The Netherlands, 2014; pp. 1–4. [CrossRef]

46. Strojan, K.; Leonardi, A.; Bregar, V.B.; Križaj, I.; Svete, J.; Pavlin, M. Dispersion of Nanoparticles in Different Media Importantly Determines the Composition of Their Protein Corona. *PLoS ONE* **2017**, *12*, e0169552. [CrossRef] [PubMed]

47. Boyles, M.S.P.; Kristl, T.; Andosch, A.; Zimmermann, M.; Tran, N.; Casals, E.; Himly, M.; Puntes, V.; Huber, C.G.; Lütz-Meindl, U.; et al. Chitosan functionalisation of gold nanoparticles encourages particle uptake and induces cytotoxicity and pro-inflammatory conditions in phagocytic cells, as well as enhancing particle interactions with serum components. *J. Nanobiotechnol.* **2015**, *13*, 84. [CrossRef]

48. McConnell, K.I.; Shamsudeen, S.; Meraz, I.M.; Mahadevan, T.S.; Ziemys, A.; Rees, P.; Summers, H.D.; Serda, R.E. Reduced Cationic Nanoparticle Cytotoxicity Based on Serum Masking of Surface Potential. *J. Biomed. Nanotechnol.* **2016**, *12*, 154–164. [CrossRef]

49. Vranic, S.; Gosens, I.; Jacobsen, N.R.; Jensen, K.A.; Bokkers, B.; Kermanizadeh, A.; Stone, V.; Baeza-Squiban, A.; Cassee, F.R.; Tran, L.; et al. Impact of serum as a dispersion agent for in vitro and in vivo toxicological assessments of TiO2 nanoparticles. *Arch. Toxicol.* **2017**, *91*, 353–363. [CrossRef]

50. Lesniak, A.; Salvati, A.; Santos-Martinez, M.J.; Radomski, M.W.; Dawson, K.A.; Åberg, C. Nanoparticle Adhesion to the Cell Membrane and Its Effect on Nanoparticle Uptake Efficiency. *J. Am. Chem. Soc.* **2013**, *135*, 1438–1444. [CrossRef]

51. Lesniak, A.; Fenaroli, F.; Monopoli, M.P.; Åberg, C.; Dawson, K.A.; Salvati, A. Effects of the Presence or Absence of a Protein Corona on Silica Nanoparticle Uptake and Impact on Cells. *ACS Nano* **2012**, *6*, 5845–5857. [CrossRef]

52. Elliott, J.T.; Rösslein, M.; Song, N.W.; Toman, B.; Kinsner-Ovaskainen, A.; Maniratanachote, R.; Salit, M.L.; Petersen, E.J.; Sequeira, F.; Romsos, E.L.; et al. Toward achieving harmonization in a nano-cytotoxicity assay measurement through an interlaboratory comparison study_suppl. *ALTEX-Altern. Anim. Exp.* **2017**, *34*, 201–218. [CrossRef]

53. Wiemann, M.; Vennemann, A.; Venzago, C.; Lindner, G.-G.; Schuster, T.B.; Krueger, N. Serum Lowers Bioactivity and Uptake of Synthetic Amorphous Silica by Alveolar Macrophages in a Particle Specific Manner. *Nanomaterials* **2021**, *11*, 628. [CrossRef]

54. Pisani, C.; Rascol, E.; Dorandeu, C.; Gaillard, J.-C.; Charnay, C.; Guari, Y.; Chopineau, J.; Armengaud, J.; Devoisselle, J.-M.; Prat, O. The species origin of the serum in the culture medium influences the in vitro toxicity of silica nanoparticles to HepG2 cells. *PLoS ONE* **2017**, *12*, e0182906. [CrossRef] [PubMed]

55. Žūkienė, R.; Snitka, V. Zinc oxide nanoparticle and bovine serum albumin interaction and nanoparticles influence on cytotoxicity in vitro. *Colloids Surf. B Biointerfaces* **2015**, *135*, 316–323. [CrossRef] [PubMed]

56. Kittler, S.; Greulich, C.; Gebauer, J.S.; Diendorf, J.; Treuel, L.; Ruiz, L.; Gonzalez-Calbet, J.M.; Vallet-Regi, M.; Zellner, R.; Köller, M.; et al. The influence of proteins on the dispersability and cell-biological activity of silver nanoparticles. *J. Mater. Chem.* **2010**, *20*, 512. [CrossRef]

57. Pyrgiotakis, G.; Blattmann, C.O.; Pratsinis, S.; Demokritou, P. Nanoparticle–Nanoparticle Interactions in Biological Media by Atomic Force Microscopy. *Langmuir* **2013**, *29*, 11385–11395. [CrossRef] [PubMed]

58. Cohen, J.; DeLoid, G.; Pyrgiotakis, G.; Demokritou, P. Interactions of engineered nanomaterials in physiological media and implications for in vitro dosimetry. *Nanotoxicology* **2013**, *7*, 417–431. [CrossRef] [PubMed]

59. Pal, A.K.; Bello, D.; Cohen, J.; Demokritou, P. Implications of in vitro dosimetry on toxicological ranking of low aspect ratio engineered nanomaterials. *Nanotoxicology* **2015**, *9*, 871–885. [CrossRef] [PubMed]

60. DeLoid, G.M.; Cohen, J.M.; Pyrgiotakis, G.; Pirela, S.V.; Pal, A.; Liu, J.; Srebric, J.; Demokritou, P. Advanced computational modeling for in vitro nanomaterial dosimetry. *Part. Fibre Toxicol.* **2015**, *12*, 32. [CrossRef]

61. Thomas, D.G.; Smith, J.N.; Thrall, B.D.; Baer, D.R.; Jolley, H.; Munusamy, P.; Kodali, V.; Demokritou, P.; Cohen, J.; Teeguarden, J.G. ISD3: A particokinetic model for predicting the combined effects of particle sedimentation, diffusion and dissolution on cellular dosimetry for in vitro systems. *Part. Fibre Toxicol.* **2018**, *15*, 6. [CrossRef]

62. Botte, E.; Vagaggini, P.; Zanoni, I.; Gardini, D.; Costa, A.L.; Ahluwalia, A.D. An integrated pipeline and multi-model graphical user interface for accurate nano-dosimetry. *bioRxiv* **2021**. [CrossRef]

63. Lison, D.; Thomassen, L.C.J.; Rabolli, V.; Gonzalez, L.; Napierska, D.; Seo, J.W.; Kirsch-Volders, M.; Hoet, P.; Kirschhock, C.E.A.; Martens, J.A. Nominal and Effective Dosimetry of Silica Nanoparticles in Cytotoxicity Assays. *Toxicol. Sci.* **2008**, *104*, 155–162. [CrossRef]

64. Keller, J.G.; Quevedo, D.F.; Faccani, L.; Costa, A.L.; Landsiedel, R.; Werle, K.; Wohlleben, W. Dosimetry in vitro—exploring the sensitivity of deposited dose predictions vs. affinity, polydispersity, freeze-thawing, and analytical methods. *Nanotoxicology* **2021**, *15*, 21–34. [CrossRef]

65. DeLoid, G.; Cohen, J.M.; Darrah, T.; Derk, R.; Rojanasakul, L.; Pyrgiotakis, G.; Wohlleben, W.; Demokritou, P. Estimating the effective density of engineered nanomaterials for in vitro dosimetry. *Nat. Commun.* **2014**, *5*, 3514. [CrossRef]

66. Ding, Y.; Weindl, P.; Lenz, A.-G.; Mayer, P.; Krebs, T.; Schmid, O. Quartz crystal microbalances (QCM) are suitable for real-time dosimetry in nanotoxicological studies using VITROCELL® cell exposure systems. *Part. Fibre Toxicol.* **2020**, *17*, 44. [CrossRef]

67. Yokel, R.A.; MacPhail, R.C. Engineered nanomaterials: Exposures, hazards, and risk prevention. *J. Occup. Med. Toxicol.* **2011**, *6*, 7. [CrossRef] [PubMed]

68. Basinas, I.; Jiménez, A.S.; Galea, K.S.; Van Tongeren, M.; Hurley, F. A Systematic Review of the Routes and Forms of Exposure to Engineered Nanomaterials. *Ann. Work. Expo. Health* **2018**, *62*, 639–662. [CrossRef]

69. Schneider, T.; Brouwer, D.H.; Koponen, I.K.; Jensen, K.A.; Fransman, W.; Van Duuren-Stuurman, B.; van Tongeren, M.; Tielemans, E. Conceptual model for assessment of inhalation exposure to manufactured nanoparticles. *J. Expo. Sci. Environ. Epidemiol.* **2011**, *21*, 450–463. [CrossRef]

70. Duncan, T.V.; Pillai, K. Release of Engineered Nanomaterials from Polymer Nanocomposites: Diffusion, Dissolution, and Desorption. *ACS Appl. Mater. Interfaces* **2015**, *7*, 2–19. [CrossRef] [PubMed]

71. Al-Kattan, A.; Wichser, A.; Vonbank, R.; Brunner, S.; Ulrich, A.; Zuin, S.; Arroyo, Y.; Golanski, L.; Nowack, B. Characterization of materials released into water from paint containing nano-SiO2. *Chemosphere* **2015**, *119*, 1314–1321. [CrossRef]

72. Murray, A.R.; Kisin, E.R.; Tkach, A.V.; Yanamala, N.; Mercer, R.; Young, S.-H.; Fadeel, B.; Kagan, V.E.; Shvedova, A.A. Factoring-in agglomeration of carbon nanotubes and nanofibers for better prediction of their toxicity versus asbestos. *Part. Fibre Toxicol.* **2012**, *9*, 10. [CrossRef]

73. Noël, A.; Charbonneau, M.; Cloutier, Y.; Tardif, R.; Truchon, G. Rat pulmonary responses to inhaled nano-TiO2: Effect of primary particle size and agglomeration state. *Part. Fibre Toxicol.* **2013**, *10*, 48. [CrossRef]

74. Murugadoss, S.; Brassinne, F.; Sebaihi, N.; Petry, J.; Cokic, S.M.; Van Landuyt, K.L.; Godderis, L.; Mast, J.; Lison, D.; Hoet, P.H.; et al. Agglomeration of titanium dioxide nanoparticles increases toxicological responses in vitro and in vivo. *Part. Fibre Toxicol.* **2020**, *17*, 10. [CrossRef]

75. Saber, A.T.; Jacobsen, N.R.; Mortensen, A.; Szarek, J.; Jackson, P.; Madsen, A.M.; Jensen, K.A.; Koponen, I.K.; Brunborg, G.; Gützkow, K.B.; et al. Nanotitanium dioxide toxicity in mouse lung is reduced in sanding dust from paint. *Part. Fibre Toxicol.* **2012**, *9*, 4. [CrossRef]

76. Pal, A.K.; Watson, C.Y.; Pirela, S.V.; Singh, D.; Chalbot, M.-C.G.; Kavouras, I.; Demokritou, P. Linking Exposures of Particles Released from Nano-Enabled Products to Toxicology: An Integrated Methodology for Particle Sampling, Extraction, Dispersion, and Dosing. *Toxicol. Sci.* **2015**, *146*, 321–333. [CrossRef] [PubMed]

77. Kim, S.; Jaques, P.A.; Chang, M.; Froines, J.R.; Sioutas, C. Versatile aerosol concentration enrichment system (VACES) for simultaneous in vivo and in vitro evaluation of toxic effects of ultrafine, fine and coarse ambient particles Part I: Development and laboratory characterization. *J. Aerosol Sci.* **2001**, *32*, 1281–1297. [CrossRef]

78. Nowack, B.; Boldrin, A.; Caballero, A.; Hansen, S.F.; Gottschalk, F.; Heggelund, L.; Hennig, M.; Mackevica, A.; Maes, H.; Navratilova, J.; et al. Meeting the Needs for Released Nanomaterials Required for Further Testing—The SUN Approach. *Environ. Sci. Technol.* **2016**, *50*, 2747–2753. [CrossRef]

79. Demokritou, P.; Lee, S.J.; Ferguson, S.T.; Koutrakis, P. A compact multistage (cascade) impactor for the characterization of atmospheric aerosols. *J. Aerosol Sci.* **2004**, *35*, 281–299. [CrossRef]

80. Bein, K.J.; Wexler, A.S. A high-efficiency, low-bias method for extracting particulate matter from filter and impactor substrates. *Atmos. Environ.* **2014**, *90*, 87–95. [CrossRef]

81. Zarcone, M.C.; Duistermaat, E.; van Schadewijk, A.; Jedynska, A.; Hiemstra, P.S.; Kooter, I.M. Cellular response of mucociliary differentiated primary bronchial epithelial cells to diesel exhaust. *Am. J. Physiol.-Lung Cell. Mol. Physiol.* **2016**, *311*, L111–L123. [CrossRef]

82. Wohlleben, W.; Brill, S.; Meier, M.W.; Mertler, M.; Cox, G.; Hirth, S.; Von Vacano, B.; Strauss, V.; Treumann, S.; Wiench, K.; et al. On the Lifecycle of Nanocomposites: Comparing Released Fragments and their In-Vivo Hazards from Three Release Mechanisms and Four Nanocomposites. *Small* **2011**, *7*, 2384–2395. [CrossRef] [PubMed]

83. Smulders, S.; Luyts, K.; Brabants, G.; Van Landuyt, K.; Kirschhock, C.E.A.; Smolders, E.; Golanski, L.; Vanoirbeek, J.; Hoet, P.H. Toxicity of Nanoparticles Embedded in Paints Compared with Pristine Nanoparticles in Mice. *Toxicol. Sci.* **2014**, *141*, 132–140. [CrossRef]

84. Smulders, S.; Luyts, K.; Brabants, G.; Golanski, L.; Martens, J.; Vanoirbeek, J.; Hoet, P.H. Toxicity of nanoparticles embedded in paints compared to pristine nanoparticles, in vitro study. *Toxicol. Lett.* **2015**, *232*, 333–339. [CrossRef]

85. Shoemaker, D.A.; Pretty, J.R.; Ramsey, D.M.; McLaurin, J.L.; Khan, A.; Teass, A.W.; Castranova, V.; Pailes, W.H.; Dalal, N.S.; Miles, P.R. Particle activity and in vivo pulmonary response to freshly milled and aged alpha-quartz. *Scand. J. Work. Environ. Health* **1995**, *21* (Suppl. 2), 15–18. [PubMed]

86. Buford, M.C.; Hamilton, R.F.; Holian, A. A comparison of dispersing media for various engineered carbon nanoparticles. *Part. Fibre Toxicol.* **2007**, *4*, 6. [CrossRef] [PubMed]

87. Polk, W.W.; Sharma, M.; Sayes, C.M.; Hotchkiss, J.A.; Clippinger, A.J. Aerosol generation and characterization of multi-walled carbon nanotubes exposed to cells cultured at the air-liquid interface. *Part. Fibre Toxicol.* **2016**, *13*, 20. [CrossRef] [PubMed]

88. Watson, C.Y.; DeLoid, G.M.; Pal, A.; Demokritou, P. Buoyant Nanoparticles: Implications for Nano-Biointeractions in Cellular Studies. *Small* **2016**, *12*, 3172–3180. [CrossRef]

89. Stock, V.; Böhmert, L.; Dönmez, M.H.; Lampen, A.; Sieg, H. An inverse cell culture model for floating plastic particles. *Anal. Biochem.* **2020**, *591*, 113545. [CrossRef]

90. Saleh, N.B.; Aich, N.; Plazas-Tuttle, J.; Lead, J.R.; Lowry, G.V. Research strategy to determine when novel nanohybrids pose unique environmental risks. *Environ. Sci. Nano* **2015**, *2*, 11–18. [CrossRef]

91. Kroll, A.; Pillukat, M.H.; Hahn, D.; Schnekenburger, J. Current in vitro methods in nanoparticle risk assessment: Limitations and challenges. *Eur. J. Pharm. Biopharm.* **2009**, *72*, 370–377. [CrossRef] [PubMed]

92. Stone, V.; Johnston, H.; Schins, R.P.F. Development of in vitro systems for nanotoxicology: Methodological considerations. *Crit. Rev. Toxicol.* **2009**, *39*, 613–626. [CrossRef] [PubMed]

93. Hillegass, J.M.; Shukla, A.; Lathrop, S.A.; MacPherson, M.B.; Fukagawa, N.K.; Mossman, B.T. Assessing nanotoxicity in cells in vitro. *Wiley Interdiscip. Rev. Nanomed. Nanobiotechnol.* **2010**, *2*, 219–231. [CrossRef]

94. Sayes, C.; Reed, K.L.; Warheit, D.B. Assessing Toxicity of Fine and Nanoparticles: Comparing In Vitro Measurements to In Vivo Pulmonary Toxicity Profiles. *Toxicol. Sci.* **2007**, *97*, 163–180. [CrossRef]

95. Guadagnini, R.; Halamoda Kenzaoui, B.; Walker, L.; Pojana, G.; Magdolenova, Z.; Bilanicova, D.; Saunders, M.; Juillerat-Jeanneret, L.; Marcomini, A.; Huk, A.; et al. Toxicity screenings of nanomaterials: Challenges due to interference with assay processes and components of classic in vitro tests. *Nanotoxicology* **2015**, *9* (Suppl. 1), 13–24. [CrossRef]

96. Ribeiro, A.R.; Leite, P.E.; Falagan-Lotsch, P.; Benetti, F.; Micheletti, C.; Budtz, H.C.; Jacobsen, N.R.; Lisboa-Filho, P.N.; Rocha, L.A.; Kühnel, D.; et al. Challenges on the toxicological predictions of engineered nanoparticles. *NanoImpact* **2017**, *8*, 59–72. [CrossRef]

97. Andraos, C.; Yu, I.J.; Gulumian, M. Interference: A Much-Neglected Aspect in High-Throughput Screening of Nanoparticles. *Int. J. Toxicol.* **2020**, *39*, 397–421. [CrossRef]

98. Savage, D.T.; Hilt, J.Z.; Dziubla, T.D. In vitro methods for assessing nanoparticle toxicity. In *Nanotoxicity*; Methods in Molecular Biology; Zhang, Q., Ed.; Humana Press: New York, NY, USA, 2019; Volume 1894, pp. 1–29. [CrossRef]

99. Kroll, A.; Pillukat, M.H.; Hahn, D.; Schnekenburger, J. Interference of engineered nanoparticles with in vitro toxicity assays. *Arch. Toxicol.* **2012**, *86*, 1123–1136. [CrossRef]

100. Piret, J.-P.; Bondarenko, O.M.; Boyles, M.S.P.; Himly, M.; Ribeiro, A.R.; Benetti, F.; Smal, C.; Lima, B.; Potthoff, A.; Simion, M.; et al. Pan-European inter-laboratory studies on a panel of in vitro cytotoxicity and pro-inflammation assays for nanoparticles. *Arch. Toxicol.* **2017**, *91*, 2315–2330. [CrossRef]

101. Han, X.; Gelein, R.; Corson, N.; Wade-Mercer, P.; Jiang, J.; Biswas, P.; Finkelstein, J.N.; Elder, A.; Oberdörster, G. Validation of an LDH assay for assessing nanoparticle toxicity. *Toxicology* **2011**, *287*, 99–104. [CrossRef]

102. Nelissen, I.; Haase, A.; Anguissola, S.; Rocks, L.; Jacobs, A.; Willems, H.; Riebeling, C.; Luch, A.; Piret, J.-P.; Toussaint, O.; et al. Improving Quality in Nanoparticle-Induced Cytotoxity Testing by a Tiered Inter-Laboratory Comparison Study. *Nanomaterials* **2020**, *10*, 1430. [CrossRef]

103. Labouta, H.I.; Sarsons, C.; Kennard, J.; Gomez-Garcia, M.J.; Villar, K.; Lee, H.; Cramb, D.T.; Rinker, K.D. Understanding and improving assays for cytotoxicity of nanoparticles: What really matters? *RSC Adv.* **2018**, *8*, 23027–23039. [CrossRef]

104. *ISO 19007:2018*; Nanotechnologies—In Vitro MTS Assay for Measuring the Cytotoxic Effect of Nanoparticles. International Organization for Standardization: Geneva, Switzerland, 2018.

105. *ISO/TS 21633:2021*; Label-Free Impedance Technology to Assess the Toxicity of Nanomaterials In Vitro. International Organization for Standardization: Geneva, Switzerland, 2021.

106. Da Luz, C.M.; Boyles, M.S.P.; Falagan-Lotsch, P.; Pereira, M.R.; Tutumi, H.R.; Santos, E.D.O.; Martins, N.B.; Himly, M.; Sommer, A.; Foissner, I.; et al. Poly-lactic acid nanoparticles (PLA-NP) promote physiological modifications in lung epithelial cells and are internalized by clathrin-coated pits and lipid rafts. *J. Nanobiotechnol.* **2017**, *15*, 11. [CrossRef]
107. NANoREG. Deliverable D 5.06. Identification and Optimization of the Most Suitable In Vitro Methodology. 2016. Available online: https://www.rivm.nl/sites/default/files/2019-01/NANoREG_D5_06_DR_Identification_and_optimization_of_the_most_suitable_in_vitro_methodology_Yo3qmgPoS9aEh1-9l2OVcA.pdf (accessed on 14 November 2022).
108. Xia, T.; Hamilton, R.F.; Bonner, J.C.; Crandall, E.D.; Elder, A.; Fazlollahi, F.; Girtsman, T.A.; Kim, K.; Mitra, S.; Ntim, S.A.; et al. Interlaboratory Evaluation of In Vitro Cytotoxicity and Inflammatory Responses to Engineered Nanomaterials: The NIEHS Nano GO Consortium. *Environ. Health Perspect.* **2013**, *121*, 683–690. [CrossRef]
109. Landsiedel, R.; Sauer, U.G.; Ma-Hock, L.; Schnekenburger, J.; Wiemann, M. Pulmonary toxicity of nanomaterials: A critical comparison of published in vitro assays and in vivo inhalation or instillation studies. *Nanomedicine* **2014**, *9*, 2557–2585. [CrossRef]
110. Sohaebuddin, S.K.; Thevenot, P.T.; Baker, D.; Eaton, J.W.; Tang, L. Nanomaterial cytotoxicity is composition, size, and cell type dependent. *Part. Fibre Toxicol.* **2010**, *7*, 22. [CrossRef]
111. Kroll, A.; Dierker, C.; Rommel, C.; Hahn, D.; Wohlleben, W.; Schulze-Isfort, C.; Göbbert, C.; Voetz, M.; Hardinghaus, F.; Schnekenburger, J. Cytotoxicity screening of 23 engineered nanomaterials using a test matrix of ten cell lines and three different assays. *Part. Fibre Toxicol.* **2011**, *8*, 9. [CrossRef]
112. Cho, W.-S.; Duffin, R.; Bradley, M.; Megson, I.L.; MacNee, W.; Lee, J.K.; Jeong, J.; Donaldson, K. Predictive value of in vitro assays depends on the mechanism of toxicity of metal oxide nanoparticles. *Part. Fibre Toxicol.* **2013**, *10*, 55. [CrossRef]
113. Han, X.; Corson, N.; Wade-Mercer, P.; Gelein, R.; Jiang, J.; Sahu, M.; Biswas, P.; Finkelstein, J.N.; Elder, A.; Oberdörster, G. Assessing the relevance of in vitro studies in nanotoxicology by examining correlations between in vitro and in vivo data. *Toxicology* **2012**, *297*, 1–9. [CrossRef] [PubMed]
114. Mannerström, M.; Zou, J.; Toimela, T.; Pyykkö, I.; Heinonen, T. The applicability of conventional cytotoxicity assays to predict safety/toxicity of mesoporous silica nanoparticles, silver and gold nanoparticles and multi-walled carbon nanotubes. *Toxicol. In Vitro* **2016**, *37*, 113–120. [CrossRef]
115. Seagrave, J.; McDonald, J.D.; Mauderly, J.L. In vitro versus in vivo exposure to combustion emissions. *Exp. Toxicol. Pathol.* **2005**, *57*, 233–238. [CrossRef] [PubMed]
116. Warheit, D.B.; Sayes, C.M.; Reed, K.L. Nanoscale and Fine Zinc Oxide Particles: Can In Vitro Assays Accurately Forecast Lung Hazards following Inhalation Exposures? *Environ. Sci. Technol.* **2009**, *43*, 7939–7945. [CrossRef]
117. Lanone, S.; Rogerieux, F.; Geys, J.; Dupont, A.; Maillot-Marechal, E.; Boczkowski, J.; Lacroix, G.; Hoet, P. Comparative toxicity of 24 manufactured nanoparticles in human alveolar epithelial and macrophage cell lines. *Part. Fibre Toxicol.* **2009**, *6*, 14. [CrossRef] [PubMed]
118. Monteiro-Riviere, N.A.; Inman, A.O.; Zhang, L.W. Limitations and relative utility of screening assays to assess engineered nanoparticle toxicity in a human cell line. *Toxicol. Appl. Pharmacol.* **2009**, *234*, 222–235. [CrossRef] [PubMed]
119. Donaldson, K.; Tran, C.L. An introduction to the short-term toxicology of respirable industrial fibres. *Mutat. Res./Fundam. Mol. Mech. Mutagen.* **2004**, *553*, 5–9. [CrossRef] [PubMed]
120. Heinrich, U.; Fuhst, R.; Rittinghausen, S.; Creutzenberg, O.; Bellmann, B.; Koch, W.; Levsen, K. Chronic inhalation exposure of Wistar rats and two different strains of mice to diesel exhaust, carbon black and titanium dioxide. *Inhal. Toxicol.* **1995**, *7*, 23. [CrossRef]
121. Driscoll, K.E.; Borm, P.J.A. Expert workshop on the hazards and risks of poorly soluble low toxicity particles. *Inhal. Toxicol.* **2020**, *32*, 53–62. [CrossRef] [PubMed]
122. Shin, H.U.; Stefaniak, A.B.; Stojilovic, N.; Chase, G.G. Comparative dissolution of electrospun Al_2O_3 nanofibres in artificial human lung fluids. *Environ. Sci. Nano* **2015**, *2*, 251–261. [CrossRef]
123. Koltermann-Jülly, J.; Keller, J.G.; Vennemann, A.; Werle, K.; Müller, P.; Ma-Hock, L.; Landsiedel, R.; Wiemann, M.; Wohlleben, W. Abiotic dissolution rates of 24 (nano)forms of 6 substances compared to macrophage-assisted dissolution and in vivo pulmonary clearance: Grouping by biodissolution and transformation. *NanoImpact* **2018**, *12*, 29–41. [CrossRef]
124. Keller, J.G.; Persson, M.; Müller, P.; Ma-Hock, L.; Werle, K.; Arts, J.; Landsiedel, R.; Wohlleben, W. Variation in dissolution behavior among different nanoforms and its implication for grouping approaches in inhalation toxicity. *NanoImpact* **2021**, *23*, 100341. [CrossRef]
125. Plumlee, G.S.; Morman, S.A.; Ziegler, T.L. The Toxicological Geochemistry of Earth Materials: An Overview of Processes and the Interdisciplinary Methods Used to Understand Them. *Rev. Miner. Geochem.* **2006**, *64*, 5–57. [CrossRef]
126. Krystek, P.; Kettler, K.; van der Wagt, B.; de Jong, W.H. Exploring influences on the cellular uptake of medium-sized silver nanoparticles into THP-1 cells. *Microchem. J.* **2015**, *120*, 45–50. [CrossRef]
127. Innes, E.; Yiu, H.H.P.; McLean, P.; Brown, W.; Boyles, M. Simulated biological fluids—A systematic review of their biological relevance and use in relation to inhalation toxicology of particles and fibres. *Crit. Rev. Toxicol.* **2021**, *51*, 217–248. [CrossRef] [PubMed]
128. Wohlleben, W.; Hellack, B.; Nickel, C.; Herrchen, M.; Hund-Rinke, K.; Kettler, K.; Riebeling, C.; Haase, A.; Funk, B.; Kühnel, D.; et al. The nanoGRAVUR framework to group (nano)materials for their occupational, consumer, environmental risks based on a harmonized set of material properties, applied to 34 case studies. *Nanoscale* **2019**, *11*, 17637–17654. [CrossRef] [PubMed]
129. Alexander, I.C.; Brown, R.C.; Jubb, G.A.; Pickering, P.; Hoskins, J.A. Durability of ceramic and novel man-made mineral fibers. *Environ. Health Perspect.* **1994**, *102*, 67–71. [CrossRef] [PubMed]

130. Christensen, V.R.; Jensen, S.L.; Guldberg, M.; Kamstrup, O. Effect of chemical composition of man-made vitreous fibers on the rate of dissolution in vitro at different pHs. *Environ. Health Perspect.* **1994**, *102*, 83–86. [CrossRef]

131. Spitler, G.; Spitz, H.; Glasser, S.; Hoffman, M.K.; Bowen, J. In Vitro Dissolution of Uranium-contaminated Soil in Simulated Lung Fluid Containing a Pulmonary Surfactant. *Health Phys.* **2015**, *108*, 336–343. [CrossRef]

132. Searl, A.; Cullen, R.T. An enzymatic tissue digestion method for fibre biopersistence studies. *Ann. Occup. Hyg.* **1997**, *41*, 721–727. [CrossRef]

133. Utembe, W.; Potgieter, K.; Stefaniak, A.B.; Gulumian, M. Dissolution and biodurability: Important parameters needed for risk assessment of nanomaterials. *Part. Fibre Toxicol.* **2015**, *12*, 11. [CrossRef]

134. Ansoborlo, E.; Hengé-Napoli, M.H.; Chazel, V.; Gibert, R.; Guilmette, R.A. Review and Critical Analysis of Available In Vitro Dissolution Tests. *Health Phys.* **1999**, *77*, 638–645. [CrossRef]

135. Farrugia, C. Flow-through dissolution testing: A comparison with stirred beaker methods. *Chronic Ill* **2002**, *6*, 17–19.

136. Braydich-Stolle, L.K.; Breitner, E.K.; Comfort, K.K.; Schlager, J.J.; Hussain, S.M. Dynamic Characteristics of Silver Nanoparticles in Physiological Fluids: Toxicological Implications. *Langmuir* **2014**, *30*, 15309–15316. [CrossRef]

137. Keller, J.; Peijnenburg, W.; Werle, K.; Landsiedel, R.; Wohlleben, W. Understanding Dissolution Rates via Continuous Flow Systems with Physiologically Relevant Metal Ion Saturation in Lysosome. *Nanomaterials* **2020**, *10*, 311. [CrossRef]

138. Murphy, F.; Dekkers, S.; Braakhuis, H.; Ma-Hock, L.; Johnston, H.; Janer, G.; di Cristo, L.; Sabella, S.; Jacobsen, N.R.; Oomen, A.G.; et al. An integrated approach to testing and assessment of high aspect ratio nanomaterials and its application for grouping based on a common mesothelioma hazard. *NanoImpact* **2021**, *22*, 100314. [CrossRef]

139. Hesterberg, T.W.; Hart, G.A. Lung Biopersistence and In Vitro Dissolution Rate Predict the Pathogenic Potential of Synthetic Vitreous Fibers. *Inhal. Toxicol.* **2000**, *12* (Suppl. 3), 91–97. [CrossRef]

140. Boyles, M.S.P.; Brown, D.; Knox, J.; Horobin, M.; Miller, M.R.; Johnston, H.J.; Stone, V. Assessing the bioactivity of crystalline silica in heated high-temperature insulation wools. *Inhal. Toxicol.* **2018**, *30*, 255–272. [CrossRef]

141. Ellingsen, D.G.; Chashchin, M.; Berlinger, B.; Fedorov, V.; Chashchin, V.; Thomassen, Y. Biological monitoring of welders' exposure to chromium, molybdenum, tungsten and vanadium. *J. Trace Elem. Med. Biol.* **2017**, *41*, 99–106. [CrossRef]

142. Gray, E.P.; Browning, C.L.; Wang, M.; Gion, K.D.; Chao, E.Y.; Koski, K.J.; Kane, A.B.; Hurt, R.H. Biodissolution and cellular response to MoO_3 nanoribbons and a new framework for early hazard screening for 2D materials. *Environ. Sci. Nano* **2018**, *5*, 2545–2559. [CrossRef]

143. Chazel, V.; Houpert, P.; Paquet, F.; Ansoborlo, E.; Hengé-Napoli, M.H. Experimental Determination of the Solubility of Industrial UF4 Particles. *Radiat. Prot. Dosim.* **2000**, *92*, 289–294. [CrossRef]

144. Chazel, V.; Gerasimo, P.; Debouis, V.; Laroche, P.; Paquet, F. Characterisation and dissolution of depleted uranium aerosols produced during impacts of kinetic energy penetrators against a tank. *Radiat. Prot. Dosim.* **2003**, *105*, 163–166. [CrossRef]

145. Stefaniak, A.B.; Guilmette, R.A.; Day, G.A.; Hoover, M.D.; Breysse, P.N.; Scripsick, R.C. Characterization of phagolysosomal simulant fluid for study of beryllium aerosol particle dissolution. *Toxicol. In Vitro* **2005**, *19*, 123–134. [CrossRef]

146. Heim, K.E.; Danzeisen, R.; Verougstraete, V.; Gaidou, F.; Brouwers, T.; Oller, A.R. Bioaccessibility of nickel and cobalt in synthetic gastric and lung fluids and its potential use in alloy classification. *Regul. Toxicol. Pharmacol.* **2020**, *110*, 104549. [CrossRef]

147. Yu, S.-J.; Chao, J.-B.; Sun, J.; Yin, Y.-G.; Liu, J.-F.; Jiang, G.-B. Quantification of the Uptake of Silver Nanoparticles and Ions to HepG2 Cells. *Environ. Sci. Technol.* **2013**, *47*, 3268–3274. [CrossRef]

148. Su, C.-K.; Sun, Y.-C. Considerations of inductively coupled plasma mass spectrometry techniques for characterizing the dissolution of metal-based nanomaterials in biological tissues. *J. Anal. At. Spectrom.* **2015**, *30*, 1689–1705. [CrossRef]

149. Jiang, X.; Miclăuş, T.; Wang, L.; Foldbjerg, R.; Sutherland, D.; Autrup, H.; Chen, C.; Beer, C. Fast intracellular dissolution and persistent cellular uptake of silver nanoparticles in CHO-K1 cells: Implication for cytotoxicity. *Nanotoxicology* **2015**, *9*, 181–189. [CrossRef]

150. Malysheva, A.; Ivask, A.; Doolette, C.L.; Voelcker, N.H.; Lombi, E. Cellular binding, uptake and biotransformation of silver nanoparticles in human T lymphocytes. *Nat. Nanotechnol.* **2021**, *16*, 926–932. [CrossRef]

151. Lehuede, P.; de Meringo, A.; Bernstein, D.M. Comparison of the chemical evolution of MMVF following inhalation exposure in rats and acellular in vitro dissolution. *Inhal. Toxicol.* **1997**, *9*, 495–523. [CrossRef]

152. Luoto, K.; Holopainen, M.; Karppinen, K.; Perander, M.; Savolainen, K. Dissolution of man-made vitreous fibers in rat alveolar macrophage culture and Gamble's saline solution: Influence of different media and chemical composition of the fibers. *Environ. Health Perspect.* **1994**, *102* (Suppl. 5), 103–107. [CrossRef]

153. Zhong, L.; Liu, X.; Hu, X.; Chen, Y.; Wang, H.; Lian, H.-Z. In vitro inhalation bioaccessibility procedures for lead in PM2.5 size fraction of soil assessed and optimized by in vivo-in vitro correlation. *J. Hazard. Mater.* **2020**, *381*, 121202. [CrossRef]

154. Guldberg, M.; Jensen, S.L.; Knudsen, T.; Steenberg, T.; Kamstrup, O. High-Alumina Low-Silica HT Stone Wool Fibers: A Chemical Compositional Range with High Biosolubility. *Regul. Toxicol. Pharmacol.* **2002**, *35*, 217–226. [CrossRef]

155. Takenaka, S.; Karg, E.; Roth, C.; Schulz, H.; Ziesenis, A.; Heinzmann, U.; Schramel, P.; Heyder, J. Pulmonary and systemic distribution of inhaled ultrafine silver particles in rats. *Environ. Health Perspect.* **2001**, *109* (Suppl. 4), 547–551. [CrossRef]

156. Ayres, J.G.; Borm, P.; Cassee, F.R.; Castranova, V.; Donaldson, K.; Ghio, A.; Harrison, R.M.; Hider, R.; Kelly, F.; Kooter, I.M.; et al. Evaluating the Toxicity of Airborne Particulate Matter and Nanoparticles by Measuring Oxidative Stress Potential—A Workshop Report and Consensus Statement. *Inhal. Toxicol.* **2008**, *20*, 75–99. [CrossRef]

157. Hellack, B.; Nickel, C.; Albrecht, C.; Kuhlbusch, T.; Boland, S.; Baeza-Squiban, A.; Wohlleben, W.; Schins, R.P.F. Analytical methods to assess the oxidative potential of nanoparticles: A review. *Environ. Sci. Nano* **2017**, *4*, 1920–1934. [CrossRef]

158. Møller, P.; Danielsen, P.H.; Karottki, D.G.; Jantzen, K.; Roursgaard, M.; Klingberg, H.; Jensen, D.M.; Vest Christophersen, D.; Hemmingsen, J.G.; Cao, Y.; et al. Oxidative stress and inflammation generated DNA damage by exposure to air pollution particles. *Mutat. Res./Rev. Mutat. Res.* **2014**, *762*, 133–166. [CrossRef]

159. Nel, A.; Xia, T.; Mädler, L.; Li, N. Toxic Potential of Materials at the Nanolevel. *Science* **2006**, *311*, 622–627. [CrossRef]

160. Møller, P.; Jacobsen, N.R.; Folkmann, J.K.; Danielsen, P.H.; Mikkelsen, L.; Hemmingsen, J.G.; Vesterdal, L.K.; Forchhammer, L.; Wallin, H.; Loft, S. Role of oxidative damage in toxicity of particulates. *Free Radic. Res.* **2010**, *44*, 1–46. [CrossRef]

161. Song, B.; Zhou, T.; Yang, W.; Liu, J.; Shao, L. Contribution of oxidative stress to TiO 2 nanoparticle-induced toxicity. *Environ. Toxicol. Pharmacol.* **2016**, *48*, 130–140. [CrossRef]

162. Bahl, A.; Hellack, B.; Wiemann, M.; Giusti, A.; Werle, K.; Haase, A.; Wohlleben, W. Nanomaterial categorization by surface reactivity: A case study comparing 35 materials with four different test methods. *NanoImpact* **2020**, *19*, 100234. [CrossRef]

163. Gandon, A.; Werle, K.; Neubauer, N.; Wohlleben, W. Surface reactivity measurements as required for grouping and read-across: An advanced FRAS protocol. *J. Phys. Conf. Ser.* **2017**, *838*, 012033. [CrossRef]

164. Lu, S.; Duffin, R.; Poland, C.; Daly, P.; Murphy, F.; Drost, E.; MacNee, W.; Stone, V.; Donaldson, K. Efficacy of Simple Short-Term In Vitro Assays for Predicting the Potential of Metal Oxide Nanoparticles to Cause Pulmonary Inflammation. *Environ. Health Perspect.* **2009**, *117*, 241–247. [CrossRef]

165. Rogers, E.J.; Hsieh, S.F.; Organti, N.; Schmidt, D.; Bello, D. A high throughput in vitro analytical approach to screen for oxidative stress potential exerted by nanomaterials using a biologically relevant matrix: Human blood serum. *Toxicol. In Vitro* **2008**, *22*, 1639–1647. [CrossRef] [PubMed]

166. Sauvain, J.-J.; Rossi, M.J.; Riediker, M. Comparison of Three Acellular Tests for Assessing the Oxidation Potential of Nanomaterials. *Aerosol Sci. Technol.* **2013**, *47*, 218–227. [CrossRef]

167. Hsieh, S.-F.; Bello, D.; Schmidt, D.F.; Pal, A.K.; Stella, A.; Isaacs, J.A.; Rogers, E.J. Mapping the Biological Oxidative Damage of Engineered Nanomaterials. *Small* **2013**, *9*, 1853–1865. [CrossRef] [PubMed]

168. Achawi, S.; Feneon, B.; Pourchez, J.; Forest, V. Assessing biological oxidative damage induced by graphene-based materials: An asset for grouping approaches using the FRAS assay. *Regul. Toxicol. Pharmacol.* **2021**, *127*, 105067. [CrossRef] [PubMed]

169. Benzie, I.F.F.; Strain, J.J. The ferric reducing ability of plasma (FRAP) as a measure of "antioxidant power": The FRAP assay. *Anal. Biochem.* **1996**, *239*, 70–76. [CrossRef] [PubMed]

170. Boyles, M.; Murphy, F.; Mueller, W.; Wohlleben, W.; Jacobsen, N.R.; Braakhuis, H.; Giusti, A.; Stone, V. Development of a standard operating procedure for the DCFH$_2$-DA acellular assessment of reactive oxygen species produced by nanomaterials. *Toxicol. Mech. Methods* **2022**, *32*, 439–452. [CrossRef]

171. Pal, A.K.; Bello, D.; Budhlall, B.; Rogers, E.; Milton, D.K. Screening for Oxidative Stress Elicited by Engineered Nanomaterials: Evaluation of Acellular DCFH Assay. *Dose-Response* **2012**, *10*, 308–330. [CrossRef]

172. Zhao, J.; Riediker, M. Detecting the oxidative reactivity of nanoparticles: A new protocol for reducing artifacts. *J. Nanopart. Res.* **2014**, *16*, 2493. [CrossRef]

173. Seleci, D.A.; Tsiliki, G.; Werle, K.; Elam, D.A.; Okpowe, O.; Seidel, K.; Bi, X.; Westerhoff, P.; Innes, E.; Boyles, M.; et al. Determining nanoform similarity via assessment of surface reactivity by abiotic and in vitro assays. *NanoImpact* **2022**, *26*, 100390. [CrossRef]

174. Janer, G.; Landsiedel, R.; Wohlleben, W. Rationale and decision rules behind the ECETOC NanoApp to support registration of sets of similar nanoforms within REACH. *Nanotoxicology* **2021**, *15*, 145–166. [CrossRef]

175. Pal, A.K.; Hsieh, S.-F.; Khatri, M.; Isaacs, J.A.; Demokritou, P.; Gaines, P.; Schmidt, D.F.; Rogers, E.J.; Bello, D. Screening for oxidative damage by engineered nanomaterials: A comparative evaluation of FRAS and DCFH. *J. Nanopart. Res.* **2014**, *16*, 2167. [CrossRef]

176. Hawkins, C.L.; Davies, M.J. Detection and characterisation of radicals in biological materials using EPR methodology. *Biochim. Biophys. Acta (BBA)—Gen. Subj.* **2014**, *1840*, 708–721. [CrossRef]

177. Kalyanaraman, B.; Darley-Usmar, V.; Davies, K.J.A.; Dennery, P.A.; Forman, H.J.; Grisham, M.B.; Mann, G.E.; Moore, K.; Roberts, L.J., II; Ischiropoulos, H. Measuring reactive oxygen and nitrogen species with fluorescent probes: Challenges and limitations. *Free Radic. Biol. Med.* **2012**, *52*, 1–6. [CrossRef]

178. Boyles, M.S.P.; Ranninger, C.; Reischl, R.; Rurik, M.; Tessadri, R.; Kohlbacher, O.; Duschl, A.; Huber, C.G. Copper oxide nanoparticle toxicity profiling using untargeted metabolomics. *Part. Fibre Toxicol.* **2016**, *13*, 49. [CrossRef]

179. Hiura, T.S.; Li, N.; Kaplan, R.; Horwitz, M.; Seagrave, J.-C.; Nel, A.E. The Role of a Mitochondrial Pathway in the Induction of Apoptosis by Chemicals Extracted from Diesel Exhaust Particles. *J. Immunol.* **2000**, *165*, 2703–2711. [CrossRef]

180. Xia, T.; Korge, P.; Weiss, J.N.; Li, N.; Venkatesen, M.I.; Sioutas, C.; Nel, A. Quinones and Aromatic Chemical Compounds in Particulate Matter Induce Mitochondrial Dysfunction: Implications for Ultrafine Particle Toxicity. *Environ. Health Perspect.* **2004**, *112*, 1347–1358. [CrossRef]

181. Čapek, J.; Roušar, T. Detection of Oxidative Stress Induced by Nanomaterials in Cells—The Roles of Reactive Oxygen Species and Glutathione. *Molecules* **2021**, *26*, 4710. [CrossRef]

182. Rushton, E.K.; Jiang, J.; Leonard, S.S.; Eberly, S.; Castranova, V.; Biswas, P.; Elder, A.; Han, X.; Gelein, R.; Finkelstein, J.; et al. Concept of Assessing Nanoparticle Hazards Considering Nanoparticle Dosemetric and Chemical/Biological Response Metrics. *J. Toxicol. Environ. Health A* **2010**, *73*, 445–461. [CrossRef]

183. Lehman, S.E.; Morris, A.S.; Mueller, P.S.; Salem, A.K.; Grassian, V.H.; Larsen, S.C. Silica nanoparticle-generated ROS as a predictor of cellular toxicity: Mechanistic insights and safety by design. *Environ. Sci. Nano* **2016**, *3*, 56–66. [CrossRef]

184. Driessen, M.D.; Mues, S.; Vennemann, A.; Hellack, B.; Bannuscher, A.; Vimalakanthan, V.; Riebeling, C.; Ossig, R.; Wiemann, M.; Schnekenburger, J.; et al. Proteomic analysis of protein carbonylation: A useful tool to unravel nanoparticle toxicity mechanisms. *Part. Fibre Toxicol.* **2015**, *12*, 36. [CrossRef]

185. Riebeling, C.; Wiemann, M.; Schnekenburger, J.; Kuhlbusch, T.A.J.; Wohlleben, W.; Luch, A.; Haase, A. A redox proteomics approach to investigate the mode of action of nanomaterials. *Toxicol. Appl. Pharmacol.* **2016**, *299*, 24–29. [CrossRef] [PubMed]

186. Reuzel, P.G.J.; Bruijntjes, J.P.; Feron, V.J.; Woutersen, R.A. Subchronic inhalation toxicity of amorphous silicas and quartz dust in rats. *Food Chem. Toxicol.* **1991**, *29*, 341–354. [CrossRef]

187. Muhle, H.; Kittel, B.; Ernst, H.; Mohr, U.; Mermelstein, R. Neoplastic lung lesions in rat after chronic exposure to crystalline silica. *Scand. J. Work Environ. Health* **1995**, *21*, 27–29.

188. Brand, W.; Peters, R.J.B.; Braakhuis, H.M.; Maślankiewicz, L.; Oomen, A.G. Possible effects of titanium dioxide particles on human liver, intestinal tissue, spleen and kidney after oral exposure. *Nanotoxicology* **2020**, *14*, 985–1007. [CrossRef]

189. Elsabahy, M.; Wooley, K.L. Cytokines as biomarkers of nanoparticle immunotoxicity. *Chem. Soc. Rev.* **2013**, *42*, 5552–5576. [CrossRef]

190. Di Ianni, E.; Erdem, J.S.; Møller, P.; Sahlgren, N.M.; Poulsen, S.S.; Knudsen, K.B.; Zienolddiny, S.; Saber, A.T.; Wallin, H.; Vogel, U.; et al. In vitro-in vivo correlations of pulmonary inflammogenicity and genotoxicity of MWCNT. *Part. Fibre Toxicol.* **2021**, *18*, 25. [CrossRef]

191. Sun, B.; Wang, X.; Ji, Z.; Li, R.; Xia, T. NLRP3 Inflammasome Activation Induced by Engineered Nanomaterials. *Small* **2013**, *9*, 1595–1607. [CrossRef]

192. Harkema, J.R.; Wagner, J.G. Pathology of the Respiratory System. In *Toxicologic Pathology for Non-Pathologists*; Humana: New York, NY, USA, 2019; pp. 311–354. [CrossRef]

193. Li, Y.; Fujita, M.; Boraschi, D. Endotoxin Contamination in Nanomaterials Leads to the Misinterpretation of Immunosafety Results. *Front. Immunol.* **2017**, *8*, 472. [CrossRef]

194. Dobrovolskaia, M.A.; Germolec, D.R.; Weaver, J.L. Evaluation of nanoparticle immunotoxicity. *Nat. Nanotechnol.* **2009**, *4*, 411–414. [CrossRef]

195. Braakhuis, H.M.; Gremmer, E.R.; Bannuscher, A.; Drasler, B.; Keshavan, S.; Barbara Rothen-Rutishauser, B.B.; Verlohner, A.; Landsiedel, R.; Meldrum, K.; Doak, S.H.; et al. Transferability and reproducibility of exposed air-liquid interface co-culture lung models. *NanoImpact*, 2022; *in preparation*.

196. Barosova, H.; Meldrum, K.; Karakocak, B.B.; Balog, S.; Doak, S.H.; Petri-Fink, A.; Clift, M.J.D.; Rothen-Rutishauser, B. Inter-laboratory variability of A549 epithelial cells grown under submerged and air-liquid interface conditions. *Toxicol. In Vitro* **2021**, *75*, 105178. [CrossRef]

197. Bessa, M.J.; Brandão, F.; Fokkens, P.; Cassee, F.R.; Salmatonidis, A.; Viana, M.; Vulpoi, A.; Simon, S.; Monfort, E.; Teixeira, J.P.; et al. Toxicity assessment of industrial engineered and airborne process-generated nanoparticles in a 3D human airway epithelial in vitro model. *Nanotoxicology* **2021**, *15*, 542–557. [CrossRef]

198. Petersen, E.J.; Sharma, M.; Clippinger, A.J.; Gordon, J.; Katz, A.; Laux, P.; Leibrock, L.B.; Luch, A.; Matheson, J.; Stucki, A.O.; et al. Use of Cause-and-Effect Analysis to Optimize the Reliability of In Vitro Inhalation Toxicity Measurements Using an Air–Liquid Interface. *Chem. Res. Toxicol.* **2021**, *34*, 1370–1385. [CrossRef]

199. Guo, C.; Buckley, A.; Marczylo, T.; Seiffert, J.; Römer, I.; Warren, J.; Hodgson, A.; Chung, K.F.; Gant, T.; Smith, R.; et al. The small airway epithelium as a target for the adverse pulmonary effects of silver nanoparticle inhalation. *Nanotoxicology* **2018**, *12*, 539–553. [CrossRef]

200. Sandberg, W.J.; Låg, M.; Holme, J.A.; Friede, B.; Gualtieri, M.; Kruszewski, M.; Schwarze, P.E.; Skuland, T.; Refsnes, M. Comparison of non-crystalline silica nanoparticles in IL-1β release from macrophages. *Part. Fibre Toxicol.* **2012**, *9*, 32. [CrossRef]

201. Svensson, C.R.; Ameer, S.S.; Ludvigsson, L.; Ali, N.; Alhamdow, A.; Messing, M.E.; Pagels, J.; Gudmundsson, A.; Bohgard, M.; Sanfins, E.; et al. Validation of an air–liquid interface toxicological set-up using Cu, Pd, and Ag well-characterized nanostructured aggregates and spheres. *J. Nanopart. Res.* **2016**, *18*, 86. [CrossRef]

202. Endes, C.; Schmid, O.; Kinnear, C.; Mueller, S.; Camarero-Espinosa, S.; Vanhecke, D.; Foster, E.J.; Petri-Fink, A.; Rothen-Rutishauser, B.; Weder, C.; et al. An in vitro testing strategy towards mimicking the inhalation of high aspect ratio nanoparticles. *Part. Fibre Toxicol.* **2014**, *11*, 40. [CrossRef]

203. Hufnagel, M.; Neuberger, R.; Wall, J.; Link, M.; Friesen, A.; Hartwig, A. Impact of Differentiated Macrophage-Like Cells on the Transcriptional Toxicity Profile of CuO Nanoparticles in Co-Cultured Lung Epithelial Cells. *Int. J. Mol. Sci.* **2021**, *22*, 5044. [CrossRef]

204. Clift, M.J.D.; Endes, C.; Vanhecke, D.; Wick, P.; Gehr, P.; Schins, R.P.F.; Petri-Fink, A.; Rothen-Rutishauser, B. A Comparative Study of Different In Vitro Lung Cell Culture Systems to Assess the Most Beneficial Tool for Screening the Potential Adverse Effects of Carbon Nanotubes. *Toxicol. Sci.* **2014**, *137*, 55–64. [CrossRef]

205. Braakhuis, H.M.; He, R.; Vandebriel, R.J.; Gremmer, E.R.; Zwart, E.; Vermeulen, J.P.; Fokkens, P.; Boere, J.; Gosens, I.; Cassee, F.R. An Air-liquid Interface Bronchial Epithelial Model for Realistic, Repeated Inhalation Exposure to Airborne Particles for Toxicity Testing. *J. Vis. Exp.* **2020**, *159*, e61210. [CrossRef]

206. Herzog, F.; Clift, M.J.D.; Piccapietra, F.; Behra, R.; Schmid, O.; Petri-Fink, A.; Rothen-Rutishauser, B. Exposure of silver-nanoparticles and silver-ions to lung cells in vitro at the air-liquid interface. *Part. Fibre Toxicol.* **2013**, *10*, 11. [CrossRef]

207. Braakhuis, H.M.; Giannakou, C.; Peijnenburg, W.J.G.M.; Vermeulen, J.; van Loveren, H.; Park, M.V.D.Z. Simplein vitromodels can predict pulmonary toxicity of silver nanoparticles. *Nanotoxicology* **2016**, *10*, 770–779. [CrossRef]

208. Li, R.; Wang, X.; Ji, Z.; Sun, B.; Zhang, H.; Chang, C.H.; Lin, S.; Meng, H.; Liao, Y.-P.; Wang, M.; et al. Surface Charge and Cellular Processing of Covalently Functionalized Multiwall Carbon Nanotubes Determine Pulmonary Toxicity. *ACS Nano* **2013**, *7*, 2352–2368. [CrossRef]

209. Wang, X.; Xia, T.; Ntim, S.A.; Ji, Z.; Lin, S.; Meng, H.; Chung, C.-H.; George, S.; Zhang, H.; Wang, M.; et al. Dispersal State of Multiwalled Carbon Nanotubes Elicits Profibrogenic Cellular Responses That Correlate with Fibrogenesis Biomarkers and Fibrosis in the Murine Lung. *ACS Nano* **2011**, *5*, 9772–9787. [CrossRef]

210. Diabaté, S.; Armand, L.; Murugadoss, S.; Dilger, M.; Fritsch-Decker, S.; Schlager, C.; Béal, D.; Arnal, M.-E.; Biola-Clier, M.; Ambrose, S.; et al. Air–Liquid Interface Exposure of Lung Epithelial Cells to Low Doses of Nanoparticles to Assess Pulmonary Adverse Effects. *Nanomaterials* **2020**, *11*, 65. [CrossRef]

211. Loret, T.; Rogerieux, F.; Trouiller, B.; Braun, A.; Egles, C.; Lacroix, G. Predicting the in vivo pulmonary toxicity induced by acute exposure to poorly soluble nanomaterials by using advanced in vitro methods. *Part. Fibre Toxicol.* **2018**, *15*, 25. [CrossRef]

212. Lenz, A.-G.; Karg, E.; Brendel, E.; Hinze-Heyn, H.; Maier, K.L.; Eickelberg, O.; Stoeger, T.; Schmid, O. Inflammatory and Oxidative Stress Responses of an Alveolar Epithelial Cell Line to Airborne Zinc Oxide Nanoparticles at the Air-Liquid Interface: A Comparison with Conventional, Submerged Cell-Culture Conditions. *BioMed Res. Int.* **2013**, *2013*, 652632. [CrossRef]

213. Loret, T.; Peyret, E.; Dubreuil, M.; Aguerre-Chariol, O.; Bressot, C.; le Bihan, O.; Amodeo, T.; Trouiller, B.; Braun, A.; Egles, C.; et al. Air-liquid interface exposure to aerosols of poorly soluble nanomaterials induces different biological activation levels compared to exposure to suspensions. *Part. Fibre Toxicol.* **2016**, *13*, 58. [CrossRef]

214. Herzog, F.; Loza, K.; Balog, S.; Clift, M.J.D.; Epple, M.; Gehr, P.; Petri-Fink, A.; Rothen-Rutishauser, B. Mimicking exposures to acute and lifetime concentrations of inhaled silver nanoparticles by two different in vitro approaches. *Beilstein J. Nanotechnol.* **2014**, *5*, 1357–1370. [CrossRef]

215. Medina-Reyes, E.I.; Delgado-Buenrostro, N.L.; Leseman, D.L.; Déciga-Alcaraz, A.; He, R.; Gremmer, E.R.; Fokkens, P.H.B.; Flores-Flores, J.O.; Cassee, F.R.; Chirino, Y.I. Differences in cytotoxicity of lung epithelial cells exposed to titanium dioxide nanofibers and nanoparticles: Comparison of air-liquid interface and submerged cell cultures. *Toxicol. In Vitro* **2020**, *65*, 104798. [CrossRef]

216. Panas, A.; Comouth, A.; Saathoff, H.; Leisner, T.; Al-Rawi, M.; Simon, M.; Seemann, G.; Dössel, O.; Mülhopt, S.; Paur, H.-R.; et al. Silica nanoparticles are less toxic to human lung cells when deposited at the air–liquid interface compared to conventional submerged exposure. *Beilstein J. Nanotechnol.* **2014**, *5*, 1590–1602. [CrossRef]

217. Öhlinger, K.; Kolesnik, T.; Meindl, C.; Gallé, B.; Absenger-Novak, M.; Kolb-Lenz, D.; Fröhlich, E. Air-liquid interface culture changes surface properties of A549 cells. *Toxicol. In Vitro* **2019**, *60*, 369–382. [CrossRef]

218. Upadhyay, S.; Palmberg, L. Air-Liquid Interface: Relevant In Vitro Models for Investigating Air Pollutant-Induced Pulmonary Toxicity. *Toxicol. Sci.* **2018**, *164*, 21–30. [CrossRef]

219. Bannuscher, A.; Schmid, O.; Drasler, B.; Rohrbasser, A.; Braakhuis, H.M.; Meldrum, K.; Zwart, E.P.; Gremmer, E.R.; Birk, B.; Rissel, M.; et al. An inter-laboratory effort to harmonize the cell-delivered in vitro dose of aerosolized materials. *NanoImpact* **2022**, *28*, 100439. [CrossRef]

220. Doak, S.H.; Clift, M.J.; Costa, A.; Delmaar, C.; Gosens, I.; Halappanavar, S.; Kelly, S.; Pejinenburg, W.J.; Rothen-Rutishauser, B.; Schins, R.P. The Road to Achieving the European Commission's Chemicals Strategy for Nanomaterial Sustainability—A PATROLS Perspective on New Approach Methodologies. *Small* **2022**, *18*, 2200231. [CrossRef]

221. Catalán, J.; Stockmann-Juvala, H.; Norppa, H. A theoretical approach for a weighted assessment of the mutagenic potential of nanomaterials. *Nanotoxicology* **2017**, *11*, 964–977. [CrossRef]

222. Kohl, Y.; Rundén-Pran, E.; Mariussen, E.; Hesler, M.; El Yamani, N.; Longhin, E.M.; Dusinska, M. Genotoxicity of Nanomaterials: Advanced In Vitro Models and High Throughput Methods for Human Hazard Assessment—A Review. *Nanomaterials* **2020**, *10*, 1911. [CrossRef]

223. Hartwig, A.; Arand, M.; Epe, B.; Guth, S.; Jahnke, G.; Lampen, A.; Martus, H.-J.; Monien, B.; Rietjens, I.M.C.M.; Schmitz-Spanke, S.; et al. Mode of action-based risk assessment of genotoxic carcinogens. *Arch. Toxicol.* **2020**, *94*, 1787–1877. [CrossRef]

224. Steiblen, G.; van Benthem, J.; Johnson, G. Strategies in genotoxicology: Acceptance of innovative scientific methods in a regulatory context and from an industrial perspective. *Mutat. Res./Genet. Toxicol. Environ. Mutagen.* **2020**, *853*, 503171. [CrossRef]

225. Catalán, J.; Norppa, H. Safety Aspects of Bio-Based Nanomaterials. *Bioengineering* **2017**, *4*, 94. [CrossRef]

226. Stone, V.; Pozzi-Mucelli, S.; Tran, L.; Aschberger, K.; Sabella, S.; Vogel, U.; Poland, C.; Balharry, D.; Fernandes, T.; Gottardo, S.; et al. ITS-NANO—Prioritising nanosafety research to develop a stakeholder driven intelligent testing strategy. *Part. Fibre Toxicol.* **2014**, *11*, 9. [CrossRef]

227. Dusinska, M.; Boland, S.; Saunders, M.; Juillerat-Jeanneret, L.; Tran, L.; Pojana, G.; Marcomini, A.; Volkovova, K.; Tulinska, J.; Knudsen, L.E.; et al. Towards an alternative testing strategy for nanomaterials used in nanomedicine: Lessons from NanoTEST. *Nanotoxicology* **2015**, *9*, 118–132. [CrossRef]

228. Kirkland, D.; Reeve, L.; Gatehouse, D.; Vanparys, P. A core in vitro genotoxicity battery comprising the Ames test plus the in vitro micronucleus test is sufficient to detect rodent carcinogens and in vivo genotoxins. *Mutat. Res./Genet. Toxicol. Environ. Mutagen.* **2011**, *721*, 27–73. [CrossRef]
229. Kirkland, D.; Aardema, M.; Henderson, L.; Müller, L. Evaluation of the ability of a battery of three in vitro genotoxicity tests to discriminate rodent carcinogens and non-carcinogens I. Sensitivity, specificity and relative predictivity. *Mutat. Res./Genet. Toxicol. Environ. Mutagen.* **2005**, *584*, 1–256. [CrossRef]
230. Doak, S.H.; Manshian, B.; Jenkins, G.J.S.; Singh, N. In vitro genotoxicity testing strategy for nanomaterials and the adaptation of current OECD guidelines. *Mutat. Res. Toxicol. Environ. Mutagen.* **2012**, *745*, 104–111. [CrossRef]
231. Clift, M.J.D.; Raemy, D.O.; Endes, C.; Ali, Z.; Lehmann, A.D.; Brandenberger, C.; Petri-Fink, A.; Wick, P.; Parak, W.J.; Gehr, P.; et al. Can the Ames test provide an insight into nano-object mutagenicity? Investigating the interaction between nano-objects and bacteria. *Nanotoxicology* **2013**, *7*, 1373–1385. [CrossRef]
232. George, J.M.; Magogotya, M.; Vetten, M.A.; Buys, A.V.; Gulumian, M. From the cover: An investigation of the genotoxicity and interference of gold nanoparticles in commonly used in vitro mutagenicity and genotoxicity assays. *Toxicol. Sci.* **2017**, *156*, 149–166.
233. Landsiedel, R.; Kapp, M.D.; Schulz, M.; Wiench, K.; Oesch, F. Genotoxicity investigations on nanomaterials: Methods, preparation and characterization of test material, potential artifacts and limitations—Many questions, some answers. *Mutat. Res./Rev. Mutat. Res.* **2009**, *681*, 241–258. [CrossRef]
234. Magdolenova, Z.; Collins, A.; Kumar, A.; Dhawan, A.; Stone, V.; Dusinska, M. Mechanisms of genotoxicity. A review of in vitro and in vivo studies with engineered nanoparticles. *Nanotoxicology* **2014**, *8*, 233–278. [CrossRef]
235. Pfuhler, S.; Elespuru, R.; Aardema, M.J.; Doak, S.H.; Maria Donner, E.; Honma, M. Genotoxicity of nanomaterials: Refining strategies and tests for hazard identification. *Environ. Mol. Mutagen.* **2013**, *54*, 229. [CrossRef]
236. Elespuru, R.; Pfuhler, S.; Aardema, M.J.; Chen, T.; Doak, S.H.; Doherty, A.; Farabaugh, C.S.; Kenny, J.; Manjanatha, M.; Mahadevan, B.; et al. Genotoxicity Assessment of Nanomaterials: Recommendations on Best Practices, Assays, and Methods. *Toxicol. Sci.* **2018**, *164*, 391–416. [CrossRef]
237. Elespuru, R.K.; Doak, S.H.; Collins, A.R.; Dusinska, M.; Pfuhler, S.; Manjanatha, M.; Cardoso, R.; Chen, C.L. Common Considerations for Genotoxicity Assessment of Nanomaterials. *Front. Toxicol.* **2022**, *4*, 859122. [CrossRef] [PubMed]
238. Fenech, M.; Morley, A.A. Cytokinesis-block micronucleus method in human lymphocytes: Effect of in vivo ageing and low dose X-irradiation. *Mutat. Res./Fundam. Mol. Mech. Mutagen.* **1986**, *161*, 193–198. [CrossRef] [PubMed]
239. Doak, S.; Griffiths, S.; Manshian, B.; Singh, N.; Williams, P.; Brown, A.; Jenkins, G. Confounding experimental considerations in nanogenotoxicology. *Mutagenesis* **2009**, *24*, 285–293. [CrossRef] [PubMed]
240. ECHA. *Guidance on Information Requirements and Chemical Safety Assessment*; Appendix R7-1 for Nanomaterials Applicable to Chapter R7a Endpoint Specific Guidance ECHA-21-H-04-EN; ECHA: Helsinki, Finland, 2021.
241. Dusinska, M.; Mariussen, E.; Rundén-Pran, E.; Hudecova, A.M.; Elje, E.; Kazimirova, A.; El Yamani, N.; Dommershausen, N.; Tharmann, J.; Fieblinger, D.; et al. In Vitro Approaches for Assessing the Genotoxicity of Nanomaterials. In *Nanotoxicity*; Methods in Molecular Biology; Springer: Cham, Switzerland, 2019; Volume 1894, pp. 83–122. [CrossRef]
242. Wills, J.W.; Hondow, N.; Thomas, A.D.; Chapman, K.E.; Fish, D.; Maffeis, T.G.; Penny, M.W.; Brown, R.A.; Jenkins, G.J.S.; Brown, A.P.; et al. Genetic toxicity assessment of engineered nanoparticles using a 3D in vitro skin model (EpiDerm™). *Part. Fibre Toxicol.* **2016**, *13*, 50. [CrossRef]
243. Conway, G.E.; Shah, U.-K.; Llewellyn, S.; Cervena, T.; Evans, S.J.; Al Ali, A.S.; Jenkins, G.J.; Clift, M.J.D.; Doak, S.H. Adaptation of the in vitro micronucleus assay for genotoxicity testing using 3D liver models supporting longer-term exposure durations. *Mutagenesis* **2020**, *35*, 319–330. [CrossRef]
244. ANSES. *Nanogenotox Final Report—Facilitating the Safety Evaluation of Manufactured Nanomaterials by Characterising Their Potential Genotoxic Hazard*; ANSES: Buenos Aires, Argentina, 2013.
245. Louro, H.; Pinhão, M.; Santos, J.; Tavares, A.; Vital, N.; Silva, M.J. Evaluation of the cytotoxic and genotoxic effects of benchmark multi-walled carbon nanotubes in relation to their physicochemical properties. *Toxicol. Lett.* **2016**, *262*, 123–134. [CrossRef]
246. Evans, S.J.; Clift, M.J.D.; Singh, N.; Wills, J.W.; Hondow, N.; Wilkinson, T.S.; Burgum, M.J.; Brown, A.P.; Jenkins, G.J.; Doak, S.H. In vitro detection of in vitro secondary mechanisms of genotoxicity induced by engineered nanomaterials. *Part. Fibre Toxicol.* **2019**, *16*, 8. [CrossRef]
247. Burgum, M.J.; Clift, M.J.D.; Evans, S.J.; Hondow, N.; Tarat, A.; Jenkins, G.J.; Doak, S.H. Few-layer graphene induces both primary and secondary genotoxicity in epithelial barrier models in vitro. *J. Nanobiotechnol.* **2021**, *19*, 24. [CrossRef]
248. SCOEL. *Methodology for Derivation of Occupational Exposure Limits of Chemical Agents*; European Commission, Ed.; SCOEL: Brussels, Belgium, 2017.
249. Decordier, I.; Papine, A.; Loock, K.V.; Plas, G.; Soussaline, F.; Kirsch-Volders, M. Automated image analysis of micronuclei by IMSTAR for biomonitoring. *Mutagenesis* **2011**, *26*, 163–168. [CrossRef]
250. Nüsse, M.; Kramer, J. Flow cytometric analysis of micronuclei found in cells after irradiation. *Cytometry* **1984**, *5*, 20–25. [CrossRef]
251. Nüsse, M.; Marx, K. Flow cytometric analysis of micronuclei in cell cultures and human lymphocytes: Advantages and disadvantages. *Mutat. Res./Genet. Toxicol. Environ. Mutagen.* **1997**, *392*, 109–115. [CrossRef] [PubMed]

252. Nüsse, M.; Recknagel, S.; Beisker, W. Micronuclei induced by 2-chlorobenzylidene malonitrile contain single chromosomes as demonstrated by the combined use of flow cytometry and immunofluorescent staining with anti-kinetochore antibodies. *Mutagenesis* **1992**, *7*, 57–67. [CrossRef] [PubMed]

253. Franz, P.; Bürkle, A.; Wick, P.; Hirsch, C. Exploring Flow Cytometry-Based Micronucleus Scoring for Reliable Nanomaterial Genotoxicity Assessment. *Chem. Res. Toxicol.* **2020**, *33*, 2538–2549. [CrossRef]

254. Azqueta, A.; Gutzkow, K.B.; Priestley, C.C.; Meier, S.; Walker, J.S.; Brunborg, G.; Collins, A.R. A comparative performance test of standard, medium- and high-throughput comet assays. *Toxicol. In Vitro* **2013**, *27*, 768–773. [CrossRef] [PubMed]

255. García-Rodríguez, A.; Rubio, L.; Vila, L.; Xamena, N.; Velázquez, A.; Marcos, R.; Hernández, A. The Comet Assay as a Tool to Detect the Genotoxic Potential of Nanomaterials. *Nanomaterials* **2019**, *9*, 1385. [CrossRef]

256. Nelson, B.C.; Wright, C.W.; Ibuki, Y.; Moreno-Villanueva, M.; Karlsson, H.L.; Hendriks, G.; Sims, C.M.; Singh, N.; Doak, S.H. Emerging Metrology for High-Throughput Nanomaterial Genotoxicology Mutagenesis Advance Access. *Mutagenesis* **2017**, *32*, 215. [CrossRef]

257. Elje, E.; Mariussen, E.; Moriones, O.H.; Bastús, N.G.; Puntes, V.; Kohl, Y.; Dusinska, M.; Rundén-Pran, E. Hepato(Geno)Toxicity Assessment of Nanoparticles in a HepG2 Liver Spheroid Model. *Nanomaterials* **2020**, *10*, 545. [CrossRef]

258. Rodriguez-Garraus, A.; Azqueta, A.; Vettorazzi, A.; de Cerain, A.L. Genotoxicity of Silver Nanoparticles. *Nanomaterials* **2020**, *10*, 251. [CrossRef]

259. Warheit, D.B.; Donner, E.M. Rationale of genotoxicity testing of nanomaterials: Regulatory requirements and appropriateness of available OECD test guidelines. *Nanotoxicology* **2010**, *4*, 409–413. [CrossRef]

260. Pfuhler, S.; van Benthem, J.; Curren, R.; Doak, S.H.; Dusinska, M.; Hayashi, M.; Heflich, R.H.; Kidd, D.; Kirkland, D.; Luan, Y.; et al. Use of in vitro 3D tissue models in genotoxicity testing: Strategic fit, validation status and way forward. Report of the working group from the 7th International Workshop on Genotoxicity Testing (IWGT). *Mutat. Res./Genet. Toxicol. Environ. Mutagen.* **2020**, *850–851*, 503135. [CrossRef]

261. Evans, S.J.; Clift, M.J.; Singh, N.; de Oliveira Mallia, J.; Burgum, M.; Wills, J.W.; Wilkinson, T.S.; Jenkins, G.J.; Doak, S.H. Critical review of the current and future challenges associated with advanced in vitro systems towards the study of nanoparticle (secondary) genotoxicity. *Mutagenesis* **2017**, *32*, 233–241. [CrossRef] [PubMed]

262. Di Ianni, E.; Jacobsen, N.R.; Vogel, U.; Møller, P. Predicting nanomaterials pulmonary toxicity in animals by cell culture models: Achievements and perspectives. *Wiley Interdiscip. Rev. Nanomed. Nanobiotechnol.* **2022**, *14*, e1794. [CrossRef]

263. Clewell, R.A.; McMullen, P.D.; Adeleye, Y.; Carmichael, P.L.; Andersen, M.E. Pathway based toxicology and fit-for-purpose assays. In *Validation of Alternative Methods for Toxicity Testing*; Springer: Cham, Switzerland, 2016; pp. 205–230.

264. Nymark, P.; Bakker, M.; Dekkers, S.; Franken, R.; Fransman, W.; García-Bilbao, A.; Greco, D.; Gulumian, M.; Hadrup, N.; Halappanavar, S.; et al. Toward Rigorous Materials Production: New Approach Methodologies Have Extensive Potential to Improve Current Safety Assessment Practices. *Small* **2020**, *16*, e1904749. [CrossRef] [PubMed]

265. OECD. *Guidance Document on the Validation and International Acceptance of New or Updated Test Methods for Hazard Assessment*; OECD Environment, Health and Safety Publications Series on Testing and Assessment; OECD: Paris, France, 2015.

266. Clift, M.J.D.; Jenkins, G.J.S.; Doak, S.H. An Alternative Perspective towards Reducing the Risk of Engineered Nanomaterials to Human Health. *Small* **2020**, *16*, e2002002. [CrossRef] [PubMed]

267. Van der Zalm, A.J.; Barroso, J.; Browne, P.; Casey, W.; Gordon, J.; Henry, T.R.; Kleinstreuer, N.C.; Lowit, A.B.; Perron, M.; Clippinger, A.J. A framework for establishing scientific confidence in new approach methodologies. *Arch. Toxicol.* **2022**, *96*, 2865–2879. [CrossRef]

268. European Commission. *Pathway to a Healthy Planet for All. EU Action Plan: 'Towards Zero Pollution for Air, Water and Soil'—Communication from the Commission to the European Parliament, the Council, the European Economic and Social Committee and the Committee of the Regions*; European Commission: Brussels, Belgium, 2021.

269. Napierska, D.; Thomassen, L.C.J.; Lison, D.; Martens, J.A.; Hoet, P.H. The nanosilica hazard: Another variable entity. *Part. Fibre Toxicol.* **2010**, *7*, 39. [CrossRef]

270. IARC. Arsenic, metals, fibres, and dusts. *IARC Monogr. Eval. Carcinog. Risks Hum.* **2012**, *100*, 11–465.

271. Höck, J.; Behra, R.; Bergamin, L.; Bourqui-Pittet, M.; Bosshard, C.; Epprecht, T.; Furrer, V.; Frey, S.; Gautschi, M.; Hofmann, H.; et al. *Guidelines on the Precautionary Matrix for Synthetic Nanomaterials*; Federal Office of Public Health: Bern, Switzerland, 2018.

272. Pavlicek, A.; Part, F.; Gressler, S.; Rose, G.; Gazsó, A.; Ehmoser, E.-K.; Huber-Humer, M. Testing the Applicability of the Safe-by-Design Concept: A Theoretical Case Study Using Polymer Nanoclay Composites for Coffee Capsules. *Sustainability* **2021**, *13*, 13951. [CrossRef]

273. OECD. Testing Programme of Manufactured Nanomaterials. Available online: https://www.oecd.org/chemicalsafety/nanosafety/testing-programme-manufactured-nanomaterials.htm (accessed on 14 November 2022).

274. NanoSafetyCluster. The Malta Initiative. Available online: https://www.nanosafetycluster.eu/international-cooperation/the-malta-initiative/ (accessed on 14 November 2022).

275. NanoHarmony. Available online: https://nanoharmony.eu/ (accessed on 14 November 2022).

276. OECD. Nanomet. Available online: https://www.oecd.org/chemicalsafety/nanomet/ (accessed on 14 November 2022).

277. Gov4Nano. Available online: https://www.gov4nano.eu/ (accessed on 14 November 2022).

278. EFSA; Andersson, N.; Arena, M.; Auteri, D.; Barmaz, S.; Grignard, E.; Kienzler, A.; Lepper, P.; Lostia, A.M.; Munn, S. Guidance for the identification of endocrine disruptors in the context of Regulations (EU) No 528/2012 and (EC) No 1107/2009. *EFSA J.* **2018**, *16*, e05311.

279. Shen, F.; Li, D.; Guo, J.; Chen, J. Mechanistic toxicity assessment of differently sized and charged polystyrene nanoparticles based on human placental cells. *Water Res.* **2022**, *223*, 118960. [CrossRef]
280. Priyam, A.; Singh, P.P.; Gehlout, S. Role of Endocrine-Disrupting Engineered Nanomaterials in the Pathogenesis of Type 2 Diabetes Mellitus. *Front. Endocrinol.* **2018**, *9*, 704. [CrossRef] [PubMed]

MDPI
St. Alban-Anlage 66
4052 Basel
Switzerland
Tel. +41 61 683 77 34
Fax +41 61 302 89 18
www.mdpi.com

Nanomaterials Editorial Office
E-mail: nanomaterials@mdpi.com
www.mdpi.com/journal/nanomaterials